Environment and Development

Environment and Development

Basic Principles, Human Activities, and Environmental Implications

Edited by

Stavros G. Poulopoulos

Kazakh-British Technical University, Almaty, Republic of Kazakhstan

Vassilis J. Inglezakis

Nazarbayev University, Astana, Republic of Kazakhstan

ELSEVIER

Amsterdam • Boston • Heidelberg • London • New York • Oxford
Paris • San Diego • San Francisco • Singapore • Sydney • Tokyo

Elsevier
Radarweg 29, PO Box 211, 1000 AE Amsterdam, Netherlands
The Boulevard, Langford Lane, Kidlington, Oxford OX5 1GB, UK
50 Hampshire Street, 5th Floor, Cambridge, MA 02139, USA

Notices
Knowledge and best practice in this field are constantly changing. As new research and experience broaden
our understanding, changes in research methods, professional practices, or medical treatment may become
necessary.

Practitioners and researchers must always rely on their own experience and knowledge in evaluating and using
any information, methods, compounds, or experiments described herein. In using such information or methods
they should be mindful of their own safety and the safety of others, including parties for whom they have a
professional responsibility.

To the fullest extent of the law, neither the Publisher nor the authors, contributors, or editors, assume any
liability for any injury and/or damage to persons or property as a matter of products liability, negligence or
otherwise, or from any use or operation of any methods, products, instructions, or ideas contained in the
material herein.

British Library Cataloguing-in-Publication Data
A catalogue record for this book is available from the British Library

Library of Congress Cataloging-in-Publication Data
A catalog record for this book is available from the Library of Congress

ISBN: 978-0-444-62733-9

For information on all Elsevier publications
visit our website at https://www.elsevier.com/

Publisher: John Fedor
Acquisition Editor: Kostas Marinakis
Editorial Project Manager: Christine McElvenny
Production Project Manager: Paul Prasad Chandramohan
Designer: Matthew Limbert

Contents

Chapter 3: Aquatic Environment 137

V.J. Inglezakis, S.G. Poulopoulos, E. Arkhangelsky, A.A. Zorpas, A.N. Menegaki

*G. Itskos, N. Nikolopoulos, D.-S. Kourkoumpas, A. Koutsianos, I. Violidakis,
P. Drosatos, P. Grammelis*

List of Contributors

E. Arkhangelsky   Nazarbayev University, Astana, Republic of Kazakhstan

M.K. Doula   Benaki Phytopathological Institute, Kifisia, Greece

I.G. Doulos   Metsovion Interdisciplinary Research Center, National Technical University of Athens, Athens, Greece

P. Drosatos   Chemical Process and Energy Resources Institute, Ptolemais, Greece

S. Folini   University of Florence, Florence, Italy

P. Grammelis   Chemical Process and Energy Resources Institute, Ptolemais, Greece

V.J. Inglezakis   Nazarbayev University, Astana, Republic of Kazakhstan

G. Itskos   Nazarbayev University, Astana, Republic of Kazakhstan

E. Katsou   Brunel University, Uxbridge, Middlesex, United Kingdom

N.M. Katsoulakos   Metsovion Interdisciplinary Research Center, National Technical University of Athens, Athens, Greece

S. Kershaw   Brunel University, Uxbridge, Middlesex, United Kingdom

V.S. Kotsios   Metsovion Interdisciplinary Research Center, National Technical University of Athens, Athens, Greece

D.-S. Kourkoumpas   Chemical Process and Energy Resources Institute, Ptolemais, Greece

A. Koutsianos   Chemical Process and Energy Resources Institute, Ptolemais, Greece

S. Malamis   National Technical University of Athens, Athens, Greece

A.N. Menegaki   Hellenic Open University, Patras, Greece; Organismos Georgikon Asfaliseon, Regional Branch of Eastern Macedonia & Thrace, Komotini, Greece

L.-M.N. Misthos   Metsovion Interdisciplinary Research Center, National Technical University of Athens, Athens, Greece

N. Nikolopoulos   Chemical Process and Energy Resources Institute, Ptolemais, Greece

S.G. Poulopoulos   Kazakh-British Technical University, Almaty, Republic of Kazakhstan

A. Sarris   Institute for Mediterranean Studies, Rethymno, Greece

D. Venetis   EPTA SA, Athens, Greece

I. Violidakis   Chemical Process and Energy Resources Institute, Ptolemais, Greece

A.A. Zorpas   Cyprus Open University, Latsia, Nicosia, Cyprus

CHAPTER 1

Introduction to Environment and Development

S.G. Poulopoulos
Kazakh-British Technical University, Almaty, Republic of Kazakhstan

Chapter Outline

1.1 Man and Environment: A Relation of Competition

The relationship between the man and the environment has been established in the early periods itself. There is a misconception that humankind used to live in harmony with nature and its other creatures in the—distant—past and that this balanced coexistence was overturned only the last 200 years. However, this is not the truth. What has really changed recently in man history is the ability to destroy natural life at devastating rates and magnitudes. But, generally, our attitude against nature and environment is not that different.

At the dawn of humankind on the face of the earth, man had to struggle against natural elements in order to survive. Man has been characterized by Benjamin Franklin and Karl Marx as a "tool-making animal" [1], because he is the only animal that manufactures various items and produces in general by transforming natural materials and shaping the environment. From the use of stone and fire at the caves of Neanderthal and of Cro-Magnon man, humanity passed in pottery and the use of metals, tamed animals, and cultivated land; built permanent settlements; and created the social and political life as we know it. In this long timeline of existence, humans never hesitated in subjugating or even

Environment and Development. http://dx.doi.org/10.1016/B978-0-444-62733-9.00001-0

destroying nature and the other living organisms, if it was considered vital in any way for survival. There is evidence that prehistoric man committed massive slaughter of animals and used extensively fire either to trap and guide wild animals to their final and fatal destination or to transform forests into land for cultivation.

For example, during the Pleistocene, the world saw a dramatic number of extinctions of very large terrestrial species [2]. Although there is an ongoing discussion on the subject, one of the hypotheses regarding the reasons behind the extinction of several megafaunal[1] species around 11000 BC proposes that the global spread of *Homo sapiens* and hunter-gatherer subsistence practices were responsible for these deaths [3]. Roberts [4] tested this hypothesis by examining extinctions in Australia. According to his measurements, the extinctions occurred around 46,400 years ago across Australia, thus ruling out any climatic impacts from the late Pleistocene as the underlying cause. Since *Homo sapiens* arrived in Australia at least 40,000 years ago [5], data concur with the human overkill hypothesis of extinction.

Speculating on the causes of this huge extinction event, the great evolutionary biologist Alfred Russel Wallace wrote [6]:

> *Looking at the whole subject again, with the much larger body*
> *of facts at our command, I am convinced that the rapidity of*
> *… the extinction of so many large Mammalia is actually due*
> *to man's agency, acting in co-operation with those general*
> *causes which at the culmination of each geological era has led*
> *to the extinction of the larger, the most*
> *specialized, or the most strangely modified forms.*

This huge loss of species is shown in Table 1.1 [8].

As mentioned previously, man used early fire to hunt animals and to achieve deforestation in order to eliminate hostile environment and create land for farming. Small-scale deforestation was practiced by some societies tens of thousands of years before the present, with some of the earliest evidence of deforestation appearing in the Mesolithic period [9]. These initial clearings were likely devised to convert closed forests into more open ecosystems, favorable to game animals. With the advent of agriculture in the mid-Holocene, greater areas were deforested, and fire became an increasingly popular method to clear land for crops. In Europe, by 7000 BC, Mesolithic hunter-gathers employed fire to create openings for red deer and wild boar.

Kaplan et al. [10] point out that mankind has transformed Europe's landscapes since the establishment of the first agricultural societies in the mid-Holocene. The most important

---

[1]The most common definition of megafauna is an animal with an adult body weight in excess of 44 kg [7].

Table 1.1: Megafaunal Extinctions (Genera) During the Last 100 ka [8].

Continent	Extinct	Living	Total	Extinct (%)
Africa	7	42	49	14.3
Europe	15	9	24	60.0
North America	33	12	45	73.3
South America	46	12	58	79.6
Australia	19	3	22	86.4

Table 1.2: Estimates of Percent Usable Land (Land Available for Clearing for Agriculture), and Percent of Forest Cover on Usable Land by Years and Each Region for Central and Western Europe [10].

Region	% Usable Land	% Forest Cover on Usable Land						
		1000 BC	500 BC	AD 1	AD 500	AD 1000	AD 1350 (Black Death)	AD 1400	AD 1850
Czechoslovakia	23.0	76.0	65.2	37.5	43.6	31.3	12.7	16.3	3.2
France	7.4	78.8	72.1	46.5	50.3	38.6	16.2	23.9	6.3
Germany	14.3	71.8	64.1	35.0	32.9	29.1	9.9	15.0	3.0
England —Wales	20.1	90.0	86.1	59.4	64.1	39.6	12.4	17.1	1.9
Ireland	30.0	64.5	68.4	69.7	50.6	38.0	13.0	19.0	0.9
Italy	10.4	69.0	51.1	30.1	47.9	40.2	22.1	30.0	7.6
Poland	9.9	95.0	91.1	75.1	69.9	46.1	22.2	24.8	4.0
Portugal	0.6	73.5	68.2	51.3	54.4	42.9	21.1	32.9	7.0
Spain	4.4	75.9	68.9	52.1	57.8	56.0	37.7	44.9	18.5
Average	**13.3**	**77.2**	**70.6**	**50.7**	**52.4**	**40.2**	**18.6**	**24.9**	**5.8**

impact on the natural environment was the clearing of forests to establish cropland and pasture, and the exploitation of forests for fuelwood and construction materials. The authors present also historical forest cover estimates as forested fractions of the usable land in each population region. The case of central and western Europe is shown in Table 1.2.

It could be even argued that anthropogenic carbon dioxide emissions per capita back then were higher than the current ones as a result of deforestation via *slash and burn*.[2] However, overpopulation nowadays does not allow any comparison in terms of absolute numbers.

This effort of humans to control and overpower environment continues even today. It has been even mentioned that despite the technological advances, we still behave as Neanderthals some 1000 years ago. But we know now that all beings live in the kingdom

---

[2]It denotes the method of agriculture in which existing vegetation is cut down and burned off before new seeds are sown, typically used as a method for clearing forest land for farming.

of nature and interact with it constantly. We have also a dialectic and continuous interaction with environment. Our actions that affect the environment are ultimately reflected on us. The influence of nature manifests itself in the air we breathe, the water we drink, the food we eat, and the flow of energy and information we consume. Despite its beauty, nature inspired fear in man in the early history of mankind and people had to fight it for survival. Today, man has become detached from nature and constitutes also a constant threat to it. At the same time, we have become a threat to our own existence since the environmental problems resulting from our actions evolve even on a planetary scale.

Although these terms have been used interchangeably, we have to make the question: What is *nature* and what is *environment*? Nature is the whole of the physical world, existing outside of any human action. Man is part of nature, but at the same time he acts upon it, thereby emancipating himself from it [11]. In Western societies, there is nostalgia for the lost paradise and the harmonic coexistence between man and nature in the past on one hand, and the urge to dominate nature on the other. Of course, the first belief is wrong as presented previously, while the second one could be catastrophic for our future. Regarding environment there are many definitions. According to that of the European Union, environment can be defined as "the whole set of elements which form the frameworks, the surroundings and the living conditions of man and society, as they are or as they are perceived" [11]. Environment includes always nature and culture. So, it can be seen as an emerging property of the man–nature relation, a field of reciprocal transformation of the human by the natural; and of the natural by the human [12].

Human activities have an adverse effect on the environment on many scales and in a variety of degrees of extent. The release of harmful compounds to the environment seems to be inevitable during chemical processes. Every day, numerous pollutants find their way into the environment through the reject of undesirable wastewater, waste gas, and liquid plus solid residues by individuals, activities, and processes. Actually, everyone has a negative and cumulative effect on the environment. We do not perceive the role that our simple daily activities play in degrading the environment. Whenever we drive our car, cook food, switch on an air conditioner, use hair spray, there is an impact on the environment. Maybe, it does not seem so important, but if you consider billions of people living on the planet and acting in similar ways, it becomes clear that the cumulative effect on the environment is tremendous. In many cases, this effect is long term, and after some point it could also become irreversible. As already mentioned, human beings interact with environment in a continuous and dynamic way all the time. Our actions create environmental problems with consequences even for the generations to come. Throughout human history, people have been threatened by various natural disasters like floods, earthquakes, volcano eruptions, etc. In recent times, technological innovations and advances have brought us new potential threats from the environment, which are now man-made and deteriorate our quality of living. Air pollution, water pollution, and soil pollution are just generic terms that include numerous environmental issues and problems that are physically harmful and stressful.

Our acting should not be seen as a number of individual actions of "frivolous" people, but they should be viewed in the context of the current economic model, where markets have a central role. The end of the Cold War provided a historic chance to reconsider our situation and direction on Earth. However, it was proved that the military threat was not the only problem. New nightmares were to come. The economy, the targets set by societies and countries, and a model based on increasing consumption set the basis and the roots of the global environmental problems arisen. In the current financial market system, which is also known as capitalism, all financial decisions are made by the market, in which buyers and sellers interact freely with minimum State or other intervention. Any economic natural resource belongs to the property of individuals or businesses, rather than to the State. The entire buying and selling scheme is based on pure competition, in which neither buyer nor seller is so strong to control the supply, demand, or price of a commodity. All buyers and sellers have absolute access to the market and enough information about the benefits and the harmful effects of economic goods, so they can make as much as possible responsible decisions.

Economists often represent the pure capitalism as a circular flow of goods and money among households and businesses operated normally, regardless of the ecosphere. As a result, the *progress* of a society has been synonym to *economic growth*, which was actually expressed through the growing demand for more services and products. So, even if it was called economic development, it meant nothing else but pure economic growth. The question arisen is obvious: since we pursuit ceaselessly economic growth, which is based on perpetual consumption of natural resources that are part of a finite ecological capital, are there any limits to this growth?

Philippe Destatte [13] provides shortly the story to answering this question. Aurelio Peccei, after founding the Club of Rome[3] in 1968, played a central role in the discussion on a better form of development through three principles for the future. He set out these principles during the following year in *The Chasm Ahead* [14]. He stated first that humanity and the global environment together are part of the same integrated macrosystem. Then, it is obvious that several parts of this macrosystem are at the risk of

---

[3]The Club of Rome is a global think tank that deals with a variety of international political issues. Founded in 1968 at Accademia dei Lincei in Rome, Italy, the Club of Rome describes itself as "a group of world citizens, sharing a common concern for the future of humanity." It consists of current and former heads of state, UN bureaucrats, high-level politicians and government officials, diplomats, scientists, economists, and business leaders from around the globe. It raised considerable public attention in 1972 with its report "The Limits to Growth" (http://www.clubofrome.org/).

being decomposed or even being completely destroyed. Finally, he argued that there is an immediate need to respond to this situation by developing a global plan. The implementation of it should be a collective obligation for all organizations with the capacity to act. Based on these principles and the methodological support of Jay Forrester and his model on dynamic systems, a research project entitled "Predicament of Mankind" was launched in 1970 at the Massachusetts Institute of Technology [15]. Two years later, the relevant report on the limits to growth was released [16]. According to the report's conclusions, as Earth is a closed system, the exponential growth of key selected variables, as population, food production, industrialization, pollution, and the use of nonrenewable resources, is clearly impossible. If the present growth trends in these variables continue in the future, it is just a matter of time (100 years) to reach the limits to growth on this planet. The ultimate consequence will be a sharp and uncontrollable decline in industrial capacity and population [16]. It is thus apparent that a collective awakening of conscience and awareness is required to avert this course to destruction. According to the report, a state of economic and ecological stability should be established so that the basic material needs of each person on Earth are satisfied and each person has an equal opportunity to realize his or her individual human potential.

However, this report was not the first to make reference to growth. Thomas Robert Malthus was the first to argue that "the power of population is indefinitely greater than the power in the earth to produce subsistence for man" back in 1798 [17,18]. His extreme ideas about population control attracted huge criticism, making him rather unpopular in his era. Malthus had foreseen that Earth's resources would have been exhausted many decades before. However, technological advances, especially in the field of agriculture, contradicted Malthus' prophecy and permitted the maintenance and feeding of a constantly increasing global population. But was it really Malthus' mistake or catastrophe had been just postponed?

Special Topic: Thomas Robert Malthus [17–24]

Malthus in a Few Words
Thomas Robert Malthus was a political economist and a precursor of demography. He was born in England in 1766 and educated at Jesus College, Cambridge, where he studied mathematics, French and English history and Literature, and was elected fellow in 1793. He served as assistant curate for short time. In 1805, he got married and was appointed as Professor of modern history and political economy at the East India College at Hertfordshire. He retained this position for the rest of his life and died in 1834.

Malthus became mainly known and stayed in the history for his work *An Essay on the Principle of Population*, which was first published anonymously in 1798, and in which Malthus warned that our society will be gradually destroyed because of the uncontrolled population, which grows geometrically, and thus it will always surpass the available food supply, which is only increased numerically. The solution proposed by Malthus can be sought in "ethical restrictions," such as

Special Topic: Thomas Robert Malthus [17−24]—cont'd

the postponement of marriage, and measures that would reduce the birth rate. These fears were calmed during the 19th century but strengthened again during the mid-20th century when the pace of population growth in the poorest regions of the planet started gaining momentum.

His Life and Work

Thomas Robert Malthus, known by middle names Robert, was born in 1766 in the Rookery, a provincial town in the South of London. He was the second son of Daniel Malthus, a noble of the province and pupil of Jean-Jacques Rousseau and David Hume, whom he knew personally. So, Robert Malthus was educated by his father and his teachers according to the perceptions of Rousseau. He entered the College of Jesus, in Cambridge, in 1784 and was ordained as a priest in the Anglican Church in 1788. He got a degree in mathematics in 1791.

He took the MA degree in 1791 and was elected a Fellow of Jesus College 2 years later. In 1789, he took orders in the Church of England and became a curate at Oakwood Chapel in the parish of Wotton, Surrey. Having been elected a Fellow of Jesus College in 1793, he enjoyed endless philosophical discussions-quarrels with his father about the possibility of the moral improvement of society and then with William Godwin and Marques de Condorcet, who persuaded him to write his ideas on paper. Finally, in 1798, Malthus issued a pamphlet known as *An Essay on the Principle of Population*.

In that famous work, Malthus argued that "the power of population is indefinitely greater than the power in the earth to produce subsistence for man." He wrote in opposition to the popular view in 18th-century Europe that saw society as improving and in principle as perfectible. He thought that the dangers of population growth precluded progress toward a utopian society. Malthus was convinced that population growth has a tendency to always exceed the food supply. Because of this trend, any effort to improve the situation of the lower classes either by increasing their income or by improving agricultural productivity would be futile, as additional livelihood would be absorbed completely by the incurred population growth. As this trend continues, according to Malthus, the possibility of moral improvement of society will always be unattainable.

In the revised and expanded edition of his Essay in 1803, Malthus focused on presenting empirical evidence to support his positions, many of which he collected himself from his travels in Germany, Russia, and Scandinavia. He also introduced the possibility of "moral restraints" as the best means of easing the poverty of the lower classes. In terms of practical politics, it meant instilling virtues of the middle class to the lower classes. He believed that this could happen with the introduction of voting rights for all, public education for the poor, with the elimination of the Poor Laws, and the establishment of a free labor market at national level. Moreover, he argued that if the poor had the chance to taste luxury, then they would require a higher standard of living before embarking on family. So, though it seems counterintuitive, Malthus suggested the possibility of "demographic transition," where sufficiently high incomes could by themselves reduce the rate of population growth.

(Continued)

Special Topic: Thomas Robert Malthus [17–24]—cont'd

The Essay converted Malthus in a celebrity intellect in a negative way. He was considered by many as a heartless monster, a false prophet, an enemy of the working class, etc. But there were also quite a few who had recognized the Essay as the first serious economic study of the welfare of the lower classes, including Karl Marx.

In 1804, Malthus got married and so he lost his scholarship at Cambridge. In 1805, he was appointed as Professor of modern history and political economy at the College of Eastern India in Hertfordshire, becoming thus the first academic Economist in England.

Malthus had shown interest in economics in 1800 when he published a pamphlet in which he had been developing a theory of endogenous money unlike the quantity theory. He argued that emerging values are followed by increases in the quantity of money supplied. Around 1810, Malthus came across with a series of leaflets of a stockbroker named David Ricardo about issues of money. Immediately, he wrote to Ricardo, and the two men began a correspondence and a friendship that would last over a decade. The relationship of Malthus–Ricardo was warm in every way except one—economics. They were found in rival camps for every topic in this field.

In 1814, Malthus took part in the debate Corn Laws. In his pamphlet, printed during the parliamentary discussion, Malthus tentatively supported the free traders. He argued that given the increasing expense of raising British corn, advantages accrued from supplementing it from cheaper foreign sources. He changed this view the following year. Foreign laws, he noted, often prohibit or raise taxes to export corn in difficult times, which meant that the power of Britain in food was captive of foreign policy.

Malthus believed that economic crises are characterized by an overall excess supply caused by insufficient consumption. Defending Corn Laws was based partly on the need for consumption of land owners to compensate for deficiencies in demand and to prevent the crisis.

The "Symposium" of Malthus: The "Sinful" Paragraph

Few have received the criticism that Malthus had sustained in his time. But what was the reason for such repulsion. It seems that the heavy criticism against him was mainly due to a paragraph that was found only in the second version of his famous work. This "sinful" paragraph that led to storm reactions is as follows:

A man who is born into a world already possessed, if he cannot get subsistence from his parents on whom he has a just demand, and if the society do not want his labour, has no claim of right to the smallest portion of food, and, in fact, has no business to be where he is. At nature's mighty feast there is no vacant cover for him. She tells him to be gone, and will quickly execute her own orders, if he does not work upon the compassion of some of her guests. If these guests get up and make room for him, other intruders immediately appear demanding the same favour. The report of a provision for all that come, fills the hall with numerous claimants. The order and harmony of the feast is disturbed, the plenty that before reigned is changed into scarcity; and the happiness of the guests is destroyed by the spectacle of misery and dependence in every part of the hall, and by the clamorous importunity of those, who are justly enraged at not finding the provision which they had been taught

---

Special Topic: Thomas Robert Malthus [17–24]—cont'd

to expect. The guests learn too late their error, in counter-acting those strict orders to all intruders, issued by the great mistress of the feast, who, wishing that all guests should have plenty, and knowing she could not provide for unlimited numbers, humanely refused to admit fresh comers when her table was already full.

Moreover, Malthus argued against the improvement of the situation of the poor, considering that a more comfortable life would lead the poor to obtain more children, which would have decreased the material resources in society, resulting in deteriorating living conditions for the entire society.

Commentary

Malthus is a person who accepted—alive and dead—intense criticism and heavy ratings as few. Nevertheless, even today his work is still referenced. Of course, Malthus could not have even conceived the current levels of the world's population, while it seems that he was wrong: advances in technology and agriculture have allowed an enormous increase in the world's population. However, Malthus was perhaps the first one who argued that there are limits to the availability and use of natural resources. In addition, the population growth today is associated with a number of significant environmental problems, either at local or planetary scale, such as global warming, water shortages, soil degradation, etc. There are scholars who state that reducing the rate of population growth could help in controlling climate change. Malthus gathered enormous criticism in his time because he went even further: he suggested specific measures to control the population and against poor people.

The example of Malthus is a serious indication of the paths in which someone can be driven without addressing the issues of development in a holistic, integrated, and interdisciplinary approach. Although Malthus foresaw that there are limits in growth, he failed in proposing sound and fair solutions. In any case, his contribution should not be forgotten. If not for anything else, at least for the following reason:

In October 1838, that is, fifteen months after I had begun my systematic inquiry, I happened to read for amusement Malthus on Population, and being well prepared to appreciate the struggle for existence which everywhere goes on from long-continued observation of the habits of animals and plants, it at once struck me that under these circumstances favorable variations would tend to be preserved, and unfavorable ones to be destroyed. The results of this would be the formation of a new species. Here, then I had at last got a theory by which to work. **Charles Darwin, from his autobiography (1876)**

In June 1972, the first United Nations Conference on the Human Environment (UNCHE) was held in Stockholm.[4] Representatives from 113 countries were present, as well as representatives from many international nongovernmental organizations, intergovernmental organizations, and many other specialized agencies. This was the first United Nations

---

[4]http://www.eoearth.org/view/article/156774/.

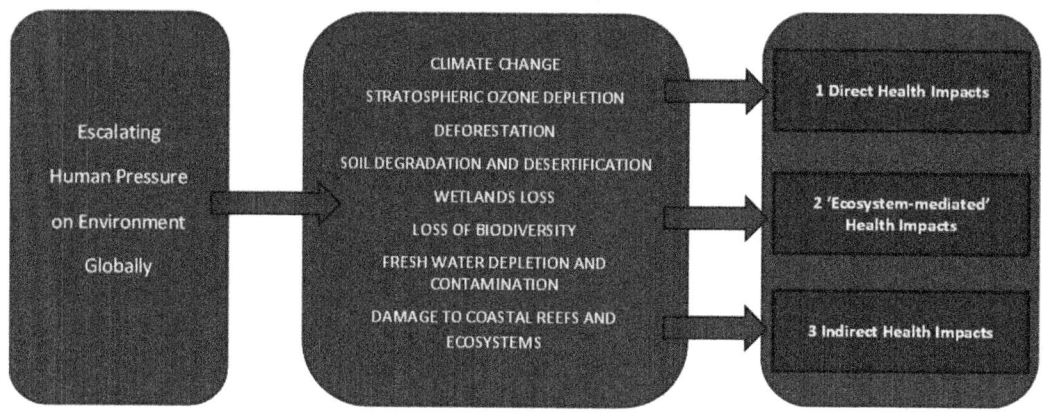

Figure 1.1
Global change. *World Health Organization, http://www.who.int/globalchange/environment/en/.*

conference on the environment as well as the first major international gathering focused on human activities in relationship to the environment, and it laid the foundation for environmental action at an international level. The conference acknowledged that the goal of reducing human impact on the environment would require extensive international cooperation, as many of the problems affecting the environment are global in nature. Following this conference, the United Nations Environmental Programme (UNEP) was launched in order to encourage United Nations agencies to integrate environmental measures into their programs.

The concerns regarding *global change*,[5] namely environmental pressures that threaten to radically alter the planet (Fig. 1.1) and put in danger the lives of many species upon it including humankind, led to the establishment in 1983 of the World Commission on Environment and Development by the UN General Assembly [25]. That body was an independent body, linked to but outside the control of governments and the UN system. It had three objectives: (1) to reexamine the critical environment and development issues and to formulate realistic proposals, (2) to propose new forms of international cooperation on these issues that would influence policies and events in the direction of needed changes, and (3) to raise the levels of understanding and commitment to action of individuals, voluntary organizations, businesses, institutes, and governments.

---

[5]According to the World Health Organization, *global change* refers to large-scale and global environmental hazards to human health and life on Earth generally. Although this term refers primarily to climate change and stratospheric ozone depletion, it may also include changes in ecosystems due to loss of biodiversity, changes in hydrological systems and the supplies of freshwater, land degradation, urbanization, and stresses on food-producing systems (http://www.who.int/globalchange/environment/en/).

In 1987, the Committee released its first report entitled "Our Common Future," known also as Brundtland Report[6] [25]. In that famous report, the first and most popular definition of "sustainable development" was given:

> *Sustainable development is development that meets the needs of the present without compromising the ability of future generations to meet their own needs.*
>
> *Ref. [25]*

According to the report, sustainable development includes two key concepts:

*   the concept of *needs*, in particular the essential needs of the world's poor to which over-riding priority should be given, and
*   the idea of limitations imposed by the state of technology and social organization on the environment's ability to meet present and future needs.

Actually, the term "sustainability" had been first used in 1713 by Carlowitz in forest sciences [26]. Many definitions and clarifications of the term have been attempted since then. The understanding of the term and the concept assigned depend on the personal perception of individuals about society and economy, the cultural and educational background, time and space, and the personal vision for a *better world*.

As the definitions of sustainability and the proposed principles have been multiplied and the term evolves continuously since Brundtland report, the models on which sustainability assessments are framed can be classified into three main categories [26]:

*   the pillar models with interacting or interdependent dimensions;
*   the human ecosystem–linked models, which are based on the concept of carrying capacity; and
*   principles of sustainability, which concentrate on the essential elements that are required for achieving sustainability.

The models in the context of the first category consider generally three dimensions of sustainability: *economic*, *social*, and *environmental*. These dimensions are assessed as being completely separate from each other, as in the triple bottom line (Fig. 1.2) and pillar (Fig. 1.3) models, or with some interaction, as in the three spheres model (Fig. 1.4). The sustainability assessment methods developed using these models are often just sets of economic, social, and environmental indicators, assessed in isolation. The more prevalent perhaps approach of "sustainable development" is the one of the three spheres, where *sustainable development* is defined as the simultaneous satisfaction of environmental

---

[6]The chairman of the World Commission on Environment and Development was Gro Harlem Brundtland, the Prime Minister of Norway, Parliamentary Leader of the Labor Party 1981–86, Member of Parliament from 1977, Minister of Environment 1974–79, and Associate Director Oslo School Health Services 1968–74 [25].

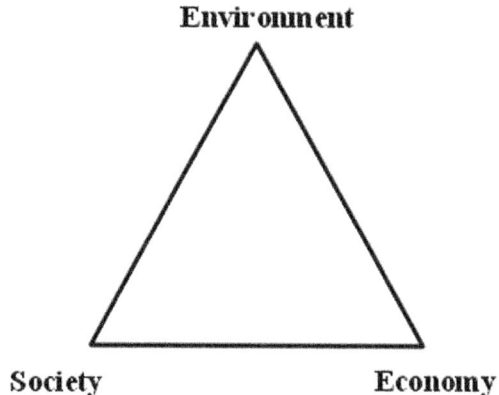

Figure 1.2
The three bottom line model of sustainability.

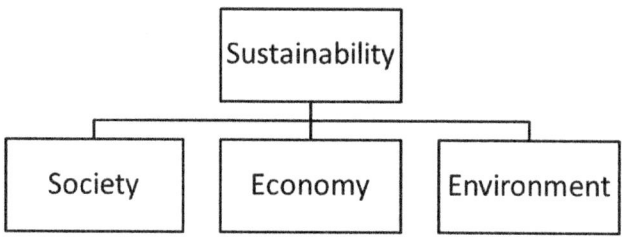

Figure 1.3
The three pillars model of sustainability.

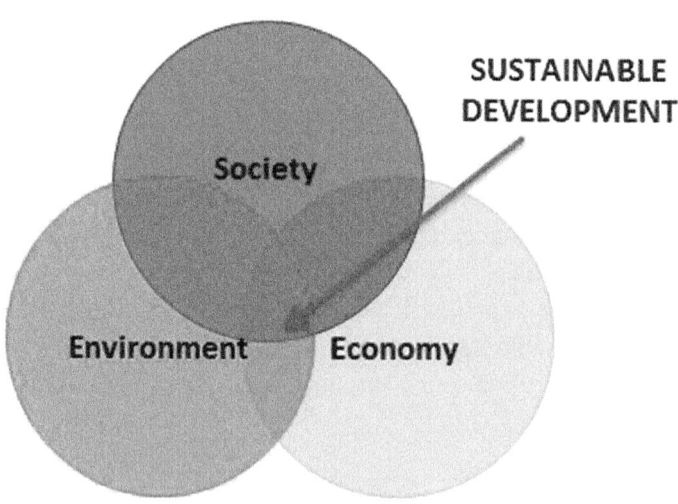

Figure 1.4
The three spheres model of sustainability.

protection, economic development, and social cohesion preservation. In many cases, these models are denoted by three E from *Environment*, *Economy*, and *Equity*.

In an effort to clarify the content of the principle of sustainable development, the Swiss "Monitoring of Sustainable Development Project" MONET tried to give a more accurate definition [27], based on the first of the 10 "Bellagio Principles",[7] which are shown in Table 1.3:

> *Sustainable development means ensuring dignified living conditions with regard to human rights by creating and maintaining the widest possible range of options for freely defining life plans. The principle of fairness among and between present and future generations should be taken into account in the use of environmental, economic and social resources.*
>
> *Putting these needs into practice entails comprehensive protection of bio-diversity in terms of ecosystem, species and genetic diversity, all of which are the vital foundations of life.*
>
> *Ref. [28]*

During the last years, other models and definitions have been proposed in order to clarify and extend the meaning of "sustainable development," such as the models of the prism and of the eggs. The first one was adapted by Valentin and Spangenberg [29] and is based on four dimensions (Fig. 1.5):

- economic dimension (man-made capital),
- environmental dimension (natural capital),
- social dimension (human capital), and
- institutional dimension (social capital).

The prism models have been criticized for paying too little concern to the environmental dimension, which is the precondition for the development of human *well-being*. Hence, a model of sustainability that would put the environment in the center should be developed. Accordingly, the International Development Research Center [30] (IDRC) proposed in 1997 the "Egg of Sustainability" (Fig. 1.6), originally designed in 1994 by the International Union for the Conservation of Nature (IUCN) [27].

Keiner [27] presents the variety of sustainable development definitions as well as alternative considerations, such as the *survivable development* and *evolutionability*.

---

[7]In 1987, the World Commission on Environment and Development called for the development of new ways to measure and assess progress toward sustainable development. In November 1996, an international group of measurement practitioners and researchers from five continents came together at the Rockefeller Foundation's Study and Conference Center in Bellagio, Italy, to review progress to date and to synthesize insights from practical ongoing efforts. The result was the principles shown in Table 1.3 that were unanimously endorsed [28].

Table 1.3: Bellagio Principles Toward Sustainable Development [27].

1. *Guiding Vision and Goals*

Assessment of progress toward sustainable development should be guided by a clear vision of sustainable development and goals that define that vision.

2. *Holistic Perspective*

Assessment of progress toward sustainable development should:

a.   include review of the whole system as well as its parts;

b.   consider the well-being of social, ecological, and economic subsystems, their state as well as the direction and rate of change of that state, of their component parts, and the interaction between parts; and

c.   consider both positive and negative consequences of human activity, in a way that reflects the costs and benefits for human and ecological systems, in monetary and nonmonetary terms.

3. *Essential Elements*

Assessment of progress toward sustainable development should:

a.   consider equity and disparity within the current population and between present and future generations, dealing with such concerns as resource use, overconsumption and poverty, human rights, and access to services, as appropriate;

b.   consider the ecological conditions on which life depends; and

c.   consider economic development and other, nonmarket activities that contribute to human/social well-being.

4. *Adequate Scope*

Assessment of progress toward sustainable development should:

a.   adopt a time horizon long enough to capture both human and ecosystem timescales thus responding to needs of future generations as well as those current to short-term decision-making;

b.   define the space of study large enough to include not only local but also long distance impacts on people and ecosystems; and

c.   build on historic and current conditions to anticipate future conditions—where we want to go, where we could go.

5. *Practical Focus*

Assessment of progress toward sustainable development should be based on:

a.   an explicit set of categories or an organizing framework that links vision and goals to indicators and assessment criteria;

b.   a limited number of key issues for analysis;

c.   a limited number of indicators or indicator combinations to provide a clearer signal of progress;

d.   standardizing measurement wherever possible to permit comparison; and

e.   comparing indicator values to targets, reference values, ranges, thresholds, or direction of trends, as appropriate.

6. *Openness*

Assessment of progress toward sustainable development should:

a.   make the methods and data that are used accessible to all; and

b.   make explicit all judgments, assumptions, and uncertainties in data and interpretations.

7. *Effective Communication*

Assessment of progress toward sustainable development should:

a.   be designed to address the needs of the audience and set of users;

b.   draw from indicators and other tools that are stimulating and serve to engage decision-makers; and

c.   aim, from the outset, for simplicity in structure and use of clear and plain language.

Table 1.3: Bellagio Principles Toward Sustainable Development [27].—cont'd

8. *Broad Participation*
Assessment of progress toward sustainable development should:
a. obtain broad representation of key grassroots, professional, technical, and social groups, including youth, women, and indigenous people—to ensure recognition of diverse and changing values; and
b. ensure the participation of decision-makers to secure a firm link to adopted policies and resulting action.

9. *Ongoing Assessment*
Assessment of progress toward sustainable development should:
a. develop a capacity for repeated measurement to determine trends;
b. be iterative, adaptive, and responsive to change and uncertainty because systems are complex and change frequently;
d. adjust goals, frameworks, and indicators as new insights are gained; and
e. promote development of collective learning and feedback to decision-making.

10. *Institutional Capacity*
Continuity of assessing progress toward sustainable development should be assured by:
a. clearly assigning responsibility and providing ongoing support in the decision-making process;
b. providing institutional capacity for data collection, maintenance, and documentation; and
c. supporting development of local assessment capacity.

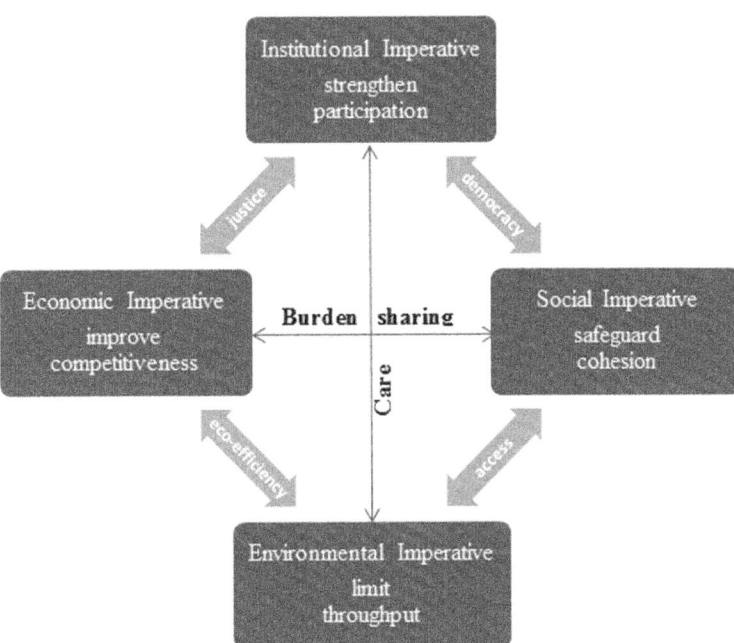

Figure 1.5
The prism of sustainability. *Adapted from Ref. [29].*

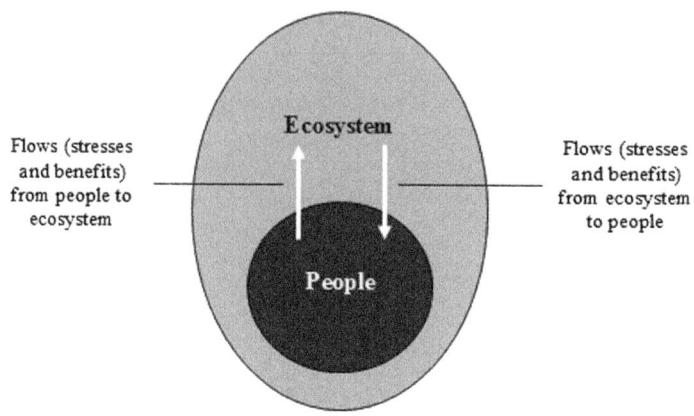

Figure 1.6
The egg of sustainability. *Adapted from Ref. [30].*

The 1992 "Earth Summit" in Rio de Janeiro, Brazil, namely the UN conference on Environment and Development (UNCED), generated not only formal endorsement for the concept of sustainable development by over 150 national governments, but also a wide range of general and particular policy initiatives under the sustainable development heading [31]. As United Nations state,[8] the Earth Summit in Rio de Janeiro was unprecedented for a UN conference, in terms of both its size and the scope of its concerns. Twenty years after the first global environment conference, the UN sought to help Governments rethink economic development and find ways to halt the destruction of irreplaceable natural resources and the pollution of the planet. Hundreds of thousands of people from all walks of life were drawn into the Rio process. They persuaded their leaders to go to Rio and join other nations in making the difficult decisions needed to ensure a healthy planet for the generations to come. In Rio, Governments, 108 of them being represented by heads of State or Government, adopted three major agreements aimed at changing the traditional approach to development:

- *Agenda 21*. It is a comprehensive program for global action in all areas of sustainable development. According to point 1.3 of the program [32], "Agenda 21 addresses the pressing problems of today and also aims at preparing the world for the challenges of the next century. It reflects a global consensus and political commitment at the highest level on development and environment cooperation. Its successful implementation is first and foremost the responsibility of Governments. National strategies, plans, policies and pro-cesses are crucial in achieving this. International cooperation should support and supple-ment such national efforts. In this context, the United Nations system has a key role to play. Other international, regional and sub regional organizations are also called upon to

---

[8]http://www.un.org/geninfo/bp/enviro.html.

contribute to this effort. The broadest public participation and the active involvement of the non-governmental organizations and other groups should also be encouraged." It contains detailed proposals for action in social and economic areas, such as combating poverty, changing patterns of production and consumption and addressing demographic dynamics, and for conserving and managing the natural resources that are the basis for life—protecting the atmosphere, oceans, and biodiversity; preventing deforestation; and promoting sustainable agriculture, for example.

- *The Rio Declaration on Environment and Development.* The Rio Declaration on Environment and Development supports Agenda 21 by defining the rights and responsibilities of States regarding these issues. Among its principles [33]:
  - Human beings are at the center of concerns for sustainable development. They are entitled to a healthy and productive life in harmony with nature.
  - Scientific uncertainty should not delay measures to prevent environmental degradation where there are threats of serious or irreversible damage.
  - States have a sovereign right to exploit their own resources but not to cause damage to the environment of other States.
  - Eradicating poverty and reducing disparities in worldwide standards of living are "indispensable" for sustainable development.
  - The full participation of women is essential for achieving sustainable development.
  - The developed countries acknowledge the responsibility that they bear in the international pursuit of sustainable development in view of the pressures their societies place on the global environment and of the technologies and financial resources they command.
- *The Statement of Forest Principles.* It was a nonlegally binding statement of principles for the sustainable management of forests, which was the first global consensus reached on forests. Among its provisions:
  - All countries, notably developed countries, should make an effort to "green the world" through reforestation and forest conservation.
  - States have a right to develop forests according to their socioeconomic needs, in keeping with national sustainable development policies.
  - Specific financial resources should be provided to develop programs that encourage economic and social substitution policies.

In addition, two legally binding Conventions that aimed at preventing both global climate change and eradication of the diversity of biological species were opened for signature at the Summit, giving high profile to these efforts:

- the United Nations Framework Convention on Climate Change and
- the Convention on Biological Diversity.

It has to be noted that many of the principles of the environmental legislation, as well as the main concepts and actions that are relevant to the environmental protection and sustainable development targets worldwide, have their roots in the texts developed in the context of Earth Summit and on Rio Declaration Principles (Table 1.4).

Another cornerstone in the sustainable development and environmental protection history was the World Summit on Sustainable Development (WSSD) that was conducted in Johannesburg in 2002. While the UNCED held in Rio provided the fundamental principles and the program of action for achieving sustainable development, the WSSD in Johannesburg reaffirmed the commitment to the Rio principles, the full implementation of Agenda 21, and the Program for the Further Implementation of Agenda [34]. The WSSD produced three types of outcomes: (1) a political declaration now known as the "Johannesburg Declaration on Sustainable Development"; (2) the "Johannesburg Plan of Implementation," a 65-page document restating existing targets eg, Millennium Declaration Goals and a limited number of new commitments; and (3) "Type II" commitments by governments and other stakeholders, including business and nongovernmental organizations [34].

Sustainable development is a very popular term, yet it has attracted a lot of criticism too. For example, Koroneos and Rokos [31] state that despite the progress that has been achieved in the 20th century, poverty and social inequality persist everywhere on the planet, including both developing and developed countries. To support their position, the authors use a number of statistics. Specifically, they report that in developing countries:

- More than a quarter of their 4.5 billion residents have no access to knowledge or to basic private and public services. Even worse, the life expectancy does not exceed 40 years!
- Roughly 1.3 billion people have no access to clean drinking water.
- Almost 840 million people are starving.

Even in the developed countries, there are a variety of regional inequalities, as expressed, for instance, from the fact that one in eight people is affected by some of the elements of human poverty, such as long-term unemployment, income below the poverty level, illiteracy, etc.

Moreover, they refer to the distribution of global income to strengthen their views. Specifically, in 1988, the richest 20% of the world's population received 82.7% of the global income, which was nearly 60 times the share of the income received by the poorest 20% of the world's population. In 1999, the richest quintile received 80 times the income earned by the poorest quintile. So, they introduced the *worth-living integrated development* to better reflect a development model that would truly combine social, economic, and environmental needs and would give also a meaning to the lives of the individuals in a society.

Table 1.4: Rio Declaration Principles [33].

No	Principle
1.	Human beings are at the center of concerns for sustainable development. They are entitled to a healthy and productive life in harmony with nature.
2.	States have, in accordance with the Charter of the United Nations and the principles of international law, the sovereign right to exploit their own resources pursuant to their own environmental and developmental policies, and the responsibility to ensure that activities within their jurisdiction or control do not cause damage to the environment of other States or of areas beyond the limits of national jurisdiction.
3.	The right to development must be fulfilled so as to equitably meet developmental and environmental needs of present and future generations.
4.	In order to achieve sustainable development, environmental protection shall constitute an integral part of the development process and cannot be considered in isolation from it.
5.	All States and all people shall cooperate in the essential task of eradicating poverty as an indispensable requirement for sustainable development, in order to decrease the disparities in standards of living and better meet the needs of the majority of the people of the world.
6.	The special situation and needs of developing countries, particularly the least developed and those most environmentally vulnerable, shall be given special priority. International actions in the field of environment and development should also address the interests and needs of all countries.
7.	States shall cooperate in a spirit of global partnership to conserve, protect, and restore the health and integrity of the Earth's ecosystem. In view of the different contributions to global environmental degradation, States have common but differentiated responsibilities. The developed countries acknowledge the responsibility that they bear in the international pursuit of sustainable development in view of the pressures their societies place on the global environment and of the technologies and financial resources they command.
8.	To achieve sustainable development and a higher quality of life for all people, States should reduce and eliminate unsustainable patterns of production and consumption and promote appropriate demographic policies.
9.	States should cooperate to strengthen endogenous capacity-building for sustainable development by improving scientific understanding through exchanges of scientific and technological knowledge, and by enhancing the development, adaptation, diffusion, and transfer of technologies, including new and innovative technologies.
10.	Environmental issues are best handled with the participation of all concerned citizens, at the relevant level. At the national level, each individual shall have appropriate access to information concerning the environment that is held by public authorities, including information on hazardous materials and activities in their communities, and the opportunity to participate in decision-making processes. States shall facilitate and encourage public awareness and participation by making information widely available. Effective access to judicial and administrative proceedings, including redress and remedy, shall be provided.
11.	States shall enact effective environmental legislation. Environmental standards, management objectives, and priorities should reflect the environmental and developmental context to which they apply. Standards applied by some countries may be inappropriate and of unwarranted economic and social cost to other countries, in particular, developing countries.
12.	States should cooperate to promote a supportive and open international economic system that would lead to economic growth and sustainable development in all countries, to better address the problems of environmental degradation. Trade policy measures for environmental purposes should not constitute a means of arbitrary or unjustifiable discrimination or a disguised restriction on international trade. Unilateral actions to deal with environmental challenges outside the

Continued

Table 1.4: Rio Declaration Principles [33].—cont'd

No	Principle
	jurisdiction of the importing country should be avoided. Environmental measures addressing transboundary or global environmental problems should, as far as possible, be based on an international consensus.
13.	States shall develop national law regarding liability and compensation for the victims of pollution and other environmental damage. States shall also cooperate in an expeditious and more determined manner to develop further international law regarding liability and compensation for adverse effects of environmental damage caused by activities within their jurisdiction or control to areas beyond their jurisdiction.
14.	States should effectively cooperate to discourage or prevent the relocation and transfer to other States of any activities and substances that cause severe environmental degradation or are found to be harmful to human health.
15.	In order to protect the environment, the precautionary approach shall be widely applied by States according to their capabilities. Where there are threats of serious or irreversible damage, lack of full scientific certainty shall not be used as a reason for postponing cost-effective measures to prevent environmental degradation.
16.	National authorities should endeavor to promote the internalization of environmental costs and the use of economic instruments, taking into account the approach that the polluter should, in principle, bear the cost of pollution, with due regard to the public interest and without distorting international trade and investment.
17.	Environmental impact assessment, as a national instrument, shall be undertaken for proposed activities that are likely to have a significant adverse impact on the environment and are subject to a decision of a competent national authority.
18.	States shall immediately notify other States of any natural disasters or other emergencies that are likely to produce sudden harmful effects on the environment of those States. Every effort shall be made by the international community to help States so afflicted.
19.	States shall provide prior and timely notification and relevant information to potentially affected States on activities that may have a significant adverse transboundary environmental effect and shall consult with those States at an early stage and in good faith.
20.	Women have a vital role in environmental management and development. Their full participation is therefore essential to achieve sustainable development.
21.	The creativity, ideals, and courage of the youth of the world should be mobilized to forge a global partnership in order to achieve sustainable development and ensure a better future for all.
22.	Indigenous people and their communities, and other local communities, have a vital role in environmental management and development because of their knowledge and traditional practices. States should recognize and duly support their identity, culture, and interests and enable their effective participation in the achievement of sustainable development.
23.	The environment and natural resources of people under oppression, domination, and occupation shall be protected.
24.	Warfare is inherently destructive of sustainable development. States shall therefore respect international law providing protection for the environment in times of armed conflict and cooperate in its further development, as necessary.
25.	Peace, development, and environmental protection are interdependent and indivisible.
26.	States shall resolve all their environmental disputes peacefully and by appropriate means in accordance with the Charter of the United Nations.
27.	States and people shall cooperate in good faith and in a spirit of partnership in the fulfillment of the principles embodied in this declaration and in the further development of international law in the field of sustainable development.

Indeed, inequalities, poverty, degraded human and natural environments are found everywhere in the world today. The *Wellbeing of Nations* surveys 180 countries in order to assess the level of sustainability globally. The Wellbeing Assessment method was developed and tested with the support of IUCN—the World Conservation Union and the International Development Research Centre (IDRC). It is intended to promote high levels of human and ecosystem well-being, demonstrate the practicality and potential of the Wellbeing Assessment method, and encourage countries, communities, and corporations to undertake their own well-being assessments. Wellbeing Assessment can be used at any level from municipality to the world, and it differs from other approaches to assessing sustainability in its dual focus on human and ecosystem well-being. Prescott-Allen further developed the method for the second phase of the IUCN/IDRC project (1997–99) and *The Wellbeing of Nations* [35]. According to the conclusions of this assessment [35]:

- No country is sustainable or even close to sustainability. The leaders (Sweden, Finland, Norway, Iceland, and Austria) and 32 other ecosystem-deficit countries (largely in Europe and North America) have high standards of living, but excessive impacts on the global environment.
- Some 27 human-deficit countries (chiefly in Africa) have fairly low demands on the ecosystem but are desperately poor.
- The remaining 116 are double-deficit countries, combining weak environmental performance and inadequate development.
- In 141 countries, ecosystem stress is higher than human well-being—a clear sign that most people's efforts to improve their lot are inefficient and overexploit the environment.
- Northern Europe is the strongest region, with 12 countries in the top 40, including the 5 leaders.
- West Asia is the weakest, with 9 countries in the bottom 40 and the most with high or extreme double deficits.

Carvahlo [36] goes one step further to suggest that within the current international political economic system, it would be nearly impossible to adopt development strategies that are conducive to truly sustainable development. The author concludes that profound changes in economic, political, and social structure are required to foster sustainable development.

In the latter approaches, the term *human well-being* is also used. Once again, defining human well-being is not easy, due to alternative views on what it means [37]. According to UNEP [37], "human well-being is the extent to which individuals have the ability and the opportunity to live the kinds of lives they have reason to value."

Human well-being incorporates personal and environmental security, access to materials for a good life, good health and good social relations, all of which are closely related to each other and underlie the freedom to make choices and take action.

Further discussion on development considerations is found in chapter "Environment and Development" of this book. The purpose of this short review was to introduce the reader to the concept of sustainable development and to highlight the fact that there is no definite interpretation of the term.

Closing this introduction, I would like to point out that despite the problems regarding human and ecosystem state worldwide, it is clear that a huge discussion is now open, the first tools and approaches have been developed, and the humankind seems able to acknowledge the mistakes of the past and make a step forward recognizing the chances but also the limitations of coexistence on Earth. Maybe, we should bear in mind that

Sustainability is more of a direction rather than a destination.

1.2 Global Environmental Problems: Our Common Future

As already presented in the previous paragraph, humankind has been trying from the dawn of history to prevail upon environment and succeeded in accomplishing this target the last 100–200 years, when humans became the dominant beings in nature and the exclusive owners of its resources. Particularly from the 1950s onward, man began to exploit natural resources in greater rates than those of replenishment. In parallel, man proceeded to large-scale interventions in the natural environment, which led to the drastic containment of ecosystems and the threat of extinction of entire species. Also, the multiple environmental pollution issues that accompany most human activities became a hard-to-solve and constant problem.

Whether it is to blame the present system of capitalism and the current way of living, the consumption patterns of the society today, the unprecedented increase of population and its concentration in urban areas, or even the very nature of man, the answer is ultimately the subject of socio-economic-political sciences. In this reading, the main objective is the description of the problems arisen and the formulation of feasible solutions from an engineering perspective.

Returning to the problems that our "development" causes over time, man brought about the pollution of seas and oceans as well as the pollution of the drinking water, both underground and surface, the overconsumption of most natural resources, the atmospheric and soil pollution as a result of the release of numerous gaseous, liquid, and solid wastes into the environment, while at the same time he caused extensive deforestation and drastically reduced the living space of most ecosystems.

Despite the severe impacts also on public health and our quality of life, mankind realized that it should take action only when global-scale problems appeared, which could extinguish life from Earth. These problems changed our view and led to international cooperation and coordinated response in order to resolve them. In the last 30 years, it has been widely—though not unanimously—accepted that man finally started disturbing the equilibrium of planetary mechanisms that are the product of the evolution of our planet during millions of years, mechanisms that are directly connected to the life on planet Earth as it stands today. These phenomena, including primarily the stratospheric ozone depletion and global warming, still threaten the planet overall, and their associated possible global-scale, long-term effects could constitute an unprecedented nightmare for humankind over periods of hundreds of years.

The first phenomenon being referred to is the destruction of ozone (O_3) in the stratosphere. Despite its small atmospheric concentration, ozone plays a vital role as it filters the solar radiation from its dangerous ultraviolet part. By the mid-80s and every September, scientists began to observe a decrease in stratospheric ozone in the Antarctic region. These observations are referred to as ozone hole. Ozone in the stratosphere is in constant dynamic equilibrium between the reaction of its formation and the reaction of its decomposition. The main elements of this equilibrium are the ultraviolet (UV) solar radiation, and molecules and oxygen atoms. But this equilibrium has been jeopardized by chlorofluorocarbons (CFCs). Let us review shortly the story of these compounds. Till 1930s, ammonia, chloromethane, carbon tetrachloride, isobutene, and propane were commonly used as refrigerants [38]. But these compounds due to some harmful properties of theirs, such as toxicity and flammability, had to be replaced by new compounds that would pose less danger during their application and use. So, in addition to having the desired thermodynamic properties, an ideal refrigerant would be nontoxic, nonflammable, completely stable inside a system, environmentally benign even with respect to decomposition products, and easy to manufacture [38]. Taking into account all these requirements, Midgley and his colleagues at General Motors suggested CFCs as clear favorites to meet all the criteria, and these compounds remained the refrigerants of choice from the 1930s until the 1990s [39]. Based on Lovelock's work, which showed that a substantial portion of the CFCs produced up to 1972 had accumulated in the atmosphere [40], Molina and Rowland propounded their hypothesis of stratospheric ozone depletion by these chemicals [41]. In the stratosphere, CFCs are exposed to UV radiation ($\lambda < 220$ nm), and subsequently they are photo-decomposed to produce active chlorine that leads to ozone destruction [39]. The discovery of the ozone hole in the mid-80s over Antarctica came to confirm all of the above.

The ozone holes are regarded by many as the precursor to an upcoming change in the planet's atmosphere. Researchers have discovered the existence of another "hole" over the Arctic zone of the planet, though less intense. These observations about ozone depletion and its potential impacts on society, including those of enhanced UVB radiation, have spurred international action to protect the ozone layer [42]. So, in the Vienna Convention for the Protection of the Ozone Layer of 1985, all Parties were called to *take appropriate measures in accordance with the provisions of the Convention and of those protocols in force to which they are party, to protect human health and the environment against adverse effects resulting or likely to result from human activities, which modify or are likely to modify the ozone layer* [43]. Due to the uncertainties back then, further research was proposed on the role of carbon monoxide, carbon dioxide, methane, nonmethane hydrocarbons, nitrous oxide, nitrogen oxides, fully halogenated alkanes, partially halogenated alkanes (both chlorine and bromine ones), hydrogen, and water, since all these compounds were thought to have the potential to modify the chemical and physical properties of the ozone layer [43]. Vienna Convention provided the framework for global cooperation on the protection of ozone layer. Following Vienna Convention, the Montreal Protocol on substances that deplete the ozone layer was signed by many industrialized countries in 1987 and came into force in 1989 [5]. Compared to Vienna Convention, it had more specific targets: to reduce and eventually eliminate emissions of ozone-depleting substances. Since then, it has been revised many times.

It is now about 40 years since stratospheric ozone depletion was predicted, and almost 30 years since the global framework convention for the protection of the ozone layer was adapted by world governments in Vienna. Since that time, evidence of ozone depletion has continued to accumulate and there is no longer any doubt that emissions of man-made chemicals (CFCs mainly) are primarily responsible for the ozone destruction [42].

The attitude of the global community toward the stratospheric ozone layer depletion can be characterized as exemplary because, despite the uncertainties and the lack of knowledge in relation with the substances and mechanisms that consumed stratospheric ozone, the international quick response and the global cooperation stopped the further destruction of ozone at the same devastating rates, allowing thus the possibility of it being restored during the 21st century. This constitutes also an excellent example of the application of the precautionary principle in environmental matters.

The second problem evolving on the planetary scale is global warming and climate change. Specifically, a number of gases in the atmosphere, carbon dioxide being the most important one, participate in the *greenhouse effect*, in which part of the radiation emitted from earth is confined in the atmosphere, heating further the surface of the planet, and ensuring average temperatures that are favorable to life. It is important to emphasize on the fact that greenhouse effect is a natural mechanism that evolved along with our planet in millions of years and which is vital to preserving life on Earth, at least as we know it today. Once again, human activities are responsible for disturbing this mechanism. Specifically, energy production is mainly based on fossil fuels and specifically on their combustion. The inevitable product of any combustion process is the release of carbon dioxide in the atmosphere. Increased concentrations of carbon dioxide in the atmosphere, along with increased levels of other greenhouse gases resulting from anthropogenic sources, enhance further the greenhouse effect, leading to the so-called global warming and ultimately to climate change [44].

Global warming and climate change will have many severe impacts on [45]:

- water resources and their management;
- ecosystems;
- food, fiber, and forest resources;
- coastal and sea-neighboring regions;
- economy, habitation, and society; and
- health.

It has been estimated by the United Nations Framework Convention on Climate Change (UNFCCC) that by 2030 the annual global cost of adapting to climate change will be approximately $49–$171 billion US dollars not including costs for ecosystem adaption that could add an additional $65–$300 billion per year [46]. The United States of America along with China and Qatar are the current top contributors of greenhouse emissions [46].

The impacts of climate change are already being witnessed globally [46]. Rising temperatures, air quality deterioration, more frequent incidences of food- and waterborne pathogens and allergens, and extreme weather events are among the current manifestation events of global warming. With the average global temperature projected to increase by 1.4–5.8°C by 2100, these events are expected to become increasingly frequent and severe [46].

Kyoto Protocol was the response of the international community to global warming in December 1997. Specifically, the Kyoto Protocol was an amendment to the United Nations Framework Convention on Climate Change (UNFCCC), which took place in the context of Rio Summit in June 1995, an international treaty on global warming. Ratifying countries committed to reducing their combined greenhouse gas levels by 5%, including carbon dioxide and five other emissions. A total of 141 countries had ratified the agreement, with the notable exceptions of the United States and Australia. The Protocol entered into force on February 16, 2005, and its emissions reduction requirements are binding on the 35 industrialized countries that have ratified it; the United States disengaged from the Protocol in 2001 and has not ratified it.

The implementation of Kyoto Protocol confronted many obstacles. For example, let us examine the position of the United States, which is the highest CO_2 emitter worldwide. The United States signed the Protocol on November 12, 1998. However, the Clinton Administration did not submit the Protocol to the Senate for advice and consent. In March

2001, the Bush Administration rejected the Kyoto Protocol. The United States continued to attend the annual conferences of the parties (COPs) to the UNFCCC, but did not participate in Kyoto Protocol—related negotiations. In February, 2002, President Bush announced a US policy for climate change that would rely on domestic, voluntary actions to reduce the "greenhouse gas intensity" (ratio of emissions to economic output) of the US economy by 18% over the next 10 years.[9]

What adds to the problem is the fact that energy demand is expected to increase year after year, and this demand will be covered by fossil fuels combustion, which means more carbon dioxide emissions released. Specifically, Solangi et al. [47] reported that world primary energy demand is projected to expand by almost 60% from 2002 to 2030, namely with an average annual increase of 1.7% per year. At the same time, fossil fuels will continue to dominate the global energy use. They will account for around 85% of the increase in world primary demand over 2002—30. And their share in total demand will increase slightly, from 80% in 2002 to 82% in 2030. The share of renewable energy sources will remain flat, at around 4%, while that of nuclear power will drop from 7% to 5% [48].

As Pacheco et al. [44] point out, despite (1) the actual risk of collective disaster, (2) the scientific consensus that anthropogenic greenhouse gas emissions perturb global climate patterns with negative consequences for many ecosystems, and (3) the predictions of early warning signals and severe climate change consequences that are already in place, *country leaders insist in discounting the severity of the problem, given the scientific uncertainty regarding the impacts of climate change.*

The difference in attitude toward global warming compared to the one against stratospheric ozone depletion could be explained by the fact that global warming is not simply connected with a purely man-made class of compounds with specific applications like CFCs, but principally with carbon dioxide, which in turn is related to energy consumption and the current way of living. So, dealing with the roots of global warming affects many economic interests worldwide.

These issues are further discussed in chapter "Atmospheric Environment". The reason why they are presented shortly in this section is to acknowledge that besides the short-term or long-term adverse effects associated with them, they helped—or forced—modern society and people, which had been disconnected from nature, to realize the importance of environment and the consequences of our actions. Moreover, since these problems are connected with compounds that are globally emitted and mixed in the atmosphere, they have consequences for everyone on Earth without exceptions. Despite the fact that the degree of guilty and the magnitude of potential impacts differ from country to country, it

---

[9]The Encyclopedia of Earth. http://www.eoearth.org/view/article/154065/.

is now universally accepted that the proper response and solution goes through international cooperation at all levels: educational, scientific, and political. In addition, these problems raised public awareness regarding environmental issues in general. It has been also obvious today that public participation is a prerequisite for coping effectively with all environmental challenges that the planet faces. For example, it is essential and of fundamental importance that people understand that every time they

- use public transport instead of vehicles, or even better bicycles or walking,
- replace an old light bulb with an energy-saving one,
- switch off completely household appliances when not using them,
- purchase products based on their environmental performance,
- gather their litter at the beach,
- separate household garbage based on the material,
- protect forests and wild life,
- invest in alternative sources of power,
- educate children to become environmentally responsible,
- do not follow unsustainable consumption patterns, trends, and practices,
- participate in environmental or development decisions that affect them,
- strongly support that development should not only rely on economic growth,

they help in saving our planet and ultimately in preserving their own status of living.

It is widely recognized that poverty, environmental degradation, and population growth are inextricably related and that none of these fundamental problems can be successfully addressed in isolation. Moreover, the measures in the list above cannot be suggested to people living in absolute poverty. How can you change a bulb, if there is no electricity at all?! The public awareness campaigns cannot be realized in such cases, at least with the same content.

So, a truly sustainable development has to address the problem of the large number of people who live in absolute poverty and they are not in the position to satisfy even the most basic of their needs. It should be also emphasized that people living in poor or developing countries have an impact on the environment too. They also contribute to the numerous pollution problems and overexploit natural resources. They are partly responsible for deforestation and thus, for climate change also, as discussed also in chapter "Atmospheric Environment."

In some cases, they could be worse emitters or wasters of resources per capita. Why so? Poverty reduces people's capacity to use resources in a sustainable manner; it intensifies pressure on the environment [25]. Most cases of extreme poverty are found in developing countries, most of them being hit by the economic stagnation of the 1980s. The woman who cooks in an earthen pot over an open fire uses perhaps eight times more energy than an affluent neighbor with a gas stove and aluminum pans. The poor who light their homes with a wick dipped in a jar of kerosene get one-fiftieth of the illumination of a 100-W electric bulb, but use just as much energy. These examples illustrate the tragic paradox of poverty. For the poor, the shortage of money is a greater limitation than the shortage of energy. They are forced to use low-quality fuels that are free and inefficient equipment because they do not have any other alternatives.

Another distinctive example refers to the tropics, which host the greatest number and diversity of species, but also host most developing nations, where population growth is fast and poverty is most widespread. If farmers in these countries are forced to continue with extensive agriculture, which is inherently unstable and leads to constant movement, then farming will tend to spread throughout the remaining wildlife environments [25]. Also, these countries base their economies on natural resources and eliminate thus, rain forests. Specifically, in Central and South America, many governments have encouraged the large-scale conversion of tropical forests to livestock ranches.

These issues were brought to light during the Earth Summit in Rio, where the world was separated into the *rich North* and the *poor South*. The central issues were as follows: Who will pay the bill for the first comprehensive attempt to repair the environmental damage caused by industrialization? And who will ensure that the poor nations of the southern hemisphere industrialize at a lower cost to the environment than the richer nations of the North did? Richer northern countries wanted to hold the poor to their pledge to pursue sustainable economic development by setting up a commission to monitor the South's strategies. But the developing world was resisting any obligation to abide by outsiders' recommendations. Poor countries wanted also to exploit northern worries about climate change, the destruction of forests, and polluted seas to extract more aid.

A necessary—but not necessarily sufficient—condition for the elimination of absolute poverty is a relatively rapid rise in per capita incomes in the Third World. It is therefore essential that the stagnant or declining growth trends of this decade be reversed. Eradication of poverty is a means of saving the environment and at the same time promotes the principles of equity, sustainable development, and human well-being.

People and countries should not stick permanently to criticizing about who is to blame for; we should keep in mind that [25]:

We will succeed or fail together.

1.3 Case Study 1: Easter Island's Human Extinction

The Easter Island case is a case of human extinction as a result of the exhaustion of natural resources due to their overexploitation by the inhabitants of the island. The history of Easter Island, its statues, and its peoples has long been shrouded in mystery. Some have even suggested that aliens planted the statues as signals to their fellow aliens to rescue them. However, new evidence based on pollen analysis supports a much simpler theory, that the Easter Island inhabitants destroyed their own society through complete deforestation. Although there are some views [49,50] that this approach has been too simplistic to explain what really happened, the narrative of Diamond [51] will be presented.

Easter Island, with an area of only 64 square miles, is the world's most isolated segment of habitable land. It lies in the Pacific Ocean more than 2000 miles west of the nearest continent (South America), 1400 miles from even the nearest habitable island (Pitcairn). When Easter Island was discovered by Europeans in 1722, it was a barren landscape with no trees over 10 ft in height. The small number of inhabitants, around 2000, lived in a state of civil disorder and everybody was thin and emaciated. Virtually no animals besides rats inhabited the island and the natives lacked sea-worthy boats. Understandably, the Europeans were mystified by the presence of great stone statues, some as high as 33 ft and weighing 82 tons. Even more impressive were the abandoned statues, as tall as 65 ft and weighing as much as 270 tons. How could such a people create, and then move such enormous structures? The answer lies in Easter islands' ecological past, when the island was not a barren place.

There is evidence that the first Polynesian colonists who stepped ashore some 1600 years ago, after a long canoe voyage from eastern Polynesia, found themselves in an untouched paradise. Easter Island of ancient times supported a subtropical forest completely with the tall Easter Island palm, a tree suitable for building homes, canoes, and latticing necessary for the construction of such statues. So, the first Polynesian colonists found themselves on an island with fertile soil, abundant food, bountiful building materials, ample lebensraum, and all the prerequisites for comfortable living. They prospered and multiplied. With the vegetation of the island, natives had fuelwood and the resources to make rope. With their sea-worthy canoes, Easter Islanders lived off a steady diet of porpoise. A complex social structure developed completely with a centralized government and religious priests.

It was that Easter Island society that built the famous statues and hauled them around the island using wooden platforms and rope constructed from the forest. The construction of these statues peaked from AD 1200 to 1500, probably when the civilization was at its greatest level. However, pollen analysis shows that at this time the tree population of the island was rapidly declining as deforestation took its toll. Eventually, Easter's growing population was cutting the forest more rapidly than the forest was regenerating. In such an isolated environment, as the forest started to decline, the inhabitants were left with no raw materials to make timber and rope to transport and erect their statues. Life must have become really uncomfortable. After some point, even wood would have become no longer available for fires [51].

Around 1400 the Easter Island palm became extinct due to overharvesting. Its capability to reproduce had become severely limited by the proliferation of rats, introduced by the islanders when they first arrived, which ate its seeds. In the years after the disappearance of the palm, ancient garbage piles reveal that porpoise bones declined sharply. Without raw materials, the inhabitants of the island could no longer build canoe to make journeys out to sea. They started eating land birds, migratory birds, and mollusks, which soon became also extinct. The islanders started to starve. With the disappearance of food surpluses, Easter Island could no longer feed the chiefs, bureaucrats, and priests who had kept a complex society running. Surviving islanders described to early European visitors how local chaos replaced centralized government and a warrior class took over from the hereditary chiefs. Survivors formed groups and bitter fighting erupted. People took to living in caves for protection against their enemies. Around 1770 rival clans started to topple each other's statues, breaking the heads off. By 1864 the last statue had been thrown down and desecrated. By the arrival of Europeans in 1722, there was almost no sign of the great civilization that once ruled the island other than the legacy of the strange statues.

Easter Island is a prime example of what widespread deforestation can do to a society. In just a few centuries, the people of Easter Island wiped out their forest, drove their plants and animals to extinction, and saw their complex society spiral into chaos and cannibalism [51].

1.4 Public Participation

Whichever interpretation and content might be given to the concept of "development," as impressive the plans we wish to apply may be, nothing can be implemented without the participation of the public. Public participation is and should be a fundamental pillar of any policies followed or choices to be made in relation with both sustainable development and environmental protection. To this end, specific steps have to be taken and dedicated laws have to be put into force so as to (1) promote public participation in environmental issues and (2) obligate authorities to integrate the results of this participation in the whole environmental planning and decision-making process.

Actually, the importance of public involvement in environmental matters was highlighted at the 1992 United Nations Conference on Environment and Development (UNCED), and particularly in principle 10 of Rio Declaration [52,53]. As stated in this fundamental principle, public participation in environmental issues consists of three pillars: (1) access to environmental information, (2) public participation in environmental decision-making, and (3) access to justice in environmental matters. The enhancement of these three pillars is considered to promote the quality, legitimacy, and effectiveness of environmental law [52].

In Europe, Aarhus Convention is a milestone regarding the promotion and enhancement of public participation in environmental issues. Specifically, on June 25, 1998, in the Danish city of Aarhus, following Rio Declaration, during the Convention on Access to Information held by the United Nations Economic Commission for Europe (UNECE), the principles of public participation in decision-making and access to justice in environmental matters were adopted.[10] It entered into force on October 30, 2001.

Specifically, the following rights of the public were introduced:

- *The Right to Access to Environmental Information.* Public authorities have to give access to the public to information regarding the state of the environment, the measures taken, human health and safety as affected by the state of the environment, or the policies implemented, when requested to do so. In addition, Aarhus Convention went one step further: Releasing information after a written demand is essential, but equally vital is collecting and disseminating information held by public authorities in a way that it is easy to access, retrieve, and understand for everyone [54]. The development of the European Pollutant Emission Register, for example, and of the subsequent European Pollutant Release and Transfer Register took place in this context.[11] This is the reason why so many registers and databases have been developed in the last decade in Europe.
- *The Right to Justice.* The public must have the right to review procedures and to challenge public decisions when its rights are accidentally, or deliberately, denied. It should be able to appeal against actions that do not comply with environmental legislation, against rejections of requests on accessing environmental information, or against failures of law in decision-making processes [54].
- *The Right to Participate.* The Aarhus Convention gives the public the right to participate in making decisions in relation with environment. Public authorities have to enable citizens and environmental nongovernmental organizations to make comments on proposals for projects, plans, and programs affecting the environment. This consultation process has to be taken into account in decision-making and all relevant information must be transparent. The Convention sets out minimum levels of opportunities for participation and the procedures that must be followed.

Gradually, the European Union issued legislation regarding each right, and all member states introduced legal acts in order to implement Aarhus Convention decisions in their territories. Mauerhofer and Larssen present the respective legislation at the EU level [52].

But Europe is not the only region to promote public engagement. Also in the United States, public participation in environmental policy-making is adopted extensively by all levels of government [55], as a result of the general trend of strengthening the engagement

---

[10]http://ec.europa.eu/environment/aarhus/.
[11]http://prtr.ec.europa.eu/.

of citizens in the public administration [56]. Consequently, there has been an extensive production of provisions and mandates for public participation throughout all levels of government, and an increased understanding of the different types of public participation available to best incorporate the public in environmental policy-making and management [55,57]. For example, the U.S. National Environmental Policy Act (NEPA), besides initiating the development of *Environmental Impact Assessment*, integrated the concept of public participation in the process [58].

Following the global tendency, China made also many steps to empower the public to participate in environmental issues by mandating disclosure of environmental information held by the government [59], although the country has been criticized for its culture of state secrecy and insufficient space for the public to participate in managing the Chinese society in general [59,60].

Generally, all over the world many actions are in place in order to promote and enhance public participation in environmental policy-making and in all levels of administration, generally. In Australia, the term *engagement* is also used. According to the guide document prepared for Consult Australia and the International Association for Public Participation Australasia[12] (IAP2) [61], "engagement is the process by which government, organizations, communities and individuals connect in the development and implementation of decisions that affect them. It is used as a tool to achieve outcomes, develop understanding, educate and/or agree to solutions on issues of concern." So, in this approach *engagement* is a broad term that may include the concept of *public participation* and can also incorporate aspects of community, stakeholder or public relations, consultation, and government and media relations.

IAP2 has developed the following core values for the practice of public participation for use in the development and implementation of engagement processes:

- Public participation is based on the belief that those who are affected by a decision have the right to be involved in the decision-making process.
- Public participation includes the promise that the public's contribution will influence the decision.
- Public participation promotes sustainable decisions by recognizing and communicating the needs and interests of all participants, including decision-makers.
- Public participation seeks out and facilitates the involvement of those potentially affected by or interested in a decision.

---

[12]IAP2 is an international association of members (with a very active Australasian Chapter) who seek to promote and improve the practice of public participation in relation to individuals, governments, institutions, and other entities that affect the public interest in nations throughout the world. In the Australasia region, the term community engagement is more frequently used to refer to public participation.

- Public participation seeks input from participants in designing how they participate.
- Public participation provides participants with the information they need to participate in a meaningful way.
- Public participation communicates to participants how their input affected the decision.

The purpose of the core values is to help make better decisions, which reflect the interests and concerns of potentially affected people and entities [61].

But why public participation is considered so important in environmental decision-making?

The importance of public participation. Public participation is essential for the protection of the environment and the promotion of sustainability in general. First of all, it satisfies and fulfills the values of democracy. People are not just subjects, but they become active citizens who take part in decisions of great importance for the environment, and their lives also. In addition, public participation utilizes the enthusiasm, knowledge, and skills of individuals and recognizes that the public has an important role to play [55].

Public participation in making decisions is vital. It can be beneficial to reaching an individual decision and may also establish democracy practices. It uses the knowledge, skills, and enthusiasm of the public to help in making a decision. Besides its added value to effectiveness, public participation facilitation is also a moral duty. Since public authorities work for and to the best interests of the public, they have to perform their tasks in a way that the public wants. So, the best way to ensure that they have the correct idea of the wishes of citizens is to engage them when making decisions. Each person has its share in protecting and enhancing the environment, and citizens know the needs of their communities through work, social interaction, and travel. That is why public involvement is a core and fundamental element of sustainable development policies. Solutions to achieve together economic, social, and environmental improvements will only be found, if everyone is involved and if the discussion is open so that new ideas and approaches can be considered [54].

Better decisions can be also taken by facilitating public participation. People can express their opinions about which solution serves the best the public, and what decisions can hold longer and more credible. Overall, the quality of life of everyone will be positively affected. As a result, relations of trust will be developed and strengthened between the public and the authorities. In this way, public participation can promote democracy. Public participation is a proof that people are valued and that their views are important. When citizens become involved in working out a mutually acceptable solution to a project or problem that affects their community and their personal lives, they mature into responsible democratic citizens and reaffirm democracy [62].

Another dimension that should not be overlooked is that without the public participation of those affected by a new legislative act, the implementation of measures decided centrally at distant think tanks and political instruments meets serious resistance and opposition in practice, especially in very small communities. For example, consider a measure decided by the bureaucrats of Brussels that affects considerably the life of the inhabitants of a small Greek island; it is very expected that the implementation will fail in case that nobody participated or had an idea about these plans from the local society. It is also very essential to add that in such small communities, most citizens and local authorities' representatives, from the baker and the teacher to the police officer, judge and mayor, have some kind of an extended family relationship. So, nobody can force anyone to act accordingly to new laws, especially when nobody participated in any level of the policy-making process.

But public participation itself cannot always lead to best results unconditionally. According to Hartley and Wood [63], there are a number of barriers that may limit the successful implementation of this principle, such as:

- poor public knowledge of planning, legal, and waste licensing issues;
- poor provision of information;
- poor access to legal advice;
- mistrust of the waste disposal industry;
- not in my back yard (NIMBY) syndrome;
- failure to influence the decision-making process;
- poor execution of participation methods; and
- regulatory constraints.

Moreover, the process public participation is not free of weaknesses like incurred costs, time-consuming procedures without tangible benefits, and a lack of a common framework to incorporate public and technical inputs [55]. In many cases, the manifestation of the concept "think globally, act locally" seems to result in a disproportional focus on local interests [55].

Specifically, public participation in environmental policy-making may be focused too much on local interests at the cost of regional and national interests merely because legitimate national stakeholders were excluded from the participation meetings [55]. Singleton [64] provided an example of this weakness: a plan to construct a dam is proposed in order to increase the power and enhance the employment and recreational opportunities of a town. However, the impacts of such a construction extend much further to the downstream inhabitants and can virtually destroy a fishing industry. As Singleton points out: "While the process is local, many of the sources of the problems it seeks to address and the constituencies it must respond to are not" [64]. Moreover, the emphasis on local interests may transform public participation from a process to protect the environment to a process to facilitate local interests and reduce social conflicts.

It is important to remember that public participation does not automatically lead to consensus. Moreover, public participation does not mean that the competent authority cannot take a decision that is unpopular. The role of public participation is to allow the public to express their opinions and for the authority to consider them in making the decision. Sometimes, the authority must take a decision that is in the interests of the wider community, but that is unpopular with the local one [54].

Finally, it should be pointed out that the proper legal framework should be developed to foster public participation. Besides legislative acts in relation with the promotion of public participation and the obligation of authorities to both disseminate environmental information and to integrate the results of people's involvement into decisions, anti-SLAPP legislation may be required. SLAPP means a *strategic lawsuit against public participation* and is a suit with a hidden purpose. Its goal is a strategic victory rather than a legal or moral one. The private citizen, faced with an alarming and costly legal defense as a result of his or her speech or other activism, will stop pursuit of public advocacy against (or oversight of) the purported polluter or developer. Other private actors may hesitate to take action in the future for the fear of being subjected to similar legal proceedings [65].

1.5 Case Study 2: European Commission's Public Consultation on Air Pollution

The purpose of this short case is to show that, especially in the territory of the European Union, the consultation process does not end with the first answers on a proposed draft legal document, but that it is rather an ongoing process providing feedback for all relevant steps and implementation actions.

The first case study to be discussed has been presented by Gemmer and Xiao [66] and concerns the public consultation on air quality held by the European Commission in December 2004 and January 2005. Two online questionnaires were developed for citizens and consumers, and businesses, and more than 11,000 answers were collected, analyzed, and used as input for the 2005 thematic strategy on air pollution. It was found that more than 80% of the population considered itself not properly informed on issues regarding air pollution, emissions, health and environmental impacts, etc. Moreover, most participants in the consultation process declared their willingness to spend substantial funding for improving air quality by taking measures in the fields of both industrial air pollution and traffic. The analysis of the answers received showed also that the framework and targets of the thematic strategy on air quality were in line with the observations and concerns of people. All the results regarding the consultation process can be obtained from TNO [67]. It is equally important that the process did not stop there. The European Commission gathered comments during all steps in the legislative process for air quality, from preparation to implementation and review. Expert groups from all stakeholders (EU Member States officials, business associations, environmental NGOs, etc.) kept meeting regularly on a 6-month basis to give feedback regarding the implementation of the existing legal framework or to propose new legislative measures. A dedicated Website was created to publicize all relevant presentations and protocols. In addition, all meetings were live streamed on the Internet and citizens were allowed to provide their views through the expert groups. On June 30, 2011, the European Commission launched again two types of online questionnaires as part of a broad consultation process on the EU Air Quality Directive: a short one for the general public and a more detailed one for experts and practitioners from all kinds of interested groups. These results can be also obtained from TNO [68].

1.6 Case Study 3: Three Cases of Environmental Activism in China

The purpose of the second case study is to show that people can change governmental decisions even under intimidation conditions.

Li et al. [59] compare three cases of environmental activism in China that changed government decisions, in a rather conservative political regime compared to the ones in Western countries. Moreover, this case study shows the importance of public participation in all stages of environmental decisions: even though the Environmental Impact Assessment Law followed the principle of public participation, the procedural rights of the public could not be taken for granted. Despite the desire of the officials and various means of oppression, such as shutdown of opposing Websites and police threatening to activists, the public strongly opposed these plans and achieved important victories, not only because citizens forced the government to change its original decisions and plans, but also because they set examples for effective public participation [59].

Specifically, the three cases of environmental activism were all targeted at projects planned to meet the goals of industrialization and urbanization. The first project concerned the construction of 13 dams on the Nu River in 2003 to provide 21 million kilowatts of energy. The second project that met strong opposition was the construction of the Xiamen PX (paraxylene) chemical plant. The Tenglong Aromatic PX (Xiamen) Corporation invested 10.8 billion yuan RMB to build the PX chemical plant in Haicang district in 2006, Xiamen, aiming at replacing imported paraxylene. The plant would add 80 billion yuan RMB worth of industrial output annually (one-fourth of Xiamen's GDP), if it started to operate in 2008. The third project referred to the plans of building the Liu Li Tun garbage incineration power plant in 2005 to convert waste to energy. All three projects were officially approved by many authorities, and the relevant environmental assessment impact studies had been also accepted.

Finally, as a result of citizens' mobilization, the second and the third constructions were relocated, whereas the first one was suspended and it is still under discussion, which showed that people have power even against conservative or even authoritative political regimes [59].

1.7 Environmental Legislation

The main environmental legislation for each sector is presented in the following chapters. In this paragraph, the purpose is not to quote an endless list of legislative acts in place globally, but to present the main underlying principles in environmental policy formulation, which constitute the common base for most legislative acts. Specifically, the main principles of the environmental legislation in the European Union is presented since it is the best structured one internationally, incorporating most of the principles derived in relevant international forums and in Rio Declaration.

European environment policy is based on the principles of *precaution*, *prevention*, and *rectifying pollution at source*, and on the "polluter pays" principle. Multiannual environmental action programs set the framework for future action in all areas of environment policy. They are embedded in horizontal strategies and taken into account in international environmental negotiations. Last but not least, implementation is crucial. So, both the integration of any European legislative act into national law and its implementation by member states in their territories are monitored centrally [69].

The European environment policy dates back to the European Council held in Paris in 1972, at which the European Heads of State and Government (in the aftermath of the first UN conference on the environment) declared the need for a community environment policy flanking economic expansion and called for an action program [69].

The Single European Act of 1987 introduced a new "Environment Title", which provided the first legal basis for a common environment policy with the aims of preserving the quality of the environment, protecting human health, and ensuring a rational use of natural resources. Subsequent Treaty revisions strengthened Europe's commitment to environmental protection and the role of the European Parliament in its development. The Treaty of Maastricht (1993) made the environment an official EU policy area, introduced the codecision procedure, and made qualified majority voting in Council the general rule. The Treaty of Amsterdam (1999) established the duty to integrate environmental protection into all EU sectorial policies with a view to promoting sustainable development. "Combating climate change" became a specific goal with the Lisbon Treaty (2009), as also sustainable development did in relations with third countries. A new legal personality enabled the European Union to conclude international agreements [69].

So, let's review the main principles of the European environmental policy [69–71].

The precautionary principle. The precautionary principle is a part of EU law, but it has been also included in virtually every recent treaty and policy document related to the protection and preservation of the environment. Although it appears in principle 15 of the 1992 Rio Declaration (*see* Table 1.4), the precautionary principle evolved out of the German sociolegal tradition. There is no universally accepted and common definition. In European Union, the precautionary principle is a risk management tool that may be invoked when there is scientific uncertainty about a suspected risk to human health or the environment emanating from a certain action or policy. For instance, to avoid damage to human health or to the environment in case of doubt about a potential dangerous effect of a product, instructions may be given to stop the distribution of this product or to remove it from the market if uncertainty persists following an objective scientific evaluation. Such measures must be nondiscriminatory and proportionate and must be reviewed once more scientific information is available. In international law also, this principle commonly states that in cases when serious harm is threatened, positive action to protect the environment should not be delayed until irrefutable scientific proof of harm is available.

The polluter pays principle. The "polluter pays" principle is implemented by the Environmental Liability Directive (ELD), which aims at preventing or otherwise remedying environmental damage to protected species and to natural habitats, water and soil. This principle is of primary significance for the effectiveness of environmental protection, as it provides a direct incentive for potential polluters not to pollute. The essence of the polluter pays principle is that the person who causes environmental damage is also responsible to take the appropriate measures to remedy it and pay for the costs. So, this principle forces the polluter to internalize the costs associated with his or her production or consumption. Moreover, it can displace other general principles like the right to property. The scope of the directive has been broadened three times to include the management of extractive waste, the operation of geological storage sites, and the safety of offshore oil and gas operations, respectively.

The principle of prevention and rectifying pollution at source. This principle is related to the polluter pays principle but is less widely recognized as it is more recent in formulation. The source principle simply states that any form of pollution should be treated as closely as possible to the source. In practice, the principle means to develop and apply environmentally friendly technologies and products. For example, air pollution should be remedied by stack scrubbers at the source, whereas water pollution should be remedied by filters at the source.

The principle of prevention. Avoiding pollution is much cheaper and feasible than spending money to rectify environmental damages resulting from inadequate technologies and products. In other words, prevention measures should be taken so as to minimize environmental damage. The difference between the principle of prevention and the precautionary principle is the evaluation of the risk threatening the environment. Precaution comes into play when the risk is high—so in fact full scientific certainty should not be required prior to the taking of remedial action.

The principle of sustainable development. Another general principle of the European environmental law is the principle of sustainability. This concept has been already extensively commented previously and has been shown that there is no universally common interpretation for it. Public participation and access to environmental information are widely regarded as very important for the implementation of the principle. Sustainable development is also an official aim of EU policy and is now listed in Article 2 of the Treaty on the European Union.

The subsidiarity principle. In contrast to the previous principles, the subsidiarity principle does not only apply to environmental policy, but to all fields of regulation. Its meaning is that only those tasks are dealt with at the higher level, which cannot be dealt with at the lower level. Since the 1992 Maastricht summit, the subsidiarity principle has become a very important topic at the European level.

The principle of integration. The environment is not only affected by environmental policy, but also by other policies and development decisions. For example, transport policy has a direct impact on the state of environment. It was thus understood that environmental protection could not be achieved, if the environmental dimension would not be integrated in other policies also. Hence, the principle of integration basically requires that environmental concerns are also taken into account in nonenvironmental policy areas. Integrating environmental concerns into other EU policy areas has become an important concept in European politics since it first arose from an initiative of the European Council held in Cardiff in 1998. The last ten years, environmental policy integration has made, for instance, significant progress in the field of energy policy, as reflected in the parallel development of the European Union's climate and energy package or in the Roadmap for moving to a competitive low-carbon economy by 2050, which looks at cost-efficient ways to make the European economy more climate-friendly and less energy-consuming.

1.8 Synopsis of the Book

In the chapter "Atmospheric Environment" of this book, the implications of human activities on the atmospheric environment are thoroughly presented. The scales of atmospheric pollution are set and the various pollutants are classified accordingly. Special emphasis is given to global environmental problems for many reasons: (1) these problems threaten life on earth as we know it on the planetary scale, (2) as they take place globally, the exact mechanisms have not been fully understood and scientific and/or political community has not reached to a common position regarding mitigating measures, (3) international cooperation is required against them, and (4) these are the problems, which led to increased public awareness and public participation to environmental issues. Moreover, these phenomena exhibit increased "popularity." Who has not heard about greenhouse effect or about stratospheric ozone depletion? In various forms, air pollution has an adverse impact on our health in the short and long term, and the problems of the greenhouse effect and the destruction of stratospheric ozone could extinguish life from the face of the Earth. All the latest data regarding the health effects of air pollution are presented. At the end, issues relevant to indoor air pollution and noise pollution are discussed.

Aquatic environment and the relevant problems are discussed in chapter "Aquatic Environment." Generally, the public is more sensitive to the pollution of aquatic environment and to issues regarding the quality of drinking water, as they have an immediate and visible impact on daily routine and recreational activities. Besides the pollution issues, the supply, use, and management of water resources are also among the topics covered. We should never forget that although water covers 71% of the planet's surface, less than 1% of it is available for human consumption, while more than 1.2 billion people in the world have no access to safe drinking water. Special emphasis is given to the configurations of plants to treat drinking water and wastewaters. To this end, many illustrations are provided. The distinctive element of the whole consideration of this chapter is that wastewater is considered as a resource and not a useless stream that has to be rejected.

Chapter "Soil Environment" deals with issues regarding soil environment. Soil pollution, erosion, and desertification are significant threats that result in both deterioration of ecosystems and in less land suitable for cultivation. It should not be overlooked that soil environment is in constant interaction with aquatic and atmospheric environment. Changes in the conditions in any of them may have a detrimental impact on the quality of the rest.

The discussion regarding greenhouse effect and atmospheric pollution would have been incomplete if the relation between energy production and consumption with environment had not been covered in chapter "Energy and the Environment." Renewable and alternative sources of power are explicitly presented as well as the promising use of hydrogen as fuel. Finally, nuclear energy along with the problems associated with its production and the available solutions are discussed.

Chapter "Extraterrestrial Environment" is devoted to a new and extremely interesting topic: Extraterrestrial Environment. Humankind has achieved in polluting also earth orbit, while there are suggestions on how to pollute outer space with nuclear waste. There are also lessons to be learnt by observing other planets in relation with greenhouse effect.

Chapter "Environment and Development," the final chapter, examines the notion of sustainable development and the main tools toward it. Alternative development concepts are also presented and the environmental implications of economic development and poverty are covered. The notions of development, sustainable development, growth, and progress are clarified.

In all chapters, the international dimensions of all issues covered are discussed, especially in relation with the policies applied worldwide. Environmental problems are dealt with in the context of development and in relation with human activities. Many case studies are provided to highlight the concepts discussed.

References

[1] S. Hook, Education for Modern Man: A New Perspective, Alfred A. Knopf, New York, 1946.
[2] V.L. Smith, Humankind in prehistory: economy, ecology, and institutions, in: T.L. Anderson, R.T. Simmons (Eds.), The Political Economy of Customs and Culture, Rowman & Littlefield Publishers, Inc., 1993, pp. 157−184.
[3] R. Gibbons, Examining the extinction of the pleistocene megafauna, Stanford Undergrad. Res. J. 3 (2004) 22−27.
[4] R.G. Roberts, et al., New ages for the last Australian megafauna: continent-wide extinction about 46,000 years Ago, Science 292 (2001) 1888−1892.
[5] D.R. Horton, Red kangaroos: last of the Australian megafauna, in: P.S. Martin, R.G. Klein (Eds.), Quaternary Extinctions: A Prehistoric Revolution, University of Arizona Press, Tucson, 1984, pp. 639−680.
[6] S.A. Elias, D.C. Schreve, Late pleistocene megafaunal extinctions, in: S.A. Elias (Ed.), Reference Module in Earth Systems and Environmental Sciences. Encyclopedia of Quaternary Science, Elsevier, 2013, pp. 700−712.
[7] P.S. Martin, Prehistoric overkill: the global model, in: P.S. Martin, R.G. Klein (Eds.), Quaternary Extinctions: A Prehistoric Revolution, University of Arizona Press, Tucson, 1984, pp. 354−403.
[8] S. Wroe, A review of terrestrial mammalian and reptilian carnivore ecology in Australian fossil faunas, and factors influencing their diversity: the myth of reptilian domination and its broader ramifications, Aust. J. Zool. 50 (2002) 1−24.
[9] T. Brown, Clearances and clearings: deforestation in Mesolithic/Neolithic Britain, Oxford J. Archaeol. 16 (2) (1997) 133−146.
[10] J.O. Kaplan, K.M. Krumhardt, N. Zimmermann, The prehistoric and preindustrial deforestation of Europe, Quat. Sci. Rev. 28 (2009) 3016−3034.
[11] P. Bourdeau, The man-nature relationship and environmental ethics, J. Environ. Radioact. 72 (2004) 9−15.
[12] F. Ost, La Nature Hors-la Loi, La Découverte, Paris, 1995.
[13] P. Destatte, Foresight: a major tool in tackling sustainable development, Technol. Forecast. Soc. Change 77 (2010) 1575−1587.
[14] A. Peccei, The Chasm Ahead, Macmillan, New York, 1969.
[15] J.W. Forrester, World Dynamics, Wright-Allen Press, Cambridge, 1971.
[16] D.H. Meadows, D.L. Meadows, J. Randers, W. Behrens, The Limits to Growth: A Report of the Club of Rome's Project on the Predicament of Mankind, Universe Books, New York, 1972.

[17] T.R. Malthus, An Essay on the Principle of Population, Oxford World's Classics Reprint, 1798.

[18] J. Avery, Malthus' Essay on the Principle of Population, University of Copenhagen, Denmark, 2005.

[19] A.J. Bennet, Environmental consequences of increasing population: some current perspectives, Agric. Ecosyst. Environ. 82 (2000) 89–95.

[20] G. Hardin, The feast of malthus – living within limits, Soc. Contract (1998) 181–187.

[21] R.L. Heilbroner, Philosophers of the Economic World, Attiki, Athens, 2000 (in Greek).

[22] D.L. Kelly, C.D. Kolstad, Malthus and climate change: betting on a stable population, J. Environ. Econo. Manag. 41 (2001) 135–161.

[23] M. Kirk, The return of Malthus? the global demographic future, 2000–2050, Futures 16 (2) (1984) 124–138.

[24] E.L. Wynder, A corner of history, Prev. Med. 4 (1975) 378–383.

[25] WCED, Report of the World Commission on Environment and Development: Our Common Future, World Commission on Environment and Development, United Nations, 1987. http://www.un-documents.net/ourcommon-future.pdf.

[26] A.M. Wallis, M.L.M. Graymore, A.J. Richards, Significance of environment in the assessment of sustainable development: the case for south west Victoria, Ecol. Econ. 70 (2011) 595–605.

[27] M. Keiner, Re-emphasizing sustainable development – the concept of 'evolutionability', Environ. Dev. Sustainability 6 (2004) 379–392.

[28] P. Hardi, T. Zdan, Assessing Sustainable Development: Principles in Practice, The International Institute for Sustainable Development, Canada, 1997. https://www.iisd.org/pdf/bellagio.pdf.

[29] A. Valentin, J.H. Spangenberg, A guide to community sustainability indicators, Environ. Impact Assess. Rev. 20 (2000) 381–392.

[30] IDRC, Assessment Tools, International Development Research Center, Ottawa, 1997.

[31] C.J. Koroneos, D. Rokos, Sustainable and integrated development – a critical analysis, Sustainability 4 (2012) 141–153.

[32] UN, Agenda 21, United Nations, Division for Sustainable Development, New York, 1992. https://sustainabledevelopment.un.org/content/documents/Agenda21.pdf.

[33] UNCED, The Rio Declaration on Environment and Development, United Nations Conference on Environment and Development 1992, Rio, 1992. http://www.unesco.org/education/nfsunesco/pdf/RIO_E.PDF.

[34] UN, Report of the World Summit on Sustainable Development, United Nations, New York, 2002. http://www.unmillenniumproject.org/documents/131302_wssd_report_reissued.pdf.

[35] R. Prescott-Allen, The Wellbeing of Nations, Island Press, Washington, 2001.

[36] G. Carvahlo, Sustainable development: is it achievable within the existing international political economy context? Sust. Dev. 9 (2001) 61–73.

[37] UNEP, Global Environment Outlook, United Nations Environment Programme, Nairobi, 2007.

[38] J.M. Calm, D.A. Didion, Trade-offs in refrigerant selections: past, present and future, Int. J. Refrig. 21 (4) (1998) 308–321.

[39] A. McCulloch, CFC and Halon replacements in the environment, J. Fluorine Chem. 100 (1999) 163–173.

[40] J.E. Lovelock, R.J. Maggs, R.J. Wade, Halogenated hydrocarbons in and over the Atlantic, Nature 241 (1973) 194–196.

[41] M.J. Molina, F.S. Rowland, Stratospheric sink for chlorofluoromethanes: chlorine atom-catalysed destruction of ozone, Nature 249 (1974) 810–812.

[42] C.V. Nolan, G.T. Amanatidis, European commission research on the fluxes and effects of environmental UVB radiation, J. Photochem. Photobiol. B 31 (1995) 3–7.

[43] UNEP, The Vienna Convention for the Protection of the Ozone Layer, Ozone Secretariat, United Nations Environment Programme, Nairobi, 2001. http://ozone.unep.org/pdfs/viennaconvention2002.pdf.

[44] J.M. Pacheco, V.V. Vasconcelos, F.C. Santos, Climate change governance, cooperation and self-organization, Phys. Life Rev. 11 (2014) 573–586.

[45] W.-T. Chen, C.-M. Shu, CO_2 reduction for a low-carbon community: a city perspective in Taiwan, Sep. Purif. Technol. 94 (2012) 154–159.

[46] E. DeNicola, P.R. Subramaniam, Environmental attitudes and political partisanship, Public Health 128 (2014) 404–409.

[47] K.H. Solangi, M.R. Islam, R. Saidur, N.M. Rahim, H. Fayaz, A review on global solar energy policy, Renew. Sust. Energy Rev. 15 (2011) 2149−2163.

[48] IEA, World Energy Outlook, International Energy Agency, Paris, 2004.

[49] T.L. Hunt, Rethinking Easter Island's ecological catastrophe, J. Archaeol. Sci. 34 (2007) 485−502.

[50] B. Pakandam, Why Easter Island Collapsed: An Answer for an Enduring Question, Working Papers No. 117/09, London School of Economics, London, 2009, http://eprints.lse.ac.uk/27864/1/WP117.pdf.

[51] J. Diamond, Easter's end, Discover 9 (1995) 62−69. http://discovermagazine.com/1995/aug/eastersend543.

[52] V. Mauerhofer, C. Larssen, Judicial perspectives from the European Union for public participation in environmental matters in east asia, Land Use Policy (2015), http://dx.doi.org/10.1016/j.landusepol.2015.06.007 in press.

[53] M. Pallemaerts, La conférence de Rio: grandeur ou décadence du droitinternational de l'environnement? Rev. b. dr. intern. 28 (1) (1995) 175−223.

[54] DETR, Public Participation in Making Local Environmental Decisions. The Aarhus Convention. Newcastle Workshop. Good Practice Handbook, Department of the Environment, Transport and the Regions, London, 2000.

[55] B. Ran, Evaluating public participation in environmental policy-making, J. US-China Public Adm. 9 (4) (2012) 407−423.

[56] T.L. Cooper, T.H. Bryer, J.W. Meek, Citizen-centered Collaborative public management, Public Adm. Rev. 66 (1) (2006) 76−88.

[57] U. Desai, Public participation in environmental policy implementation: case of Surface Mining Control and Reclamation Act, Am. Rev. Public Adm. 19 (1) (1989) 49−65.

[58] A.N. Glucker, P.P.J. Driessen, A. Kolhoff, H.A.C. Runhaar, Public participation in environmental impact assessment: why, who and how? Environ. Impact Assess. Rev. 43 (2013) 104−111.

[59] W. Li, J. Liu, D. Li, Getting their voices heard: three cases of public participation in environmental protection in China, J. Environ. Manage. 98 (2012) 65−72.

[60] G. Wu, China in 2010: dilemmas of "scientific development", Asian Surv. 51 (1) (2011) 18−32.

[61] IAP2 AUSTRALASIA, Valuing Better Engagement Proudly Supported by: An Economic Framework to Quantify the Value of Stakeholder Engagement for Infrastructure Delivery, International Association for Public Participation2 Australasia, Wollongong, 2015. http://www.consultaustralia.com.au/docs/default-source/infrastructure/engagement/valuing-better-engagement--economic-framework.pdf?sfvrsn=2.

[62] T. Webler, H. Kastenholz, O. Renn, Public participation in impact assessment: a social learning perspective, Environ. Impact Assess. Rev. 15 (1995) 443−463.

[63] N. Hartley, C. Wood, Public participation in environmental impact assessment − implementing the Aarhus Convention, Environ. Impact Assess. Rev. 25 (2005) 319−340.

[64] S. Singleton, Collaborative environmental planning in the American west: the good, the bad and the ugly, Environ. Polit. 11 (3) (2002) 54−75.

[65] C.S. Norman, Anti-slapp legislation and environmental protection in the USA: an overview of direct and indirect effects, RECIEL 19 (1) (2010) 28−34.

[66] M. Gemmer, B. Xiao, Air quality legislation and standards in the European Union: background, status and public participation, Adv. Clim. Change Res. 4 (1) (2013) 50−59.

[67] TNO, Public Views on Air Pollution in the European Union: Results of the European Commission's Public Consultation on Air Pollution, TNO-report B&O-A R2005/100, TNO Built Environment and Geosciences, 2005.

[68] TNO, Survey of Views of Stakeholders, Experts and Citizens on the Review of the EU Air Policy Part I: Main Results, TNO report TNO-060-UT-2012−00714, TNO Innovation for Life, 2012.

[69] T. Ohliger, Environmental Policy: General Principles and Basic Framework, Fact Sheets on the European Union, European Parliament, Brussels, 2015. http://www.europarl.europa.eu/ftu/pdf/en/FTU_5.4.1.pdf.

[70] E. Engle, General principles of european environmental law, Pennsylvania State Environmental Law Reviews 17 (2008) 215−224.

[71] H. von Seht, H.E. Ott, EU Environmental Principles: Implementation in Germany, Wuppertal Papers Nr. 105, Wuppertal Institute for Climate, Environment and Energy, Climate Policy Division, Wuppertal, 2000.

CHAPTER 2

Atmospheric Environment

S.G. Poulopoulos

Kazakh-British Technical University, Almaty, Republic of Kazakhstan

Chapter Outline

Environment and Development. http://dx.doi.org/10.1016/B978-0-444-62733-9.00002-2

2.1 Introduction: The Stratification of Atmosphere

Earth is surrounded by a thin layer of gases that is called atmosphere. Although it is difficult to define a specific thickness of this layer, it is characterized thin because it is about 100–150 km above the sea level and the average radius of earth is 6370 km. Yet, this thin layer affects numerous processes that take place on earth.

Its existence sustains the life on earth—at least as we know it—for a number of reasons:

- It keeps a global surface average temperature of 15°C that is favorable for the life forms evolved on earth. So, although extreme temperatures ranging from −89°C (at Vostok, Antarctica) to 58°C (at Al Aziziyah, Libya) have been recorded, Earth is a friendly planet in terms of climate variation and temperature, as a result of the role of carbon dioxide contained in it and the greenhouse effect mechanism [1].
- It contains today about 1.20×10^{18} kg of oxygen (about 21% in volume), a level probably reached not later than 25 million years ago. Respiration of humans and animal life is absolutely depended on this oxygen [2].
- It acts as a filter for dangerous radiation from sun. Atmosphere is the transparent layer through which, solar radiation passes and reaches the surface of Earth, land or water. Solar radiation is the only source to supply energy for the most vital process on earth that supports all other life: photosynthesis. Atmosphere filtrates 95% of the harmful ultraviolet part of solar radiation via ozone layer [3].
- The continuous interaction of the gases contained in the atmosphere with solar radiation regulates the flow of energy through the climate system, which makes the composition of the atmosphere a key determinant of Earth's climate [1].
- Atmosphere as a part of biosphere interacts continuously with hydrosphere and lithosphere. Its components participate in most critical elements cycles, such as carbon, oxygen, nitrogen, and water cycle. This is the reason why Earth's atmosphere can be described as being in a state of dynamic equilibrium. All the time, without any interruption, substances enter and leave the atmosphere, forming different compounds at different times and places [4].
- Outer space is full of potentially dangerous objects like asteroids, comets, meteoroids, and even man-made satellites. Although most of them are not on a path to collision with earth, some occasionally pose a threat to our planet. Atmosphere burns all these objects before they getting crashed on Earth's surface, protecting thus again life on Earth.

It is also noteworthy that atmosphere does not only protect life on earth, but it seems to have played also an important role in the origin of it. Particularly, the redox state of Earth's early atmosphere may have controlled greatly the efficiency of abiotic synthesis of biologically important organic compounds [5,6]. Atmosphere is not also static, but changes with Earth's aging. Hashimoto et al. [5] provide a short description of the evolution of atmospheric composition.

It is obvious that the composition of the atmosphere is very important. Dry atmospheric air is composed mostly of molecular nitrogen ($\sim 78\%$) and molecular oxygen ($\sim 21\%$). Inert gas argon ($\sim 1\%$) is the next most abundant gas in the atmosphere. These gases make up more than 99.9% of the mass of dry air, whereas the ratio of the number of molecules of each is nearly constant up to a height of about 80 or 90 km. The atmospheric gases that are important for the absorption and emission of radiant energy comprise less than 1% of the atmosphere's mass: water vapor, carbon dioxide, and ozone followed by methane, nitrous oxide, and a number of other minor species in order of importance for surface temperature [1]. The composition of atmospheric air is presented in Table 2.1.

Almost 50% of the atmospheric mass can be found within 5.5 km from the sea level; 90% is within about 16 km, and 99.9% is below 50 km. However, it is still possible to detect air at altitudes as high as 500 km [7].

Earth's atmosphere can be stratified, meaning that it can be considered to be composed of a number of layers. The way in which these layers are defined depends on the property considered. The vertical distribution of temperature, pressure, density, and composition of the atmosphere constitutes atmospheric structure. So, layers defined by different types of properties can overlap. For example, if the concentration of atmospheric gases is taken into account, then atmosphere is divided into [3]:

Table 2.1: Near Sea Level Concentration of Atmospheric Gases [2].

Gas	Chemical Symbol	Concentration (ppmv)
Nitrogen	N_2	780,840
Oxygen	O_2	209,460
Argon	Ar	9340
Carbon dioxide	CO_2	384
Neon	Ne	18.18
Helium	He	5.24
Methane	CH_4	1.774
Krypton	Kr	1.14
Hydrogen	H_2	0.56
Nitrous oxide	N_2O	0.320
Xenon	Xe	0.09
Ozone	O_3	0.01−0.10

- *Homosphere*: It is the lower part from the surface of the earth to a height of 80—100 km above the earth. In this layer, gases are more or less uniform in their chemical composition, as a result of bulk motion of large volumes of air that effectively mixes the components of atmosphere.
- *Heterosphere*: Starting from the upper portion of homosphere, it extends to a height of 500 km above the earth. In this layer, chemical composition changes considerably with height since molecular diffusion dominates mixing. Specifically, the concentration of gases decreases with altitude.
- *Exosphere*: The part of atmosphere above about 500 km is called exosphere, because molecules with an upward trajectory, given sufficient velocities, may escape the atmosphere without colliding with other molecules.

However, the most important property that varies with height in our atmosphere is temperature, and as a result the most common classification of atmosphere's layers is based on its vertical distribution. Specifically, we may break up atmosphere into four major layers: troposphere, stratosphere, mesosphere, and thermosphere, with exosphere being again the exterior layer considered, as shown in Fig. 2.1.

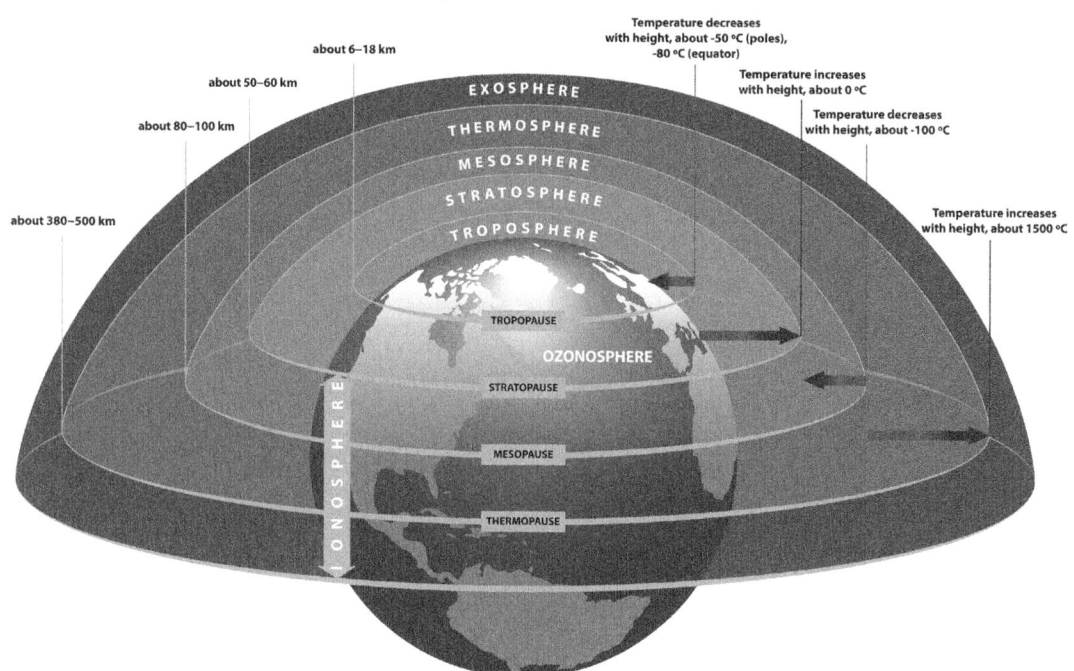

Figure 2.1
The stratification of atmosphere. *Used under license from Shutterstock.com/Image ID:115547989.*

This subdivision of the atmosphere is based on the lapse rate $\gamma = -dT/dz$ (°C/1000 m) because it characterizes the meteorological properties of a layer (mixing, separation, convection, etc.) in the simplest way. So, the limits of each boundary are easily decided by its change of sign or by a sudden change in its absolute value [8]. The classification of atmospheric layers is as follows [1−3,7,8]:

Troposphere: The first layer concerns the air that we breathe. Here, all gas emissions resulting from human activities are primarily released. It contains almost 70% of the total mass of the atmosphere, ranging from 6 to 18 km, depending on whether we are on the poles or at the equator. Weather events are also revealed in this layer. It is characterized by temperature decrease with height and violent and turbulent movement of air, which results in almost uniform concentrations of gases. The temperature decreases due to the fact that the troposphere is heated primarily by convection from the surface of the Earth. Convective heating is more effective near the surface, and consequently, its effectiveness decreases as the height increases. The tropospheric average lapse rate is −6.4°C/km, but the lapse rate varies with altitude, season, and latitude. The upper part of troposphere is called *tropopause*. Tropopause is the point where temperature does not decrease with height.

Stratosphere: It is the layer above the troposphere, which extends up to a height of about 50 km. In the stratosphere, the temperature, after an initial invariance with height in the lower section, increases with height, creating thus a stable atmosphere with very small vertical mixing trends. This temperature increase with altitude, contrary to what happens in the troposphere, prevents vertical winds, and as a result only horizontal winds that flow almost parallel to the earth's surface are observed. Thus, turbulence is absent here and a very calm atmosphere is created in this layer. The increase of temperature with height is owed to the presence of ozone that absorbs solar ultraviolet radiation in this layer; 90% of ozone is found here. The concentration of ozone is 3 ppmv at 20 km altitude; 8−10 ppmv at 35 km, and 2 ppmv at 50 km. Above stratosphere, there is a small layer called *stratopause*, where temperature neither decreases nor increases with height up to some level.

Mesosphere: It is the layer above the stratopause, starting from about 52 km and reaching the height of 80−100 km. The mesosphere temperature decreases with height. Its upper boundary is the coldest point of the whole atmosphere. It is characterized by strong vertical mixing. At the end of mesosphere, *mesopause* is found, where temperature neither decreases nor increases with height. The temperature at the height of mesopause drops to almost −90°C.

Thermosphere: The uppermost layer in our atmosphere is called thermosphere. It extends up to the *thermopause*, somewhere between 500 and 1000 km above the surface of the Earth. The position of thermopause changes radically and depends on the amount of the solar radiation that falls on it. The temperature of this layer increases with altitude, as a result of solar radiation absorption, and it is reported to take values as high as 1000°C. The density and the pressure in the thermosphere are extremely low. The pressure can be one million times less than the one on the surface of the earth.

The part above thermosphere is again called *exosphere*. The upper mesosphere and lower thermosphere constitute the *ionosphere*. The name derives from the strong creation of ions through photoionization of the air molecules.

Mesosphere and thermosphere are fields of interest to scientists involved in space travel because a spacecraft has to cross these layers on the way to the moon or other planets. Also, satellites are also found here in orbit around Earth, and therefore these layers are also important to radio communications.

Stratosphere is important to aviation since it is the common travel zone of airplanes that fly at these heights to take benefit from the absence of turbulence in this layer. Stratosphere is also connected with global environmental problems since ozone depletion takes place in it. Moreover, pollution from dispersed particles resulting from volcanic eruptions or atomic bomb tests may also take place in this part of the atmosphere.

Finally, troposphere is also directly connected or related to daily life since most air pollution effects are present in the first atmospheric layer.

2.2 Scales of Atmospheric Pollution

In the previous paragraph, the role of atmosphere in both life origin and life-sustaining mechanisms on Earth was highlighted. This critical role is based on many, multilevel, different—but interacting at the same time—mechanisms and processes, which have been developed in millions and billions of years to the present form and are constantly at a dynamic equilibrium. However, numerous human activities result in the release of many harmful substances in the environment, polluting thus atmosphere, besides soil and water.

Although all environmental problems are important and should be confronted adequately and promptly, there are some mechanisms in operation in the atmospheric environment, the overturn of which could pose a threat to all life forms on earth. It is well known also today that atmospheric problems take place on many time and place scales.

So, air pollution is a problem that manifests itself at both local and global scale. Its duration also variates considerably depending on the mechanism affected: it could be a few minutes or even decades and hundreds of years. For example, predictions for the global warming potential (GWP) by the Intergovernmental Panel on Climate Change

(IPCC) show that the warming effect of CO_2 has several timescales: 20, 100, and 500 years [9]. For purposes of studying, air pollution is commonly categorized according to the levels at which it appears:

- *Local level.* Air pollution at this level concerns generally a region within a maximum 5-km radius. It is characterized by high concentrations of specific pollutants and may be related to emissions released by a large emitter like a chemical plant, or the result of minor releases from many small emitters like cars. High buildings and the terrain can also contribute to high local concentrations of pollutants.
- *Urban level.* This level is related to air pollution observed in cities. Although it is or may be related to local level problems, the term is used to describe types of air pollution with specific characteristics like winter-type smog and photochemical smog [10]. It is recognized today that exposition to this level of air pollution has a negative impact on health and average life span of inhabitants of large cities [11].
- *Regional level.* Regional air pollution affects large areas of 50–1000 km magnitude. It is related to the transport and dispersion or even transformation of urban pollutants over large areas by means of winds and sun, such as tropospheric ozone and acidifying substances like SO_2 and NO_2 [12].
- *Transboundary level.* This level and its air pollution phenomena overlaps with regional level problems; however, the extent regards regions of 1000 km and the "exchange" of air pollution between countries or even continents. For example, Japan and Canada "import" various types of air pollution from China and the United States, respectively. This level of air pollution is another proof of the fact that pollution is a problem that needs intergovernmental cooperation in order to be adequately addressed [13].
- *Global level.* As discussed in the first chapter, global air pollution is related to two famous problems that are responsible to a great extent for the increased public awareness about environmental matters: global warming and stratospheric ozone depletion. These phenomena could lead to a drastic global-scale change in the climate or atmosphere that could ultimately have a tremendous impact on life on Earth [14]. Regarding ozone depletion, the rapid evolvement of ozone depletion and its obvious tremendous potential effects led considerably very quickly to the decision to ban and replace all chlorofluorocarbons (CFCs) involved in ozone destruction. It has to be noted that resolving this problem is generally easier compared to the greenhouse effect, since it is related with compounds "supported" by fewer financial interests. In contrast, carbon dioxide releases and the related greenhouse effect are connected with the current way of living, in the Western world at least, with carbon and fossil fuels interests, with poverty, and with many uncertainties in the relevant research. But it is well understood that only global-scale cooperation could lead to feasible solutions.

2.3 Ambient Air Pollution and Health Effects

Human well-being and sustainability cannot be considered or attempted outside the context of the environment component. Thus, clean air is an integral prerequisite for achieving or at least approaching sustainable development. However, most production or even daily activities like industrial manufacture, agriculture, transportation, cooking, even smoking, etc. are accompanied by the release of numerous compounds into the atmosphere. We should also keep in mind that any process related to energy production, transformation, transfer, and finally use has an inevitable result: air pollution. Every time we switch on the lights or a device, somewhere combustion of a carbon form takes place releasing harmful gases or carbon dioxide into the atmosphere. It has to be noted that although CO_2 cannot be considered as a pollutant in the strict sense, since everybody exhales carbon dioxide, we should not forget that it is the most important contributor to global warming. Air pollution is a major factor that affects human health negatively, especially in large cities.

Air pollution is not a new phenomenon; in medieval times, the burning of coal was forbidden in London while Parliament was in session. However, nowadays air pollution problems have dramatically increased in both intensity and scale due to the increase in emissions since the Industrial Revolution.

As new important data are being accumulated day after day about the potential adverse health effects resulting from air pollution, it is worth presenting some of them. In Table 2.2, the main air pollutants present at the urban scale along with their health effects are shown [15—17].

On March 25, 2014, the World Health Organization (WHO) announced that air pollution is now the world's largest single environmental health risk, resulting in 7 million premature deaths in 2012 worldwide, namely one in eight of total global deaths.[1] That finding doubled previous estimates of the size of the challenge! Air pollution is the biggest environmental health problem and affects everyone from developing and developed countries, rich and poor. For those speculating about many health issues nowadays despite medicine's progress, or those who like to back up conspiracy theories, the message should be repeated in the most clear way:

> …*millions of people in both developing and developed countries die prematurely every year because of long-term exposure to air pollutants. The health of many more is seriously affected.*

Ref. [18]

---

[1]   http://www.who.int/mediacentre/news/releases/2014/air-pollution/en/.

Table 2.2: The Main Health Effects of the Most Important Air Pollutants [15–17].

Pollutant	Main Health Effects
Sulfur oxides (SO_x)	Can cause or aggravate cardiovascular and lung diseases, heart attacks, and arrhythmias. Can cause cancer. May lead to atherosclerosis, adverse birth outcomes, and childhood respiratory disease. The outcome can be premature death.
Nitrogen dioxide (NO_2)	Exposure to NO_2 is associated with increased all-cause, cardiovascular and respiratory mortality and respiratory morbidity.
Particulate matter (PM)	Can cause or aggravate cardiovascular and lung diseases, heart attacks, and arrhythmias. Can cause cancer. May lead to atherosclerosis, adverse birth outcomes, and childhood respiratory disease. The outcome can be premature death.
PAHs, in particular Benzo-a-pyrene (BaP)	Can cause cancer.
Carbon monoxide (CO)	Interferes with oxygen transport by blood, resulting in the reduction of oxygen supply to the heart (chronic anoxia), heart and brain damage, impaired perception.
Ozone (O_3) and other photochemical oxidants	Pain on deep breathing, irritation and inflammation of lungs, heart stress or failure. Can decrease lung function. Can lead to premature mortality.
Benzene (C_6H_6)	Can cause cancer.
1,3-Butadiene	Can cause cancer.
Lead (Pb)	Can affect almost every organ and system, especially the nervous and cardiovascular systems. It may also have adverse cognitive effects in children and lead to increased blood pressure in adults.
Arsenic (As)	Inorganic arsenic is a human carcinogen. The critical effect of inhalation of inorganic arsenic is considered to be lung cancer.
Cadmium (Cd)	Cadmium and cadmium compounds are carcinogenic. Inhalation is a minor part of total exposure, but ambient levels are important for deposition in soil and, thereby, dietary intake.
Mercury (Hg)	Can affect the liver, the kidneys, and the digestive and respiratory systems. It may also affect the central nervous system adversely.
Nickel (Ni)	Several nickel compounds are classified as human carcinogens.

Data gathered during the last ten years show that there is a connection between both indoor and outdoor air pollution exposure and various diseases, which is stronger than previously believed. The range of possible health effects includes cardiovascular diseases, such as strokes and ischemic heart disease, as well as—the forbidden word in many places in the world—cancer. This is in addition to air pollution's role in the development of respiratory diseases, including acute respiratory infections and chronic obstructive pulmonary diseases.

According to WHO [19], taking into account the health effects of only outdoor (or ambient) air pollution, 3.7 million deaths globally can be connected to this type of environmental problem in 2012. Regarding the regional distribution of the burden, about 88% of these deaths occur in low- and middle-income countries, which, however, represent 82% of the world population. The regions that bear most of these devastating consequences are the Western Pacific and South East Asian regions, 1.67 million and 936,000 deaths, respectively. Roughly 236,000 deaths correspond to the Eastern Mediterranean region, 200,000 in Europe, 176,000 in Africa, and 58,000 in the Americas. The remaining deaths occur in high-income countries of Europe (280,000), Americas (94,000), Western Pacific (67,000), and Eastern Mediterranean (14,000). The results are depicted in Fig. 2.2 [19].

127,000 of these deaths are related to acute lower respiratory infection; 227,000 to lung cancer, 389,000 to chronic obstructive pulmonary disease; 1,485,000 to stroke; and 1,505,000 to ischemic heart disease, as shown in Fig. 2.3 [19].

Previous corresponding estimates for deaths related to outdoor air pollution revealed only (!) 1.3 million deaths in 2008. WHO attributes this huge difference from 2008 to 2012 to (1) additional evidence, available on the relationship between exposure and health outcomes and the use of integrated exposure-response functions, (2) an increase in

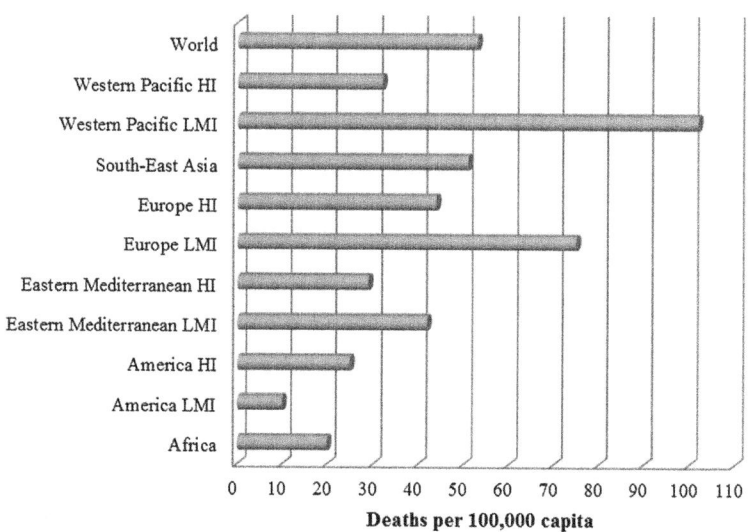

Figure 2.2

Deaths per capita attributable to ambient air pollution in 2012, by region (*LMI*, low- and middle-income; *HI*, high-income). *Adapted from Ref. [19].*

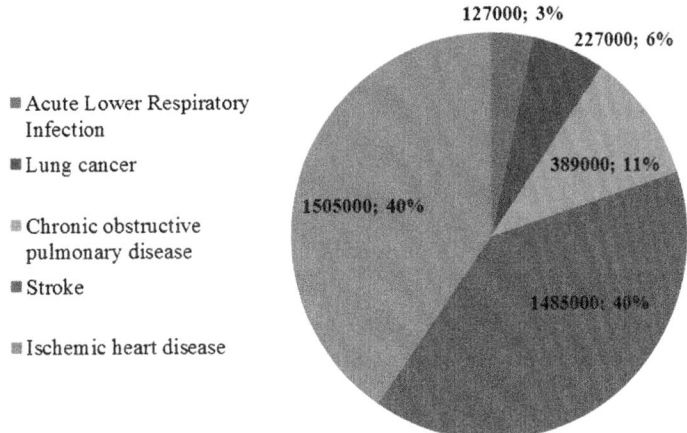

Figure 2.3

Deaths attributable to ambient air pollution in 2012, by disease. *Adapted from Ref. [19].*

noncommunicable diseases, (3) the inclusion of the rural population in the study for 2012, whereas the previous estimate for 2008 only covered the urban population, and (4) the use of a lower counterfactual, ie, the baseline exposure against which the effect of air pollution was measured.

For those that are more sensitive to financial figures, the relevant UNEP (United Nations Environment Programme) report adds also that

> *The cost of air pollution to the world's most advanced economies plus India and China is estimated to be US$3.5 trillion per year in lives lost and ill health.*
>
> *In OECD countries the monetary impact of death and illness due to outdoor air pollution in 2010 is estimated to have been US$1.7 trillion.*
>
> *Ref. [18]*

If not reported by global well-esteem organizations, the numbers above are too high to believe. But this is the reality; the quality of the atmospheric environment has an imminent connection with our health and a direct impact on the life expectancy, which in turn may be expressed in huge economic and social costs. The main health effects along with their respective air pollutants are presented in Fig. 2.4 [20].

In addition, air pollutants contribute to atmospheric problems such as acidification and global climate change, which have impacts on crop productivity, forest growth, biodiversity, buildings, and cultural monuments. There is a variety also at the levels of which these problems appear: starting from photochemical smog at local and urban level, ozone formation and acid rain at regional and transboundary level, to the global warming and ozone layer depletion at global level.

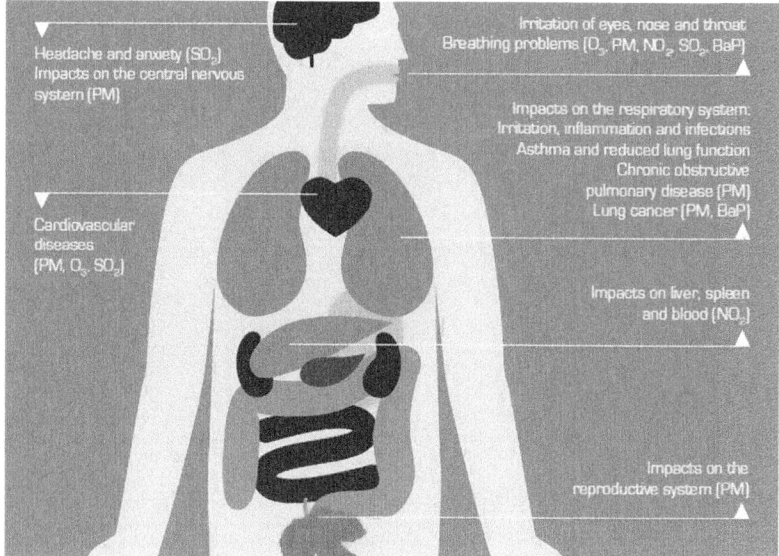

Figure 2.4
Health impacts of air pollution. *Reproduced from EEA Signals 2013.*
Every breath we take. © EEA, 2013.

Table 2.3: Air Pollution Problems in Association With the Major Pollutants [19].

Pollutant	Smog	Urban Air Quality	Acid Deposition	Global Warming	Ozone Depletion	Health
Ozone	×	×		×		×
Sulfur dioxide	×	×	×	×		
Carbon monoxide	×	×		×		
Carbon dioxide			×	×		
CFCs				×	×	
Nitrogen oxides	×	×	×	×	×	
Volatile organic compounds	×	×		×		
Toxics[a]	×	×				×
Particles	×	×	×	×		×
Total reduced sulfur compounds		×				×

[a]Toxic metals and organic compounds.

These problems are discussed in detail in the next pages. The main air pollution problems in association with major pollutants are presented in Table 2.3 [21].

As far as air pollution terminology is concerned, air or atmospheric pollution is divided into *outdoor* or *ambient air pollution*, also referred to as simply air pollution hereinafter, which denotes problems taking place from troposphere up to the stratosphere, and indoor air pollution, which refers to air quality issues in closed space indoors, in apartments, offices, etc. Indoor air pollution is under investigation mainly in the last 10 years, and it has been found to have also a number of impacts in our quality of life, as already discussed previously. Therefore, a separate paragraph is devoted to this type of air pollution, while *only issues related to outdoor air pollution are covered in the context of Section 2.3.*

Air pollutants are also classified into two broad categories: *primary* and *secondary* ones. Primary pollutants are those emitted directly into the air by automotive vehicles or industrial activities, such as particulate matter (PM) and carbon monoxide, whereas the term secondary pollutants refers to compounds not directly emitted into the atmosphere, but being formed there as a result of a variety of chemical and/or photochemical reactions among numerous primary pollutants and elements present in the atmosphere. Ozone formation in smog constitutes the most characteristic example of the latter category. In many cases, it is more difficult to deal with secondary pollutants since they are not directly correlated with sources, but with a number of parallel atmospheric reactions that are difficult to control [21].

Sources of air pollutants can be also classified into two broad categories: *natural sources* like volcanos and *anthropogenic sources*. The latter can be further divided into *mobile sources* like cars and any means of transport and *stationary sources* like industrial facilities (Fig. 2.5). This discrimination is very important since relevant tools, legislative measures, and technological solutions depend on the type of anthropogenic sources.

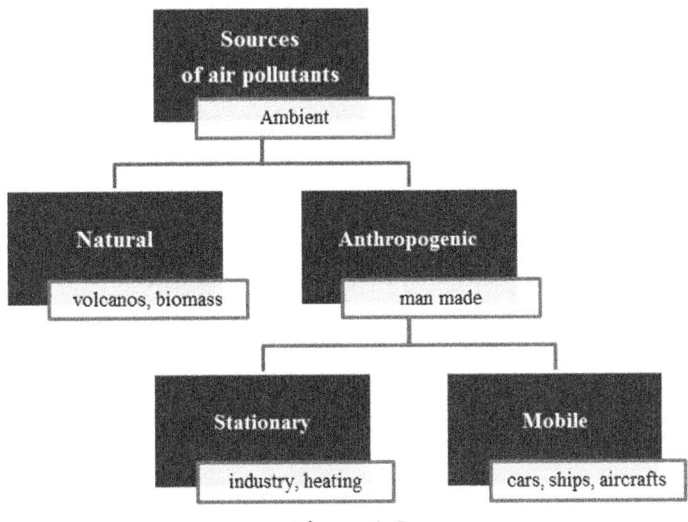

Figure 2.5
Classification of sources of ambient air pollution.

The main sources of air pollutants include agriculture, energy plants, road transport, and industry. Agriculture, for example, is connected with the release of acid compounds into the atmosphere that might lead to acid rain formation with a dramatic impact on lakes, rivers, and forests.

Sources, pollutants, and types of air pollution are further discussed based on the scale that they appear.

2.3.1 Local Level

Local air pollution refers to elevated ambient concentrations of compounds over a small area, caused by a large local emitter releasing these compounds into the atmosphere or many small emitters gathered like cars. Automobile vehicles stuck in traffic, for instance, may cause high levels of carbon monoxide affecting a specific street. Exposure to urban air pollutants at the roadside constitutes a threat to public health [22]. High buildings along with local meteorological conditions may worsen the case for inhabitants [23]. Even in a small village, the concentration of livestock farms and activities can result in local air pollution. The emissions of a variety of compounds like sulfur dioxide, nitrogen oxides, carbon monoxide, fine PM, organic compounds like benzene, toluene and polycyclic aromatic hydrocarbons (PAHs), and heavy metals in PM (lead, cadmium), can cause local concentrations to reach levels that are harmful to human health. For example, Lu et al. [24] reported that short-term exposures of aerodynamic diameters less than 10 μm (PM_{10}), nitrogen dioxide (NO_2), and sulfur dioxide (SO_2) were positively associated with increases in mortality risks in the city of Nanjing in China. They also found that *years of life* lost were significantly elevated corresponding to an increase in current ambient concentrations of PM_{10}, NO_2, and SO_2.

Besides the compounds mentioned above, which are typical primary compounds, some compounds like nitrogen oxides and volatile organic compounds (VOCs) present in stagnant air can react by means of solar radiation to form secondary pollutants like ozone and other oxidants. Ozone is a serious concern in many cities worldwide, and it is a difficult-to-treat problem as it is connected to a complex reaction network taking place in the atmosphere.

It is very common in large cities, one air quality—monitoring station to report elevated concentrations of target compounds in the atmosphere, while another one to measure normal ones within the limits set, despite its vicinity to the first one. Most city centers experience incidents of local air pollution frequently in contrast to areas outside the limits of the center, yet in the same city.

All reports on air pollution in the 19th and early 20th centuries indicate that air pollution events were mostly local ones, being concentrated in or around industrial centers and large cities. Even the infamous environmental disasters in the area of Liege in the 1930s—the first recorded occurrence of death by air pollution—or in London in

December 1952, were essentially local phenomena. In the London smog episode, presented also as case study in Section 2.3.5, sulfur dioxide and sulfuric acid were accumulated in stagnant air at dangerous levels, resulting in ~ 4000 deaths from lung and heart conditions [21].

Since sources and air pollutants at local level are essentially the same as the ones at urban level, these topics are discussed in detail in the next paragraph.

2.3.2 Urban Level

Besides the local incidents discussed in the previous paragraph, there are three major types of air pollution found in urban areas [21]:

- High annual average concentration levels of various pollutants, eg, benzene, lead, sulfur dioxide (SO_2), and PM. As in the case of air pollution at the local level, this type of pollution is linked to specific pollutants resulting from either large industrial and power plants or automotive vehicles.
- Winter-type smog, characterized by high concentrations of SO_2 and PM that arise mainly from the combustion of coal and fuels with a high content of sulfur. This kind of pollution occurs in urban areas with many power plants or industrial units clustered together, where low temperatures and mist are observed in the year. It has been also termed "industrial pollution." Central heating systems are also major contributors, especially in low-income or developing countries.
- Summer-type smog, characterized by high concentrations of carbon monoxide, VOCs, and nitrogen oxides (NO_x). It is also called "photochemical smog" or "LA smog," since it first appeared in Los Angeles. This type of pollution is closely connected to automotive vehicles, and its formation is favored by sunlight and high temperatures.

Primary air pollutants such as nitric oxide (NO), carbon monoxide (CO), and sulfur dioxide (SO_2), which are emitted directly into the atmosphere from various sources, are involved in the first two types of air pollution, whereas the third type is associated with secondary pollutants, primarily ozone, resulting from atmospheric reactions among primary pollutants by means of solar irradiation and heat, constituting thus a complicated issue to cope with. This ozone is also referred to as *ground-level ozone* (or "bad" *ozone*) and its role should not be confused with the one at the stratosphere. The measures to be taken against pollution from primary pollutants are more straightforward since these compounds are directly linked to their corresponding sources to be targeted. The numerous health effects of these compounds and the effect on life expectancy have been already summarized in Paragraph 2.3.

According to the U.S. Environmental Protection Agency (EPA),[2] the main urban air pollutants include PM, ground-level ozone, carbon monoxide, sulfur oxides, nitrogen oxides, and lead. A small description of these compounds is given in the following text regarding their definition, major sources, and main health and environmental effects [25], as well as the suggested guidelines for maximum concentrations in the atmosphere by the World Health Organization where applicable.[3]

Non-Methane Volatile Organic Compounds (NMVOCs). The term nonmethane volatile organic compounds (NMVOCs) is used to describe a collection of organic compounds that differ widely in their chemical composition, but display similar behavior in the atmosphere. They include nonmethane hydrocarbons (NMHC) and oxygenated NMHC (eg, alcohols, aldehydes, and organic acids), which have short atmospheric lifetimes (fractions of a day to months) and small direct impact on radiative forcing. NMVOCs are emitted into the atmosphere from a large number of sources including combustion activities, solvent use, and production processes. They contribute to the formation of ground-level ozone. In addition, certain NMVOC species or species groups such as benzene and 1,3-butadiene are hazardous to human health. Given their short lifetimes and geographically varying sources, it is not possible to derive a global atmospheric burden or mean abundance for most volatile organic compounds (VOCs) from current measurements. VOC abundances are generally concentrated very close to their sources.

Carbon Monoxide. Carbon monoxide (CO) is a colorless, odorless, and poisonous gas. After inhalation, carbon monoxide combines with hemoglobin to form carboxyhemoglobin (HbCO) in the blood. This prevents hemoglobin from carrying oxygen to the tissues, effectively reducing the oxygen-carrying capacity of the blood, leading to hypoxia. Additionally, myoglobin and mitochondrial cytochrome oxidase are thought to be adversely affected. Carboxyhemoglobin can revert to hemoglobin, but the recovery takes time because the HbCO complex is fairly stable. Young children, the elderly, and people with heart or respiratory problems are especially susceptible to carbon monoxide effects. Inhalation, especially during exercise, can lead to chest pains, reduce exercise capacity, and may affect visual perception, the physical skills, as well as the ability to learn and perform complex tasks. It has been given also a special role to ozone formation.

It is a product of incomplete combustion of hydrocarbons, and its main sources in urban centers are the exhaust emissions from automotive vehicles. It is the pollutant with the highest concentrations in the lower layers of the atmosphere. Its formation is favored during starting of vehicles, when the air supply is limited and the engine and catalytic converter temperatures are still relatively low. Poorly maintained vehicles increase also its emissions. Carbon monoxide emissions are also increased during winter months when low temperatures prevail.

The U.S. National Institute for Occupational Safety and Health recommends an 8-h TWA limit of 35 ppm with a 200-ppm ceiling.

---

[2] http://www3.epa.gov/airquality/urbanair/.
[3] http://www.who.int/mediacentre/factsheets/fs313/en/.

—cont'd

Nitrogen Oxides. Among the seven nitrogen oxides (NO, NO_2, NO_3, N_2O, N_2O_3, N_2O_4, N_2O_5), the most important ones are nitric oxide (NO) and nitrogen dioxide (NO_2), which are together referred to as NO_x. 90% of them are produced from natural sources, while anthropogenic emissions are related to combustion processes. They may be formed in two ways:

- Thermal NO_x: They constitute the result of oxidation reaction between the atmospheric molecular nitrogen with the coexisting oxygen at high temperatures in the combustion chamber (>700°C).
- Fuel NO_x: They are the product of the oxidation of nitrogen compounds present in the fuel.

Most NO_x emissions are in the form of NO, since NO_2 formed during combustion is typically less than 0.5% of total NO_x. Nitric oxide (primary pollutant) oxidizes easily in air to nitrogen dioxide, which in turn reacts with volatile hydrocarbons by means of solar light to create ozone. Moreover, nitrogen dioxide may react with hydroxyl radicals in the atmosphere to form nitric acid, one of the major components of acid rain. Generally, nitrogen oxides are not considered hazardous to health at concentrations that occur in the atmosphere, but they exhibit very strong photochemical action. It should be noted that nitrous oxide (N_2O) is also present in the atmosphere, which is one of the most active compounds in global warming. It is mainly emitted by natural sources.

WHO guideline values
NO_2
40 $\mu g/m^3$ annual mean
200 $\mu g/m^3$ 1-h mean

Ozone (O_3). It is found naturally in the stratosphere, where it creates the "ozone layer", which protects life on Earth by filtering out sunlight from ultraviolet light. In contrast, ozone in the low layers of the atmosphere, called *ground-level* or *tropospheric ozone*, is a dangerous pollutant. It is the major component of photochemical smog in urban areas and a major problem associated with air quality. It may cause many irritating symptoms such as cough and eye irritation. It can destroy the tissues of the lung, and cause respiratory diseases or make people more susceptible to respiratory infections. Children and people with weakened immune system are particularly vulnerable to the adverse effects of ozone. It also may cause damage to plastic materials, textile materials, foliage trees, and plants. High levels of ozone can stop the development of plantations and cause extensive damage to rural farming and forests. It is considered responsible for 90% of the total damage caused to agricultural production that is related to air pollution.

Ozone is a secondary pollutant, as it is not released directly into the air, but it is formed when hydrocarbons, carbon monoxide, and nitrogen oxides react in the presence of sunlight and heat. The rate of formation of these reactions depends on both the temperature and the intensity of sunlight. So, alarming levels of ozone in the atmosphere usually occur during summer months. It should be noted that ground-level ozone unfortunately does not contribute to an increase of the "good" ozone in the stratosphere, which protects the Earth from the harmful ultraviolet radiation of the Sun.

WHO guideline values
O_3
100 $\mu g/m^3$ 8-h mean

(*Continued*)

—cont'd

Particulate Matter (PM). The soot or smoke is another type of air pollutant, which is generally referred to as particulate matter. It results in hazing in urban areas and constitutes a health risk as it enters the body through the respiratory system. It is noteworthy that during the presence of these particles in the atmosphere, a lot of low-volatility compounds, which are formed secondarily in the gas phase, are condensed on them. As a result, inhaled particles form a complex mixture of hazardous-to-health chemical compounds. So, the major components of PM are sulfate, nitrates, ammonia, sodium chloride, black carbon, mineral dust, and water. PM affects more people than any other pollutant. The most health-damaging particles are those with a diameter of 10 μm or less ($\leq PM_{10}$), which can penetrate and lodge deep inside the lungs. Commonly, they are divided into PM_{10} and $PM_{2.5}$. This pollutant is emitted primarily from diesel engines and it is associated with the type of flame in a process.

WHO guideline values
$PM_{2.5}$
10 $\mu g/m^3$ annual mean
25 $\mu g/m^3$ 24-h mean

PM_{10}
20 $\mu g/m^3$ annual mean
50 $\mu g/m^3$ 24-h mean

Sulfur Oxides (SO_x). Sulfur oxides may be formed during the combustion of the sulfur contained in the fuel either as element or in organic and inorganic compounds. They are corrosive and dangerous to both health and the environment because of the role they play in creating acid rain. Particularly, SO_2 is a colorless gas with a sharp odor. It can affect the respiratory system and the functions of the lungs, and cause irritation of the eyes. Inflammation of the respiratory tract causes coughing, mucous secretion, aggravation of asthma, and chronic bronchitis and makes people more prone to infections of the respiratory tract. Hospital admissions for cardiac disease and mortality increase on days with higher SO_2 levels. When SO_2 combines with water, it forms sulfuric acid; this is the main component of acid rain, which is a cause of deforestation and a threat to natural life. Restricting sulfur in gasoline has drastically reduced the emissions of sulfur oxides from vehicles, but it is still a concern for low-income and developing countries where low-quality fuels, containing high-sulfur amounts, are used for transport or heating purposes.

WHO guideline values
SO_2
20 $\mu g/m^3$ 24-h mean
500 $\mu g/m^3$ 10-min mean

Lead (Pb). Lead, because of its multiplicity of uses, is present in air, dust, soil, and water. Lead enters the body mainly by ingestion or inhalation. It is emitted in the atmosphere mainly by vehicles powered by gasoline with lead in the form of tetra-ethyl lead ($(C_2H_5)_4Pb$), which is used as an antiknocking compound.

Lead affects practically all body systems. Most toxic exposures occur at chronic low levels and can result in reductions in intelligence quotient (IQ), increased blood pressure, and a range

—cont'd

of behavioral and developmental effects. The range and extent of adverse health effects has been appreciated only—relatively—recently. Furthermore, lead is now understood to be toxic, especially to children, at levels that previously were thought to be safe. In more severe cases of poisoning, adverse health effects include gastrointestinal symptoms, anemia, neurological damage, and renal impairment. Its serious effects on human health in combination with its poisoning effect on automotive catalytic converters led to its gradual phase out by the gasoline in most developed countries. Specifically, currently about 60 countries have phased out leaded petrol and ~85% of petrol sold worldwide is lead-free. However, it is still a concern in many developing countries where leaded gasoline is still in use. Other important lead sources are more difficult to control, such as leaded kitchenware ceramics, water pipes, and house paints.

There is no known safe blood lead concentration. But it is known that as lead exposure increases, the range and severity of symptoms and effects also increases. Even blood lead concentrations as low as 5 μg/dL, which once was thought to be a "safe level," may result in decreased intelligence in children, behavioral difficulties, and learning problems.

The main health effects that have been associated with air pollution have been earlier presented in Paragraph 2.3. The reader may retrieve more information on the subject via relevant reports of international organizations [18,19,26].

The Organization for Economic Co-Operation and Development (OECD) has developed an extensive database[4] (OECD.Stat) with the annual releases of many pollutants for many countries around the world. Pollutants are related to one or more of the following anthropogenic sources:

1. Total mobile sources
   a. Road transport
   b. Other mobile sources
2. Total stationary sources
   a. Power stations
   c. Combustion
      i. Industrial combustion
      ii. Other combustion
   d. Industrial processes
   e. Solvents

---

[4] http://stats.oecd.org/#.

  f. Miscellaneous
      i. Agriculture
     ii. Energy
    iii. Waste

In Figs. 2.6—2.9, the annual emissions from mobile and stationary sources of nitrogen oxides, sulfur oxides, nonmethane volatile organic compounds, and carbon monoxide are shown, respectively, for a number of countries or regions around the world (*source*: OECD.Stat).

It is obvious that most releases are decreased over time as a result of both more stringent environmental requirements and technological advancements. However, some challenges still remain as in the case of NMVOCs and CO from stationary sources in the United States.

In Fig. 2.10, the releases of nitrogen oxides from various sources are shown in the case of the US mobile sources and particularly road transport are the main emitters followed by

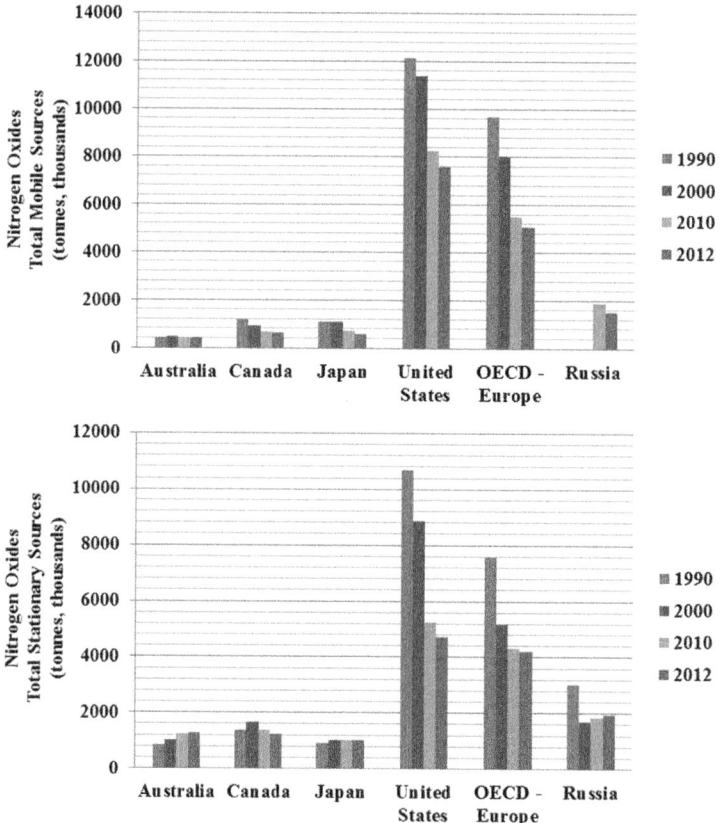

Figure 2.6

Nitrogen oxides releases from mobile and stationary sources in 1990—2012 around the world. *Data extracted on 01 Nov 2015 from OECD.Stat.*

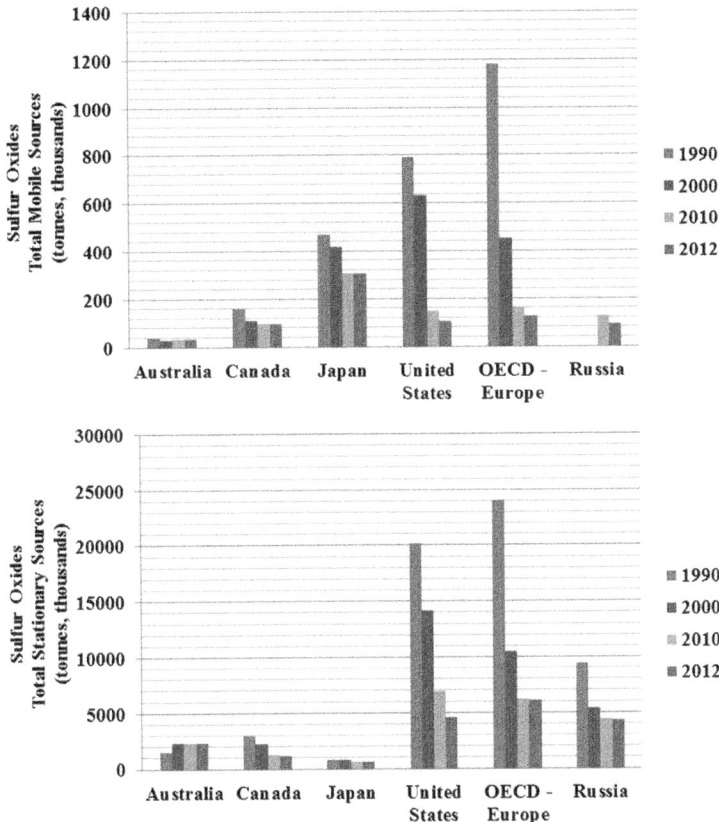

Figure 2.7

Sulfur oxides releases from mobile and stationary sources in 1990–2012 around the world. *Data extracted on 01 Nov 2015 from OECD.Stat.*

power station and combustion. In the case of $PM_{2.5}$ and PM_{10}, industrial processes and combustion processes are the major contributors (Fig. 2.11).

Information about sources and pollutants in Europe is summarized in the relevant report of the European Environment Agency (EEA) [20]. Specifically, in Europe more than 40% of nitric oxide emissions result from road transport, while around 60% of the sulfur oxides come from energy production and distribution in the EEA member and cooperating countries. Commercial, government, and public buildings and households contribute to around half of the $PM_{2.5}$ and carbon monoxide emissions. Agriculture is another important contributor, since it is responsible for around 90% of ammonia emissions and 80% of total methane emissions (Fig. 2.12). Other methane sources include waste (landfills), coal mining and long-distance gas transmission. Methane is a very active greenhouse gas (GHG).

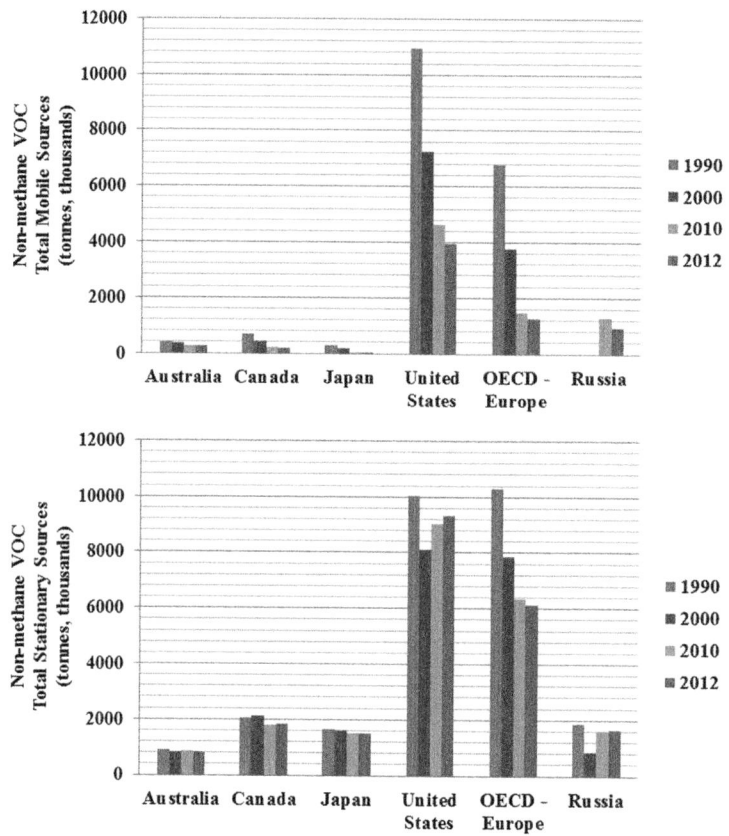

Figure 2.8

Nonmethane volatile organic compounds releases from mobile and stationary sources in 1990—2012 around the world. *Data extracted on 01 Nov 2015 from OECD.Stat.*

Overall, the air quality in Europe has improved significantly in the last 60 years due to accumulated scientific knowledge, demands by the public, and legislation strengthening. The concentrations of many air pollutants, including sulfur dioxide, carbon monoxide, and benzene, have decreased significantly, whereas lead concentrations have dropped sharply below the limits set by legislation. However, the desired air quality standards have not been achieved, since PM and ozone keep posing serious risks to human health and the environment [17,20]. Most urban regions in the world have an ozone problem: Europe, North America, China, India, and Japan. Even the rapidly developing countries such as Brazil (where biomass burning and vehicles release ozone precursor gases) have ozone problem [20]. Furthermore, large cities in China, which have been changed rapidly over the last three decades, face substantial air pollution. Therefore, Chinese cities constitute the subject of many works related to air pollution issues [24]. According to WHO, annual PM_{10} levels are estimated to increase by 6% during the recent 3-year periods (2009—12 or earlier period) globally, as assessed in cities present in its databases and weighted by regional urban population.

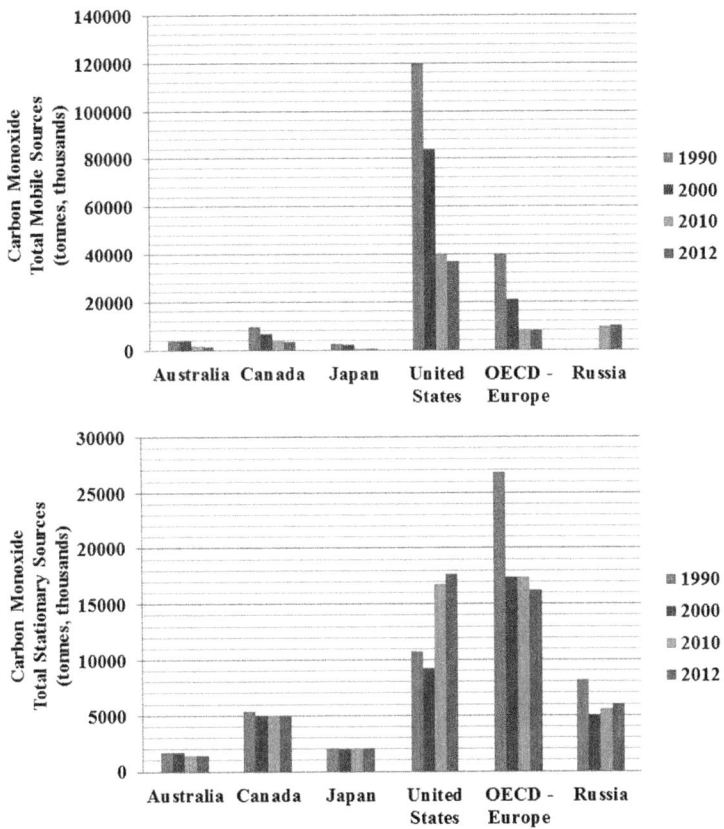

Figure 2.9

Carbon monoxide releases from mobile and stationary sources in 1990–2012 around the world. *Data extracted on 01 Nov 2015 from OECD.Stat.*

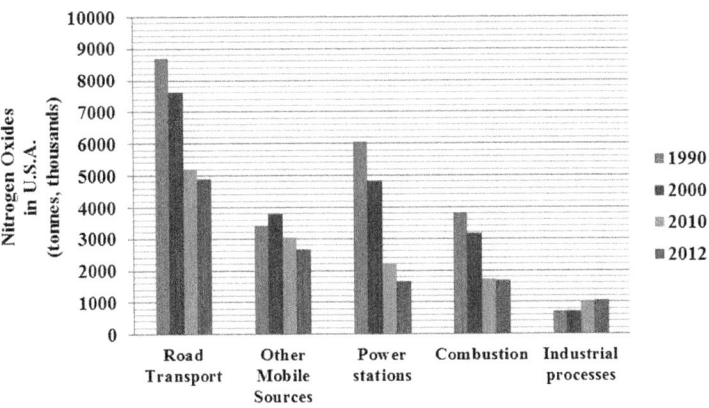

Figure 2.10

Nitrogen oxides releases from various sources in 1990–2012 in the United States. *Data extracted on 01 Nov 2015 from OECD.Stat.*

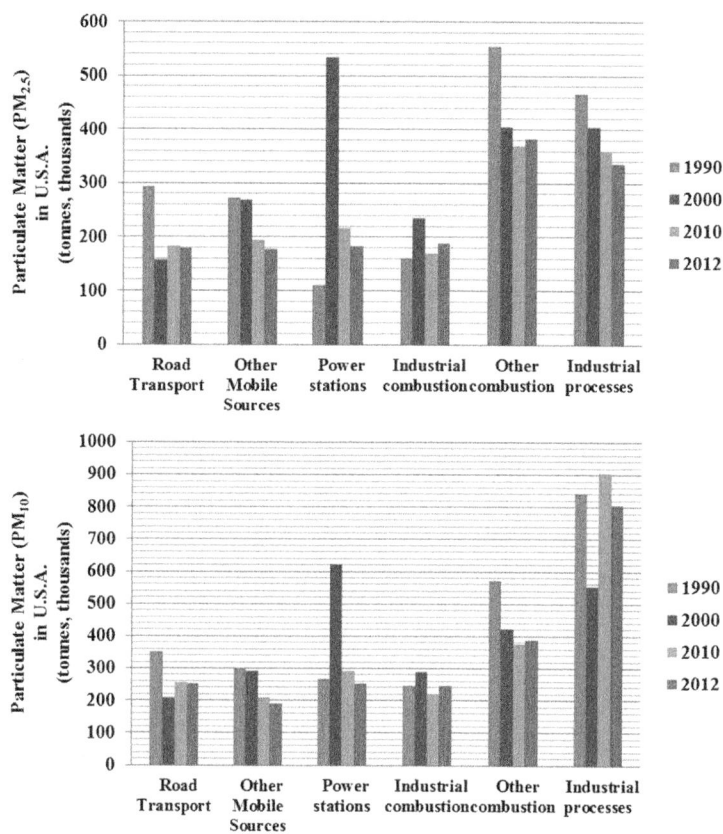

Figure 2.11

Particulate matter releases as $PM_{2.5}$ or PM_{10} from various sources in 1990–2012 in the United States. *Data extracted on 01 Nov 2015 from OECD.Stat.*

Rodríguez et al. [27] presented evidence that urban fragmentation is correlated with higher concentration of NO_2 and PM_{10}, when controlling for socioeconomic factors and climate conditions. Moreover, they suggested that cities with larger artificial areas may experience higher NO_2 and PM_{10} concentrations, whereas densely populated cities are associated with higher SO_2 concentrations. Their results showed that NO_2, SO_2, and PM_{10} are affected differently by urban characteristics due to differences in emission sources. They proposed that concerns about increased levels of NO_2 and PM_{10} concentrations, stemming from further expansion of urban areas in Europe, could be partially addressed by spatial policies aiming at the reduction of urban fragmentation. Continuous urban areas enhance connectivity, reduce mobility needs and car dependency, and facilitate the use of nonmotorized modes of transport such as biking and walking.

Sources of air pollution in Europe
Air pollution is not the same everywhere. Different pollutants are released into the atmosphere from a wide range of sources, including industry, transport, agriculture, waste management and households. Certain air pollutants are also released from natural sources.

1 / Around 90 % of ammonia emissions and 80 % of methane emissions come from **agricultural activities.**

2 / Some 60 % of sulphur oxides come from **energy production and distribution.**

3 / Many **natural phenomena,** including volcanic eruptions and sand storms, release air pollutants into the atmosphere.

4 / **Waste (landfills), coal mining and long-distance gas transmission** are sources of methane.

5 / More than 40 % of emissions of nitrogen oxides come from **road transport.**

Almost 40 % of primary $PM_{2.5}$ emissions come from transport.

6 / **Fuel combustion** is a key contributor to air pollution — from road transport, households to energy use and production.

Businesses, public buildings and households contribute to around half of the $PM_{2.5}$ and carbon monoxide emissions.

Figure 2.12
Sources of air pollution in Europe. *Reproduced from EEA Signals 2013. Every breath we take. © EEA, 2013.*

2.3.3 Regional and Transboundary Level

Air pollution and the associated impacts on the environment are discussed together for regional and transboundary scales, since they are essentially the same issues taking place at areas of different magnitudes.

Air pollution at these levels can be attributed to two mechanisms [21]:

• transport and dispersion of urban pollutants over large areas, and
• transport and transformation of primary pollutants into secondary ones.

Polluted air masses containing ozone, sulfur (S), and nitrogen (N) compounds can travel over long distances from urban centers to rural areas, or even cross-national borders and damage other countries' resources: their freshwaters, forests, grasslands, or cultural heritage (ie, buildings, monuments, etc.). Three kinds of environmental issues are connected to these mechanisms: spread of *tropospheric ozone, acidification*, and *eutrophication*, all of them having serious consequences to crops, aquatic life, and forests (Fig. 2.13) [28,29].

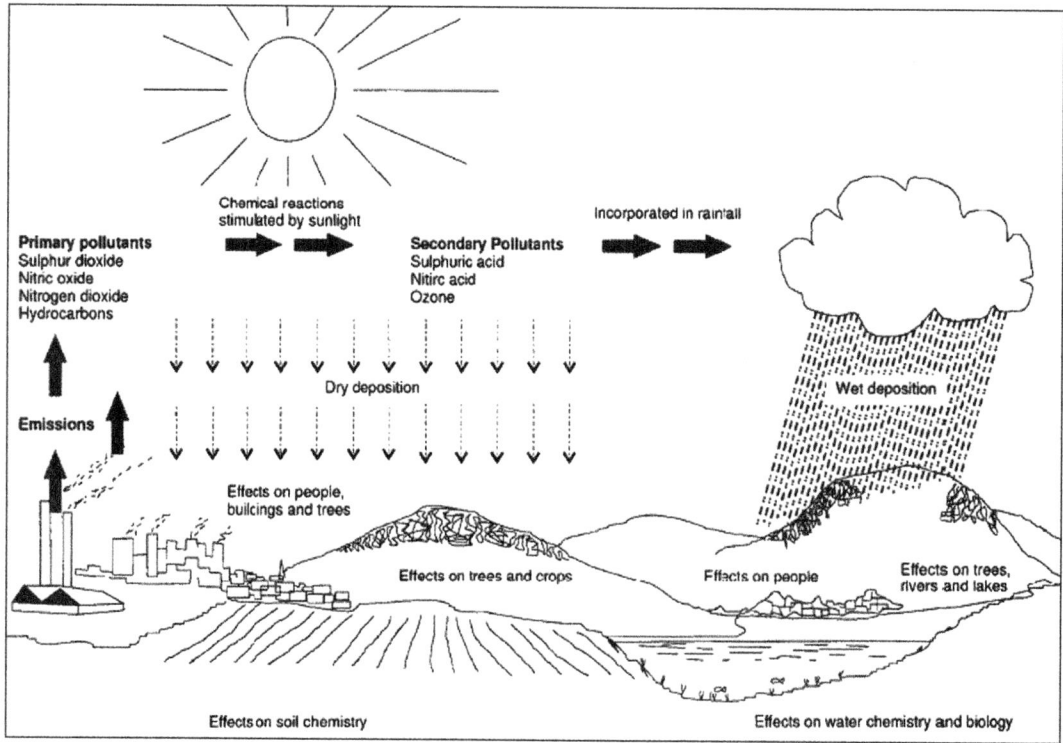

Figure 2.13
Transport, dispersion, transformation, and deposition of air pollutants at the regional—transboundary level. *Reproduced from Ref. [29].*

Tropospheric or ground-level ozone is formed in urban areas via the reaction of NO_x with NMVOC by means of solar light and transferred by the wind to rural areas. It can be also formed via atmospheric reactions during the transfer of NO_x and NMVOC from cities to rural areas by means of beneficial meteorological conditions. As discussed earlier, ozone has adverse impacts on both human health and ecosystems [29]. It is known to cause leaf injury in plants, including crops and trees, significantly reducing plant growth and crop yield, and causes some materials, particularly organic ones such as paints and rubber, to disintegrate [28]. About 90% of the adverse impact of air pollution on agriculture is attributed to ozone [21].

Furthermore, *acidification* may occur when air pollutants emitted to the atmosphere undergo dispersion, and many of them subsequently may be involved in a number of (photo)chemical reactions. Some compounds, after a certain lifetime, are deposited on earth by one of the two processes: *wet deposition* or *dry deposition* [30]. Wet deposition occurs when pollutants are emitted into the atmosphere and oxidized to form an acid, as in the case of sulfur dioxide. Specifically, SO_2 emitted when sulfur-containing fuels are burnt may be oxidized to sulfuric acid (H_2SO_4) in the atmosphere. The acid falls then to earth as acidic precipitation (as rain, snow, hail, or dew). This phenomenon is widely known as *acid rain*. On the other hand, dry deposition occurs when the acids are transformed chemically into gases and salts and then deposit, for example, on vegetation surfaces. They can subsequently be washed off into soils by rain. Acidification affects fish populations and forest soils in sensitive areas of Europe and North America, and in addition, it causes corrosion of buildings and monuments [28]. Sulfur oxides (SO_2) and nitrogen oxides (NO_x) have been identified as the main compounds involved, while ammonia (NH_3) released from agricultural activities adds also to the problem. The lifetime of SO_2 in the air is a few days, whereas the lifetime of NO_x ranges from a few hours to 1−2 weeks, depending on whether it is dispersed close to the Earth's surface, where it can react with other pollutants, or in higher layers of the atmosphere. Airborne NH_3 is usually relatively short-lived (1−5 days or less) [30].

Finally, in sensitive areas, high levels of nitrogen deposition from nitrogen oxide and ammonia emissions result in *eutrophication*. The increase of this plant nutrient in natural ecosystems results in both excessive growth of specific plant species and disappearance of others. Moreover, in coastal and inland waters, the excessive increase of algae due to eutrophication leads to oxygen depletion, which in turn has dramatic impacts on plants, fish, and the rest life forms. Excess nitrogen deposition directly increases nitrate concentrations in groundwater normally used for drinking, making it thus harmful to consumers, and also causes nitrogen to leach from soils, increasing the acidification of surface and groundwaters [28].

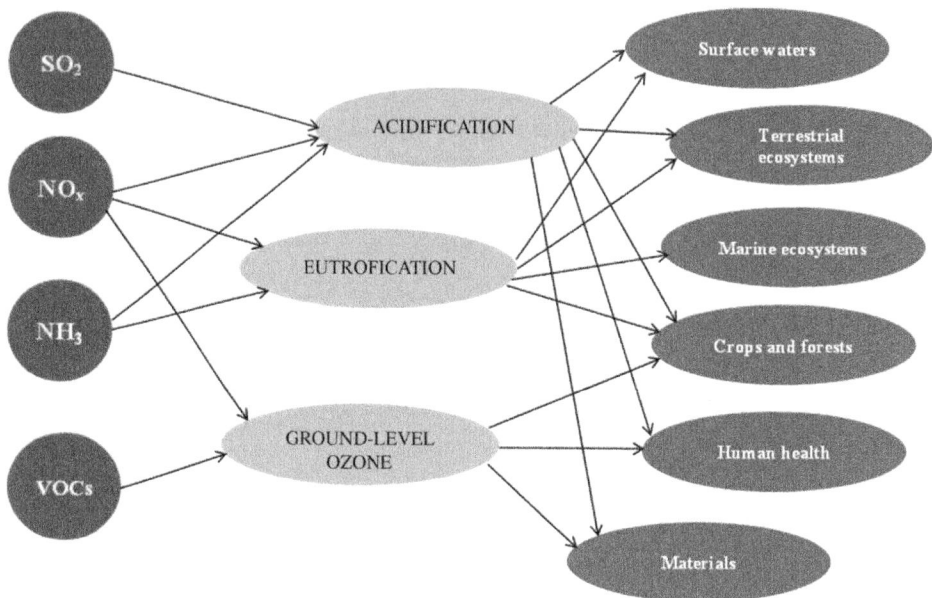

Figure 2.14
Regional—transboundary level air pollutants, associated problems, and impacts.
Adapted from Ref. [28].

Air pollutants along with the relevant environmental phenomena and the corresponding consequences to environment and human health are depicted in Fig. 2.14.

The problems that appear on a regional level may also be exchanged between adjacent countries. Transboundary air pollution is related to the transfer of air pollution from one country to another. For example, Greece exchanges ozone with Italy, whereas Japan and Canada "import" various types of air pollution from China and the United States, respectively. Moreover, the transboundary transport of acidifying pollutants has led to a marked change in the acidity of lakes and streams in Scandinavia, as observed in the 1960s [29]. Air pollution problems seem also to travel across different continents. Jaffe et al. [31] showed that emissions from East Asia significantly impact the air arriving to North America. Holloway et al. [32] reported that also North America "exports" air pollutants to Europe. A number of toxic substances are capable of intercontinental atmospheric transport, including mercury, toxaphene, hexachlorobenzene, and polychlorinated biphenyls [33]. An extensive review on transboundary air pollution in the United Kingdom is given by the Centre for Ecology & Hydrology [34]. It is apparent that international cooperation at all levels is required for an effective response to these issues and prevention of negative effects on the environment like the one shown in Fig. 2.15.

Figure 2.15
Deforestation as a result of an acid rain episode. *Source: Pixabay.com.*

2.3.4 Global Level

Stratospheric ozone depletion

As discussed in Section 2.1, most of the atmosphere's ozone is found within the stratosphere. This ozone layer has an essential-to-life role: it absorbs much of the incoming solar ultraviolet radiation, providing thus vital protection from this radiation to all organisms living at Earth's surface. It should not be confused with the ground-level ozone, which is a harmful secondary air pollutant resulting indirectly from human activities. Stratospheric ozone has been created in the atmosphere gradually and can be considered as a product of life on Earth, which began around 3.5 billion years ago. McMichael et al. [35] describe shortly the history of ozone layer; until a half billion years ago, living organisms could not inhabit the land surface and life was restricted to the world's oceans and deep waters in general, which offered protection from the harmful unfiltered solar ultraviolet radiation. About 2 billion years ago, ozone (O_3) started being formed gradually in the atmosphere, as a result of the first photosynthesizing organisms' action that emitted oxygen (O_2). Around 400 million years ago, aqueous plants were able to migrate onto the now-protected land and evolve into terrestrial plants. That was the beginning of the food chain on earth; animal life fed on plants followed and life evolved via several evolutionary paths, through herbivorous and carnivorous dinosaurs, mammals, and omnivorous humans. Today, all species on Earth enjoy the protection of this ozone layer formed on the stratosphere that acts as a shield for Earth against solar ultraviolet radiation.

The Antarctic ozone hole is one of the most striking examples of the impact of human activities on the atmosphere. In 1985, the first so-called *ozone hole* was discovered [36].

Back then, the mechanisms behind ozone depletion were not known, and a number of competing theories were employed to solve the "mystery" [37]. However, Rowland and Molina had warned world community since 1974 about the detrimental action of CFCs in the stratosphere [38]. Their story is interesting[5]: Frank Sherwood Rowland was born in Ohio, the United States, in 1927. He obtained his BA from Ohio Wesleyan University in 1948. He went on to do his MS and then PhD at the University of Chicago. In 1964, he became Professor of Chemistry at the University of California, Irvine, where he still works. José Mario Molina-Pasquel Henríquez was born in Mexico City in 1943. He studied chemical engineering at the National Autonomous University of Mexico in 1965, and then did his postgraduation from the Albert Ludwigs University of Freiburg, West Germany, in 1967. He got a PhD in Chemistry from University of California, Berkeley, in 1972. He now works at the University of California, San Diego. Molina worked as a postdoctoral scholar in Rowland's laboratory in 1974. They together heard a lecture, in which James Lovelock's data on CFCs in the atmosphere were presented. Intrigued, they started further investigations. They discovered that CFCs decompose in sunlight to release chlorine atoms, which subsequently convert ozone to oxygen, and then they can attack other ozone molecules. A single atom can destroy millions of ozone molecules before it is neutralized (Fig. 2.16). Molina and Rowland's findings were published in 1974 [38] and shocked the entire world. For historical reasons let us read the abstract of their famous publication:

Chlorofluoromethanes are being added to the environment in steadily increasing amounts. These compounds are chemically inert and may remain in the atmosphere for 40–150 years, and concentrations can be expected to reach 10 to 30 times present levels. Photodissociation of the chlorofluoromethanes in the stratosphere produces significant amounts of chlorine atoms, and leads to the destruction of atmospheric ozone.

Ref. [38]

Their findings were later confirmed by scientists around the world and led to the Montreal Protocol[6] of 1987 that banned CFCs around the world. They received the Nobel Prize for Chemistry in 1995 for these findings. Before this "happy end", of course, as in the case of global warming and climate change, there were some disputes; Kelly reported in 1976 that "The primary conclusion of this study is that neither the longest available stratospheric ozone record (from Arosa, Switzerland) nor the most complete set of hemispheric records (1957–74) provides evidence that the ozone has been affected or reduced by the inputs of man" [39]. Today, however, a lot of information has been accumulated about the fundamental mechanisms triggering ozone

---

[5]   http://humantouchofchemistry.com/frank-rowland-and-mario-molina.htm.

[6]   The Montreal Protocol on Substances that Deplete the Ozone Layer is an international treaty designed to protect the ozone layer by phasing out the production of numerous substances that are responsible for ozone depletion. It is further discussed in Section 2.3.7.

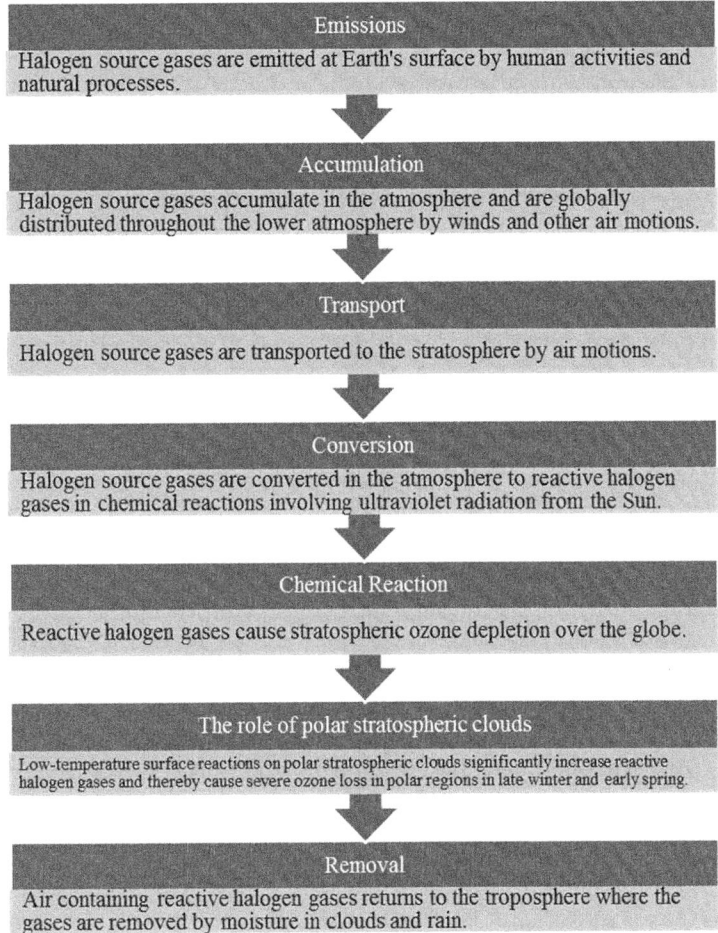

Emissions
Halogen source gases are emitted at Earth's surface by human activities and natural processes.

Accumulation
Halogen source gases accumulate in the atmosphere and are globally distributed throughout the lower atmosphere by winds and other air motions.

Transport
Halogen source gases are transported to the stratosphere by air motions.

Conversion
Halogen source gases are converted in the atmosphere to reactive halogen gases in chemical reactions involving ultraviolet radiation from the Sun.

Chemical Reaction
Reactive halogen gases cause stratospheric ozone depletion over the globe.

The role of polar stratospheric clouds
Low-temperature surface reactions on polar stratospheric clouds significantly increase reactive halogen gases and thereby cause severe ozone loss in polar regions in late winter and early spring.

Removal
Air containing reactive halogen gases returns to the troposphere where the gases are removed by moisture in clouds and rain.

Figure 2.16
The steps of stratospheric ozone depletion.

destruction, and man-made halogen compounds have been identified as the ultimate cause of stratospheric ozone loss, although there are still some points open to discussion [40].

According to Grooβ and Müller [41], Antarctic ozone loss is caused by a chain of events starting in autumn with cooling of the polar stratosphere and formation of the polar vortex. The polar vortex is a strong stratospheric wind jet that isolates the polar air mass from midlatitudes. In winter, when stratospheric temperatures are low enough, polar stratospheric clouds are formed that are composed of water ice, nitric acid hydrates, and supercooled liquid ternary $HSO_4/HNO_3/H_2O$ droplets. These polar stratospheric clouds along with cold binary sulfate particles provide the surfaces that foster the heterogeneous reactions through which the ozone-destroying form of chlorine is produced from the so-called reservoir species (mostly HCl and $ClONO_2$). Ozone depletion takes place in catalytic cycles; in the Antarctic stratosphere mainly through a cycle involving chlorine peroxide (ClOOCl). This ozone-depleting process is terminated in spring either after the temperature rise above the threshold value for efficient heterogeneous chlorine activation ($\approx 195K$) and the halogen reservoir species reform, or when extremely low mixing ratios of ozone are reached. It has to be noted that the problem is not now limited only to the atmosphere above Antarctica, but it is spread all over the world with generally lower stratospheric concentrations of ozone. Observations of ozone and of chlorine-related trace gases near 40 km provide evidence that gas phase chemistry has indeed currently depleted about 10% of the stratospheric ozone [37]. Due to the fact that the use of CFCs has been considerably decreased worldwide, ozone layer is expected to recover gradually during the 21st century [37].

In 2010, Lu suggested that besides the photochemical model for ozone depletion, the cosmic ray—driven electron reaction model could be more effective in the interpretation of ozone depletion observations [42]. The latter mechanism drastically differs from the photochemical model for stratospheric ozone depletion. While the typical photochemical model assumes that the sunlight photolysis of CFCs in the upper tropical stratosphere, air transport, and the subsequent heterogeneous chemical reactions of transported inorganic halogens on ice surfaces in polar stratospheric clouds are the three major processes for the activation of halogenated compounds into photoactive halogens, the cosmic ray—driven electron reaction model holds that the in situ cosmic ray—driven electron-induced reaction of halogenated molecules including organic and inorganic molecules (CFCs, HCl, $ClONO_2$, etc.) adsorbed or trapped at polar stratospheric clouds ice in the winter polar stratosphere is the key step for forming the photoactive halogen [42]. However, Grooβ and Müller rejected that possibility [41].

The halogen source gases, often referred to as ozone-depleting substances (ODSs), include manufactured chemicals, which are released into the atmosphere by a variety of applications, such as refrigeration, air conditioning, and foam blowing. CFCs are an important example of chlorine-containing gases. The low reactivity of these manufactured halogenated gases is one property that makes them well suited for specialized applications such as refrigeration; however, it is also the property that allows these compounds to accumulate, exhibiting high lifetimes in the atmosphere ($\approx 50-500$ years). For example, CFC-11 has a lifetime of about 50 years in the atmosphere, whereas this number in the case of CFC-115 becomes 500 years [37]. As halocarbons return toward preindustrial levels, future evolution of ozone will depend on nitrous oxide (N_2O), which is another ozone-destroying compound [43]. Some other gases, like methane and carbon dioxide, have also an indirect role on ozone evolution. For example, the CH_4 increase during the 20th century reduced the ozone losses owing to halocarbon increases, and the N_2O chemical destruction of O_3 was buffered by CO_2 thermal effects in the middle stratosphere.

The destruction of stratospheric ozone has numerous effects on many levels and aspects, of variable severity, whereas lately it has been recognized that it is interconnected also with climate change. UNEP has examined the variety of ozone depletion impacts in two extensive reports (over 300 pages each) published in 2010 and 2015 [44,45]. Only some key-points of the executive summary of the first one are presented here [44]:

Ozone depletion and climate change
* There are strong interactions between ozone depletion and changes in climate induced by increasing GHGs. Ozone depletion affects climate, and climate change affects ozone. The successful implementation of the Montreal Protocol has had a marked effect on climate change. Calculations show that the phaseout of CFCs reduced Earth's warming effect far more than the measures taken under the Kyoto protocol for the reduction of GHGs.
* The Montreal Protocol is working, but it will take several decades for ozone to return to 1980 levels.
* Without the Montreal Protocol, peak values of sun burning UV radiation could have tripled by 2065 at midnorthern latitudes.

Human health
* The incidences of cataract and skin cancers continue to rise in many countries, with significant societal impacts and costs to health-care systems. In some regions the incidence of melanoma in children and young people is no longer increasing, or increasing incidence is confined to less lethal forms.

Terrestrial ecosystems
* In areas where substantial ozone depletion has occurred, results from a wide range of field studies suggest that increased UV-B radiation reduces terrestrial plant productivity by about 6%.
* UV radiation promotes the breakdown of dead plant material and consequently carbon loss to the atmosphere.

Aquatic ecosystems
* Detrimental effects of solar UV-B radiation have been demonstrated for many aquatic organisms.
* Climate change will alter the exposure of aquatic organisms to solar UV radiation by influencing their depth distribution as well as the transparency of the water.

Biogeochemical cycles
* There are interactions between the effects of solar UV radiation and climate change on the processes that drive the carbon cycle. These interactions could

accelerate the rate of atmospheric CO_2 increase and subsequent global warming beyond current predictions.

- Feedbacks involving GHGs other than CO_2 are increasing due to interactive effects of UV radiation and climate change. For example, increases in oxygen-deficient regions of the ocean caused by climate change enhance emissions of nitrous oxide, an important greenhouse and ozone-depleting gas.

Air quality

- Ultraviolet radiation is one of the controlling factors for the formation of photochemical smog, which includes tropospheric ozone and aerosols; it also initiates the production of hydroxyl radicals, which control the amount of many climate- and ozone-relevant gases in the atmosphere.
- Numerical models predict that future changes in UV radiation and climate will modify the trends and geographic distribution of hydroxyl radicals, thus affecting urban and regional photochemical smog formation, as well as the abundance of several GHGs.

Materials

- Increased ambient temperature accelerates the UV-induced degradation of plastics and wood, thus shortening their useful outdoor lifetimes.

Global warming and climate change

Greenhouse effect: When we feel cold in a winter night, we use a blanket to keep our bodies at the desired temperature. The blanket is not a heating body, since it does not produce any energy itself; instead, it traps the heat released by the body, preventing it thus from being wasted in the cold environment. In the same sense, carbon dioxide is the blanket of our planet. It traps the long wave radiation given off by Earth. The oxygen atoms in CO_2 molecule vibrate around the carbon atom in the center with a frequency that corresponds to some of the infrared wavelengths of the long wave radiation. So, when the frequency of the radiation from the Earth's surface and the atmosphere coincides with the frequency of carbon dioxide vibration, the radiation is absorbed by the CO_2 molecule, and then is subsequently converted to heat by collision with other air molecules, before being given back to the surface [14]. Owing to this mechanism, there is excessive heat energy on the planet, since the outgoing long wave radiation is reduced by CO_2, and not as much heat escapes to balance the net solar radiation reaching the Earth. It has to be emphasized that this is a natural mechanism keeping the surface of Earth at temperatures favorable to life on average (Fig. 2.17).

Global warming: Human activities once again disturbed the biogeochemical cycle of carbon and influenced the concentrations of carbon dioxide in the atmosphere that are directly connected to the current extent of the greenhouse effect, by releasing continuously large amounts of carbon dioxide in the atmosphere through the combustion of fossil fuels.

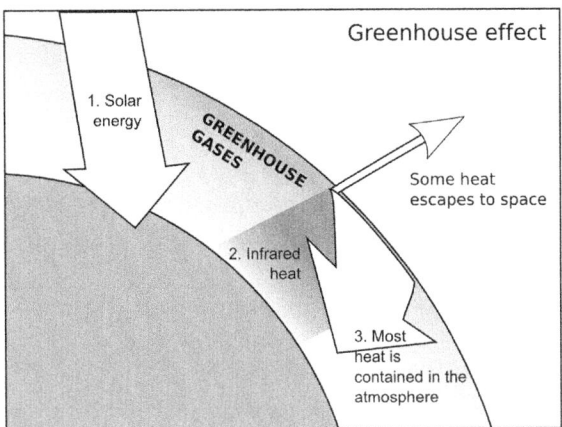

Figure 2.17
The greenhouse effect. *Source: Pixabay.com.*

But CO_2 is not the only compound of concern. There are several other gases that participate in global warming and thus, they are termed greenhouse gases. Kyoto Protocol[7] focuses on the emissions of six gases[8] [46]:

- *Carbon dioxide (CO_2)*. As already mentioned, burning fossil fuels, namely coal, natural gas, and oil, release carbon dioxide into the atmosphere. Burning also solid waste as well as trees and wood products adds to these releases. Chemical industry also (eg, manufacture of cement) is responsible for a part of CO_2. Finally, since carbon dioxide is removed from the atmosphere (or as it is called "sequestered") when it is absorbed by plants as part of the biological carbon cycle, *deforestation is also a contributor to the problem*.
- *Methane (CH_4)*. Methane is released into the atmosphere during the production and transport of natural gas. It is emitted also from livestock and other agricultural practices. Municipal solid waste landfills are also responsible for methane emissions.
- *Nitrous oxide (N_2O)*. Agricultural and industrial activities, as well as combustion of fossil fuels and solid waste, are the main anthropogenic sources of nitrous oxide emissions into the atmosphere.
- *Hydrofluorocarbons (HFCs)*. They are used as refrigerants, aerosol propellants, solvents, and fire retardants, with the first use being their major emissions source, for example, in air conditioning systems in both vehicles and buildings. These compounds were developed as a replacement for CFCs and hydro chlorofluorocarbons (HCFCs), because they

---

[7] The Kyoto Protocol is an international treaty, which extends the 1992 United Nations Framework Convention on Climate Change (UNFCCC) that commits State Parties to reduce greenhouse gas emissions, based on the premise that (1) global warming exists and (2) man-made CO_2 emissions have caused it.
[8] http://www3.epa.gov/climatechange/ghgemissions/gases.html.

do not deplete the stratospheric ozone layer. Unfortunately, they proved to be very potent GHGs with long atmospheric lifetimes and high GWP. They are released into the atmosphere through leaks, servicing, and disposal of equipment in which they are used.

- *Perfluorocarbons (PFCs)*. They are produced as by-products of various industrial processes associated with aluminum production and the manufacturing of semiconductors. They exhibit generally high atmospheric lifetimes and high GWPs.
- *Sulfur hexafluoride (SF_6)*. It is used in magnesium processing and semiconductor manufacturing, as well as a tracer gas for leak detection. It finds application in electrical transmission equipment, including circuit breakers.

Carbon dioxide, methane and nitrous oxide are the most important GHGs, since they account for 98% GHG emissions covered by the Kyoto Protocol. Particularly, CO_2 emissions account for 70% of global GHG emissions [46]. The annual emissions for CO_2, CH_4, and N_2O and for HFCs, PFCs and SF_6 per each source in the United States are presented in Tables 2.4–2.7, respectively [47].

Carbon dioxide was the major GHG emitted by human activities in the United States, accounting for about 82.5% of total GHG emissions. The largest source for both carbon dioxide and total GHG emissions was the combustion of fossil fuels. Methane emissions have decreased by 14.6% since 1990; their main sources are enteric fermentation associated with domestic livestock, natural gas systems, and decomposition of wastes in landfills, whereas in the case of nitrous oxide, agricultural soil management, manure management, mobile source fuel combustion, and stationary fuel combustion were the major sources. Emissions of HFC-23 during the production of HCFC-22 were the primary contributors to aggregate HFC emissions. PFC emissions resulted from primary aluminum production and semiconductor manufacturing as a by-product, whereas electrical transmission and distribution systems accounted for most SF_6 emissions [47].

The largest emitters (countries) of carbon dioxide in the atmosphere are shown in Table 2.8 [48].

The effect of each gas on global warming is the result of the combination of three factors [47]:

- *Concentration*. Larger emissions of GHGs lead to higher concentrations in the atmosphere. GHG concentrations are measured in parts per million for carbon dioxide, parts per billion for methane and nitrous oxide, and in parts per trillion for the rest. It has to be noted that while CO_2, CH_4, and N_2O result also from natural sources, the rest GHGs are man-made compounds, found in the atmosphere only through human activities. This is reflected on the scales of units used to express their concentrations.
- *Lifetime*. Each of these gases can remain in the atmosphere for different amounts of time, ranging from a few years to thousands of years.

Table 2.4: Recent Trends in US CO_2 Emissions and Sinks (MMT CO_2 Equivalent) [47].

Source	1990	2005	2010	2013
CO_2	5123.7	6134.0	5704.5	5505.2
Fossil fuel combustion	4740.7	5747.7	5367.1	5157.7
Electricity generation	1820.8	2400.9	2258.4	2039.8
Transportation	1493.8	1887.8	1732.0	1718.4
Industrial	842.5	827.8	775.7	817.3
Residential	338.3	357.8	334.7	329.6
Commercial	217.4	223.5	220.2	220.7
US territories	27.9	49.9	46.2	32.0
Nonenergy use of fuels	117.7	138.9	114.6	119.8
Iron and steel production and metallurgical coke production	99.8	66.7	55.7	52.3
Natural gas systems	37.6	30.0	32.3	37.8
Cement production	33.3	45.9	31.3	36.1
Petrochemical production	21.6	28.1	27.4	26.5
Lime production	11.7	14.6	13.4	14.1
Ammonia production	13.0	9.2	9.2	10.2
Incineration of waste	8.0	12.5	11.0	10.1
Petroleum systems	4.4	4.9	4.2	6.0
Liming of agricultural soils	4.7	4.3	4.8	5.9
Urea consumption of nonagricultural purposes	3.8	3.7	4.7	4.7
Other process uses of carbonates	4.9	6.3	9.6	4.4
Urea fertilization	2.4	3.5	3.8	4.0
Aluminum production	6.8	4.1	2.7	3.3
Soda ash production and consumption	2.7	2.9	2.6	2.7
Ferroalloy production	2.2	1.4	1.7	1.8
Titanium dioxide production	1.2	1.8	1.8	1.6
Zinc production	0.6	1.0	1.2	1.4
Phosphoric acid production	1.6	1.4	1.1	1.2
Glass production	1.5	1.9	1.5	1.2
Carbon dioxide consumption	1.5	1.4	1.2	0.9
Peatlands remaining peatlands	1.1	1.1	1.0	0.8
Lead production	0.5	0.6	0.5	0.5
Silicon carbide production and consumption	0.4	0.2	0.2	0.2
Magnesium production and processing	+	+	+	+
Land-use, land-use change, and forestry (sink)[a]	(775.8)	(911.9)	(871.6)	(881.7)
Wood biomass and ethanol consumption[b]	219.4	229.8	265.1	283.3
International bunker fuels[c]	103.5	113.1	117.0	99.8

Note: Emission values are presented in CO_2 equivalent mass units using IPCC AR4 GWP values; +, does not exceed 0.05 MMT CO_2 Eq.
[a]Parentheses indicate negative values or sequestration. Sinks (ie, CO_2 removals) are only included in the Net Emissions total. Refer to Table 2.8 for a breakout of emissions and removals for Land Use, Land-Use Change, and Forestry by gas and source category.
[b]Emissions from Wood Biomass and Ethanol Consumption are not included specifically in summing energy sector totals. Net carbon fluxes from changes in biogenic carbon reservoirs are accounted for in the estimates for Land Use, Land-Use Change, and Forestry.
[c]Emissions from International Bunker Fuels are not included in totals.

Table 2.5: Recent Trends in US CH$_4$ Emissions and Sinks (MMT CO$_2$ Equivalent) [47].

Source	1990	2005	2010	2013
CH$_4$	745.5	707.8	667.2	636.3
Enteric fermentation	164.2	168.9	171.1	164.5
Natural gas systems	179.1	176.3	159.6	157.4
Landfills	186.2	165.5	121.8	114.6
Coal mining	96.5	64.1	82.3	64.6
Manure management	37.2	56.3	60.9	61.4
Petroleum systems	31.5	23.5	21.3	25.2
Wastewater treatment	15.7	15.9	15.5	15.0
Rice cultivation	9.2	8.9	11.1	8.3
Stationary combustion	8.5	7.4	7.1	8.0
Abandoned underground coal mines	7.2	6.6	6.6	6.2
Forest fires	2.5	8.3	4.7	5.8
Mobile combustion	5.6	3.0	2.3	2.1
Composting	0.4	1.9	1.8	2.0
Iron and steel production and metallurgical coke production	1.1	0.9	0.6	0.7
Field burning of agricultural residues	0.3	0.2	0.3	0.3
Petrochemical production	0.2	0.1	0.1	0.1
Ferroalloy production	+	+	+	+
Silicon carbide production and consumption	+	+	+	+
Peatlands remaining peatlands	+	+	+	+
Incineration of waste	+	+	+	+
International bunker fuels[a]	0.2	0.1	0.1	0.1

Note: Emission values are presented in CO$_2$ equivalent mass units using IPCC AR4 GWP values; +, does not exceed 0.05 MMT CO$_2$ Eq.
[a]Emissions from International Bunker Fuels are not included in totals.

- *Global Warming Potential (GWP)*. The GWP is an indicator calculated to show how strongly a GHG absorbs energy. Gases with higher GWPs contribute more to global warming. The GWP of a GHG is defined as the ratio of the time-integrated radiative forcing from the instantaneous release of 1 kg of a trace substance relative to that of 1 kg of a reference gas [49].

These factors for each GHG are presented in Table 2.9 [47].

Climate change and consequences: Global warming has manifold effects both regionally and internationally. The resulting climate change is the most pressing global problem today since it could lead to tremendous, long-lasting, and life-changing events all over the world. The fact that the associated phenomena are based on planetary mechanisms and interactions not fully understood so far introduces many uncertainties in the relevant research. For example, Buchwitz et al. [50] emphasize on the fact that despite the importance of carbon dioxide in global warming, the knowledge regarding its sources and sinks is still incomplete [51,52]. Moreover, despite the efforts to decrease CO$_2$ emissions,

Table 2.6: Recent Trends in US N_2O Emissions and Sinks (MMT CO_2 Equivalent) [47].

Source	1990	2005	2010	2013
N_2O	329.9	355.9	360.1	355.2
Agricultural soil management	224.0	243.6	264.3	263.7
Stationary combustion	11.9	20.2	22.2	22.9
Mobile combustion	41.2	38.1	23.7	18.4
Manure management	13.8	16.4	17.1	17.3
Nitric acid production	12.1	11.3	11.5	10.7
Wastewater treatment	3.4	4.3	4.7	4.9
N_2O from product uses	4.2	4.2	4.2	4.2
Adipic acid production	15.2	7.1	4.2	4.0
Forest fires	1.7	5.5	3.1	3.8
Settlement soils	1.4	2.3	2.4	2.4
Composing	0.3	1.7	1.6	1.8
Forest soils	0.1	0.5	0.5	0.5
Incineration of waste	0.5	0.4	0.3	0.3
Semiconductor manufacture	+	0.1	0.1	0.2
Field burning of agricultural residues	0.1	0.1	0.1	0.1
Peatlands remaining peatlands	+	+	+	+
International Bunker Fuels[a]	0.9	1.0	1.0	0.9

Note: Emission values are presented in CO_2 equivalent mass units using IPCC AR4 GWP values; +, does not exceed 0.05 MMT CO_2 Eq.
[a]Emissions from International Bunker Fuels are not included in totals.

Table 2.7: Recent Trends in US HFCs, PFCs, and SF_6 Emissions and Sinks (MMT CO_2 Equivalent) [47].

Source	1990	2005	2010	2013
HFCs	**46.6**	**131.4**	**152.6**	**163.0**
Substitution of ozone-depleting substances[a]	0.3	111.1	144.4	158.6
HCFC-22 production	46.1	20.0	8.0	4.1
Semiconductor manufacture	0.2	0.2	0.2	0.2
Magnesium production and processing	0.0	0.0	+	0.1
PFCs	**24.3**	**6.6**	**4.4**	**5.8**
Aluminum production	21.5	3.4	1.9	3.0
Semiconductor manufacture	2.8	3.2	2.6	2.9
SF_6	**31.1**	**14.0**	**9.5**	**6.9**
Electrical transmission and distribution	25.4	10.6	7.0	5.1
Magnesium production and processing	5.2	2.7	2.1	1.4
Semiconductor manufacture	0.5	0.7	0.4	0.4

Note: Emission values are presented in CO_2 equivalent mass units using IPCC AR4 GWP values; +, does not exceed 0.05 MMT CO_2 Eq.
[a]Small amounts of PFC emissions also result from this source.

Table 2.8: Per Capita Metric Tons of
Carbon Dioxide Emissions [48].

Country	2009
Australia	18.4
Brazil	1.9
Canada	15.2
China	5.8
Germany	9
New Zealand	7.4
United Kingdom	7.7
United States	17.3

Table 2.9: The Main Greenhouse Gases Along With Their Concentration, Lifetime, and Global
Warming Potential in the Atmosphere [47].

Greenhouse Gas	Formula	Concentration		Lifetime in Atmosphere (years)	Global Warming Potential (100-year)
		1750	1998		
Carbon dioxide	CO_2	280[a]	365[a]	50−200	1
Methane	CH_4	700[b]	1745[b]	12	28−36
Nitrous oxide	N_2O	270[b]	314[b]	114	298
Hydrofluorocarbons	HFCs	0[c]	≈ppt	1−270	12−14,800
Perfluorocarbons	PFCs	0[c]	≈ppt	2600−50,000	7390−12,200
Sulfur hexafluoride	SF_6	0[c]	4.2[c]	3200	22,800

[a]ppm.
[b]ppb.
[c]ppt.

the concentration of carbon dioxide in the atmosphere still increases at a rate of approximately 2 ppm/year [53]. So, the predictions about the future climate of our planet remain insecure [50,54]. The same applies also to methane. Its atmospheric abundance increased till 2000, then remained almost constant during 2000−06, and it started rising again in recent years [55,56]. Unfortunately, it is not clear why these trends were observed for CH_4 before 2007 or why it started to increase again at an approximate rate of 7−8 ppb/year [50,53]. But it could take many decades till we achieve a complete understanding of all complicated and interconnected mechanisms on our planet that regulate its climate and many life-supporting processes.

Another distinctive example of uncertainty in relation with sources and sinks of carbon dioxide is the case of oceans. The significance of the role that oceans play in regulating Earth's climate is described in detail in the work of De Suarez et al. [57]. Specifically, oceans play a complicated, multilevel, and vital role in preserving life on Earth. They generate half of the global oxygen and constitute the largest active carbon sink since they absorb a great part of man-produced carbon dioxide. Besides regulating climate and temperature constantly, they are also an important economic resource and a provider of environmental services to humankind all over the world. In addition, they offer a stable life-support system to all species in the entire biosphere. Since oceans are in constant and dialectic interaction with the atmosphere of Earth, they regulate global climate and temperature; however, at the same time ocean warming is a major driver of climate change. Increasing ocean acidification, an effect resulting from increased carbon dioxide levels in the atmosphere, has serious impacts on coral reefs integrity and marine invertebrates, changing thus the structure and nature of ocean ecosystems. The oceans offer an important service as a sink of carbon dioxide to humanity, preventing so humankind from some of the long-term, devastating results of global warming. It has been found that out of all the biological carbon captured in the world, over half is captured by marine living organisms, hence the term "blue carbon". However, increasing ocean acidification could lead to less blue carbon absorbed by oceans. So, there are feedbacks and interactions that cannot be currently entirely predicted.

Hemming et al. [58] reported also that uncertainties in vegetation and carbon cycle responses to climate change reflect hugely different implications for food supply, forestry, and carbon cycle feedback. Additionally, the effects of clouds, black carbon, aerosols, and ocean heat content changes add to the uncertainties involved. One of the factors that complicates even more our attempts to evaluate climate sensitivity is that there appear to be complex natural fluctuations in Earth's climate, whether being internally or externally forced is unknown [59].

Then, what should we do? The case of stratospheric ozone depletion is a good example, where the prompt global action against ODSs, which probably saved life on our planet, following the precautionary principle and despite the fact that back in 1985 the relevant underlying mechanism had not been fully known and described, is an excellent example of what our attitude and response toward climate change should be.

To understand the risks involved, it is essential at this point to examine the existing or potential consequences of global warming and climate change. All effects and associated future risks are presented in full detail in the relevant extensive report of the IPCC [60]. The most important of the current impacts owing to climate change along with the corresponding confidence are summarized in paragraph SPM 1.3 (p. 6) of this report:

In recent decades, changes in climate have caused impacts on natural and human systems on all continents and across the oceans. Impacts are due to observed climate change, irrespective of its cause, indicating the sensitivity of natural and human systems to changing climate.

Evidence of observed climate change impacts is strongest and most comprehensive for natural systems. In many regions, changing precipitation or melting snow and ice are altering hydrological systems, affecting water resources in terms of quantity and quality (medium confidence). Many terrestrial, freshwater and marine species have shifted their geographic ranges, seasonal activities, migration patterns, abundances and species inter-actions in response to ongoing climate change (high confidence). Some impacts on human systems have also been attributed to climate change, with a major or minor contribution of climate change distinguishable from other influences. Assessment of many studies covering a wide range of regions and crops shows that negative impacts of climate change on crop yields have been more common than positive impacts (high confidence). Some impacts of ocean acidification on marine organisms have been attributed to human influence (medium confidence).

Ref. [60]

The report includes also future and long-term threats; however, this is a very modest statement about climate change by all means. Ryu and Jacobson [61] suggest that positive feedback linking GHG emissions, Arctic climate change, and global warming may be stronger than previously realized. Based on previous works [54,62,63], they reported that global surface temperatures could increase from 1 to 6°C by 2100 as a result of global warming. This temperature rise could be even greater for the Arctic, where sea ice reduction, organic matter decomposition in lakes, and thawed permafrost, as well as other positive feedbacks, can potentially amplify the global trend. The researchers suggested that collapse of much larger continental ice sheets could have sustained sizable carbon dioxide evasion fluxes, perhaps sufficient to influence atmospheric CO_2 levels over glacial–interglacial timescales.

Brysse et al. [64] examined in their work whether IPCC underestimates generally global warming effects or not, and whether this global-scale issue is just a hoax. The authors concluded that evidence from recent analyses suggests that scientists may have underestimated the magnitude and rate of expected impacts of anthropogenic climate change. They further suggested that this underestimation reflects a systematic bias, which they labeled "erring on the side of least drama (ESLD)". They explained this as an effort of scientists to avoid drama; as they pointed out "the scientific community may be biasing its own work—a bias that needs to be appreciated because it could prevent the full recognition, articulation, and acknowledgment of dramatic natural phenomena that may, in fact, be occurring."

The U.S. National Aeronautics and Space Administration (NASA) summarizes the current and future trends regarding global warming effects in the United States both nationally and regionally, based on the "Climate Change Impacts in the United States" report [65]:

- *Change will continue through this century and beyond.* Global climate is projected to continue to change over this century and beyond. The magnitude of climate change beyond the next few decades depends primarily on the amount of heat-trapping gases emitted globally and how sensitive the Earth's climate is to those emissions.

- *Temperatures will continue to rise.* Because human-induced warming is superimposed on a naturally varying climate, the temperature rise has not been, and will not be, uniform or smooth across the country or over time.
- *Frost-free season (and growing season) will lengthen.* The length of the frost-free season (and the corresponding growing season) has been increasing nationally since the 1980s, with the largest increases occurring in the western United States, affecting ecosystems and agriculture. Across the United States, the growing season is projected to continue to lengthen. In a future in which heat-trapping gas emissions continue to grow, increases of a month or more in the lengths of the frost-free and growing seasons are projected across most of the United States by the end of the century, with slightly smaller increases in the northern Great Plains. The largest increases in the frost-free season (more than 8 weeks) are projected for the western United States, particularly in high elevation and coastal areas. The increases will be considerably smaller if heat-trapping gas emissions are reduced.
- *Changes in precipitation patterns.* Average US precipitation has increased since 1900, but some areas have had increases greater than the national average, and some areas have had decreases. More winter and spring precipitation is projected for the northern United States, and less for the Southwest, over this century. Projections of future climate over the United States suggest that the recent trend toward increased heavy precipitation events will continue. This trend is projected to occur even in regions where total precipitation is expected to decrease, such as the Southwest.
- *More droughts and heat waves.* Droughts in the Southwest and heat waves (periods of abnormally hot weather lasting days to weeks) everywhere are projected to become more intense, and cold waves less intense. Summer temperatures are projected to continue rising, and a reduction of soil moisture, which exacerbates heat waves, is projected for much of the western and central United States in summer. By the end of this century, what have been once-in-20-year extreme heat days (one-day events) are projected to occur every 2 or 3 years over most of the nation.
- *Hurricanes will become stronger and more intense.* The intensity, frequency, and duration of North Atlantic hurricanes, as well as the frequency of the strongest (Category 4 and 5) hurricanes, have all increased since the early 1980s. The relative contributions of human and natural causes to these increases are still uncertain. Hurricane-associated storm intensity and rainfall rates are projected to increase as the climate continues to warm.
- *Sea level will rise 1–4 feet (0.3–1.2 m) by 2100.* Global sea level has risen by about 8 in. since reliable record keeping began in 1880. It is projected to rise another 1–4 ft by 2100. This is the result of added water from melting land ice and the expansion of seawater as it warms. In the next several decades, storm surges and high tides could combine with sea level rise and land subsidence to further increase flooding in many of

these regions. Sea level rise will not stop in 2100 because the oceans take a very long time to respond to warmer conditions at the Earth's surface. Ocean waters will therefore continue to warm and sea level will continue to rise for many centuries at rates equal to or higher than those of the current century.

*   *Arctic likely to become ice-free.* The Arctic Ocean is expected to become essentially ice-free in summer before midcentury.
*   *US regional effects.*

    Northeast. Heat waves, heavy downpours, and sea level rise pose growing challenges to many aspects of life in the Northeast. Infrastructure, agriculture, fisheries, and ecosystems will be increasingly compromised. Many states and cities are beginning to incorporate climate change into their planning.

    Northwest. Changes in the timing of streamflow reduce water supplies for competing demands. Sea level rise, erosion, inundation, risks to infrastructure, and increasing ocean acidity pose major threats. Increasing wildfire, insect outbreaks, and tree diseases are causing widespread tree die-off.

    Southeast. Sea level rise poses widespread and continuing threats to the region's economy and environment. Extreme heat will affect health, energy, agriculture, and more. Decreased water availability will have economic and environmental impacts.

    Midwest. Extreme heat, heavy downpours, and flooding will affect infrastructure, health, agriculture, forestry, transportation, air and water quality, and more. Climate change will also exacerbate a range of risks to the Great Lakes.

    Southwest. Increased heat, drought, and insect outbreaks, all linked to climate change, have increased wildfires. Declining water supplies, reduced agricultural yields, health impacts in cities due to heat, and flooding and erosion in coastal areas are additional concerns.

This list of potential consequences is not exhaustive. Numerous other chain effects could be triggered from global warming! For example, Bezirtzoglou et al. [66] relate climate change with infectious diseases. Specifically, they point out that climate affects mainly the range of infectious diseases, whereas weather affects the timing and intensity of outbreaks. The ranges of several vectorborne diseases or their vectors are already changing in altitude due to warming. Moreover, more intense weather events can bring about conditions conductive to outbreaks of infectious diseases: Heavy rains leave insect breeding sites, drive rodents from burrows, and contaminate clean water systems. The incidences of mosquito-borne parasitic and viral diseases are among those diseases that are the most sensitive to climate. Climate change affects disease transmission by shifting the vector's geographic range and by shortening the pathogen incubation period. Climate-related increases in temperature in sea surface and level would lead to higher incidence of waterborne infectious and toxin-related illnesses, such as cholera and seafood intoxication. Food security is another sector to be affected by global warming. Sommer has reported, for instance, that global warming and related climate change may pose a major challenge to agriculture and rural livelihoods in Central Asia, with its five countries Kazakhstan, Uzbekistan, Kyrgyzstan, Tajikistan, and Turkmenistan [67]. The impact of climate change on the hydrological cycle is also discussed in the next chapter on aquatic environment. Shortly, global warming is another water stressor. Biodiversity is also considerably affected by climate change, and the list continues!

Mitigation and adaptation: It has been discussed that global warming and climate change affects and is affected at the same time by complex mechanisms taking place on the planetary scale, mechanisms not entirely understood so far. Many uncertainties are involved. Even if we succeeded in stabilizing the global average surface temperature, it would not be given that all aspects and mechanisms of the climate system would be alleviated. Soil carbon, biomass, ice sheets, ocean temperatures, and associated sea level rise all have their own intrinsic long timescales, which will result in changes lasting even hundreds of years after global surface temperature is stabilized [60]. Moreover, many of the GHGs involved have very long lifetimes in the atmosphere, reaching even hundreds of years in some cases. It is therefore evident that many aspects of environmental change and related effects will proceed for a considerable length of time, even for centuries, regardless of our efforts to reduce the man-made emissions of GHGs.

Of course, as long as global warming is allowed to evolve and extend, the risk of experiencing violent and/or irreversible global changes is increased. It is almost positive that the average sea level will continue to rise globally for many centuries beyond 2100, the exact rise being associated with the future GHG emissions. The threshold for the loss of the Greenland ice sheet over a millennium or more, and an associated sea level rise of up to 7 m, is greater than about 1°C (low confidence) but less than about 4°C (medium confidence) of global warming with respect to preindustrial temperatures [60].

These facts impose our response against global warming in two directions: *adaptation* and *mitigation*.

In general, climate policy has mostly focused on *mitigation*, namely the reduction of GHG emissions and/or the enhancement of sinks—with instruments such as the Kyoto Protocol. Considerable reductions of GHG emissions over the next decades can reduce climate risks in the 21st century and beyond, increase prospects for effective adaptation, reduce the costs and challenges of mitigation in the longer term, and contribute to climate-resilient pathways for sustainable development. Without more measures in the mitigation direction to decrease the anthropogenic emissions of GHGs beyond those in place today, and even with adaptation, warming by the end of the 21st century will lead to a very high risk of severe, broad impacts all over the world, impacts that will be impossible to reverse. Mitigation is also accompanied by a number of co-benefits and of risks due to adverse side effects, but in any case these risks are not of the same extent, severity, and irreversible nature as those stemming from climate change [60].

While there is a wide consensus among climate experts and policy makers that mitigation of climate change is and should remain the prime focus of climate policy, it is increasingly recognized that *adaptation* to climate change has become unavoidable. Even under optimistic assumptions for the success of present day mitigation efforts and policies, human activity is likely to lead to further climate change with possibly severe impacts [60]. As also discussed, long timescales of mechanisms involved in climate regulation as well as long lifetimes of some GHGs in the atmosphere will delay the effectiveness of the mitigation measures implemented. So, it is a necessity to deal with the adverse effects connected to climate change for the near future. The Stern review noted that adaptation is the only response available for the impacts that will occur over the next several decades before mitigation measures can have an effect [68].

It is therefore obvious that mitigation and adaptation are complementary approaches for reducing risks of climate change impacts over different timescales. Mitigation, in the near term and through the century, can substantially reduce climate change impacts in the latter decades of the 21st century and beyond. Benefits from adaptation can already be realized in addressing current risks, and they can be also realized in the future for addressing emerging risks [60].

Global warming and climate change are global-scale issues that can be confronted only through collective action and international cooperation. After all, wherever the emissions are released, they accumulate over time and mix globally, affecting the entire planet without any discrimination. Effective mitigation will not be achieved if individual agents advance their own interests independently. In addition, the effectiveness of adaptation can be enhanced through complementary actions across all levels, including international cooperation. The evidence suggests that outcomes seen as equitable can lead to more effective cooperation [60,68].

According to the Stern review, GHG emissions can be cut in four ways [68]:

- reducing demand for emissions-intensive goods and services;
- increased efficiency, which can save both money and emissions;
- action on nonenergy emissions, such as avoiding deforestation; and
- switching to lower-carbon technologies for power, heat, and transport.

Regarding the third action on the list, and as far as *deforestation* is concerned, it should not be forgotten that it *is responsible for more than 18% of global emissions*, a share greater than the one for the global transport sector [68]. Hence, action to preserve the remaining areas of natural forest is needed urgently.

The policy to reduce emissions should be based on three essential elements: *carbon pricing*, *technology policy*, and *removal of barriers to behavioral change*.

In relation with adaptation, there are four key areas [68]:

- High-quality climate information and tools for risk management will help to drive efficient markets. Improved regional climate predictions will be critical, particularly for rainfall and storm patterns.
- Land-use planning and performance standards should encourage both private and public investment in buildings and other long-lived infrastructure to take account of climate change.
- Governments can contribute through long-term polices for climate-sensitive public goods, including natural resources protection, coastal protection, and emergency preparedness.
- A financial safety net may be required for the poorest in society, who are likely to be the most vulnerable to the impacts and least able to afford protection.

Sustainable development and equity provide a basis for assessing climate policies. Sustainability and well-being cannot be achieved without either addressing the impacts of climate change or eliminating anthropogenic GHG emissions. Poverty eradication is also a prerequisite since equity cannot be established among people with different means of living, and poverty itself has an impact on global warming and air quality, mainly via deforestation and use of low-quality carbon fuels. The contribution to GHG emissions, and thus to global warming, differs from one country to another. Moreover, countries also cope with different challenges, having at the same time different capacities to address mitigation and adaptation. Most of the populations that are most vulnerable to climate change have contributed and contribute little to GHG emissions. Delaying mitigation responses just shifts burdens from the present to the future generations, and insufficient adaptation responses to emerging impacts are already eroding the basis for sustainable development. Comprehensive strategies in response to climate change that are consistent with sustainable development take into account the co-benefits, adverse side effects, and risks that may arise from both adaptation and mitigation options [60].

2.3.5 Case Studies

Four case studies are presented to underscore the effects of air pollution. The first three are about respective notorious disasters caused either by severe events of urban air pollution or by large-scale industrial accidents. The purpose is to show how the air we breathe can become lethal and spread the dangers associated with industrial accidents over large areas. The fourth case study included aims at highlighting the effects of global warming on coral reefs.

The Great Smog of December 1952

The *Great Smog* of 1952 was a severe air pollution event that affected London during December 1952 [69]. Specifically, a dense fog covered Greater London between the 5 and 8 December 1952, accompanied by a sudden rise in mortality that exceeded by far anything previously recorded during similar and frequent incidents of smog. According to estimates of the Ministry of Health's committee, between 3500 and 4000 more people died than it would be expected under normal conditions [70].

Although the deadly smog was exceptional in density and duration, the death toll only became evident after the event, because of the fact that such air pollution events were not unusual in London back then [70]. During the incident, both smoke and sulfur dioxide levels reached exceptional concentrations. The weather also played a significant role in the tragic effects of this air pollution event. On December 4, 1952, an anticyclone settled over a windless London, causing a temperature inversion with cold, stagnant air trapped under a layer of warm air. The resultant fog, mixed with chimney smoke, particulates such as those from vehicle exhausts, and other pollutants such as sulfur dioxide, formed a persistent smog, which covered the whole city the following day. The presence of tarry particles of soot gave the smog its yellow-black color, hence the nickname "peasouper." The absence of significant wind prevented its dispersion and allowed an unprecedented accumulation of pollutants. The relevant report of the City Hall of London provides us with the details [70]: The previous December, the mean smoke concentration across 12 sites ranged in $0.12-0.44$ mg/m^3. In 1952, the wind speed dropped during the afternoon of Thursday 4 December and patches of fog had begun to appear by 6 pm. Air pollution measurements were taken by the London County Council at its headquarters, County Hall in Lambeth, later to be the offices of the Greater London Council. By noon the next day, smoke concentrations at County Hall had risen from 0.49 mg/m^3 to 2.46 mg/m^3. They continued to rise to 4.46 mg/m^3 on both 7 and 8 December. Sulfur dioxide (SO$_2$) followed a similar pattern with concentrations rising from 0.41 mg/m^3 on 4 December to 2.15 mg/m^3 on the 5th and to 3.83 mg/m^3 on both the 7 and 8 December. The concentrations of both smoke and SO$_2$ dropped sharply to 1.22 mg/m^3 and 1.35 mg/m^3, respectively, as wind speeds rose and the smog cleared on Tuesday 9 December.

The first casualties of the smog were cattle at the Smithfield Show. Then, an Aberdeen Angus died, 12 other cattle had to be slaughtered, 60 needed major veterinary treatment, and about a 100 more needed minor attention. All were relatively young and in prime condition. At Sadler's Wells, the opera La Traviata had to be abandoned after the first act because the theater was so full of smog. A number of stories on the smog appeared in the newspapers, with The Times covering the high economic cost of the smog. However, it was the coroners, pathologists, and the Registrars of Deaths who first became aware of the exceptional effects on the people of London. The number of deaths attributable to the smog was similar to the numbers caused by the cholera epidemic of 1854 and the influenza epidemic of 1918 [70]!

Britain's economic situation in 1952 was extremely tenuous; for this reason, low-quality, high-sulfur coal was burned in London to permit the export of the more valuable low-residue, low-sulfur coal. Thus, the economic situation contributed indirectly to the degree of pollution [71].

These tragic public events in London half a century ago made people realize that polluted air could not only cause an immediate increase in deaths and illnesses, but it could also result in longer-term and more subtle effects [69].

Seveso accident

In July 1976, at a chemical plant in the suburbs of Milan, Italy, an accident occurred that caused the release of dioxin, which is a deadly poison. The Seveso disaster obtained its name from the city of Seveso (population: 17,000 in 1976), which was the community most affected.[9] The details of the accident are presented in the relevant article of Fortunati [72].

Shortly, operators stopped a batch operation in an incorrect way, disregarding operating instructions. As a result, the temperature rose, a large amount of dioxin was generated by a runaway reaction, a rupture disk operated, and contents including dioxin diffused into the atmosphere. The soil of 1800 ha of land was contaminated, and 220,000 persons were injured, suffering from sequelae. There was no correct knowledge among researchers and factory managers of a runaway reaction of the reactor, and the generation of dioxin from a runaway reaction. The damage spread because administrative authorities were not informed until the existence of high-density dioxin was reconfirmed.

Specifically, at a plant which produced 2,4,5-trichlorophenol by a batch operation in the suburbs of Milan, Italy, the batch operation was finished with a different method from the one described in the operation manual. Therefore, in a distillation vessel that was used also as a reactor, a runaway reaction occurred, and dioxin was generated in large quantities. As a result of the runaway reaction, a rupture disk operated, and reaction products including dioxin were discharged. Carried by the wind, the reaction products, including dioxin, spread over a wide area. When dioxin fell to the earth by rain, there was a large scale of dioxin poisoning and people suffered from sequelae. Extensive soil pollution was caused.

In a laboratory, the existence of dioxin was confirmed 5 days later, but the local government was not informed immediately. After five more days, the existence of dioxin

---

9 *source*: https://en.wikipedia.org/wiki/Seveso_disaster; http://greenliving.about.com/od/greenprograms/a/Seveso-TCDD.htm; http://www.sozogaku.com/fkd/en/hfen/HC1300002.pdf.

was confirmed again by the same laboratory. The local government was informed by the laboratory about the dioxin. Due to the delay in communication, the surrounding inhabitants were more exposed to dioxin, and the damage spread.

The response to the Seveso accident was widely criticized as slow and bungled. Several days passed before it was announced that a gas containing dioxin had been released from the facility; evacuation of the worst-affected areas took several more days. Seveso and its residents continue to function as a kind of "living laboratory" into the effects of dioxin exposure on people and animals. Throughout Europe, the name Seveso is now associated with tough regulations that require any facilities storing, manufacturing, or handling hazardous materials to inform local authorities and communities about the nature of their facility, and to create and publicize measures to prevent and respond to any accident that may occur.

Bhopal gas tragedy

The Bhopal disaster, also referred to as the *Bhopal gas tragedy*, was a gas leak incident in India, which is considered the world's worst industrial disaster.[10] All the details of the accident can be found in the relevant ICMR document [73].

The night of December 2–3, 1984 in Bhopal, India, was one of those nights where the week winds kept changing direction. Under the dear dark sky, the key units of the Union Carbide India Limited facility, one of the largest employers in the city, were quietly waiting to be dismantled and shipped to another developing country. The Union Carbide plant had once been port of an ambitious Indian plan to achieve self-agricultural production by increasing the national production of pesticides, but the plan was severely curtailed by the crop failures and famine that spread across India in the early 1090s. At 11 pm, while most of Bhopal's 900,000 inhabitants were sleeping, one operator of the plant noticed a small leak as well as elevated pressure inside storage tank 610, which contained methyl isocyanate (MIC), a highly reactive chemical used as an intermediate in the production of the insecticide Sevin. The leak had been created by a strong exothermic reaction resulting from mixing of 1 ton of water normally used for cleaning internal pipes with 40 tons of MIC contained in the tank. Because of the fact that the coolant for the refrigeration unit had been previously used in another part of the plant, tank 610 could not be cooled quickly. Therefore, pressure and heat continued to build inside the tank, and the tank continued to leak. Both the vent gas scrubber and the gas flare system, two safety devices designed to neutralize potential toxic discharges from the tank before they escaped into the atmosphere, had been turned off several weeks before. At around 1 am, a loud rumbling echoed around the plant as the safety valve of the tank gave way. Nearly 40 tons

---

[10] *source*: https://en.wikipedia.org/wiki/Bhopal_disaster; http://www.bhopal.com/.

of MIC gas were released into the morning air of Bhopal. It did not take long for the plume carried by the changing winds to spread over a large area.

At least 3800 people died immediately, killed in their sleep. Local hospitals were soon overwhelmed with the thousands of injured people. The crisis was further deepened by a lack of knowledge of which gas was specifically involved, and hence what the appropriate course of treatment should be. Estimates of the number of people killed in the first few days by the plume from the Union Carbide plant are as high as 10,000, with 15,000–20,000 premature deaths reportedly occurring in the subsequent two decades. The Indian government reported that more than half a million people were exposed to the gas. The greatest impact was on the densely populated poor neighborhoods immediately surrounding the plant.

The reader can retrieve shocking photos relevant to this disaster from the India Today news Website.[11]

Global warming and coral reefs

Ocean warming is one of the multiple effects of global warming. Moreover, the dissolved carbon dioxide in the oceans results in ocean acidification. These effects have detrimental impacts on the integrity of coral reefs. They are described in the work of De Suarez et al. [57]. According to the authors, the levels of atmospheric carbon dioxide of 450 ppm are widely considered to be unsustainable. Long-term levels of atmospheric carbon dioxide of no more than 350 ppm should be sought by humanity; otherwise, many ecosystems such as coral reefs will be fundamentally changed.

Specifically, since the 1980s, corals have undergone unprecedented high temperature mass bleaching and mortality. Locations, intensity, and severity of bleaching are predictable using a sea surface temperature "Hotspot" method. For example, current observations using the Hotspot method indicate large patches of warm waters around the Seychelles with on-the-ground confirmation of coral bleaching in many areas. The recurrence of coral bleaching, since a devastating event in 1998, is another proof that global warming is taking place. Without prompt and adequate action, coastal areas will be at stake around the world.

The authors point out that with the current trends in GHG emissions, *it won't take more than 20–40 years for the remaining reefs to be lost forever to coral bleaching* as a result of the combined effect of ocean warming and ocean acidification. Even the most optimistically low future atmospheric CO_2 concentrations could be high enough to cause carbonate coral reef ecosystems to dissolve, large areas of polar waters to become corrosive to the shells of some key marine species, and marine ecosystems to become nearly unrecognizable.

---

[11] http://indiatoday.intoday.in/gallery/bhopal-gas-tragedy-30th-anniversary-warren-anderson/1/13577.html.

The trends indicated by measured temperature records suggest that negative effects on coral reef ecosystems will be much more immense than predicted by current models of climate change.

2.3.6 Air Pollution Abatement Techniques

The minimization of the releases into the environment can be largely achieved by [74].

- process-integrated techniques, such as emissions reuse and pollution prevention measures and
- end-of-pipe treatment (individual and/or central facilities).

The first approach may involve cleaner synthesis processes, improved technology, recycling of residues, improved use of catalysts, and generally, every technique integrated into the process that leads to less waste, whereas the second one refers to the treatment of tailpipe emissions, which are inevitably produced by a chemical process [21]. Both approaches have to be combined so that our releases into the environment are as decreased and harmless as possible.

An important categorization for the selection of the appropriate minimization technology is the one into *stationary* and *mobile* sources, as discussed previously. Stationary sources include mainly industrial facilities and central heating systems in households, where mobile sources include all transport means.

Gas emissions can be also divided into *ducted* and *diffuse* emissions. Only ducted emissions can be treated effectively. As far as diffuse emissions are concerned, the objective of waste gas management is their prevention and/or minimization [74].

In the context of this chapter, special interest is given on the end-of-pipe techniques for the treatment of industrial ducted emissions, since for the transport sector the solutions for tailpipe exhaust emissions minimization are very specific and determinate, such as three-way catalysts for gasoline-fueled cars.

The most important *ducted* gas emissions in the chemical industry, as presented in the Best Available Techniques (BAT) Reference Document for Common Waste water and Waste Gas Treatment/Management Systems in the Chemical Sector released by the European Commission [74], include:

- process emissions released through a vent pipe by the process equipment and inherent to the running of the plant;
- flue gases from energy-providing units, such as process furnaces, steam boilers, combined heat and power units, gas turbines, and gas engines;

- waste gases from emission control equipment, such as filters, incinerators/oxidizers, or absorbers, likely to contain unabated pollutants or pollutants generated in the abatement system;
- tail gases from reaction vessels and condensers;
- waste gases from catalyst regeneration;
- waste gases from solvent regeneration;
- waste gases from vents from storage and handling (transfers, loading, and unloading) of products, raw materials, and intermediates;
- waste gases from purge vents or preheating equipment, which are used only in start-up or shutdown operations;
- discharges from safety relief devices (eg, safety vents, safety valves);
- exhaust from general ventilation systems; and
- exhaust from vents from captured diffuse sources, eg, diffuse sources installed within an enclosure or building.

The main air pollutants from chemical processes and energy supply are [74]:

- carbon dioxide;
- sulfur oxides (SO_2, SO_3) and other sulfur compounds (H_2S, CS_2, COS);
- nitrogen oxides (NO_x, N_2O) and other nitrogen compounds (NH_3, HCN);
- halogens and their compounds (Cl_2, Br_2, HF, HCl, HBr);
- incomplete combustion compounds, such as CO and $CXHY$;
- volatile organic compounds and organosilicon compounds; and
- particulate matter (such as dust, soot, alkali, and heavy metals).

As already demonstrated, carbon dioxide is not a pollutant in the strict sense of the term, but in any case it is the most important GHG. Ultimately, waste-treatment techniques are classified by the type of contaminant. The main techniques concerning waste gas treatment are the following [21]:

VOCs and inorganic compounds: membrane separation, condensation, adsorption, wet scrubbing, biofiltration, bioscrubbing, biotrickling, thermal oxidation, catalytic oxidation, and flaring.

Particulate matter: separator, cyclone, electrostatic precipitator, wet dust scrubber, fabric filter, catalytic filtration, two-stage dust filter, absolute filter, high-efficiency air filter, and mist filter.

Gaseous pollutants in combustion exhaust gases: dry sorbent injection, semidry sorbent injection, wet sorbent injection, selective noncatalytic reduction of NO_x (SNCR), selective catalytic reduction of NO_x (SCR).

Table 2.10: Evaluation of Alternative Treatment Processes Used to Control Industrial Gas Emissions [21].

Case	Activated Carbons	Thermal Oxidation	Scrubbers	Particulate Filters	Catalytic oxidation
Low VOC levels	×		×		
High VOC levels		×	×		×
Continuous load	×	×	×	×	×
Intermittent loads	×			×	
Halogenated organics	×				
$T > 65°C$		×	×		×
$T < 65°C$	×		×	×	
High flows	×		×		
Low flows	×	×	×		×
High humidity		×		×	×
Inorganic particles				×	

VOC, volatile organic compound.

In Table 2.10, the conditions for the application of some treatment processes are shown, as reported by Inglezakis and Poulopoulos [21].

The BAT reference document divides the selection of techniques into those by the pollutant to be removed and those by the waste gas flow, as presented in Tables 2.11 and 2.12 [74].

2.3.7 Air Pollution Legislation

As already discussed, air pollution is related to many pollutants and problems that manifest themselves at various scales. It is thus expected that the relevant legislation is extensive, especially if someone attempts to illustrate all legislative variations around the world. Hence, the main principles, trends, and reference documents are presented hereinafter based on the air pollution problem type. Emphasis is put on European legislation, which can be characterized as the most progressive one in the world and on international agreements.

Air legislation in Europe

The European legislation and tools in relation with ambient air quality is representative of most similar policies around the world. The European Environment Agency includes a short synopsis of the European legislation in the report "Every Breath We Take" [20]. Action takes place on many directions in order to ensure clean air for all citizens [20].

Targeting pollutants: The European Union sets legally binding and nonbinding limits for its territory for specific air pollutants. Accordingly, the European Law has set standards for

Table 2.11: Selection of Techniques for Waste Gas Emission Reduction by Pollutant to be Removed [74].

Technique	Dry Matter	Wet Matter	Inorganic Particulates	Organic Particulates	Inorganic Gaseous or Vaporous Components	Organic gaseous or vaporous components	Odors
Recovery and Abatement for VOCs and Inorganic Compounds							
Membrane separation (pre)						×	
Condensation (pre)					(×)	×	
Cryocondensation (pre, FT)					(×)	×	(×)
Adsorption (FT, pol)					×	×	×
Wet gas scrubber (water) (FT)	(×)	(×)	(×)	(×)	×	×	×
Wet gas scrubber (alkaline) (FT)	(×)	(×)	(×)	(×)	×	×	×
Wet gas scrubber (alkaline oxidation) (FT)	(×)	(×)	(×)	(×)			×
Wet gas scrubber (acidic) (FT)	(×)	(×)	(×)	(×)	×	×	×
Abatement for VOCs and Inorganic Compounds							
Biofiltration (FT)					×	×	×
Bioscrubbing (FT)					×	×	×
Biotrickling (FT)					×	×	×
Moving-bed trickling filter (FT)					×	×	×
Thermal oxidation (FT)				×		×	×
Catalytic oxidation (FT)						×	×
Ionization (FT)						×	×
Photo/UV oxidation (FT)						×	×
Recovery and Abatement for Particulates							
Settling chamber/gravitational separator (pre)	×	×	×	×			
Cyclone (pre)	×	×	×	×			
Electrostatic precipitator (FT)	×	×	×	×	(×)	(×)	
Wet dust scrubber (FT)	×	×	×	×			
Fabric filter (FT)	×		×	×			

Continued

Table 2.11: Selection of Techniques for Waste Gas Emission Reduction by Pollutant to be Removed [74].—cont'd

Technique	Dry Matter	Wet Matter	Inorganic Particulates	Organic Particulates	Inorganic Gaseous or Vaporous Components	Organic gaseous or vaporous components	Odors
Ceramic and metal filter (FT)	×		×	×			
Catalytic filtration (FT)	×	×	×	×		×	
Two-stage dust filter (pol)	×		×	×			
Absolute (HEPA) filter (pol)	×		×	×			
High-efficiency air filter (pol)		×					
Mist filter (pre, pol)		×			(×)		
Recovery and Abatement for Inorganic Compounds							
Dry alkali injection (FT)					×		
Semidry alkali injection (FT)					×		
Wet lime injection (FT)					×		
SNCR (FT)					×		
SCR (FT)					×		
NSCR (FT)					×		
Wet gas scrubber for NO$_x$ (FT)					×	(×)	
Flaring							
Flaring (FT)						×	×

NB: ×, primary application; (pre), mainly used as a pretreatment; (FT), treatment technique used as a final treatment technique; (pol), mainly as a polishing technique after application of a standard technique; VOC, volatile organic compound.

Table 2.12: Selection of Techniques for Waste Gas Emission Reduction by Waste Gas Flow Rate [74].

Technique	100 (Nm³/h)	1000 (Nm³/h)	10,000 (Nm³/h)	100,000 (Nm³/h)
Recovery and Abatement for VOCs and Inorganic Compounds				
Membrane separation				
Condensation	×	×	××	×
Cryocondensation	×	×		
Adsorption	×	××	××	×
Wet gas scrubber (water)	×	×	××	××
Wet gas scrubber (alkaline)	×	×	××	××
Wet gas scrubber (alkaline-oxidation)	×	×	××	×
Wet gas scrubber (acidic)	×	×	××	××
Abatement for VOCs and Inorganic Compounds				
Biofiltration	×	××	××	××
Bioscrubbing	×	×	×	×
Biotrickling	×	×	×	×
Moving-bed trickling filter		×	×	
Thermal oxidation		×	××	
Catalytic oxidation		×	××	
Ionization	×	×	×	×
Photo/UV oxidation		×	×	
Recovery and Abatement for Particulates				
Settling chamber/gravitational separator	×	×	××	××
Cyclone	×	××	××	×
Electrostatic precipitator			×	×
Wet dust scrubber		×	××	××
Fabric filter	×	×	××	××
Ceramic filter		××	×	×
Metal filter		NI		
Catalytic filtration	×	×	×	
Two-stage dust filter		×	×	
Absolute (HEPA) filter	×	×		
High-efficiency air filter	××	××	×	
Mist filter		×	××	××
Recovery and Abatement for Inorganic Compounds				
Dry alkali injection			××	×
Semidry alkali injection			×	××
Wet lime injection		×	×	×
SNCR	×	×	×	×
SCR		×	××	××
NSCR			×	
Wet gas scrubber for NO$_x$			×	
Flaring				
Flaring	×	×	×	×

NB: *NI*, no information provided; ×, application; ××, most common applications; *VOC*, volatile organic compound.

PM of certain sizes, ozone, sulfur dioxide, nitrogen oxides, lead, and other pollutants that may cause either adverse health effects or damage to ecosystems. The European directives currently regulating ambient air concentrations of the main pollutants include the Directive 2008/50/EC on ambient air quality and cleaner air for Europe, as well as the Directive 2004/107/EC relating to arsenic, cadmium, mercury, nickel, and polycyclic aromatic hydrocarbons in ambient air. In the case of noncompliance with these air quality limit and target values, air quality management plans must be developed and implemented in the areas where noncompliance is observed [17].

Targeting emissions: In addition, the Gothenburg Protocol to the United Nations Economic Commission for Europe's Convention on Long-range Transboundary Air Pollution (LRTAP) and the EU National Emission Ceilings Directive (2001/81/EC) both set annual emissions limits for European countries on air pollutants, including those pollutants responsible for acidification, eutrophication, and ground-level ozone pollution. So, member states are responsible for implementing the measures that are required to ensure that their emissions do not exceed on annual basis the ceiling set for each pollutant [20].

Targeting sectors: Besides the air quality standards described above, European legislation aims also at regulating the particular sectors that constitute the main air pollution sources. Particularly, the exhaust emissions from automotive vehicles have been regulated through a number of performance and fuel standards, including the Directive 98/70/EC relating to the quality of petrol and diesel fuels and vehicle emission standards, known as the Euro standards. The Euro 5 and 6 standards cover emissions from light vehicles including passenger cars, vans, and commercial vehicles. Industry is naturally another sector under strict regulation. Specifically, the Industrial Emissions Directive 2010/75/EU and the Directive 2001/80/EC on the limitation of emissions of certain pollutants into the air from large combustion plants apply. Other international agreements on the emissions of air pollutants are also in place; for instance, sulfur dioxide emissions from shipping are regulated through the International Maritime Organization's 1973 Convention for the Prevention of Pollution from Ships (MARPOL), with its additional protocols [17,20].

It is therefore apparent that a pollutant is usually regulated by more than one piece of legislation. A nonexhaustive list of the main legislative documents that are applicable in EU regarding air pollutant emissions (either directly or indirectly by regulating emissions of precursor gases) and ambient concentrations of air pollutants is given in Table 2.13 [17].

Transboundary air pollution

The *Convention on Long-range Transboundary Air Pollution* was the first international legally binding instrument to deal with problems of air pollution on an international basis. It was signed in 1979 and entered into force in 1983. Since then, it has been extended by eight specific protocols. The Convention is one of the central means for protecting our

Table 2.13: Legislation in Europe Regulating Emissions and Ambient Concentrations of Air Pollutants [17].

	Policies	PM	O_3	NO_2 NO_x NH_3	SO_2 SO_x	CO	Heavy Metals	BaP PAHs	VOC
Directives for ambient air quality	2008/50/EC	PM	O_3	NO_2	SO_2	CO	Pb		Benzene
	2004/107/EC						As, Cd, Hg, Ni	BaP	
Directives for emissions of air pollutants	2001/81/EC	(a)	(b)	NO_x, NH_3	SO_2				NMVOC
	2010/75/EU	PM	(b)	NO_x, NH_3	SO_2	CO	Cd, Tl, Hg, Sb, As, Pb, Cr, Co, Cu, Mn, Ni, V		VOC
	Euro standards on road vehicle emissions	PM	(b)	NO_x		CO			VOC, NMVOC
	94/63/EC	(a)	(b)						VOC
	2009/126/EC	(a)	(b)						VOC
	1999/13/EC	(a)	(b)						VOC
	91/676/EEC			NH_3					
Directives for fuel quality	1999/32/EC	(a)							
	2003/17/EC	(a)	(b)				Pb		
International conventions	MARPOL 73/78	PM	(b)	NO_x	SO_x	CO		PAHs	Benzene, VOC
	LRTAP	PM (a)	(b)	NO_2, NH_3	SO_2	CO	Cd, Hg, Pb	BaP	NMVOC

Note: (a) Directives and conventions limiting emissions of PM precursors, such as SO_2, NO_x, NH_3, and VOC, indirectly aim to reduce particulate matter ambient air concentrations. (b) Directives and conventions limiting emissions of O_3 precursors, such as NO_x, VOC, and CO, indirectly aim to reduce troposphere O_3 concentrations.

environment. It has substantially contributed to the development of international environmental law and has created the essential framework for controlling and reducing the damage to human health and the environment caused by transboundary air pollution. It is a successful example of what can be achieved through intergovernmental cooperation.[12]

Particularly, the *Protocol to Abate Acidification, Eutrophication and Ground-level Ozone* is probably the most sophisticated environmental agreement so far and marks a leap forward in international law-making. It also initiates a new phase in the life of the Convention, featuring increased emphasis on implementation, compliance, review and extension of existing protocols [28]. It was adopted in Gothenburg, Sweden, on November 30, 1999. The Protocol sets national emission ceilings for 2010 up to 2020 for four pollutants: sulfur dioxide, nitrogen oxides, VOCs, and ammonia. It thus builds on the previous protocols that addressed sulfur emissions (1985 Protocol; 1994 Protocol), VOCs, and NO_x. The national emission ceilings set were decided after an overall assessment of the scientific evidence on pollution effects and the abatement techniques available.

Furthermore, the Protocol defines strict emission limits for certain sources, such as combustion plants, electricity production, dry cleaning, cars and lorries, and it necessitates the employment of the best available techniques to cut down the emissions of interest. VOC emissions from products like paints or aerosols have also to be reduced. Finally, farmers have to take specific measures to control ammonia emissions. Guidance documents adopted together with the Protocol provide a wide range of abatement techniques and economic instruments for the reduction of emissions in the relevant sectors, including transport [28].

Stratospheric ozone depletion

In searching for a new refrigerant, the requirements for any new candidate compound included low boiling point, low toxicity, and to be generally nonreactive. From this point of view, synthesis of CFCs was a great success; they fulfilled all criteria and found thus many applications worldwide as refrigerants in refrigerators, in aerosol spray cans, as cleaning solvents, and as foam-blowing agents [75]. Fears of ozone depletion due to human activities first emerged in the late 1960s. A decade of denial and debate followed with eventual acceptance by scientists and policy makers that ozone depletion was likely to occur and would represent a global environmental crisis [35]. After the work of Molina and Rowland in the 1970s [38], which showed that chlorine from CFCs could destroy ozone, the use of CFCs in spray cans (aerosols) for hairspray, deodorants, and paints, was restricted in the United States, Canada, Norway, and Sweden [75]. The growing use of CFCs and the hypothesis that CFCs cause ozone depletion, since back then the knowledge on the depletion mechanism had not been fully understood, prompted many nations of the

---

[12] http://www.unece.org/env/lrtap/lrtap_h1.html.

world to sign the Vienna Convention for the Protection of the Ozone Layer in 1985. The Vienna Convention did not impose any restrictions on the use of CFCs. However, it recognized the need to further study the ozone layer and the effect of CFCs on it, as well as the fact that it was an international issue that could be resolved only by the international community. Thus, it set the stage and provided the framework for the Montreal Protocol of 1987 [75]. After the first satellite observations of ozone in the stratosphere in the early 1980s and the discovery of the ozone hole [37], the Montreal Protocol of 1987 on Substances that Deplete the Ozone layer was adopted, widely ratified, and the phasing out of major ozone-destroying gases began. The Montreal Protocol entered into force in 1989 and aimed at protecting the stratospheric ozone layer by phasing out more than 200 substances, including CFCs, halons, hydrofluorocarbons (HCFCs), hydrobromo-fluorocarbons (HBFCs), carbon tetrachloride (CTC), trichloroethane (TCA), hydrochloromethane (BCM), and methylbromide (MB). Most known substances with significant ozone-depleting potential (ODP) are covered by the Montreal Protocol [76]. The countries that signed the original document in 1987 agreed to freeze CFC production and use at the 1986 rates by the year 1989, and to cut CFC production and use by 50% over the next 10 years. In 1990, London amendments were added to the Montreal Protocol with the provision to totally phase out ozone the original CFCs listed by 2000, following new scientific evidence on the relation between ozone destruction and chlorine from CFCs. After the London amendments were adopted, the chlorine and bromine atmospheric levels were expected to peak by 2020 and not return to preozone hole levels until the end of the 21st century [75]. More amendments have taken place since then; the text of the Protocol along with its amendments and the current ratification are available via the United Nations Environment Programme Website at www.unep.ch/ozone.

Within the European Union, the use and trade in controlled substances is regulated by Regulation (EC) 1005/2009 (known as the ODS Regulation) [76,77]. All companies that are involved in the production, import, or export into the European Union, as well as feedstock users, process agent users, and destruction facilities are required to report their activities in relation with controlled substances annually. The ODS Regulation includes also five additional substances that are not covered by the Montreal Protocol.

It is noteworthy that the Montreal Protocol has successfully brought about international cooperation to address a serious global environmental hazard. Moreover, scientific advancements and policy decisions progressed in parallel and in coordination. Finally, it should be noted that the Montreal Protocol case is a successful example of the application of the precautionary principle in environmental policy [75,78,79].

Global warming and climate change

The *Kyoto Protocol* is an international agreement linked to the United Nations Framework Convention on Climate Change, which commits its Parties by setting internationally binding emission reduction targets for GHGs. Recognizing that developed countries are principally responsible for the current high levels of GHG emissions in the atmosphere as a result of more than 150 years of industrial activity, the Protocol places a heavier burden on developed nations under the principle of "common but differentiated responsibilities".[13]

It was adopted in Kyoto, Japan, on December 11, 1997, and entered into force on February 16, 2005. The detailed rules for the implementation of the Protocol were adopted at COP 7 in Marrakesh, Morocco, in 2001, and are referred to as the "Marrakesh Accords." Its first commitment period started in 2008 and ended in 2012.

Specifically, under the Protocol and during the first commitment period, the industrialized countries had committed to reduce, during the period 2008−12, the emissions of six gases responsible for global warming, namely carbon dioxide, methane, nitrous oxide, hydrofluorocarbons, perfluorocarbons, and hexafluoride sulfur, at least by 5% compared to 1990 levels. In this context, EU member states pledged to reduce their emissions by 8% during this period.

In Doha, Qatar, on December 8, 2012, the Kyoto Protocol was amended in order to include[13]:

- new commitments of its Parties for the period 2013−20;
- a revised list of GHGs to be reported in the second commitment period; and
- amendments to several articles of the Kyoto Protocol that required to be updated for the second commitment period.

During the second commitment period, Parties committed to reduce the GHG emissions by at least 18% below 1990 levels in the 8-year period from 2013 to 2020; however, the composition of Parties in the second commitment period is different from the first.

The Protocol offers three market-based mechanisms to its Parties to achieve their targets. Specifically, the Kyoto mechanisms are[13]:

- *International Emissions Trading*. Emissions trading, as set out in Article 17 of the Kyoto Protocol, allows countries that have emission units to spare—emissions permitted them but not "used"—to sell this excess capacity to countries that are over their targets. Thus, a new commodity was created in the form of emission reductions or removals. Since carbon dioxide is the principal GHG, people speak simply of trading in carbon. Carbon is now tracked and traded like any other commodity. This is known as the "carbon market".
  - *Clean Development Mechanism*. This mechanism, as defined in Article 12 of the Protocol, allows a country with an emission-reduction or emission-limitation commitment under the Kyoto Protocol to implement an emission-reduction project

---

[13] http://unfccc.int/kyoto_protocol/items/2830.php.

in developing countries. Such projects can earn saleable certified emission reduction credits, each equivalent to 1 ton of carbon dioxide, which can be counted toward meeting Kyoto targets.

- *Joint implementation.* The joint implementation mechanism, as defined in Article 6 of the Kyoto Protocol, allows a country with an emission reduction or limitation commitment under the Kyoto Protocol to earn emission reduction units from an emission-reduction or emission-removal project in another Party, each equivalent to 1 ton of carbon dioxide, which can be counted toward meeting its Kyoto target.

The Kyoto Protocol can be considered as an important first step toward a really worldwide emission reduction system that will stabilize GHG emissions and can provide the architecture for the future international agreement on climate change.

2.4 Indoor Air Pollution

2.4.1 Description

Indoor air pollution refers to the deterioration of air quality indoors as a result of the presence of chemical, biological and physical pollutants in the indoor air [80]. As it is the case with the ambient air pollution, it is related with many adverse health effects, which can be either immediate or long term. In poor and developing countries, the main source of indoor air pollution is connected with low-quality combustion technologies for cooking or heating that produce biomass smoke, which contains suspended particulate matter (SPM), nitrogen dioxide (NO_2), sulfur dioxide (SO_2), carbon monoxide (CO), formaldehyde, and polycyclic aromatic hydrocarbons (PAHs). In industrialized countries, in addition to nitrogen dioxide, carbon monoxide and formaldehyde, radon, asbestos, mercury, human-made mineral fibers, VOCs, allergens, tobacco smoke, bacteria, and viruses are among the pollutants found in the atmospheric environment indoors.

These pollutants may stem from sources outdoors like an industrial facility nearby or be related to activities taking place inside the building. In many cases, building materials, paintings, and special materials for applications like insulation, may release undesirable compounds in households, offices, etc. According to the California Air Resources Board (ARB),[14] which is part of the California Environmental Protection Agency, many pollutants may build up rapidly indoors, reaching concentrations higher than the ones occurring outside. Though it may seem strange, this is more frequent in newly constructed buildings in developed countries, where tighter construction standards and elimination of indoor air renewal techniques, as a result of requirements related to energy losses minimization, prevent particles or other compounds from escaping the building. For example, buildings in Sweden are responsible for 40% of energy nationally. Directive

---

[14] http://www.arb.ca.gov/research/indoor/rediap.htm.

2010/31/EU of the European Parliament states that from January 1, 2021 all new buildings in the European Union should become "nearly zero energy buildings." Passive buildings utilize a number of technologies, such as efficient insulation, advanced window technology, airtightness, and heat recovery techniques, in order to significantly reduce energy consumption. However, these measures aiming at saving energy in buildings may result in deteriorated indoor air quality, often associated with "sick building" syndrome symptoms [81,82]. The term *sick building* is used for buildings whose tenants experience unexplained symptoms such as respiratory allergies, skin diseases, skin irritation, headaches, nausea, and lethargy. In any case, the pollutants accumulated indoors can cause a variety of health problems and can be even fatal at high levels. Sadly to say, in Greece of 2015, deaths related to breathing poisonous gases resulting from inappropriate combustion of materials used for heating purposes have been considerably increased due to the current economic situation.

Although the indoor environment quality has a significant impact on modern life, the relevant research does not count many decades. Yet, there is an increased interest in the quality of the air indoors nowadays. This is partly attributed to the fact that today most activities occur indoors. We live, work, exercise, and entertain ourselves in closed spaces. It is estimated that currently 90% of the population in developed countries spends its time indoors. Most of this time spent indoors occurs at home (67% on average for the French population) [83,84]. For some parts of the population, such as infants and the elderly, the time spent indoors can be even greater. Another distinctive characteristic is that exposure to some pollutants, such as tobacco smoke and radon, occurs almost exclusively indoors.

2.4.2 Sources of Indoor Air Pollution

Indoor air pollution can be caused by tobacco smoke, carbon monoxide, or nitrogen oxides from unvented or faulty gas appliances, particles from wood-burning stoves, fireplaces and aerosol sprays, and biological agents, such as pet dander, dust, and mold. Asbestos, lead, and radon originating from building materials are particularly dangerous indoor pollutants that can cause brain damage and cancer.

Poverty is also a factor of indoor air pollution. According to WHO,[15] around 3 billion people heat their homes and cook food using solid fuels (ie, wood, charcoal, coal, dung, crop wastes) on open fires or traditional stoves. Such inefficient cooking and heating practices produce high levels of household (indoor) air pollution, which includes a range of health-damaging pollutants, such as fine particles and carbon monoxide.

Generally, the determination of air quality indoors is a complex task, which is based on the identification of the activities that release gas pollutants and the interaction between a

---

[15] http://www.who.int/mediacentre/factsheets/fs292/en/.

number of factors that regulate the production or the elimination of pollutants indoors. The main factors affecting indoor air quality are:

- *Ambient air quality*: The air quality in an internal space may vary with time, depending on changes in the composition of the external air intake and on the response rate, which in turn is influenced by the permeability of the building structure, the nature of pollutants, etc. Shortly, air pollutants produced outdoors by automotive vehicles, industrial facilities, etc., such as nitrogen oxides, sulfur oxides, lead, VOCs, ozone, and PM, can be transferred indoors through doors, windows, etc. Specifically, according to EPA,[16] outdoor air can enter and leave a building via infiltration, natural ventilation, and mechanical ventilation. In infiltration, outdoor air flows into buildings through openings, joints and cracks in walls, floors, ceilings, windows and doors. In natural ventilation, air moves through open windows and doors. Air movement associated with infiltration and natural ventilation is caused by air temperature differences between indoors and outdoors and by wind. Finally, there is a number of mechanical ventilation devices, from outdoor-vented fans that intermittently remove air from a single room like bathrooms and kitchen, to air handling systems that use fans and duct work to continuously remove indoor air and distribute filtered and conditioned outdoor air to strategic points throughout the house. The rate at which outdoor air replaces indoor air is described as the air exchange rate. When there is little infiltration and natural or mechanical ventilation, the air exchange rate is low and pollutant levels can increase.
- *Internal production of pollutants*: Indoor air quality is mainly determined by the concentration of pollutants present and the thermal conditions prevailing in a closed space. The main internal sources of pollutants are:
  - building materials, paints and furniture used inside the building, and
  - activities of people indoors (use of gas in the kitchen, cooking, cleaning, smoking, use of consumer products, and even the movement of people in the area causing the suspension of solid particles).

The categorization of the most important sources of air pollution indoors is shown in Fig. 2.18.

According to EPA, the sources of indoor air pollution are classified as:

- Fuel-burning combustion appliances
- Tobacco products
- Building materials and furnishings as diverse as:
  - Deteriorated asbestos-containing insulation
  - Newly installed flooring, upholstery, or carpet
  - Cabinetry or furniture made of certain pressed wood products

---

[16] http://www2.epa.gov/indoor-air-quality-iaq/introduction-indoor-air-quality.

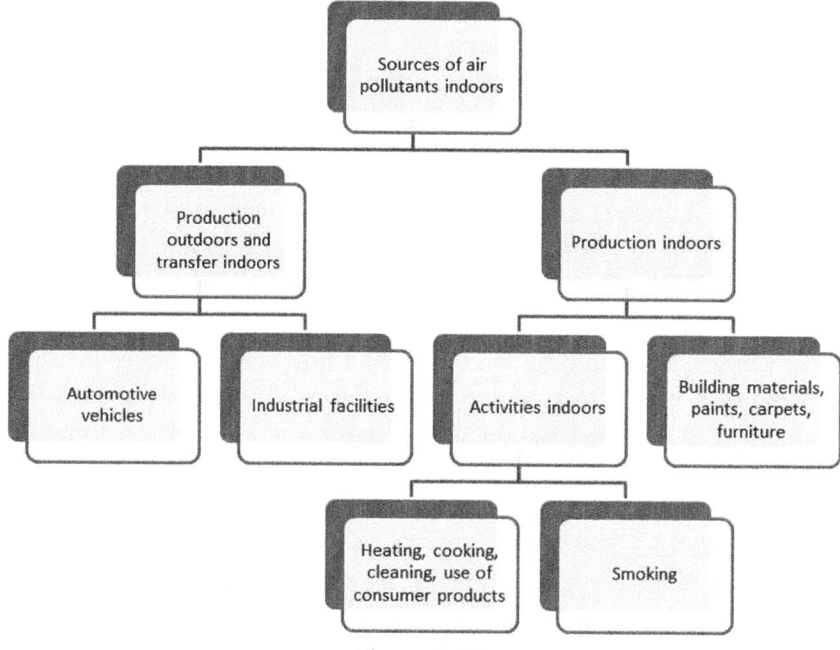

Figure 2.18
Sources of indoor air pollution.

- Products for household cleaning and maintenance, personal care, or hobbies
- Central heating and cooling systems and humidification devices
- Excess moisture
- Outdoor sources such as:
  - Radon
  - Pesticides
  - Outdoor air pollution

The problem of indoor air pollution has been intensified after 1980, when energy-saving measures started being applied. Efforts to reduce energy consumption in buildings led to the proliferation of systems for internal air recycling, reducing natural ventilation and creating watertight and thermally sealed buildings. The reduced use of natural ventilation has resulted in an increase of pollutants in the indoor air.

2.4.3 Indoor Air Pollutants

Depending on their source, air pollutants found in the atmospheric environment indoors are classified into three categories; the first category includes the pollutants that originate from the external environment, the second one the pollutants produced in both the inside and outside environment of buildings, and the third category concerns the pollutants generated mainly in the interior of buildings.

The main pollutants emanating from the external environment are:

- Sulfur oxides (gases, particles)
- Ozone
- Lead, manganese
- Calcium, chlorine, cadmium, Silicon
- Volatile organic compounds

Pollutants produced inside and outside of buildings include:

- Nitrogen oxides
- Carbon monoxide, carbon dioxide
- Particulate matter
- Organics
- Allergens
- Microorganisms

Pollutants generated mainly inside buildings:

- Radon
- Formaldehyde
- Asbestos, synthetic fibers
- Organics
- Ammonia
- Polycyclic aromatic hydrocarbons, nicotine, acrolein, etc.
- Mercury
- Aerosols

The concentrations of pollutants that come mainly from the external atmospheric environment are generally lower indoors. Particularly, SO_2 and O_3 are found in significantly lower abundancies indoors because of their high chemical activity. Sulfur dioxide is a by-product of combustion processes, while ozone is a secondary pollutant produced by photochemical reactions in the atmosphere. Lead and manganese are produced by vehicle emissions. Nitrogen oxides, carbon monoxide, and PM are produced during cooking with gas, heating, and smoking. Their levels indoors are higher than those in the external environment. Nitrogen dioxide is produced by the metabolic activity of man, while steam results from biological activity and evaporation. Organic compounds in the atmospheric environment are produced from natural sources, petrochemical solvents, and evaporation of fuels. Organic compounds indoors and outdoors are produced by insecticides, metabolic activities, painting, and combustion and exhaust processes. The organic compounds that are produced only inside buildings derive from adhesives, solvents, cosmetics, and cooking. Therefore, their concentrations indoors can be much higher than those outside. The same applies for formaldehyde, which is released mainly from insulating materials, carpets, and synthetic wood furniture. Radon and formaldehyde are two of the most dangerous pollutants indoors for the human health. Asbestos and synthetic fiber emissions, which are known to cause cancer, are related to insulating and fire protection materials. Polycyclic aromatic hydrocarbons, acrolein, and nicotine are present in cigarette smoke and they are found in much higher concentrations indoors than outdoors. Mercury comes from fungicides, paints, and broken thermometers, while ammonia and aerosols are released from consumer products. Finally, microorganisms come from animals, humans, and plants, while allergenic substances are usually of plant origin; they enter the interior air through either ventilation systems or people's clothes. Therefore, attention should be given to the type of vegetation that exists around the buildings in such a way as to avoid the production and transfer of pollen. Some organizations suggest plants fertilized by birds or insects, instead of those being fertilized by the wind. Mao et al. [85] reported that high-rise residential buildings where doors and windows are generally closed exhibit higher infection risks associated with indoor airborne diseases transmission. They attributed this to the fact that during cold seasons, when airborne pathogens are prevalent, the outdoor temperature is low and the wind is strong. The diseases spread inside these buildings not only horizontally but also vertically through elevator or stairwell shafts due to the stack and/or wind effect. As a result of the relative importance of stack and wind effects in a building, different pressure profiles may be developed, which could lead to different indoor airflow movements and gaseous pollutant transport patterns.

WHO [86,87] on the other hand has divided air pollutants into two categories, as shown in Table 2.14. The first group includes pollutants for which WHO guidelines for indoor air were needed, while the second group includes pollutants of potential interest.

According to EPA, the most important air pollutants indoors are the following:

- Asbestos
- Biological pollutants
- Carbon monoxide (CO)
- Formaldehyde/pressed wood products
- Lead (Pb)
- Nitrogen dioxide (NO_2)
- Pesticides
- Radon (Rn)
- Respirable particles
- Secondhand smoke/environmental tobacco smoke
- Volatile organic compounds

Table 2.14: Pollutants Considered for Inclusion in the WHO Indoor Air Quality Guidelines [87].

Group 1. Development of Guidelines Recommended	Group 2. Current Evidence Uncertain or Not Sufficient for Guidelines
Benzene	Acetaldehyde
Carbon monoxide	Asbestos
Formaldehyde	Biocides, pesticides
Naphthalene	Flame retardants
Nitrogen dioxide	Glycol ethers
Particulate matter ($PM_{2.5}$ and PM_{10})	Hexane
Polycyclic aromatic hydrocarbons, especially	Nitric oxide
benzo-[α]-pyrene	Ozone
Radon	Phthalates
Trichloroethylene	Styrene
Tetrachloroethylene	Toluene
	Xylenes

Indoor air pollutants and linking sources in a typical household are shown in Fig. 2.19 [88]. In Table 2.15, the categories of major indoor air pollutants along with their sources and health effects are shown, as proposed by the University of Kentucky [89].

Exposure to high concentrations of these pollutants can have a direct toxic effect, while exposure to lower concentrations over a long period of time can cause chronic diseases, even cancer. The relative importance of any single source depends on how much of a given pollutant it emits and how hazardous these emissions are. In some cases, factors like how old the source is and whether it is properly maintained are also significant.

Most of the pollutants mentioned above are discussed in the part related to ambient air pollution or are generally known even to nonexperts, with one exception: Radon. This is the reason for presenting more information on this pollutant. Also, cigarette smoke composition and related health effects are discussed since this habit is directly related to increased cancer cases and deaths worldwide.

Radon

Radon is a radioactive gas formed from the radioactive decay of uranium. It is released from the ground in areas with uranium-containing soils and rocks. It is generally abundant in granitic areas [90]. So, the most important pathway for human exposure is permeation of radon gas into buildings, but also radon from water, outdoor air, and construction materials can contribute to the total exposure [89]. Hence, radon exposure depends strongly on regional variations.

Radon decays to radon daughters, some of which emit alpha radiation. Alpha-emitting radon daughters are adsorbed on dust particles which, when inhaled, are trapped in the

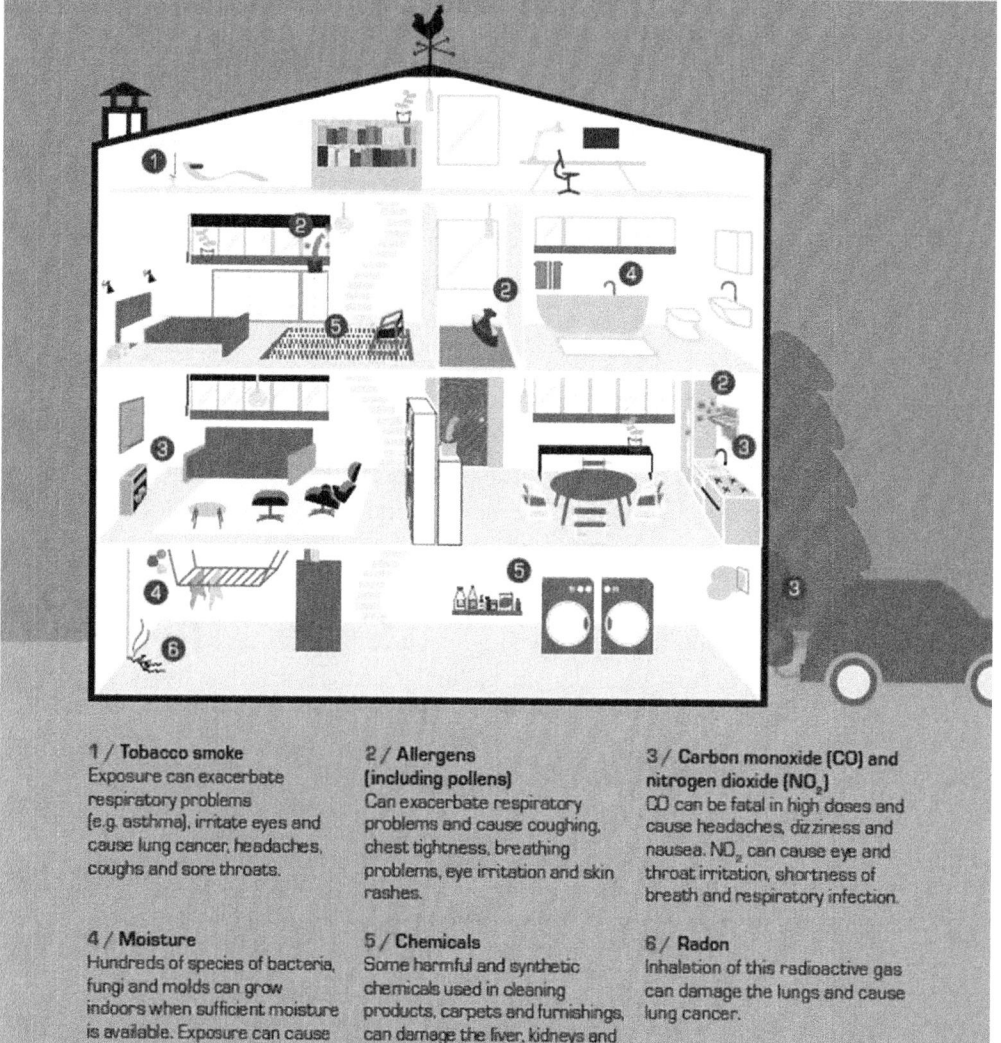

Indoor air pollution
We spend a large part of our time indoors — in our homes, workplaces, schools or shops. Certain air pollutants can exist in high concentrations in indoor spaces and can trigger health problems.

1 / Tobacco smoke
Exposure can exacerbate respiratory problems (e.g. asthma), irritate eyes and cause lung cancer, headaches, coughs and sore throats.

2 / Allergens (including pollens)
Can exacerbate respiratory problems and cause coughing, chest tightness, breathing problems, eye irritation and skin rashes.

3 / Carbon monoxide (CO) and nitrogen dioxide (NO_2)
CO can be fatal in high doses and cause headaches, dizziness and nausea. NO_2 can cause eye and throat irritation, shortness of breath and respiratory infection.

4 / Moisture
Hundreds of species of bacteria, fungi and molds can grow indoors when sufficient moisture is available. Exposure can cause respiratory problems, allergies and asthma, and affect the immune system.

5 / Chemicals
Some harmful and synthetic chemicals used in cleaning products, carpets and furnishings, can damage the liver, kidneys and nervous system, cause cancer, headaches and nausea, and irritate the eyes, nose and throat.

6 / Radon
Inhalation of this radioactive gas can damage the lungs and cause lung cancer.

Figure 2.19
Indoor air pollution in a household. *Reproduced from EEA Signals 2013. Every breath we take. © EEA, 2013.*

Table 2.15: Major Indoor Air Pollutants, Sources, and Health Effects [89].

Pollutant	Major Sources Indoors	Potential Health Effects
Radon Colorless, tasteless, and odorless gas that comes from the radioactive decay of uranium or radium.	Earth and rock under buildings; Some earth-derived building materials; Groundwater.	Lung cancer.
Biological contaminants Molds, mildews and fungi, bacteria, viruses, dust, mites.	House dust; Infected humans or animals; Poorly maintained humidifiers, dehumidifiers, and air conditioners; Wet or moist surfaces; Carpets and home furnishings.	Allergies and asthma; Headaches; Eye, nose and throat irritation; Colds, flu, and pneumonia.
Carbon monoxide Colorless, odorless gas produced by incomplete combustion of all carbon fuels.	Heating equipment; Wood or coal stoves; Fireplaces; Ovens; Charcoal grills; Engines; Tobacco smoke.	Headaches, drowsiness, dizziness; Impairment of human respiration, vision and brain functioning, nausea, mental confusion; Very high levels can cause death.
Nitrogen oxides and sulfur dioxide Gases formed by incomplete combustion of all carbon fuels.	Heating equipment; Wood or coal stoves; Fireplaces; Ovens; Charcoal grills; Engines; Tobacco smoke.	Damage to respiratory tract and lungs (NO_2); Irritation of eyes, nose and respiratory (SO_2).
Respirable suspended particulates Particles small enough to inhale that come in a variety of sizes, shapes, and levels of toxicity.	Wood-burning stoves, fireplaces; Unvented kerosene space heaters; Gas-fired ranges, furnaces, water heaters; Vacuum cleaning and house dust; Tobacco smoke; Soap powders, pollen, lint, dust, cleaning and cooking sprays.	Eye, nose, and throat irritation; Respiratory infections and bronchitis; Emphysema; Lung cancer.
Environmental tobacco smoke Secondhand smoke exhaled by smokers.	Cigarettes; Cigars; Pipes.	Eye, nose and throat irritation; Respiratory irritation; bronchitis and pneumonia; Increased risk of emphysema, lung cancer, and heart disease.
Asbestos A natural mineral fiber used in various building materials. (All homes more than about 20 years old are likely to have some asbestos)	Damaged or deteriorating ceiling, wall, and pipe insulation; Vinyl-asbestos floor material; Fireproof gaskets in heat shields, wood stoves, and furnaces; Acoustical materials; Thermal insulation; Exterior siding.	No immediate symptoms; Chest, abdominal and lung cancers and asbestosis.

Continued

Table 2.15: Major Indoor Air Pollutants, Sources, and Health Effects [89].—cont'd

Pollutant	Major Sources Indoors	Potential Health Effects
Volatile Organic Chemicals Airborne chemicals contained in many household products.	Aerosol sprays, hair sprays, perfumes, solvents, glues, cleaning agents, fabric softeners, pesticides, paints, moth repellents, deodorizers, and other household products; Dry-cleaned clothing; Moth balls; Tobacco smoke.	Eye, nose, throat irritation; Headaches; Loss of coordination; Confusion; Damage to liver, kidneys, and brain; Various types of cancer.
Formaldehyde Pungent gas released into air.	Pressed wood products; Urea-formaldhyde foam wall Insulation; Carpets, draperies, furniture Fabrics; Paper products, glues, adhesives; Some personal care products; Tobacco smoke.	Allergic reactions; Eye, nose and throat irritation; Headaches; Nausea, dizziness, coughing; Cancer is a possibility; Sensitivity varies widely.
Lead Natural element once used as a component in gasoline, house paint, solder and water pipes.	Household dust from lead paint; Lead-based paint; Water from lead or lead-soldered pipes or brass fixtures; Soil near highways, lead industries; Hobbies such as working with stained glass and target shooting; Lead-glazed ceramic ware; Some folk medicines.	Damage to brain, kidneys, and nervous system; Behavioral and learning Problems; Slowed growth; Anemia; Hearing loss; Large doses can be fatal.

lungs and may cause gene damage, mutations, and finally cancer [89]. After smoking, radon is considered to be the second cause of lung cancer. The fact that most of the radon-induced lung cancer cases are observed among smokers shows a strong combined adverse health effect of smoking and radon [89]. There is unfortunately another strong evidence for radon's detrimental effects to health; according to EPA estimates, radon is the number one cause of lung cancer among *nonsmokers*. It is responsible for about 21,000 lung cancer deaths every year in the United States. About 2900 of these deaths occur among people who have never smoked.

Cigarette smoke

There are over 4000 chemicals in the smoke from cigarettes, and at least 69 of these chemicals are known to cause cancer. EPA defines *secondhand smoke* as a mixture of the smoke given off by the burning of tobacco products, such as cigarettes, cigars, or pipes, and the smoke exhaled by smokers. Secondhand smoke is also called *environmental tobacco smoke* and exposure to it is sometimes called involuntary or *passive smoking*.

Secondhand smoke contains more than 7000 substances, several of which are known to cause cancer in humans or animals. EPA has concluded that exposure to secondhand smoke can cause lung cancer in adults who do not smoke, and according to its estimates, exposure to secondhand smoke causes ~3000 lung cancer deaths per year in nonsmokers. Exposure to secondhand smoke has also been shown in a number of studies to increase the risk of heart disease and stroke. Instead of more comments, Fig. 2.20—although not complete—is very characteristic of smoke composition and relevant health risks.

Nowadays, there is a lot of marketing and promotion of e-cigarettes, which are supposed to be a healthy (?) alternative to conventional cigarettes. However, Offermann [91] reported that e-cigarettes emit many harmful chemicals into the air and need to be regulated in the same manner as for tobacco smoking. Consumers should be warned that, while the health risks associated with the usage of e-cigarettes appear to be less than those associated with tobacco smoking, there remain substantial health risks associated with the use of e-cigarettes.

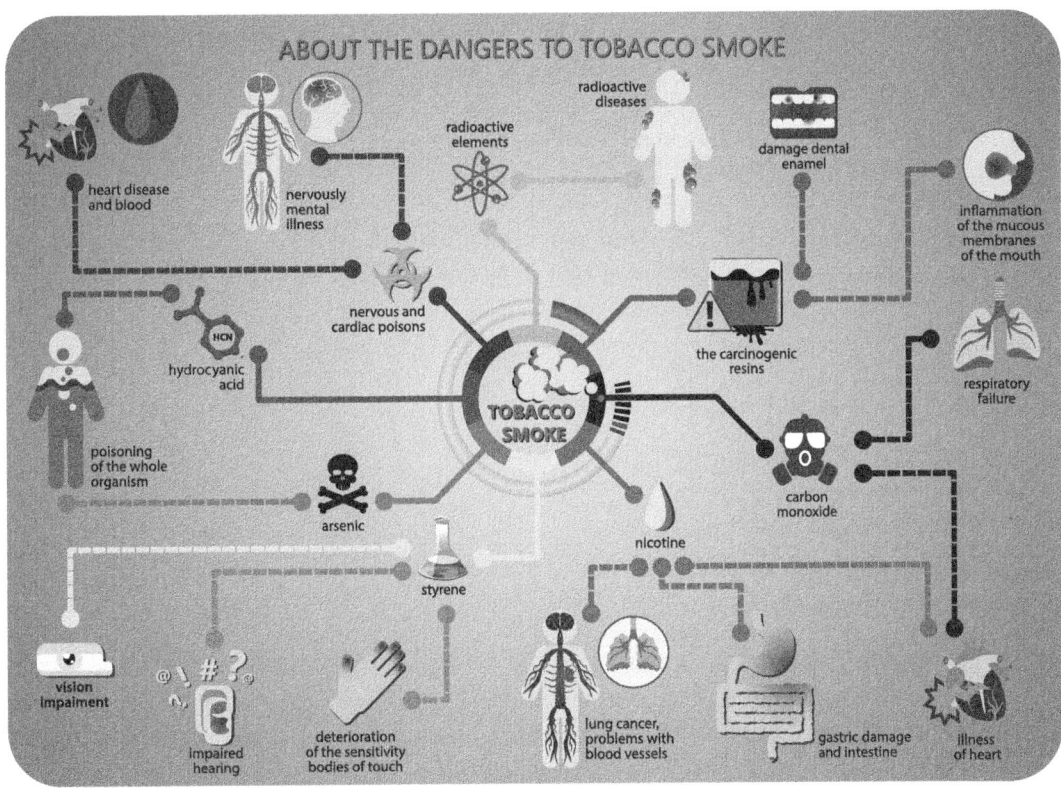

Figure 2.20
Cigarette smoke composition and associated health risks. *Used under license from Shutterstock.com/Image ID:255520378.*

2.4.4 Indoor Air Pollution and Health Effects

The U.S. EPA states that there are immediate and long-term effects of indoor air pollution. Intermediate health effects refer to health effects, which may appear shortly after a single exposure or repeated exposures to a pollutant, such as irritation of the eyes, nose and throat, headaches, dizziness, and fatigue. Such immediate effects are usually short term and treatable. Other health effects may show up either years after exposure has occurred or only after long or repeated periods of exposure. These effects, which include some respiratory diseases, heart disease and cancer, can be severely debilitating or fatal. It is therefore evident that we should make substantial efforts to improve the indoor air quality in our houses even if symptoms are still not noticeable.

According to Bruce et al. [92], the air pollution in households resulting from the combustion of solid and other low-quality fuels is responsible for a very substantial public health burden, affecting mainly low- and middle-income households. In the Global Burden of Disease study that was released in 2010 [93], cooking with solid fuels, such as wood, dung, crop wastes, charcoal, and coal, was estimated to have caused 3.5 million premature deaths in 2010, with a further 0.5 million outdoor air pollution deaths being attributed to emissions from household cooking. But the death toll is not restricted to these numbers. In 2014, WHO updated its estimates for 2012, reporting that around 4.3 million deaths globally can be connected to poor quality of the air indoors in low-and middle-income countries [94].

The numbers are dreadful: *1.69 million deaths* in the South East Asia, *1.62 million deaths* in the Western Pacific region, *600,000* deaths in Africa, *200,000* in the Eastern Mediterranean region, *99,000* in Europe and *81,000* in the Americas; the remaining *19,000* deaths occurred in high-income countries (Fig. 2.21).

As shown in Fig. 2.22, among these deaths [94]:

- 12% was due to pneumonia,
- 34% from stroke,
- 26% from ischemic heart disease,
- 22% from chronic obstructive pulmonary disease, and
- 6% from lung cancer.

This large increase in burden in comparison with previous observations has been attributed to (1) additional health outcomes such as cerebrovascular diseases and ischemic heart disease included in the analysis of WHO, (2) additional evidence that has become available on the relationship between exposure and health outcomes and the use of integrated exposure-response functions, and (3) an increase in noncommunicable diseases.

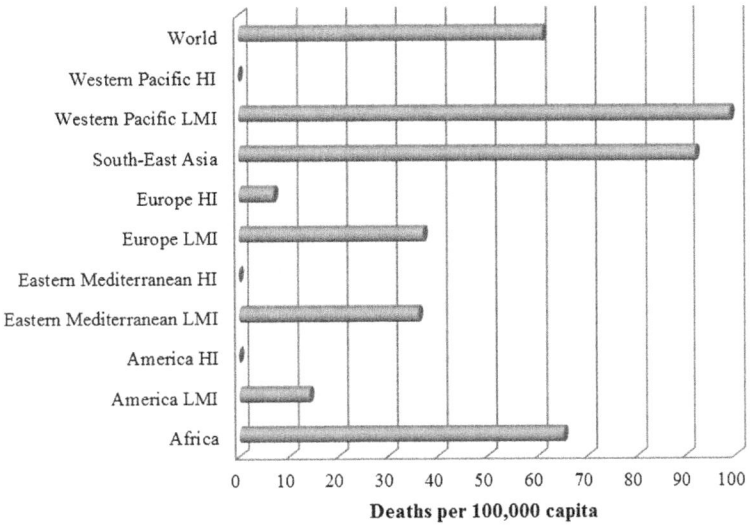

Figure 2.21
Deaths per capita attributable to household air pollution in 2012, by region
(*LMI*, low- and middle-income; *HI*, high-income). *Adapted from Ref. [94].*

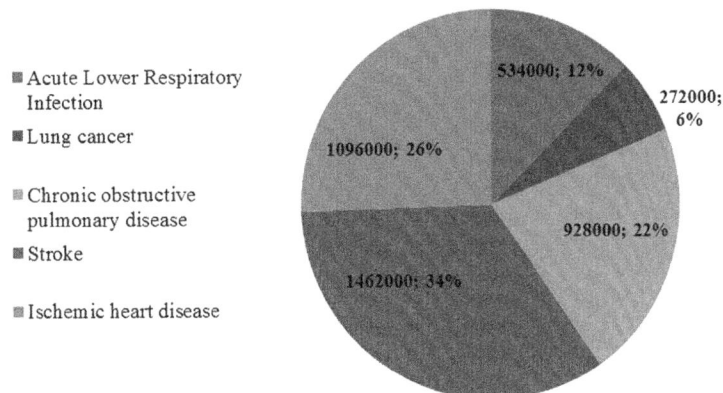

Figure 2.22
Deaths attributable to household air pollution in 2012, by disease. *Adapted from Ref. [94].*

Although women are more exposed to indoor air pollution than men due to their greater involvement in daily cooking activities, which in turn means higher relative risk to experience adverse health effects, the absolute burden is larger in men due to larger underlying disease rates in them [93,95].

2.4.5 Policy for Indoor Air Pollution

OECD suggests the following measures for improving the policy design for environmentally sustainable buildings:

General policy framework

- Establish a national strategy for improving the environmental performance of the building sector.
- Establish a framework to regularly monitor the environmental performance of the building sector.
- Develop a close partnership between government and industry for the support of R&D and technology diffusion.

Policy instruments for preventing indoor air pollution

- Improve the quality of building materials by implementing instruments that target building material manufacturers.
- Avoid providing misleading information to consumers.
- Establish a framework to identify newly emerging indoor health problems.

Besides some laws against smoking in various public places and workplaces, and some regulations about building materials, the quality of the air indoors has not been regulated yet [20].

The easiest way to avoid high concentrations of air pollutants indoors is to design in such a way that allows the proper ventilation and renewal of the air. Such a design for a given indoor space depends largely on the numbers of people to use it, the type of the use (for example, for sleeping or working or being used for patients in a hospital), and whether there is already a pollution background for the specific building.

2.5 Noise

2.5.1 Noise Pollution

Noise is a major environmental issue, particularly in urban areas, affecting a large number of people. There is not a single interpretation of what noise or noise pollution exactly is. The word noise originates from the Latin word *nausea* [96]. According to the U.S. Environmental Protection Agency,[17] the traditional definition of *noise* is "unwanted or disturbing sound," namely the sound that either inhibits to some extent normal activities like sleeping, studying, and conversation, or disrupts the quality of life of a person. The World Health Organization (WHO) has excluded the noise resulting from industrial

---

[17] http://www2.epa.gov/clean-air-act-overview/title-iv-noise-pollution.

activities in its definition of *environmental noise* as "noise emitted from all sources except for noise at the industrial workplace" [97]. In European Union, the relevant Directive 2002/49/EC on the management of environmental noise defines *environmental noise* as "unwanted or harmful outdoor sound created by human activities, including noise from road, rail, airports and from industrial sites" [98]. The Organization for Economic Co-operation and Development (OECD) provides a similar definition for *noise pollution*: "Noise pollution is sound at excessive levels that may be detrimental to human health" [80].

The fact that noise has not attracted as much attention as the other types of air pollution does not mean that there are no effects on our health, ranging from mild to severe ones. It has to be noted that regarding the potential health effects of noise, most people in the past used to believe that these included only hearing ability. However, in recent years, evidence has accumulated regarding the health effects of environmental noise. For example, well-designed epidemiological studies have found cardiovascular diseases to be consistently associated with exposure to environmental noise [5]. Today, the noise-associated health effects are internationally categorized in three general types:

- noise adversely affects the human hearing system;
- noise adversely affects the mental and physical health, given its contribution to the development of stress; and
- noise has a decisive impact on people already suffering from a disease or abnormal physiology.

The estimation of the *environmental burden of disease* (EBD) due to environmental noise requires a quantitative risk assessment approach. The process of risk assessment of environmental noise requires information about the [99]:

- nature of the health effects of noise;
- levels of exposure at which health effects begin to occur and how the extent of each effect changes with increasing noise levels; and
- number of people exposed to these hazardous levels of noise.

The EBD is expressed as disability-adjusted life years (DALYs). DALYs are the sum of the potential years of life lost due to premature death and the equivalent years of "healthy" life lost by virtue of being in states of poor health or disability [99].

Besides possible health effects, noise may disturb a person to a level that will not allow him/her to perform adequately common activities like reading, sleeping, communicating, or even sleeping. As it is the case with many environmental problems, it affects the most sensitive groups like children and the elderly people, especially when another disease is also present.

2.5.2 Noise Measurement

Sound or noise is the result of fluctuations or oscillations in atmospheric pressure that excite the ear mechanism and evoke the sensation of hearing. The human ear responds to changes in sound pressure over a very wide range—the loudest sound pressure to which the human ear responds is 10 million times greater than the softest. A logarithmic unit is thus used in order to reduce this wide range and provide a more appropriate way of comparing sounds. At the same time, to avoid a very compressed scale, a factor of 10 is introduced, giving rise to the decibel unit. So, sound intensity is measured in decibels (dB); the unit A-weighted dB (dBA) is used to indicate how a given sound is perceived by human ear. Zero dBA corresponds to the point at which a sound can be heard by a person. Generally, a change of 1 dBA or less in the level of a sound is undetectable, while a 3-to-5-dBA change corresponds to small but noticeable change in loudness. A 10-dBA change is generally accepted to correspond to an approximate doubling or halving in loudness. A soft whisper at 1 m equals to 30 dBA, a busy freeway at 20 m is around 80 dBA, and a chain saw can reach 110 dBA or more at operating distance. Brief exposure to sound levels exceeding 120 dBA without hearing protection may even cause physical pain [100]. In Fig. 2.23, the intensity of the sound caused by a number of activities is presented.

2.5.3 Noise Health Effects

Roughly 30 million workers are estimated to be exposed to hazardous sound levels on the job in the United States [100]. According to the U.S. National Institute for Occupational Safety and Health (NIOSH), *hazardous noise* is defined as the sound that exceeds the time-weighted average of 85 dBA, meaning the average noise exposure measured over a typical 8-h work day. Construction, transportation, agriculture, and mining are characteristic examples of industries with workers exposed to such noise levels [100]. In 2001, it was estimated that 12.5% of American children between the ages of 6−19 years had impaired hearing in one or both ears [101]. But noise pollution is not a "privilege" of the United States only. In Japan, for instance, continuous exposure to noise resulting from public loudspeaker messages and other forms of city noise have forced many Tokyo citizens to wear earplugs daily to decrease discomfort. In Europe, about 65% of the population is exposed to ambient sound at levels above 55 dBA, while about 17% is exposed to levels above 65 dBA, according to the European Environment Agency [100].

Besides hearing impairment issues that could be caused by chronic exposure to noise, WHO has documented five categories of adverse health effects of noise pollution on humans in the report "Burden of disease from environmental noise" [99], where a quantification of healthy life years lost in Europe has been attempted. The main health effects as presented in this report are summarized below [99].

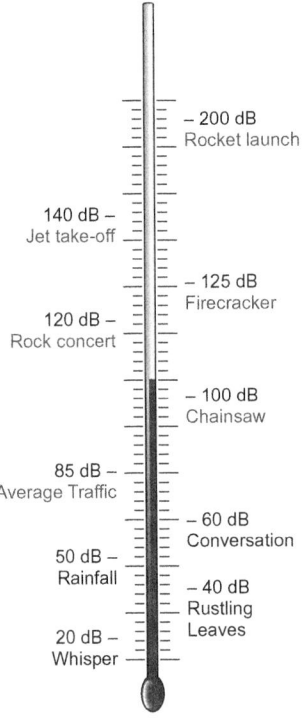

Figure 2.23
Noise thermometer. *Adapted with permission from Sight and Hearing Association.*

Cardiovascular diseases. There is an accumulation of evidence from epidemiological studies recently that shows a linkage between road traffic and aircraft noise and hypertension and ischemic heart disease. Moreover, road traffic noise has been connected to increased risk of ischemic heart disease, including myocardial infarction. High blood pressure, a common concern nowadays, is also increased by exposure to road traffic and aircraft noise. Based on the exposure data from the noise maps of EU Member States, it has been estimated that the EBD from environmental noise is ~61,000 (!) years for ischemic heart disease in high-income European countries.

Cognitive impairment in children. It is defined as the "reduction in cognitive ability in school-age children that occurs while the noise exposure persists and will persist for some time after the cessation of the noise exposure." The estimated DALYs for the high-income European countries are 45,000 years for children aged 7—19 years.

Sleep disturbance. It can be measured electrophysiologically or by self-reporting in epidemiological studies using survey questionnaires. The latter is the most cost-effective method, where "self-reported sleep disturbance" is the most easily measurable outcome indicator, because electrophysiological measurements are costly and difficult to carry out on large samples. Conservative estimates using exposure data from noise maps give a total of 903,000 DALYs lost from noise-induced sleep disturbance for the EU population living in towns with more than 50,000 inhabitants.

Tinnitus. It refers to the sensation of sound in the absence of an external sound source. Tinnitus caused by excessive noise exposure has long been described; 50–90% of patients with chronic noise trauma report tinnitus. It may further cause a number of adverse health effects like sleep disturbance, cognitive effects, anxiety, psychological distress, depression, communication problems, frustration, irritability, tension, inability to work, reduced efficiency, and restricted participation in social life. DALYs for noise-induced tinnitus were estimated to be 22,000 years for the adult population in high-income European countries.

Annoyance. According to WHO, health is not just the absence of disease, but rather a state of complete physical, mental and social well-being. Therefore, a high level of annoyance caused by environmental noise should be considered as one of the environmental health burdens. Conservative estimates using exposure data from noise maps give a total of 654,000 DALYs lost from noise-induced annoyance for the EU population living in towns with more than 50,000 inhabitants.

Goines and Hagler [101] provide a comprehensive review of all noise-related health effects, whereas Hammer et al. [102] relate noise pollution to a wide variety of adverse health effects, including sleep disturbance, annoyance, noise-induced hearing loss (NIHL), cardiovascular disease, endocrine effects, and increased incidence of diabetes. The authors focus on several highly prevalent health effects: sleep disruption and heart disease, stress, annoyance, and NIHL (Fig. 2.24) [102].

It is noteworthy to point out that noise acts as a nonspecific biologic stressor, eliciting reactions that prepare the body for a fight or fight response [103,104]. As a result, noise can

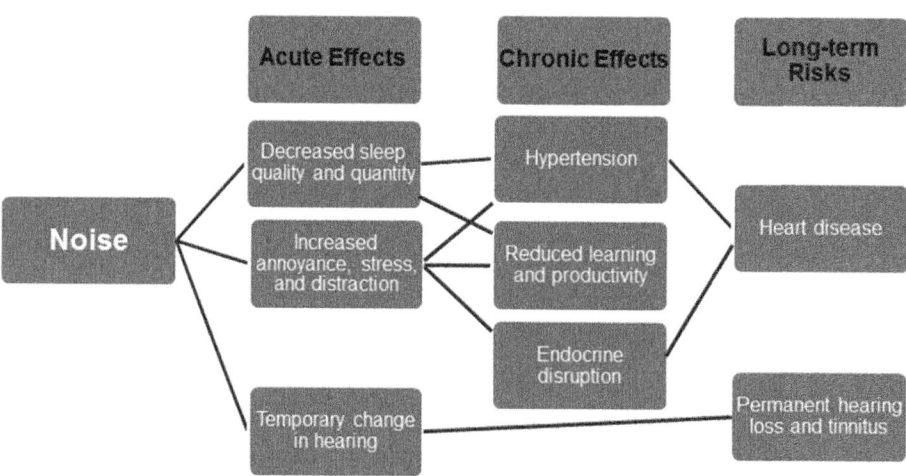

Figure 2.24
Noise possible health effects. *Adapted from Ref. [102].*

trigger both endocrine and autonomic nervous system responses that affect the cardiovascular system and thus, it may be a risk factor for cardiovascular disease [101,104,105].

The numbers are devastating; according to the report "Noise in Europe" released by the European Environment Agency in 2014 [106]:

* Environmental noise causes at least 10,000 cases of premature death in Europe each year.
* Almost 20 million adults are annoyed and a further 8 million suffer sleep disturbance due to environmental noise.
* Over 900,000 cases of hypertension are caused by environmental noise each year.
* Noise pollution causes 43,000 hospital admissions in Europe per year.

As illustrated in the same report, potential noise health effects can be seen as pyramid in relation with people affected and severity of effects (Fig. 2.25) [106].

2.5.4 Noise Effects on Wildlife

Despite the numerous possible adverse health effects of noise, this type of pollution has been underestimated, as deduced also from our own experience. The case is even worse

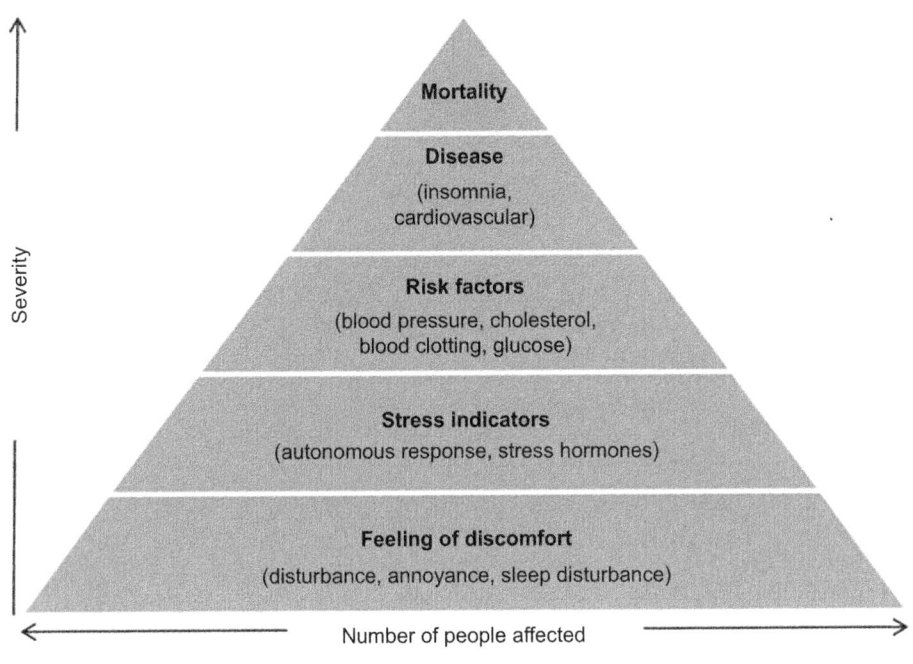

Figure 2.25
Noise health effects pyramid. *Adapted from Ref. [106].*

regarding the impact of noise on the wildlife, where more research is still needed. However, it is already known that chronic and frequent noise interferes with the ability of many natural species to detect important sounds, whereas intermittent and unpredictable noise is often perceived as a threat [107].

Francis et al. [108] state that although acoustic signals may be adapted to natural sources of ambient sounds like wind, noise is often so loud that it overlaps frequencies used by many species for communication. Moreover, it has been shown that predators avoid noisy areas, presumably because noise impairs predators' abilities to locate preys [109]. In a subsequent work, Francis et al. [110] reported that besides altering the behavior and distribution of birds and other vertebrates, noise might change critical ecological services in general. The authors found out that noise pollution may have an impact on pollination, seed dispersal and seedling establishment, within a study system that isolated the effects of noise from confounding stimuli common to human-altered landscapes.

In 2013, Francis and Barber [107] conducted a review of substantial literature detailing the impacts of noise on wildlife and provided a conceptual framework to guide future research. They reported that most noise-related impacts appear to involve behavioral responses across four categories:

- changes in temporal patterns;
- alterations in spatial distributions or movements;
- decreases in foraging or provisioning efficiency coupled with increased vigilance and antipredator behavior; and
- changes in mate attraction and territorial defense.

They demonstrated that these disturbance-, distraction-, and masking-mediated behavioral changes could directly impact individual survival and fitness or lead to physiological stress that may then compromise fitness.

It is thus apparent that noise is a type of pollution with many effects on both the quality of our lives and on the wildlife as well. Therefore, it should be dealt with the same importance as any other pollution type.

2.5.5 Noise Sources

The most significant sources of noise that are responsible for the degradation of the quality of the acoustic environment are the following:

- all means of transport;
- industrial and craft facilities;
- recreation and entertainment facilities;
- household appliances.

It is easily understood that most problems related to noise are found in urban areas and especially large urban centers, where most daily activities are concentrated in a limited area. For example, in Athens, the capital of Greece, 40% of the population of the country is gathered along with 50% of country's industrial and artisanal activity, 55% of vehicles, and 70% of services. As a result, noise is an integral part of the daily routine. Even in the night, when it seems that absolute peace prevails, there is a constant noise background originating from the traffic in large road axes crossing a city that never rests.

Besides urban areas, tourist regions also experience noise problems. Owing to many contrasting activities present in such areas, it is common to see on one hand some inhabitants welcoming noisy recreational activities that are profitable to them, and on the other hand people against the same activities as these produce undesirable constant sound, especially in the night, which affects both the quality of the services provided and the quality of life of the residents in these areas.

According to Hamer et al. [102], the main sources of noise in the United States include road and rail traffic, air transportation, and occupational and industrial activities. Additional individual-level exposures include amplified music and recreational activities including concerts and sporting events. Firearms constitute also additional individual kinds of exposure to noise. Harmful sound levels can be also experienced during the use of personal music players [111]. Despite the fact that the sound from recreational activities and music cannot be considered strictly as "noise" in the sense of being unwanted sound, it cannot be overlooked that *undesirable adverse health effects are possible even from desirable sounds.*

In Europe, road traffic is the major source of environmental noise, with an estimation of 125 million people being affected by noise levels greater than 55 dB [106].

2.5.6 Noise Control

Reducing and managing noise is based on the concept:

Source—Path—Receiver

Specifically, the minimization of noise pollution requires alteration or modification of any or all of the following [112]:

- modifying the source to reduce its noise output;
- altering or controlling the transmission path and the environment to reduce the noise level reaching the listener; and
- providing the receiver with personal protective equipment.

Control at the source. It refers to actions aiming at reducing the disturbance caused by mechanical devices or processes that radiate noise or vibratory energy. The solutions may involve (1) the requirement of noise level specifications written on any product so that the consumer or the buyer in general can select the quieter equipment, (2) process substitution by a quieter one, (3) machine substitution, and (4) efficient systems design, including reducing impact forces, speeds and pressures, frictional resistance, radiating area and noise leakage, and isolating and damping vibrating elements.

Control in the transmission path. Sometimes, it is impossible to control noise at the source even if noise specifications are applied. Traffic is a distinctive example of this. Increasing numbers of automobiles cause cumulatively higher noise levels despite the achievements of car industry in reducing the noise emitted by single cars. Moreover, large road axes constructed in, around, or outside large urban centers diffuse the traffic noise to the residential areas lying on both sides of these roads. In these cases, it is important to isolate the transmission path of the noise. This can be done by several ways: (1) absorbing the sound along the path, (2) deflecting the sound in some other direction by placing a reflecting barrier in its path, and (3) containing the sound by placing the source inside a sound-insulating box or enclosure.

Protect the receiver of the sound. When previous techniques cannot deal with the problem successfully, then the last resort is to protect the receiver of the sound. This includes (1) ear-protection equipment like earplugs, and (2) individual enclosures or noise shelters for workers in noisy working environments.

Besides noise control through the techniques presented above, specific policy and planning measures may confront noise pollution. These may include [106]:

- access controls to manage the relationship between residents and traffic;
- systems engineering to reduce traffic;
- management including pedestrianization, parking and loading controls, delivery time windows, etc.;
- integration of different traditional transport modes in the mobility policy such as bike sharing, car sharing, and ride sharing;
- supporting modal shift to an increased share of walking and cycling, and the development of a good and accessible public transport network;
- development of a sustainable urban mobility plan;
- land-use planning, to avoid noisy and opposing activities in the same area; and/or
- construction of large transport infrastructures like airports far from residential areas.

The type of measures to be implemented in Europe to reduce noise levels resulting from transport are presented in the following box [106].

Noise Action Plans (as Proposed in Ref. [106])

The type of measures planned in action plans of the first reporting round is very much linked to the noise source.

In the case of urban areas, information is currently available for 40% of the cities. Groups of actions referring to land use and urban planning are the predominant ones. This kind of action is presented in 23% of all actions plans related to agglomerations. Followed by measures related to traffic management (20%) and others (17%), this last one includes measures related to increasing public awareness, avoiding the generation of additional traffic and promoting public transport, and encouraging cycling and walking. The high percentage of measures related to traffic and transport in actions plans linked to agglomerations underlines the importance of these topics as noise sources inside agglomerations.

In the case of major roads, the actions that predominate are those related to measures on the propagation path (32%), at the receiver (23%), on traffic management (22%), and on land use and urban planning (12%).

Major railways differ from major roads, although propagation path (22%) and land use and urban planning (22%) are also included inside the most predominant actions, and measures at the receiver are presented in less than 15% of the actions plans related to railways. Measures of traffic management presented in major roads are replaced by other specific railways management actions (17%) such as tram track improvements.

In the case of major airports, the actions highlighted are those considered as operational (presented in 27% of the actions plans related to major airports) followed by measures at the receiver (19%).

Noise control techniques and policy measures should be accompanied by actions targeting individuals with the aim of:

- informing the public in a clear, simple and uniform way;
- combating misleading advertising for products;
- changing standards and methods of behavior and choice; and
- increasing public awareness against noise.

2.5.7 Noise Legislation

Currently, there is not any international law for controlling noise pollution. The European Commission developed a new framework for noise policy based on shared responsibility between the European Union and national and local governments. It included a comprehensive set of measures to improve the accuracy and standardization of data to help improve the coherency of different actions [99]:

- The creation of a Noise Expert Network with the mission to assist the Commission in the development of its noise policy;
- EU Directive 2002/49/EC on the management of environmental noise [98];
- The follow-up and further development of existing EU legislation relating to sources of noise such as motor vehicles, aircraft, and railway rolling stock, and the provision of financial support to noise-related studies and research projects.

Directive 2002/49/EC is the main legislation document related to noise in Europe. Its principles share many common elements with the other environmental policy Directives, such as [99]:

- Enforcing Member States to declare competent authorities with the task to monitor noise and produce strategic noise maps for major roads, railways, civil airports, and urban agglomerations;
- Introducing harmonized noise indicators;
- Informing the public about noise pollution along with its effects and the mitigation measures to be implemented in order to increase public participation and awareness in line with the principles of the Aarhus Convention;
- Addressing local noise issues by requiring competent authorities to draw up action plans to reduce noise where necessary;
- Developing a long-term EU strategy, including objectives to reduce the number of people affected by noise, and providing a framework for developing existing EU policy on noise reduction from sources.

In the context of the actions above, NOISE was created. NOISE stands for the Noise Observation and Information Service for Europe maintained by the European Environment Agency (EEA) and the European Topic Centre for Air Pollution and Climate Change Mitigation (ETC-ACM) on behalf of the European Commission. It contains data related to strategic noise maps delivered in accordance with European Directive 2002/49/EC relating to the assessment and management of environmental noise, which is also known as the Environmental Noise Directive (END). NOISE database[18] covers the data reported by member states and member countries corresponding to 2005 and 2007 deliveries, and to 2010 (2008) and 2012 deliveries, updated up to June 10, 2014 for both noise exposure data and for noise contour maps.

The report released by Noise Free America in 2010 offers a short history of noise legislation in the United States [113]. Specifically, the Noise Control Act of 1972

---

[18] http://noise.eionet.europa.eu/.

(42 U.S.C. 4901 to 4918) established a national policy to promote a healthy, noise-free environment for all Americans. The Act requires the coordination of federal research and activities in noise control, authorizes the establishment of federal noise emission standards for commercial products, and authorizes the dissemination of information to the public regarding the noise characteristics of commercial products. To enforce the Noise Control Act, the Office of Noise Abatement and Control (ONAC) was created within EPA, but in 1981, the director of ONAC was informed that the Office of Management and Budget was eliminating all funding for ONAC. Even though ONAC's funding was eliminated, the Noise Control Act of 1972 remains in effect and EPA remains legally responsible for enforcing the relevant provisions. But, as stated in the relevant Website of EPA,[19] "EPA phased out the office's funding in 1982 as part of a shift in federal noise control policy to transfer the primary responsibility of regulating noise to state and local governments. However, the Noise Control Act of 1972 and the Quiet Communities Act of 1978 were never rescinded by Congress and remain in effect today, although essentially unfunded." In any case, the Act provides for:

- Identification of major noise sources;
- Noise emission standards for products distributed in commerce;
- Labeling;
- Quiet communities, research, and public information;
- Development of low-noise-emission products;
- Motor carrier noise emission standards.

It is obvious that the noise pollution issue has not been a priority for policy makers so far and that it has not been approached like air and water pollution. The case is even worse in developing countries where little emphasis is placed on noise pollution. In many cases, noise pollution is addressed in general environmental or labor laws. Moreover, noise pollution is often regulated and monitored by governmental bodies that do not focus specifically on the environment [114].

References

[1] D.L. Hartman, Global physical climatology, in: R. Dmowska, J.R. Holton (Eds.), International Geophysics Series, vol. 56, Academic Press, San Diego, 1994.

[2] T.W. Schlatter, Atmospheric Composition and Vertical Structure, U.S. Department of Commerce, National Oceanic and Atmospheric Administration, Earth System Research Laboratory, Global Systems Division, USA, 2009. http://ruc.noaa.gov/AMB_Publications_bj/2009%20Schlatter_Atmospheric%20Composition%20and%20Vertical%20Structure_eae319MS-1.pdf.

[3] M.N. Rao, H.V.N. Rao, Air Pollution, Tata McGraw-Hill Publishing Co. Ltd, New Delhi, 1989.

---

[19] http://www2.epa.gov/aboutepa/epa-history-noise-and-noise-control-act.

[4] P. Falkowski, R.J. Scholes, E. Boyle, J. Canadell, D. Canfield, J. Elser, N. Gruber, K. Hibbard, P. Högberg, S. Linder, F.T. MacKenzie, Moore, T. Pedersen, Y. Rosenthal, S. Seitzinger, V. Smetacek, W. Steffen, The global carbon cycle: a test of our knowledge of earth as a system, Science 290 (2000) 291–296.

[5] G.L. Hashimoto, Y. Abe, S. Sugita, The chemical composition of the early terrestrial atmosphere: formation of a reducing atmosphere from CI-like material, J. Geophys. Res. 112 (2007) 1–12.

[6] A. Bar-Nu, S. Chang, Photochemical reactions of water and carbon monoxide in earth's primitive atmosphere, J. Geophys. Res. 88 (11) (1983) 6662–6672.

[7] C.D. Ahrens, Meteorology Today: An Introduction to Weather, Climate, and the Environment, ninth ed., Cengage Learning Belmont, 2008.

[8] H. Flohn, R. Penndorf, The stratification of the atmosphere, Bull. Am. Meteorol. Soc. 31 (3) (1950) 71–78.

[9] IPPC, Climate Change 2014 Synthesis Report. Summary for Policymakers, Intergovernmental Panel on Climate Change, Geneva, 2014.

[10] A.G. Barnett, L.D. Knibbs, Higher fuel prices are associated with lower air pollution levels, Environ. Int. 66 (2014) 88–91.

[11] B. Ostro, Outdoor Air Pollution. Assessing the Environmental Burden of Disease at National and Local Levels, World Health Organization, Protection of the Human Environment, Geneva, 2004.

[12] R.B. Husar, D.E. Patterson, Regional scale air pollution: sources and effects, Ann. N.Y. Acad. Sci. 338 (1) (2008) 399–417.

[13] F. DiGiovanni, P. Fellin, Transboundary air pollution, in environmental monitoring, in: H.I. Inyang, J.L. Daniels (Eds.), Encyclopedia of Life Support Systems (EOLSS), Developed under the Auspices of the UNESCO, Eolss Publishers, Oxford, 2006.

[14] V. Ramanathan, Y. Feng, Air pollution, greenhouse gases and climate change: global and regional perspectives, Atmos. Environ. 43 (2009) 37–50.

[15] POST, Air Quality in the UK, Postnote No 188, Parliamentary Office of Science and Technology, London, 2002.

[16] UNEP, The World Environment 1972–1992, United Nations Environment Programme, Chapman & Hall, London, 1992.

[17] EEA, Air Quality in Europe − 2014 Report, EEA Report No 5/2014, European Environment Agency, Copenhagen, 2014.

[18] UNEP, UNEP Year Book 2014 Emerging Issues Update. Air Pollution: World's Worst Environmental Health Risk, United Nations Environment Programme, Environment for Development, Nairobi, 2014. www.unep.org/yearbook/2014/PDF/chapt7.pdf.

[19] WHO, Burden of Disease from Ambient Air Pollution for 2012, World Health Organization, Geneva, 2014. http://www.who.int/phe/health_topics/outdoorair/databases/FINAL_HAP_AAP_BoD_24March2014.pdf?ua=1.

[20] EEA, Every Breath We Take, Improving Air Quality in Europe, EEA Signals 2013, European Environment Agency, Copenhagen, 2013. http://www.eea.europa.eu/publications/eea-signals-2013/at_download/file.

[21] V.J. Inglezakis, S.G. Poulopoulos, Adsorption, Ion Exchange and Catalysis, Design of Operations and Environmental Applications, first ed., Elsevier, Amsterdam, 2007.

[22] P. Brimblecombe, Z. Ning, Effect of road blockages on local air pollution during the Hong Kong protests and its implications for air quality management, Sci. Total Environ. 536 (2015) 443–448.

[23] M.S. Alam, I.J. Keyte, J. Yin, C. Stark, A.M. Jones, M. Harrison, Diurnal variability of polycyclic aromatic compound (PAC) concentrations: relationship with meteorological conditions and inferred sources, Atmos. Environ. 122 (2015) 427–438.

[24] F. Lu, L. Zhou, Y. Xuc, T. Zheng, Y. Guo, G.A. Wellenius, B.A. Bassig, X. Chen, H. Wang, X. Zheng, Short-term effects of air pollution on daily mortality and years of life lost in Nanjing, China, Sci. Total Environ. 536 (2015) 123–129.

[25] S.G. Poulopoulos, Catalytic Oxidation of Oxygenated Compounds in Exhaust Emissions (Ph.D. thesis), National Technical University of Athens, 2004, http://hdl.handle.net/10442/hedi/16814 (in Greek).

[26] WHO, Review of Evidence on Health Aspects of Air Pollution — REVIHAAP Project, World Health Organization Technical Report, WHO European Centre for Environment and Health, Bonn, 2013.

[27] M.C. Rodríguez, L. Dupont-Courtade, W. Oueslati, Air pollution and urban structure linkages: evidence from European cities, Renewable Sustainable Energy Rev. 53 (2016) 1—9.

[28] UNECE, Protocol to Abate Acidification, Eutrophication and Ground-level Ozone, United Nations Economic Commission for Europe, Gothenburg, 1999.

[29] H.M. ApSimon, R.F. Warren, Transboundary air pollution in europe, Energy Policy 24 (7) (1996) 631—640.

[30] EEA, Effects of Air Pollution on European Ecosystems, EEA Technical report No 11/2014, European Environment Agency, Copenhagen, 2014.

[31] D. Jaffe, T. Anderson, D. Covert, R. Kotchenruther, B. Trost, J. Danielson, W. Simpson, T. Berntsen, S. Karlsdottir, D. Blake, J. Harris, G. Carmichael, I. Uno, Transport of Asian air pollution to North America, Geophys. Res. Lett. 26 (1999) 711—714.

[32] T. Holloway, A. Fiore, M.G. Hastings, Intercontinental transport of air pollution: will emerging science lead to a new hemispheric treaty? Environ. Sci. Technol. 37 (2003) 4535—4542.

[33] IAQAB, Summary of Critical Air Quality Issues in the Transboundary Region, International Air Quality Advisory Board, 2004. http://www.ijc.org/files/publications/ID1539.pdf.

[34] RoTAP, Review of Transboundary Air Pollution: Acidification, Eutrophication, Ground Level Ozone and Heavy Metals in the UK, Contract Report to the Department for Environment, Food and Rural Affairs. Centre for Ecology & Hydrology, Midlothian, 2012.

[35] A.J. McMichael, R. Lucas, A.-L. Ponsonby, S.J. Edwards, Stratospheric ozone depletion, ultraviolet radiation and health, in: A.J. McMichael, D.H. Campbell-Lendrum, C.F. Corvalan, et al. (Eds.), Climate Change and Human Health — risks and Responses, World Health Organization, Geneva, 2003, pp. 159—180.

[36] J.C. Farman, B.G. Gardiner, J.D. Shanklin, Large losses of total ozone in Antarctica reveal seasonal ClO_x/NO_x interaction, Nature 315 (1985) 207—210.

[37] S. Solomon, Stratospheric ozone depletion: a review of concepts and history, Rev. Geophys. 37 (1999) 275—316.

[38] M. Molina, F. Rowland, Stratospheric sink for chlorofluoromethanes: chlorine atom catalysed destruction of ozone, Nature 249 (1974) 810—812.

[39] G.R. Kelly, The integrity of the ozone layer, Atmos. Environ. 10 (8) (1976) 677—680.

[40] R. Müller, A brief history of stratospheric ozone research, Meteorol. Z. 18 (2009) 3—24.

[41] J.-U. Grooß, R. Müller, Do cosmic-ray-driven electron-induced reactions impact stratospheric ozone depletion and global climate change? Atmos. Environ. 45 (2011) 3508—3514.

[42] Q.-B. Lu, Cosmic-ray-driven electron-induced reactions of halogenated molecules adsorbed on ice surfaces: implications for atmospheric ozone depletion and global climate change, Phys. Rep. 487 (2010) 141—167.

[43] R.W. Portmann, J.S. Daniel, A.R. Ravishankara, Stratospheric ozone depletion due to nitrous oxide: influences of other gases, Philos. Trans. B 367 (2012) 1256—1264.

[44] UNEP, Environmental Effects of Ozone Depletion: 2010 Assessment, United Nations Environment Programme, Nairobi, 2010.

[45] UNEP, Environmental Effects of Ozone Depletion: 2014 Assessment, United Nations Environment Programme, Nairobi, 2015.

[46] V. Marchal, et al., Climate change, in: OECD Environmental Outlook to 2050: The Consequences of Inaction, OECD Publishing, 2012.

[47] EPA, Inventory of U.S. Greenhouse Gas Emissions and Sinks: 1990—2013, U.S. Environmental Protection Agency, Washington, 2015.

[48] D.K. Bird, K. Haynes, R. Honert, J. McAneney, W. Poorting, Nuclear power in Australia: a comparative analysis of public opinion regarding climate change and the Fukushima disaster, Energy Policy 65 (2014) 644−653.

[49] IPCC, 2013 Supplement to the 2006 IPCC Guidelines for National Greenhouse Gas Inventories: Wetlands, Intergovernmental Panel on Climate Change, Switzerland, 2013.

[50] M. Buchwitz, et al., The Greenhouse Gas Climate Change Initiative (GHG-CCI): comparison and quality assessment of near-surface-sensitive satellite-derived CO_2 and CH_4 global data sets, Remote Sens. Environ. 162 (2013) 344−362.

[51] J.G. Canadell, P. Ciais, S. Dhakal, H. Dolman, P. Friedlingstein, K.R. Gurney, et al., Interactions of the carbon cycle, human activity, and the climate system: a research portfolio, Curr. Opin. Environ. Sustainability 2 (2010) 301−311.

[52] B.B. Stephens, K.R. Gurney, P.P. Tans, C. Sweeney, W. Peters, L. Bruhwiler, et al., Weak northern and strong tropical land carbon uptake from vertical profiles of atmospheric CO_2, Science 316 (2007) 1732−1735.

[53] O. Schneising, M. Buchwitz, M. Reuter, J. Heymann, H. Bovensmann, J.P. Burrows, Long-term analysis of carbon dioxide and methane column-averaged mole fractions retrieved from SCIAMACHY, Atmos. Chem. Phys. 11 (2011) 2881−2892.

[54] S. Solomon, D. Qin, M. Manning, Z. Chen, M. Marquis, K.B. Averyt, et al., Climate change 2007: the physical science basis, in: Contribution of Working Group I to the Fourth Assessment Report of the Intergovernmental Panel on Climate Change (IPCC), Cambridge University Press Cambridge, 2007.

[55] E.J. Dlugokencky, L. Bruhwiler, J.W.C. White, L.K. Emmons, P.C. Novelli, S.A. Montzka, et al., Observational constraints on recent increases in the atmospheric CH_4 burden, Geophys. Res. Lett. 36 (18) (2009) L18803.

[56] C. Frankenberg, I. Aben, P. Bergamaschi, E.J. Dlugokencky, R. van Hees, S. Houweling, et al., Global column-averaged methane mixing ratios from 2003 to 2009 as derived from SCIAMACHY: trends and variability, J. Geophys. Res. 116 (D4) (2011) D04302.

[57] J.M. De Suarez, B. Cicin-Sain, K. Wowk, R. Payet, O. Hoegh-Guldberg, Ensuring survival: oceans, climate and security, Ocean Coastal Manage. 90 (2014) 27−37.

[58] D. Hemming, R. Betts, M. Collins, Sensitivity and uncertainty of modelled terrestrial net primary productivity to doubled CO_2 and associated climate change for a relatively large perturbed physics ensemble, Agric. For. Meteorol. 170 (2013) 79−88.

[59] C. Loehle, A minimal model for estimating climate sensitivity, Ecol. Modell. 276 (2014) 80−84.

[60] IPCC, Climate change 2014: synthesis report, Contribution of Working Groups I, II and III to the Fifth Assessment Report of the Intergovernmental Panel on Climate Change, Geneva, 2014.

[61] J.-S. Ryu, A.D. Jacobson, CO_2 evasion from the Greenland Ice Sheet: a new carbon-climate feedback, Chem. Geol. 320−321 (2012) 80−95.

[62] S.A. Zimov, E.A.G. Schuur, F.S. Chapin, Permafrost and the global carbon budget, Science 312 (2006) 1612−1613.

[63] M.C. Serreze, A.P. Barrett, J.C. Stroeve, D.N. Kindig, M.M. Holland, The emergence of surface-based Arctic amplification, Cryosphere 3 (2009) 11−19.

[64] K. Brysse, N. Oreskes, J. O'Reilly, M. Oppenheimer, Climate change prediction: erring on the side of least drama? Global Environ. Change 23 (1) (2013) 327−337.

[65] J.M. Melillo, T.C. Richmond, G.W. Yohe, Climate Change Impacts in the United States: The Third National Climate Assessment, U.S. Global Change Research Program, Washington, 2014.

[66] C. Bezirtzoglou, K. Dekas, E. Charvalos, Climate changes, environment and infection: facts, scenarios and growing awareness from the public health community within Europe, Anaerobe 17 (2011) 337−340.

[67] R. Sommer, Impact of climate change on wheat productivity in Central Asia, Agric. Ecosyst. Environ. 178 (2013) 78−99.

[68] N. Stern, The Economics of Climate Change. The Stern Review, Cambridge University Press, London, 2007.

[69] D.L. Davis, M.L. Bell, T. Fletcher, A look back at the London smog of 1952 and the half century since, Environ. Health Perspect. 110 (12) (2002) A734.

[70] GLA, Fifty Years on. The Struggle for Air Quality in London since the Great Smog of December 1952, Greater London Authority, City Hall, London, 2012.

[71] D.V. Bates, A half century later: recollections of the London fog, Environ. Health Perspect. 110 (12) (2002) A735.

[72] C.U. Fortunati, The seveso accident, Chemosphere 14 (6−7) (1985) 729−737.

[73] ICMR, The Bhopal gas tragedy, No 702-006-1, ICFAI Center for Management Research, Telangana, India, 2002. http://www.econ.upf.edu/~lemenestrel/IMG/pdf/bhopal_gas_tragedy_dutta.pdf.

[74] EC, Best Available Techniques (BAT) Reference Document for Common Waste Water and Waste Gas Treatment/Management Systems in the Chemical Sector, Final Draft, European Commission, Institute for Prospective Technological Studies, Seville, 2014. http://eippcb.jrc.ec.europa.eu/reference/BREF/CWW_Final_Draft_07_2014.pdf.

[75] A.M. Middlebrook, M.A. Tolbert, Stratospheric Ozone Depletion, Global Change Instruction Program, University Science Books, Sausalito, California, 2000.

[76] EEA, Ozone-depleting Substances 2014, EEA Technical report No 10/2015, European Environment Agency, Copenhagen, 2015.

[77] EC, Regulation (EC) No 1005/2009 of the European Parliament and of the Council of 16 September 2009 on Substances that Deplete the Ozone Layer, European Commission, O.J. L 286, October 31, 2009, pp. 1−30.

[78] C.S. Norman, S.J. DeCanio, L. Fan, The Montreal Protocol at 20: ongoing opportunities for integration with climate protection, Global Environ. Change 18 (2008) 330−340.

[79] J.R. Jacobs, The precautionary principle as a provisional instrument in environmental policy: the Montreal Protocol case study, Environ. Sci. Policy 37 (2014) 161−171.

[80] UN, Glossary of Environment Statistics, Studies in Methods, Series F, No. 67, United Nations, New York, 1997.

[81] S. Langer, G. Bekoe, E. Bloom, A. Widheden, L. Ekberg, Indoor air quality in passive and conventional new houses in Sweden, Build. Environ. 93 (2015) 92−100.

[82] W.J. Fisk, A.G. Mirer, L.J. Mendell, Quantitative relationship of sick building syndrome symptoms with ventilation rates, Indoor Air 19 (2009) 159−165.

[83] W. Wei, O. Ramalho, C. Mandin, Indoor air quality requirements in green building certifications, Build. Environ. 92 (2015) 10−19.

[84] S. Kirchner, Quality of Indoor Air, Quality of Life, a Decade of Research to Breathe Better, Breathe Easier, CSTB ed., French Indoor Air Quality Observatory, Marne-La-Vallée, 2013.

[85] J. Mao, W. Yang, N. Gao, The Transport of Gaseous Pollutants Due to Stack and Wind Effect in High-rise Residential Buildings, Building and Environment 94 (2) (2015) 543−557.

[86] WHO, Development of WHO Guidelines for Indoor Air Quality. Report on a Working Group Meeting, Bonn, Germany, 23−24 October 2006, World Health Organization, Regional Office for Europe, Copenhagen, 2006. http://www.euro.who.int/__data/assets/pdf_file/0007/78613/AIQIAQ_mtgrep_Bonn_Oct06.pdf.

[87] WHO, WHO Guidelines for Indoor Air Quality: Selected Pollutants, World Health Organization, Regional Office for Europe, Copenhagen, 2010.

[88] EEA, Environment and Human Health, EEA Report no 3/2013, European Environment Agency, Copenhagen, 2013.

[89] CER, Common Indoor Air Pollutants: Sources and Health Impacts, IAQ Fact Sheet 2, Cooperative Extension Service, University of Kentucky, Kentucky, 2000. http://www2.ca.uky.edu/HES/fcs/FACTSHTS/HF-LRA.161.PDF.

[90] J. Madureira, I. Paciência, J. Rufo, A. Moreira, E. de Oliveira Fernandes, A. Pereira, Radon in indoor air of primary schools: determinant factors, their variability and effective dose, Environ. Geochem. Health (2015) 1−11.

[91] F.J. Offermann, Chemical emissions from e-cigarettes: direct and indirect (passive) exposures, Build. Environ. 93 (2015) 101−105.

[92] N. Bruce, D. Pope, E. Rehfuess, K. Balakrishnan, H. Adair-Rohani, C. Dora, WHO indoor air quality guidelines on household fuel combustion: strategy implications of new evidence on interventions and exposure risk functions, Atmos. Environ. 106 (2015) 451−457.

[93] K.R. Smith, N.G. Bruce, K. Balakrishnan, H. Adair-Rohani, J. Balmes, Z. Chafe, et al., Millions dead: how do we know and what does it mean? Methods used in the comparative risk assessment of household air pollution, Ann. Rev. Public Health 35 (2014) 185−206.

[94] WHO, Deaths from Household Air Pollution, 2012, World Health Organization, Geneva, 2014.

[95] K. Balakrishnan, S. Ghosh, B. Ganguli, S. Sambandam, N. Bruce, D.F. Barnes, K.R. Smith, State and national household concentrations of $PM_{2.5}$ from solid cookfuel use: results from measurements and modeling in India for estimation of the global burden of disease, Environ. Health 12 (2013) 77.

[96] R.D. Gupta, Environment Pollution: Hazards and Control, Concept Publishing Company, New Delhi, India, 2006.

[97] WHO, Guidelines for Community Noise, World Health Organization, Geneva, 1999. http://www.who.int/docstore/peh/noise/guidelines2.html.

[98] EU, Directive 2002/49/EC of the European Parliament and of the Council of 25 June 2002 Relating to the Assessment and Management of Environmental Noise, European Commission, Official Journal of the European Communities, O.J. L. 189, 2002, pp. 12−25.

[99] WHO, Burden of Disease from Environmental Noise. Quantification of Healthy Life Years Lost in Europe, World Health Organization, Regional Office for Europe, Bonn, 2011.

[100] R. Chepesiuk, Decibel hell: the effects of living in a noisy world, Environ. Health Perspect. 113 (1) (2005) 34−41.

[101] L. Goines, L. Hagler, Noise pollution: a modern plague, South. Med. J. 100 (3) (2007) 287−294.

[102] M.S. Hammer, T.K. Swinburn, R.L. Neitzel, Environmental noise pollution in the United States: developing an effective public health response, Environ. Health Perspect. 122 (2) (2014) 115−119.

[103] H. Ising, B. Kruppa, Health effects caused by noise: evidence from the literature from the past 25 years, Noise Health 6 (2004) 5−13.

[104] W. Babisch, Noise and health, Environ. Health Perspect. 113 (2005) 14−15.

[105] G.W. Evans, S.J. Lepore, Non-auditory effects of noise on children; a critical review, Child. Environ. 10 (1993) 42−72.

[106] EEA, Noise in Europe, European Environment Agency, Copenhagen, 2014.

[107] C.D. Francis, J.R. Barber, A framework for understanding noise impacts on wildlife: an urgent conservation priority, Front. Ecol. Environ. 11 (6) (2013) 305−313.

[108] C.D. Francis, C.P. Ortega, A. Cruz, Different behavioural responses to anthropogenic noise by two closely related passerine birds, Biol. Lett. 7 (6) (2011) 850−852.

[109] C.D. Francis, C.P. Ortega, A. Cruz, Noise pollution changes avian communities and species interactions, Curr. Biol. 19 (2009) 1415−1419.

[110] C.D. Francis, N.J. Kleist, C.P. Ortega, A. Cruz, Noise pollution alters ecological services: enhanced pollination and disrupted seed dispersal, Proc. R. Soc. B 279 (1739) (2012) 1−9.

[111] H.A. Breinbauer, J.L. Anabalón, D. Gutierrez, R. Cárcamo, C. Olivares, J. Caro, Output capabilities of personal music players and assessment of preferred listening levels of test subjects: outlining recommendations for preventing music-induced hearing loss, Laryngoscope 122 (11) (2012) 2549−2556.

[112] D.H.F. Liu, H.C. Roberts, Noise pollution, in: D.H.F. Liu, B.G. Liptak (Eds.), Environmental Engineers' Handbook, CRC Press LLC, 1999.

[113] NFA, The American Noise Pollution Epidemic: The Pressing Need to Reestablish the Office of Noise Abatement and Control, Noise Free America, New York, 2010. https://www.noisefree.org/ONAC_2010.pdf.

[114] U.S. International Trade Commission, Air and Noise Pollution Abatement Services: An Examination of U.S. and Foreign Markets, Investigation No. 332−461, Publication 3761, Washington, 2005.

CHAPTER 3

Aquatic Environment

V.J. Inglezakis
Nazarbayev University, Astana, Republic of Kazakhstan

S.G. Poulopoulos
Kazakh-British Technical University, Almaty, Republic of Kazakhstan

E. Arkhangelsky
Nazarbayev University, Astana, Republic of Kazakhstan

A.A. Zorpas
Cyprus Open University, Latsia, Nicosia, Cyprus

A.N. Menegaki
Hellenic Open University, Patras, Greece; Organismos Georgikon Asfaliseon, Regional Branch of Eastern Macedonia & Thrace, Komotini, Greece

Chapter Outline

Environment and Development. http://dx.doi.org/10.1016/B978-0-444-62733-9.00003-4

3.1 Introduction: The Importance of Water

Earth is called as the "blue planet" because of the fact that 71% of its surface is covered by water, which is mainly salty in the oceans. Water is not just another substance; it is connected with the development and support of all life on our planet. A tree contains 60% water, most animals are composed of about 65% water, while our bodies contain around 55% water [1]. It plays manifold roles: it regulates the climate on Earth via oceans as discussed in the previous chapter, it accepts and disperses contaminants, it is vital for most life forms as drinking water, and besides all other precious values, clean water is a prerequisite for both economic and sustainable development of mankind [2]. Apart from drinking it to survive, people use water in so many ways, including cooking, washing their bodies, washing clothes, washing cooking and eating utensils, keeping houses and communities clean, recreational activities like swimming and fishing, keeping plants alive in gardens and parks, etc. Freshwater is consumed every day to cover food demands; for domestic use; in agriculture which is a huge consumer; in construction, transport, and chemical industry; and many more other human activities. It has to be emphasized that the production of any type of goods is based on the supply of clean water. The production of food is also entirely depended on clean water availability. To eradicate hunger, spread of diseases, and poverty in our world, the starting point, the most important requirement, and the most critical factor is to ensure access to clean drinking water to populations!

These numerous uses of water and its importance for life on Earth are owed to its unique properties, which in turn are related to the molecule structure of water and the hydrogen bonding between water molecules.[1,2]

Water is the only substance that exists naturally on Earth in all three physical states of matter, gas, liquid, and solid, changing from one form to another. This means that liquid water absorbs solar energy to evaporate into the atmosphere. As water vapor rises in the atmosphere, it cools and condenses into tiny liquid droplets that scatter light and become visible as clouds. Under proper conditions, these droplets further combine and become heavy enough to precipitate (fall out) as drops of liquid or, if the air is cold enough, as flakes of solid, thus returning to the surface of the Earth to continue this cycle of water (hydrological cycle) between its condensed and vapor phases. This procedure is essential for climate regulation and for purifying and supplying us with fresh clean water. It is also apparent that water is a connecting link between atmospheric, aquatic, and soil environments.

The great majority of liquids contract under decreasing temperatures and they reach a maximum density when they solidify. Water, however, exhibits a unique behavior. As it cools down, it contracts till the temperature of 4°C, where it starts to expand before being frozen at 0°C. Then, ice is formed, which is less dense than water, allowing it thus to float on water. This function is of unique importance since it allows aquatic life to survive in the unfrozen liquid below the surface of frozen lakes or rivers during winter seasons.

Water has a remarkably high boiling point for a substance with such a small molecule. If hydrogen bonding did not exist, taking into account the molar mass of water, it should be boiling at −90°C instead of 100°C (under 1 atm). So, it is able to create and participate in the hydrological cycle, and distribute heat on Earth.

Water has an unusually high specific heat, meaning that it takes more energy to raise the temperature of 1 g of it by 1°C, than any other liquid. As a result, water must absorb relatively high amounts of heat energy to raise its temperature. This property is responsible for the ocean's ability to act as a thermal reservoir that moderates variations in the Earth's temperature from day to night and from winter to summer, making thus the planet suitable for its diverse inhabitants.

---

[1] http://www.eoearth.org/view/article/153627/.
[2] http://scifun.chem.wisc.edu/chemweek/PDF/COW-Water-Jan2011.pdf.

Many substances can be dissolved in water. This enables the transfer of dissolved nutrients in the tissues of living organisms, or the dispersion and elimination of waste products by humans.

Water has the second highest surface tension of all common liquids; only mercury has a higher one. Surface tension is the property that allows water to stride and proceed from the root to the leaves of a plant.

Despite these unique features and its importance, water does not take the respect it deserves from humans. Numerous compounds are released daily into water receivers, leading to various pollution problems that deteriorate water quality, making it inappropriate for use, whether it was destined to be used as drinking water or for recreational activities. Besides pollution problems, water quantity is another issue of global interest. Since water has manifold value for people and sustainable development, we have to ensure that enough quantities of freshwater are available to cover the ever-increasing demand.

So, human activities have serious impacts on water quality and availability. Water is polluted or exhausted every day all over the world. In addition, man is responsible for large-scale phenomena like climate change that affect severely the hydrological cycle and thus the availability of water globally, since freshwater is purified and reallocated through the hydrological cycle in nature. Another stressor is the ever-increasing world population that requires access to clean drinking water. The situation is expected to worsen in the near future, especially in densely populated or industrial areas. These areas consume large amounts of freshwater and at the same time produce and release large amounts of wastewater into the environment. Water management is not an easy task, because it is a multiparameter problem, as it will be discussed in the next paragraphs.

3.2 The Distribution of Freshwater

Water is extremely vital for life, and additionally it is considered to be a crucial resource for economic development, let alone the fact that it plays an important role in the climate regulation cycle. Therefore, the protection and sound management of water resources, including freshwater, seawater, salty water, drinking water, and bathing waters, is considered to be one of the cornerstones of environmental protection. This is why the EU's water policy for more than 30 years has been focused on that direction and the Water Framework Directive (WFD) was developed [3].

The importance of freshwater to our life support systems has been widely recognized, as it can be seen clearly in the international context (eg, Agenda 21, World Water Fora, the Millennium Ecosystem Assessment, and the World Water Development Report).

Freshwater is indispensable for all forms of life and is needed, in large quantities, for almost all human activities. However, freshwater is considered to be limited in the entitled world. Although 71% of the Earth surface is water-covered, most of this water is salty in the oceans that hold about 96.5% of all Earth's water. The remaining is freshwater; however, not even that small amount can be easily either accessible or exploited, because it is partly stored as ice on the poles and on mountaintops. Specifically, freshwater exists in the air as water vapor, in rivers and lakes, in ice caps and glaciers, in the ground as soil moisture, and in aquifers [4].

It is projected that 96% of the world's total water supply (1338 million cubic kilometers) is saline (salty). Regarding freshwater, more than 68% of it is locked up in ice and glaciers, while 30% is found as underground water. Rivers are the source of most of the fresh surface water that people use, but they only constitute about 1250 km^3, about 1/10,000th of 1% of total water (Table 3.1) [5].

The global water distribution according to Shiklomanov is shown in Fig. 3.1 [5].

Nixon et al. have reported a similar distribution as depicted in Fig. 3.2 [4,6].

Generally, freshwater availability in a country is determined by climate conditions, geomorphology, land uses, and transboundary water flows (ie, external inflows). The case of Europe is interesting due to the existence of many transboundary water resources. There are significant differences among countries, with France, Germany, Italy, Sweden, and the

Table 3.1: Estimation of Global Water Distribution [5].

Water Source	Water Volume (km^3)	Percentage of Freshwater	Percentage of Total Water
Oceans, seas, and bays	1,338,000,000	—	96.54
Ice caps, glaciers, and permanent snow	24,064,000	68.7	1.74
Groundwater	23,400,000	—	1.69
Fresh	10,530,000	30.1	0.76
Saline	12,870,000	—	0.93
Soil moisture	16,500	0.05	0.001
Ground ice and permafrost	300,000	0.86	0.022
Lakes	176,400	—	0.013
Fresh	91,000	0.26	0.007
Saline	85,400	—	0.006
Atmosphere	12,900	0.04	0.001
Swamp water	11,470	0.03	0.0008
Rivers	2120	0.006	0.0002
Biological water	1120	0.003	0.0001

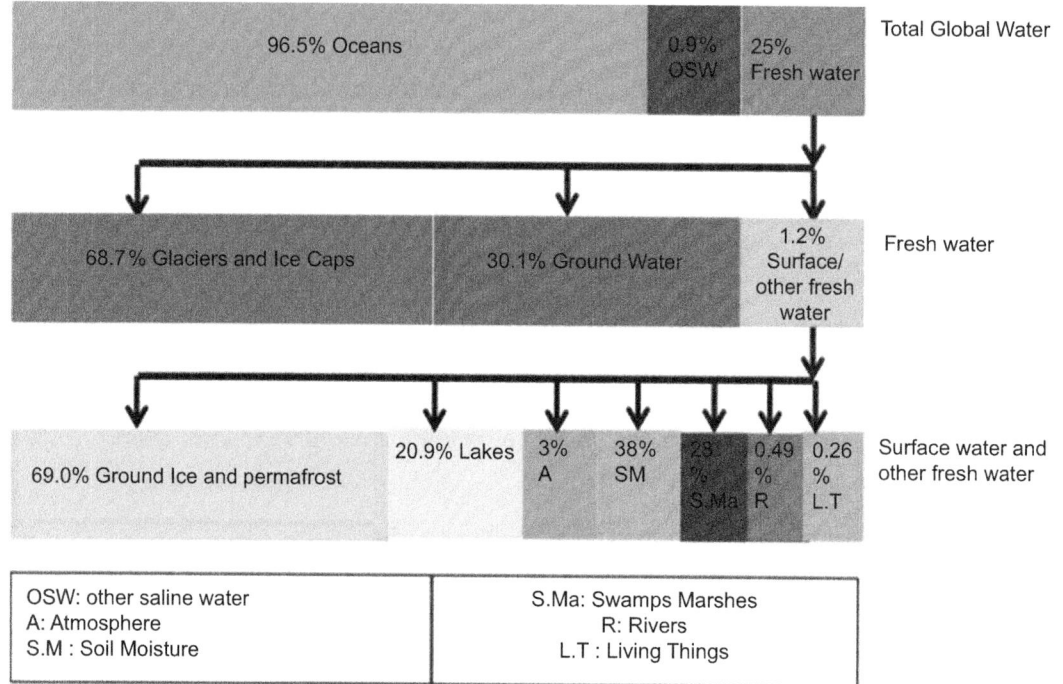

Figure 3.1
Global water distribution according to Shiklomanov [5].

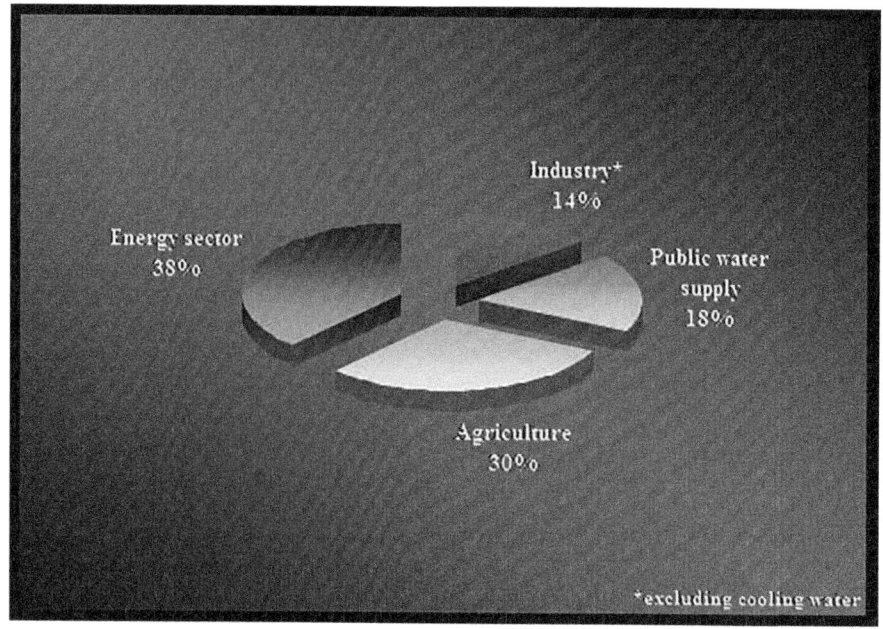

Figure 3.2
Global water distribution according to Nixon et al. [4,6].

United Kingdom being the Member States with the highest amount of freshwater resources, with a long-term annual average between 164,300 and 188,000 million cubic meters (Table 3.2).[3]

Several countries receive a significant proportion of their freshwater resources as external inflow (Fig. 3.3). For instance Hungary, Serbia, and the Netherlands are the countries with the highest dependency on transboundary water resources, as external inflow accounts for 93.5%, 92.7%, and 8.8% of their total freshwater resources, respectively. Serbia, Hungary, and Bulgaria receive the highest external flow (162,600, 108,900, and 89,100 million cubic meters, respectively; Table 3.2). Important transboundary rivers are the Danube (EU Member States sharing the basin: Austria, Bulgaria, Croatia, the Czech Republic, Germany, Hungary, Italy, Poland, Romania, Slovenia, Slovakia), the Elbe (EU Member States sharing the basin: Austria, the Czech Republic, Germany, Poland), the Meuse (EU Member States sharing the basin: Belgium, Germany, France, Luxembourg, the Netherlands), and the Rhine (EU Member States sharing the basin: Austria, Belgium, Germany, France, Italy, Luxembourg, the Netherlands) [7].

A significant water-related indicator is the freshwater resources per inhabitant. Between the EU-28 Member States, Croatia, Finland, and Sweden recorded the highest freshwater annual resources per inhabitant (around 20,000 m^3 or more). In contrast, relatively low levels per inhabitant (below 3000 m^3) were recorded in the six most populated Member States (France, Italy, the United Kingdom, Spain, Germany, and Poland). An area is considered to experience water stress when annual water resources drop below 1700 m^3 per person [8]. Poland, the Czech Republic, Cyprus, and Malta present the lowest values with between 200 and 1600 m^3 per person.

3.3 Conventional and Nonconventional Sources of Drinking Water

The sources of drinking water can be generally classified as conventional and nonconventional. There are two main *conventional* drinking water resources: *surface water* and *groundwater*. Surface waters are lying above Earth's surface, forming rivers, lakes, streams, ponds, etc. On the other hand, groundwater is the water found underground in the cracks and spaces in soil, sand, and rock. It is stored in and moves slowly through geologic formations of soil, sand, and rocks called *aquifers*.

Rain, snow, and atmospheric water are also frequently used for drinking purposes. These sources, however, depend on environmental conditions, and the associated water production capacity is rather low. Solutions like iceberg towing and artificial rain creation in order to produce drinking water are rather distant as well as extremely difficult and uncertain solutions.

---

[3] http://ec.europa.eu/eurostat/web/main/home.

Table 3.2: Freshwater Resources—Long-Term Annual Average (1000 million cubic meters).

Countries	Precipitation (P)	Actual Evapotranspiration (AE)	Internal Flow (IF) = (P) − (AE)	External Inflow (EI)	Freshwater Resources (FWR) = (IF) + (EI)	Outflow
Belgium	28.9	16.6	12.3	7.6	19.9	15.6
Bulgaria	69.8	52.3	18.1	89.1	107.2	108.5
Czech Republic	54.7	39.4	15.2	0.7	16.0	16.0
Denmark	38.5	22.1	16.3	0.0	16.3	1.9
Germany	307	190	117	75	188	182
Estonia	29					
Ireland	80	32.5	47.5	3.5	51	
Greece	115	55	60	12	72	
Spain	346.5	235.4	111.1	0.0	111.1	111.1
France	500.8	320.8	175.3	11	186.3	168
Croatia	65.7	40.1	23			
Italy	241.1	155.8	167	8	175	155
Cyprus	3	2.7	0.3		0.3	0.1
Latvia	42.7	25.8	16.9	16.8	33.7	32.9
Lithuania	44	28.5	15.5	9	24.5	25.9
Luxembourg	2	1.1	0.9	0.7	1.6	1.6
Hungary	55.7	48.2	7.5	108.9	116.4	115.7
Malta	150.4	72.5	0.1		0.1	
Netherlands	31.6	21.3	8.5	81.2	89.7	86.3
Austria	98	43	55	29	84	84
Poland	193.1	138.3	54.8	8.3	63.1	63.1
Portugal	82.2	43.6	38.6	35	73.6	34
Romania	154	114.6	39.4	2.9	42.3	17.9
Slovenia	31.7	13.2	18.6	13.5	32.1	32.3
Slovakia	37.4	24.3	13.1	67.3	80.3	81.7
Finland	222	115	107	3.2	110	110
Sweden	342.2	169.4	172.5	13.7	186.2	186.2
United Kingdom	275	117.2	157.9	6.4	164.3	164.3
Iceland	200	30	170		170	170
Norway	470.7	112	371.8	12.2	384	384
Switzerland	61.6	21.6	40.7	12.8	53.5	53.5
FYR of Macedonia	19.5			1		6.3
Serbia	56.1	43.3	12.8	162.6	175.4	175.4
Turkey	503.1	275.7	227.4	6.9	234.3	178

Source: Eurostat.

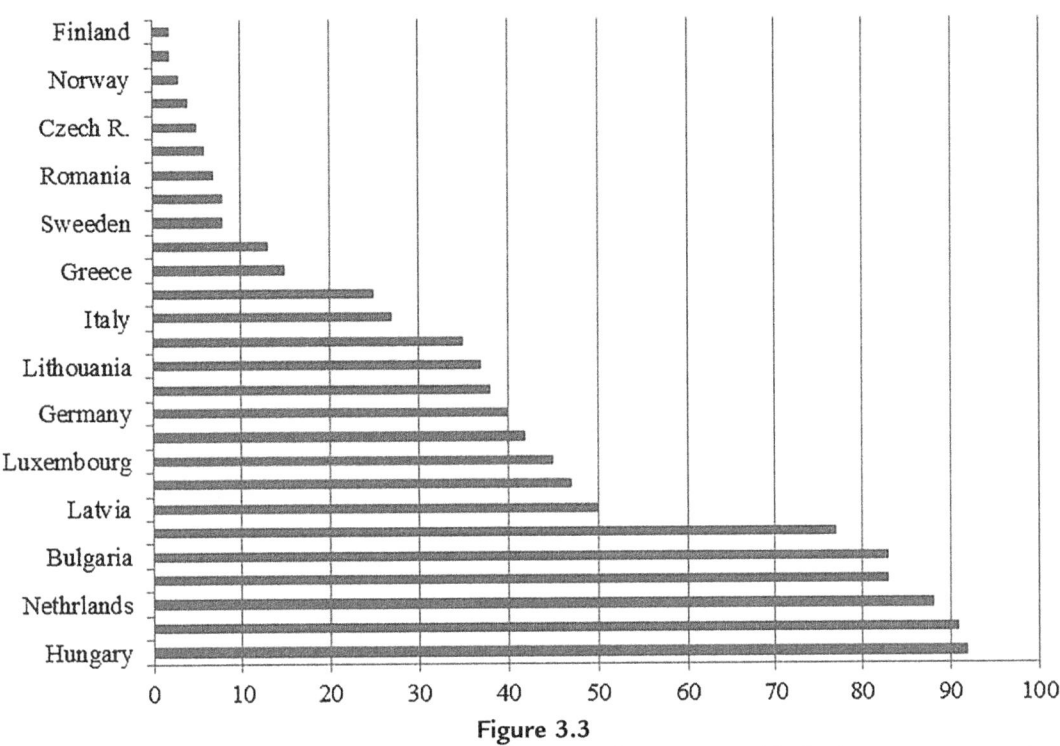

Figure 3.3

Proportion of external inflow from neighboring territories to renewable freshwater resources in percentage (long-term average). *Source: Eurostat.*

When conventional sources of drinking water are not available or are in shortage, *nonconventional water resources* may be sought: *desalinated seawater* and *brackish water*. The desalination of salty water is discussed in detail in Section 3.11.2. Another alternative potential source of potable water is reclaimed wastewater. For example, reclaimed wastewater is used for the production of drinking water in Singapore, Namibia, and Belgium [9]. Recently, reclaimed water is attracting vast attention around the world and is a potential candidate as potable water source for the upcoming decades. Water reclamation is also discussed in detail in Sections 3.11.3 and 3.11.4.

In order to use a water resource as drinking water, it should meet strict drinking water regulations, which set legal limits on the levels of certain contaminants in drinking water. The U.S. Environmental Protection Agency (EPA)[4] has set maximum concentration standards for 90 pollutants that can be roughly divided into six main groups, ie, microorganisms, radionuclides, inorganics, organics (volatile and synthetic), disinfectants, and disinfection by-products.

[4] http://water.epa.gov/.

Microorganisms coming from human and animal fecal waste are pathogenic bacteria, protozoa, and viruses. They cause headaches, gastrointestinal illness, such as diarrhea, vomiting, cramps, and pneumonia. For people with weak immune systems these diseases can be severe or even fatal. EPA has set zero limits for pathogenic organisms, such as the poliovirus, coxsackievirus, echovirus, and other enteroviruses, such as *Cryptosporidium*, *Giardia lamblia*, *Legionella*, and *Escherichia coli*. The second pollutant group is represented by radionuclides that are naturally occurring in certain rock types and are associated with cancer. Maximum Contaminant Level (MCL) for radium, uranium, alpha and beta emitters is set to 5 pCi/L, 30 µg/L, 15 pCi/L, 4 mrem/year, respectively. Inorganic pollutants of concern include antimony, arsenic, asbestos, barium, beryllium, cadmium, chromium, copper, cyanide, fluoride, lead, mercury, nitrate, nitrite, selenium, and thallium. They may reach water bodies through erosion of natural deposits, runoff from landfills/croplands, corrosion of plumbing systems, and industrial discharge. MCL for these contaminants range from 0 to 10 mg/L. Inorganic impurities can have the following health effects: gastrointestinal distress, bone disease, kidney damage, liver problems, hair and nail loss, high blood pressure, delay in physical and mental development, increased risk of getting cancer, and many more adverse health effects, including also death. Atrazine, alachlor, benzene, carbon tetrachloride, chlorobenzene, carbofuran, chlordane, dichloromethane, endrin, ethylbenzene, styrene, toluene, etc. are related to organic contaminants. EPA's limits, sources of discharge, and health effects for these pollutants are similar (same order of magnitude) to the ones for the inorganic group compounds. The next two classes are related to disinfection processes. The Maximum Residual Disinfectant Level Goal (MRDLG) for both chlorine and chloramines is 4 mg/L, and 0.8 mg/L for chlorine dioxide. When disinfectants are added to water that already contains naturally occurring organic matter, they can form disinfection by-products like bromates, chlorate, haloacetic acid, and trihalomethanes; their MCL values lie in between 0.01 and 1 mg/L. Both disinfectants and disinfection by-products may cause anemia, irritations, etc., and they may even affect nervous system and increase cancer risk.

3.4 Hydrological Cycle

Hydrological cycle is also known as the "water cycle"; it is the normal water recycling system on Earth (Fig. 3.4). Due to solar radiation, water evaporates, generally from the sea, lakes, etc. Water also evaporates from plant leaves through the mechanism of *transpiration*. As the steam rises in the atmosphere, it is being cooled, condensed, and returned to the land and the sea as precipitation. Precipitation falls on the earth as surface water and shapes the surface, creating thus streams of water that result in lakes and rivers. A part of the water precipitating penetrates the ground and moves downward through the incisions, forming aquifers. Finally, a part of the surface and

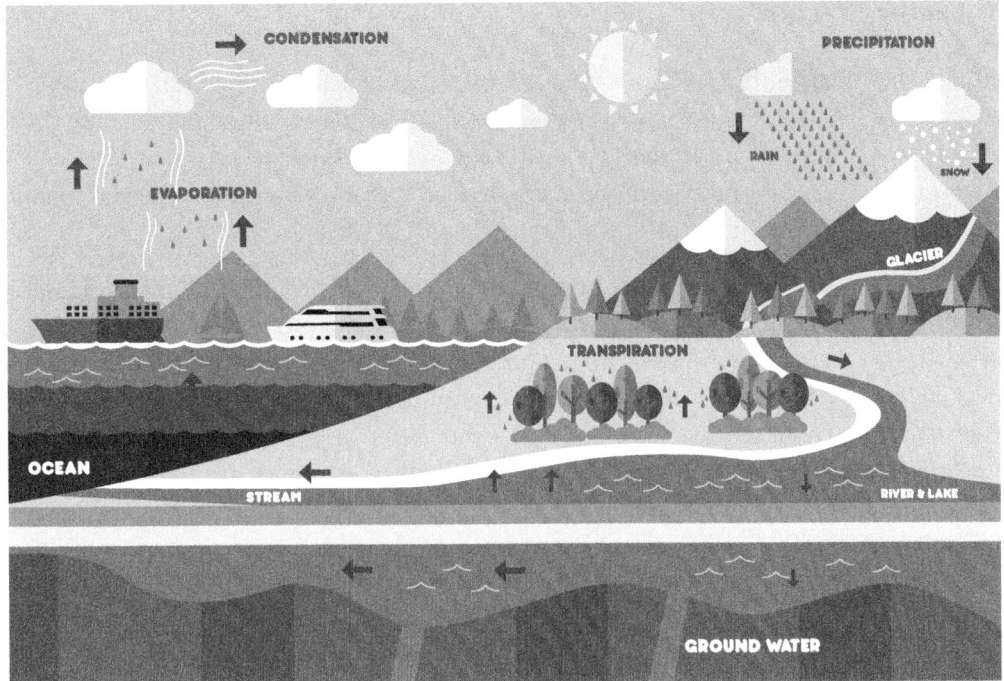

Figure 3.4
The hydrological cycle. *Used under license from Shutterstock.com/Image ID:236708653.*

underground water leads to sea. During this trip, water is converted in all phases: gas, liquid, and solid. As mentioned above, water always changes states between liquid, vapor, and ice, with these processes happening in the blink of an eye and over millions of years.

The hydrological cycle is intimately linked with changes in the atmospheric temperature and radiation balance. Warming of the climate system in recent decades is unequivocal, as it is now evident from observations of increases in global average air and ocean temperatures, widespread melting of snow and ice, and rising of the sea level globally.

It is expected that the hydrological cycle will be affected from global warming due to the enhanced greenhouse effect [10]. The hydrological cycle may be strengthened with more precipitation and more evaporation, but the extra precipitation will be unequally distributed around the globe. It is expected that some areas of the world may see significant reductions in precipitation or even more major variations in the timing of wet and dry seasons. Many aspects of the economy, environment, and society are dependent upon water resources, and changes in the hydrological resource base have the potential to severely impact upon environmental quality, economic development, and social well-being [11].

3.5 Water Stressors

In the previous chapter, the problems related with the atmospheric environment were discussed in detail. The main concerns were associated with air quality and the emissions of gas pollutants into the air. In the case of the aquatic environment, however, the problems that require our proper response are divided into two dimensions: water pollution (quality matter) and management of freshwater resources (quantity matter).

In Section 3.2, it was shown that freshwater is relatively limited, taking into account that it covers 71% of the surface of our planet. To understand the pressures and challenges associated with water, it is required at this point to present also its main uses for people. In Europe, 38% of the freshwater is used for energy production, 30% in agriculture, 14% in industry, and the rest 18% is used as public water supply [6]. However, the water use varies dramatically from one part of the world to another. In Egypt, for example, 98% of water is used for irrigation leaving only ~27 L/capita per day for domestic use. On a global basis, about 70% of freshwater is currently used for crop irrigation, around 20% for industrial purposes, and 10% for domestic purposes [12]. Another dimension that adds to the problems regarding water is that it is not equally distributed in the world. Asia, for example, has 60% of the world's population, but only 36% of the world's water [13].

Zimmerman et al. [13] examined the individual and integrated effects of several important stressors on the global water resource: population and consumption, demographic and land-use changes, urbanization, and of course climate change, all of which can contribute to changes in quality, quantity, and availability of water. The main conclusions are shortly as follows [13]:

- *Population and Consumption.* Current population is 6.7 billion with an average annual growth rate of 1.4%. Inevitably, the freshwater per capita is going to decrease.
- *Urbanization.* Only half of the increase of the 100 largest urban areas in the United States from 1970 to 1990 was because of population growth. 21 of the world's 33 megacities are located in coastal areas. So, the demand for freshwater and clean water generally for recreational activities is going to increase considerably for specific areas.
- *Economic growth.* It is always accompanied by increasing demands for clean water. By 2025, water withdrawals are predicted to increase from current levels by 50% in developing countries and 18% in developed countries. The potential impact includes water used for energy production and increased use in both the nonconsumptive domestic sector and the consumptive industrial sector. Moreover, one-half of the jobs worldwide are associated with water-dependent resources, such as fisheries, forests, and agriculture.

- *Land use.* Current methods of urbanization result in loss of forested riparian cover along streams and increased flooding, sedimentation, dredging and filling of wetlands, and eutrophication. Approximately 50% of precipitation recharges to groundwater in a natural system, whereas only 15% recharges in a highly urbanized environment. Human changes in land cover increase runoff and sediment loading.
- *Climate change.* This stressor will result in greater variability in precipitation, and runoff will affect erosion and sedimentation, which is already affected by land use.

Irina Bokova, Director-General of UNESCO, summarizes the issues that have to be dealt with water quality and quantity in a very clear and alarming way:

> *Freshwater is a cross-cutting issue that is central to all development efforts. It faces rising challenges across the world — from urbanization and overconsumption, from underinvestment and lack of capacity, from poor management and waste, from the demands of agriculture, energy and food production. Freshwater is not being used sustainably according to needs and demands. Accurate information remains disparate, and management is fragmented. In this context, the future is increasingly uncertain, and risks are set to deepen. If we fail today to make water an instrument of peace, it might become tomorrow a major source of conflict. More than ever, we need Integrated Water Resources Management to provide coherent leadership. We need better information gathering and sharing on the state of freshwater, on the nature of demand and its use. We need better systems for measurement and control at the local, national and global levels. We must start early, by building water issues into education. We need also for governments, the private sector and civil society to work more closely together and to integrate water as an intrinsic part of their decision-making.*
>
> *Ref. [8]*

The topics that are relevant to the effect of climate change and water pollution on water distribution and quality are further discussed in the next paragraphs.

3.6 Climate Change and Freshwater Pressure

Freshwater, climate, socioeconomic, and biophysical systems are interconnected, and each one affects the others in many ways. Anthropogenic climate change adds a major pressure to nations that are already confronting the issue of sustainable freshwater use. The challenges related to freshwater are having too little water or too much water, and having too much pollution. Each of these difficulties may be intensified by climate change. Freshwater-related issues play a pivotal role among the key regional and sectoral vulnerabilities. Consequently, the relationship between freshwater resources and climate change is of primary concern and deserves special attention. So far, water resource issues have not been adequately addressed in climate change analyses and climate policy formulations. Likewise, in several cases, climate change problems have not been adequately dealt with in water resources analyses, management, and policy formulation.

Water availability and quality will be the main pressures on, and issues for, the society and the environment under climate change; hence, it is necessary to improve our understanding of the parameters involved. In several countries, water use has increased over the last decades due to changes in lifestyle, population, and economic growth as well as due to expanded water supply systems. As mentioned, irrigation water use (mainly for agricultural purposes) is globally the most important consumer. Irrigation accounts for about 70% of total water withdrawals worldwide and for more than 90% of consumptive water use (ie, the water volume that is not available for reuse downstream). Irrigation generates about 40% of the total agricultural output [14]. The area of global irrigated land has increased approximately linearly since 1960 at a rate of roughly 2% per annum, from 140 million hectares in 1961/63 to 270 million hectares in 1997/99, representing about 18% of today's total cultivated land [15].

Generally, global climate changes predict warmer and drier conditions for much of the northwest European Union [16]; however these phenomena mask regional and seasonal shifts in temperature and precipitation. According to Weyhenmeyer et al. [17], freshwaters are affected by large-scale climate drivers.

The temperature of water in lakes, rivers, and generally in freshwaters is strongly related to the air temperature. According to Durance and Ormerod [18] and Davidson and Boet [19], rising temperatures affect macroinvertebrate communities in streams, salmonid fish, and other fish communities. Moreover, changes in climate have a strong impact on the hydrological cycle operation, and particularly affect precipitation. Specifically, high temperatures lead to increasing evapotranspiration rates, which could offset any net increase in precipitation. According to Arnell [20], mean summer flows regarding precipitation in rivers in southeast England might be 30% lower than in the period of 1961−90. Also, longer and more frequent periods of drought, combined with more intense rainfall and runoff activities, are known to act as stressors of in-stream ecology [21]. Those changes may be related with land use practices to drive catchment and subcatchment responses of freshwater ecosystems.

The Mediterranean region is an interesting example, because it is a sensitive and vulnerable area to these effects. Specifically, Mediterranean region lies in a transition zone between the arid climate of North Africa and the temperate and rainy climate of central Europe, and it is affected by interactions between mid-latitude and tropical processes. Because of these features, even relatively minor modifications of the general circulation, eg, shifts in the location of mid-latitude storm tracks or subtropical high pressure cells, can lead to substantial changes in the Mediterranean climate. This makes the area a potentially vulnerable region to climatic changes as induced, for example, by increasing concentrations of greenhouse gases [22]. Indeed, the Mediterranean region has shown large climate shifts in the past [23], and it has been identified as one of the most prominent "hot spots" in future climate change projections [24]. The Mediterranean climate is mild and wet during the winter, and hot and dry during the summer. Winter climate is mostly dominated by the westward movement of storms originating over the Atlantic and impinging upon the western European coasts. The winter Mediterranean climate, and most importantly precipitation, is thus affected by the North Atlantic Oscillation (NAO) over its western areas [25] and the East Atlantic (EA) and other patterns over its northern and eastern areas [26]. Over the Eastern Mediterranean, the El Nino Southern Oscillation (ENSO) has also been suggested to significantly affect winter rainfall variability along with spring and fall precipitation over Iberia and Northwestern Africa [27]. Mediterranean storms (in addition to Atlantics) can be produced internally to the region in correspondence of cyclogenetic areas such as the lee of the Alps, the Gulf of Lyon, and the Gulf of Genoa [28]. In summer, high pressure and descending motions dominate over the region, leading to dry conditions particularly over the southern Mediterranean region. The Mediterranean summer climate variability has been found to be connected with both the Asian and African monsoons, and with strong geopotential blocking anomalies over central Europe [26]. Moreover, the climate of the Mediterranean region is affected by local processes induced by the complex physiography of the region and the presence of a large body of water (the Mediterranean Sea). The Alpine chain, for example, is a robust factor in modifying traveling synoptic and mesoscale systems, and the Mediterranean Sea is an important source of moisture and energy for storms [28]. The coastline, the topography which is too complicated, and the vegetation cover of the region as well, are well known to modulate the regional climate signal at small spatial scales [29].

Summarizing, the water cycle is a delicate balance among precipitation, evaporation, and all the steps in between. According to the U.S. Environmental Protection Agency,[5] changes in the amount of rain falling during storms provide evidence that the water cycle is already being affected. Warmer winter temperatures seem to cause more precipitation to fall as rain rather than as snow. Furthermore, rising temperatures cause snow to begin melting earlier in the year. This alters the timing of streamflow in rivers whose sources are in mountainous areas. Generally, climate change affects the hydrological cycle and the distribution of water in so many ways; it makes freshwater environments more susceptible to toxic algal blooms and can lead to proliferation of harmful microbes and bacteria, it causes ocean acidification, and it has an impact on both water quality and quantity.

3.7 Water Pollution

Although water pollution is a common topic in many books and articles, there is not a widely accepted definition about it. There could be many reasons for this. The lack of a

---

[5] http://water.epa.gov/scitech/climatechange/Water-Impacts-of-Climate-Change.cfm.

unanimous definition could be explained by the fact that most people have an idea about what is meant by "water pollution." Another reason could be that a strict definition could be considered superfluous. Moreover, there are many definitions found in many official documents, which may differ in both the details and fundamental aspects. This could be attributed to the fact that these definitions have been used in different contexts and served different purposes [30].

According to Inglezakis and Poulopoulos [4], "water-quality deterioration can be attributed to water pollution or contamination. Water pollution is generally defined as any physical, chemical, or biological alteration in water quality that has a negative impact on living organisms."

Searching the term through the site of the European Environment Agency (EEA), the following definition provided by Wikipedia is given[6]: "Pollution is the introduction of substances or energy into the environment, resulting in deleterious effects of such a nature as to endanger human health, harm living resources and ecosystems, and impair or interfere with amenities and other legitimate uses of the environment".

According to the Water Framework Directive [3], "Pollution means the direct or indirect introduction, as a result of human activity, of substances or heat into the air, water or land which may be harmful to human health or the quality of aquatic ecosystems or terrestrial ecosystems directly depending on aquatic ecosystems, which result in damage to material property, or which impair or interfere with amenities and other legitimate uses of the environment."

In the stricter sense, pollution can be defined as the transfer of any substance to the environment. However, there is a tolerance limit for each pollutant, since zero-level pollution is economically and technically unpractical [4]. So, in these cases the term contamination is more appropriate. According to the Food and Agriculture Organization of the United Nations,[7] contamination is the presence of elevated concentrations of substances in the environment above the natural background level for the area and for the organism.

Contamination of water by physical and bacteriological agents, whether it is drinking water, ice water, or harbor water, may be detected by laboratory tests. Test results are usually expressed in parts per million (ppm) or parts per billion (ppb) for physical parameters, and bacterial counts per 100 mL for organisms. For both types of contaminants, maximum levels are usually stipulated and these levels may differ from country to country.

---

[6] http://www.eea.europa.eu/themes/water/wise-help-centre/glossary-definitions/pollution.
[7] http://www.fao.org/docrep/x5624e/x5624e04.htm#1.1.

There is a great variety in the classification of the types of water pollution in the relevant literature, depending on the subject of categorization. An important discrimination for study, technological solutions, or law purposes refers to the kind of the source. Specifically, the sources of water pollution are divided into

- *point sources*, such as chemical industries and human communities, and
- *nonpoint or diffuse sources*, such as agricultural activities and landfill leachates.

Almost all human activities have adverse impacts on water. Water quality is influenced by both direct point source and diffuse pollution, which come from urban and rural populations, industrial emissions, and farming. Diffuse pollution from farming, and point source pollution from sewage treatment and industrial discharge are principal sources. For agriculture, the key pollutants include nutrients, pesticides, sediments, and fecal microbes. Oxygen-consuming substances and hazardous chemicals are more associated with point source discharges. Point sources are mainly responsible for the pollution of surface waters (rivers, lakes, seas), whereas nonpoint sources mainly contribute to the pollution of groundwater resources. Moreover, releases from point sources can be treated by wastewater treatment plants, whereas nonpoint source releases can only be minimized.

So, depending on the type of source, we divide water pollution into *point source pollution* and *diffuse pollution*.

If the specific source is taken into account, one can describe water pollution as *industrial pollution* if it is related to industrial activities, *sewage pollution* if it is connected to sewage disposals from urban and rural areas, or *agricultural pollution* if it is about nutrients, pesticides, or other chemicals used in agricultural activities.

Although not so common, water pollution can be also termed by the extent it appears. So, the term *transboundary pollution* is also used to describe polluted waters traveling through rivers or oceans from one country or even continent to another.

Depending on the water receiver, water pollution can be characterized as

- *surface water pollution*, if it refers to the pollution of lakes, rivers, oceans, and any surface waters in general, and
- *groundwater pollution*, if it refers to the pollution of water that is held in underground rock structures known as aquifers.

The latter one is by far less obvious and in many cases, more difficult to treat.

Whatever the context and the terminology we may use, it is equally or even more important to classify the types of water pollutants, since their knowledge will also indicate the technological solution to be implemented. So, the various types of water pollutants can be classified into the following major categories [4].

Thermal pollution. The discharge of warm wastewaters into a surface receiver may lead to the increase in temperature of waters, which in turn will result in the decrease in the oxygen concentration in water. A lot of aquatic species are sensitive to oxygen concentration in waters, and such variations could cause the immediate elimination of the most sensitive aquatic life. Temperature changes may also cause changes in the reproductive periods of fishes, growth of parasites and diseases, or even thermal shock to the animals found in the thermal plume.

Organic pollutants. They are further categorized into oxygen-demanding wastes and persistent organic chemicals.

> *Oxygen-demanding wastes.* The release of biodegradable organic compounds into water bodies results in the decrease in the oxygen dissolved in water due to its consumption by the aquatic microorganisms that decay the organic pollutants. A minimum of 6 mg of oxygen per liter of water is essential to support aquatic life. A few species, like carp, can survive in low-oxygen waters. Each biodegradable waste is characterized by the biological oxygen demand (BOD), which is a measure of the amount of dissolved oxygen needed by aquatic microorganisms for the degradation of waste. They are found in water bodies mainly by industrial processes and agricultural activities.
> *Persistent organic chemicals (POPs).* This term is used for synthetic organic compounds that show great resistance and high life spans in the environment, thus constituting a long-term danger to life. Dioxins, polychlorinated biphenyls (PCBs), and pesticides (DDT and others) are man-made compounds that remain intact for months in the environment. Consequently, people and animals at the top of the food chain eventually consume food containing these compounds. DDT, a popular compound that helped in the elimination of malaria, was proved to have many adverse effects on natural life.
> *Oils.* These are complex mixtures of hydrocarbons that are generally degradable under bacterial action, with the biodegradation rate being dependent on the oil. Oil spills, leaks from oil pipes, and wastewaters from industrial production and refineries are the main sources of oil into water.

Nutrients and agriculture runoff. The excess presence of plant nutrients, like nitrates and phosphorous, in the environment through agricultural activities directly, or indirectly through agriculture runoff, may cause the problem of eutrophication. *Eutrophication* is the rapid depletion of dissolved oxygen in a body of water because of an increase in biological productivity. Moreover, high nitrogen levels in the water supply cause a potential health risk, especially to infants less than 6 months. This is when the methemoglobin results in a decrease in the oxygen-carrying capacity of the blood (blue baby disease) as nitrate ions in the blood readily oxidize ferrous ions in the hemoglobin.

Inorganic pollutants. Metals, nonmetals, and acids/bases released by human activities severely deteriorate water quality, since they are toxic even at concentrations of parts per million. Particularly, heavy metals are extremely dangerous to human health and aquatic life, since they are accumulated in the environment and the food chain.

Pathogens. Pathogenic microorganisms, like viruses, bacteria, and protozoans, are released into water bodies through the discharge of sewage wastes, and wastewaters from animal industries like slaughterhouses. They are responsible for waterborne diseases, such as cholera, typhoid, dysentery, polio, and infectious hepatitis in humans.

Suspended solids and sediments. Another type of pollution involves the disruption of sediments (fine-grained powders) that flow from rivers into the sea. They comprise of silt, sand, and minerals eroded from land. These appear in the water through the surface runoff during rainy season and through municipal sewers. This can lead to the siltation and reduces storage capacities of reservoirs. Moreover, during construction work, soil, rock, and other fine powders sometimes enter nearby rivers in large quantities, causing it to become turbid. The extra sediment can block the gills of fish, effectively suffocating them.

Radioactive pollutants. They may originate from the mining and processing of ores; use of radioactive isotopes in research, agriculture, medical, and industrial activities; radioactive discharges from nuclear power plants and nuclear reactors; and finally the testing and/or use of nuclear weapons. These isotopes are toxic to all life forms; they accumulate in the bones and teeth and can cause serious disorders.

It is useful to recall that while road transport and combustion installations, mainly of the energy sector, are the most important sources of air pollutants, agriculture and metal industry activities constitute the major polluters for water bodies [4].

3.8 International Water Policy and Legislation

3.8.1 An Overview

Until recently, water-related laws around the world were people-focused, and water was seen as a commodity intended for human consumption and related domestic needs, as well as for economic exploitation [31]. However, the rapid growth of the world population combined with the climate change, the pollution of surface and underground waters, and the salinization of coastal waters, affected the availability of, and the demand for, freshwater. United Nations' Agenda 21, reviewed during Rio+20 summit held in 1992, called for improved integrated water resource management and as a result, gradually water-related law internationally evolved beyond the anthropocentric approach and expanded to embrace broader environmental protection issues.

The trend of greening water laws has influenced national legislation, and an increasing number of countries incorporate the environmental protection aspect when drafting or reviewing their water-related legislation [31]. The extent to which water management is regulated varies widely among different countries, ranging from no water pollution control regulations to a comprehensive policy framework and regulations [32]. The last 20 years several countries passed innovative legislation or amended the existing one in order to fundamentally rework their approaches to water management. Common principles include decentralized water decision making, increased stakeholder participation, and user-pay and polluter-pay principles.

Over the last four decades the European Union and its Member States have successively implemented a number of directives along with the relevant national implementation measures to ensure sustainable water management. Indeed, the water sector was one of the first environmental sectors that was reshaped in the 1970s and 1980s, and a large number of legislative acts were issued. The legislation of this period was characterized by a primarily regulatory approach that resulted in a highly fragmented and rather ineffective water legislation framework (Fig. 3.5). These legislative acts either lay down environmental quality standards (EQSs) for specific types of water, like the Surface Water, Bathing Water and Drinking Water Directives, or establish emission limit values (ELVs) for specific water uses, like the Groundwater Directive [33,34]. In the 1990s, under the pressure of the increasing eutrophication of sea and freshwaters, two new legal instruments were adopted, focusing on the sources of pollution: urban wastewater and nitrates coming from agricultural activities. Urban Wastewater Treatment Directive, UWWTD (91/271/EEC), became obligatory even in the smallest settlement, and Nitrate Directive (91/676/EEC) introduced measures for protecting water bodies from nitrates pollution coming from agricultural sources (Fig. 3.6). The UWWTD sets the requirements for the development of sewerage systems and wastewater treatment plants as well as quality targets for the treated effluent disposed to water recipients. The Integrated Pollution Prevention and Control Directive, IPPC (96/61/EC), enhanced emissions control while Seveso II Directive (96/82/EEC) contains important aspects of water protection. Water Framework Directive (2000/60/EC) was adopted in 2000, designed in the view of integrated water management, and its main target is to achieve a

Figure 3.5
Evolution of water legislation in the European Union.

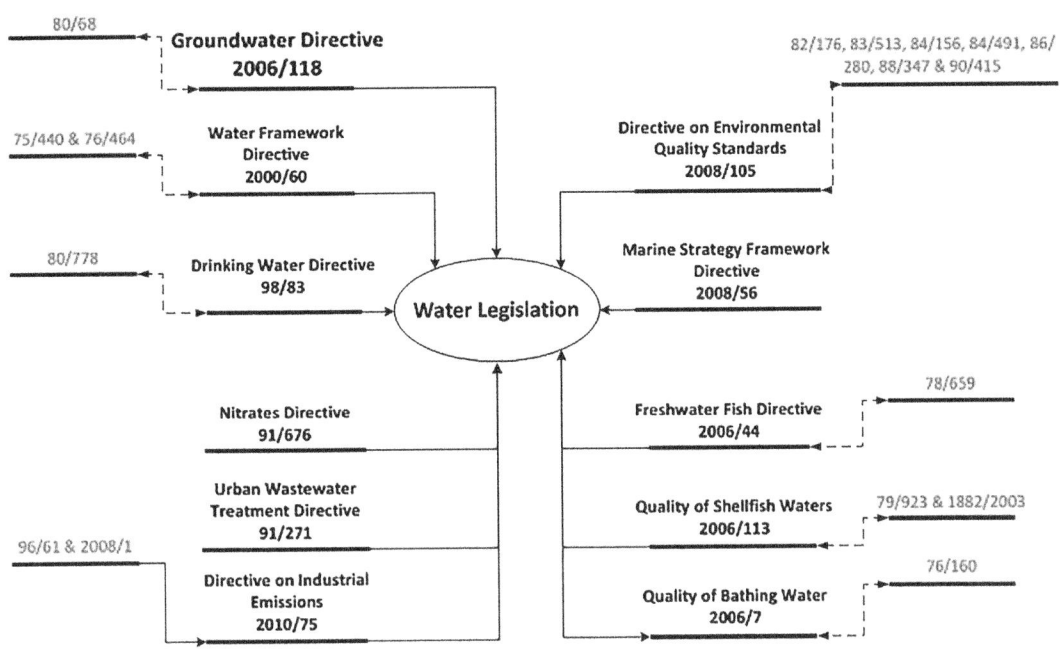

Figure 3.6

Overview of EU water legislation (Repealed legislative acts are those connected with *dotted lines* to the legislation currently in force.).

"good status" of all aquatic systems, surface waters (rivers and lakes), groundwater, and coastal waters. Furthermore, the WFD tries to reconcile the conflicting approaches of ELVs and EQSs [33]. Finally, Groundwater Directive entered into force in 2007 aiming to prevent the pollution of groundwater from agricultural residues such as pesticides and other harmful chemicals [34].

In the United States, the Clean Water Act (CWA) establishes the basic structure for regulating discharges of pollutants into waters and quality standards for surface waters. The basis of the CWA was enacted in 1948 and was called the Federal Water Pollution Control Act, which was significantly reorganized and expanded in 1972. Prior to this act, no national policy existed in the United States calling for the protection of the quality of water resources, and emphasis up to that time was focused primarily on regulating the quality of drinking water [35]. Under the CWA, pollution control programs are implemented by setting wastewater standards for industry. Furthermore, US legislation sets water quality standards for all contaminants in surface waters [36]. In addition to the CWA, the Safe Drinking Water Act (SDWA), passed in 1974, establishes stringent standards for drinking water quality [35]. Lately, in 2008, the Environmental Protection Agency and the U.S. Army Corps of Engineers issued revised regulations governing compensatory mitigation for authorized impacts to wetlands, streams, and other waters under Section 404 of the Clean Water Act. It is interesting to note that while drinking water quality in the European Union and the United States requires legal compliance with specific standards, in many other countries these standards are expressed as guidelines or targets rather than requirements.

China has also modernized its water legislation and has enacted several laws and regulations directly relevant to water resource management including water quality and emissions standards. Water Law of the People's Republic of China was originally introduced in 1988 and was significantly amended in 2002. Its purpose is to provide for the development, utilization, and protection of water resources, as well as the prevention and control of water hazards. As it is the case in similar modern acts worldwide, the overall planning has to be made on the basis of river basins or regions [37]. The Law on Prevention and Control of Water Pollution was previously adopted in 1984 and revised in 1996, and most recently in 2008, following the severe water pollution across China. This law aims at the prevention and control of pollution of surface water bodies and groundwater. Among several new concepts, the new law upgrades the importance of public participation in environmental protection [38]. Finally, China adopted its own drinking water standard enacted by Ministry of Environmental Protection in 2002.

Climate change and water management are recognized as two of the most important public policy issues that Australia faces [39]. Historically, most public focus on environmental water management has occurred in the Murray-Darling Basin, largely prompted by the need to address the overallocation of water associated with irrigation development [40]. Following the international trends, Australia commenced in 2007 a reform of its water management legislation, passing the Commonwealth Water Act, which has been amended several times in the period 2008–13. Australia's water reform has been closely tied to increasing the efficiency of water use, largely through a water-rights market, and has created a new federal repository of water monitoring and measurement information [32,41]. Furthermore, the National Health and Medical Research Council has developed the Australian Drinking Water Guidelines, whereas there are several other state-level acts, such as the Water and Sewerage Act 2000, which makes provision in relation with the supply of plumbing or sanitary drainage services. Finally, Australia has recently become the second country after South Africa to regulate the plantation forest water use [42].

Since 1992, Russia has enacted an integrated environmental legislation that in many cases meets or exceeds commonly accepted international standards, but at the same time the country faces challenges in the enforcement of this legislation. In 2006, Russia rewrote its water code (Russian Federation Water Code No. 174-Ф3) to focus on integrated regional water management [41]. The code's founding principles are that protection of water bodies takes priority over use, and that utilization is prioritized toward drinking and other domestic purposes. Some of the code's innovations include its river basin approach, the introduction of integrated water basin management schemes, as well as the civil society involvement in decision making [32,43].

It is true that water issues do not only concern local, regional, and national authorities; water traverses borders and thus becomes an international matter also. Worldwide, there are over 260 watercourses and more than 270 groundwater basins shared by two or more sovereign States. The governance of international freshwater is a complicated task that involves an enormous number of stakeholders and instruments [44]. Approximately 400 international regional and basin agreements exist, most of which deal with surface water and groundwater [44]. Up to date, the only global water course agreement in force is the United Nations Watercourses Convention finalized in 1997, and it entered into force in 2014. Its purpose as defined in Article 1 is to protect, preserve, and manage international watercourses and their waters for nonnavigational purposes. The convention serves as a legal umbrella that can "supplement, facilitate, and sustain transboundary water cooperation at all levels." [44]. Despite the fact that no international guidelines exist for ecosystem water quality, there are several international water quality guidelines [32]:

*   World Health Organization (WHO) guidelines for drinking water contaminant levels;
*   WHO guidelines for the safe use of wastewater, excreta, and graywater in agriculture and aquaculture;
*   Food and Agriculture Organization (FAO) wastewater quality guidelines for agricultural use;
*   FAO guidelines for water quality for livestock and poultry;
*   International Standards Organization (ISO). Out of a total of more than 19,000 International Standards, more than 550 are related to water, 260 of which on water quality. For instance, standard 13.060 on Water quality including toxicity, biodegradability, protection against pollution, related installations, and equipment.

3.8.2 Environmental Quality Standards and Emission Limit Values

Environmental Quality Standard (EQS) is a concept for which there is no uniform definition in the legislative systems around the world. The term EQS is mostly used in Europe, while in the United States and Canada the terms Ambient Water Quality Criteria and Water Quality Guidelines, respectively, are used [45]. In any case, when set in legislation, they are legally binding limits and are translated into concentrations of individual substances. EQS is an environmental medium quality standard for specific substances, which sets concentration thresholds below which, no adverse impact on the medium occurs, and which takes explicit account of available dilution at different discharge locations [45]. EQSs concern the environmental medium without any reference to the sources of emissions and should take into account the specific local environmental conditions. Another term used is emission limit value (ELV) referring to the "occurrence of pollutants in the environment regardless the source and the manner of distribution" [46]. The quality objectives may be substantiated at various levels. For example, there are "desirable levels" (long-term objectives), "maximum tolerable levels" (short-term objectives), and "intervention levels" (alarm function). When intervention levels are exceeded, remediation actions should be implemented. The setting of standards follows a multistep process, which incorporates priority setting, risk assessment, risk management, and public consultation. The process for setting standards also recognizes that some contaminants can move easily through the natural environment, persist for long periods of time, and/or accumulate in the food chain (bioaccumulation).

The term "emission" is generally defined as the direct or indirect release of substances, vibrations, heat or noise, from point or diffuse sources, into the air, water, or land. The Threshold Limit Values (TLV) are developed as guidelines to assist in the control of health hazards and are not developed for use as legal standards. On the other hand, an emission limit value (ELV) is usually expressed as legally binding minimum standard. It may be expressed either as concentration of a substance in the effluent or as a load (mass) discharged per unit of production, and may not be exceeded during one or more periods of time. ELVs are normally applied at the point where the emissions leave the installation, any dilution being disregarded when determining them. Furthermore, ELVs should take into account the local environmental conditions and thus, the relevant EQSs. ELVs can be set for the specific industrial sector, the specific environmental medium, or both.

EQSs values are lower than ELVs, and ELV/EQS ratio varies between 2 and 1000 [47]. This ratio describes the necessary dilution that must be attained in the physical environment in order to avoid environmental damage. In this approach, ELV is considered to protect against acute effects while EQS is supposed to prevent long-term impacts [47]. Another way to prescribe ELVs is through the concept of the Best Available Techniques (BATs) [45]. BAT means the best available technologies that can protect the environment and are economically justified.

Summarizing, EQSs concern the environmental medium (ambient standard) while ELVs the discharge at the point where the emissions leave the installation (effluent standard). Thus, EQS is biologically rather than technologically based [45]. Furthermore, ELV takes into account the overall load discharged from pollution sources, but it does not consider the response of the recipient. On the contrary, EQS does not permit restriction on the overall load discharged from the pollution sources, but it is an explicit recognition of the available dilution. Thus, the combined approach is the optimum choice, as it is done, for example, in the Water Framework Directive in the European Union.

3.9 Wastewater as a Resource

3.9.1 How a Problem Becomes Part of the Solution

Population and economic growth, dependency on high-water-demand agriculture, fast urbanization, and water pollution pose serious stress on the limited clean water availability, and the need for alternative water sources is imminent. The overall conclusion of a large number of studies is that major part of the world population, up to two-thirds mostly found in the developing world, will be affected by water scarcity over the next decades [31,48,49]. Even in Europe, a region of plenty of water resources compared to other parts of the world, it has been estimated that approximately half of the countries, representing almost 70% of the total population, are facing water stress issues [50]. However, it is not always easy to determine whether water is scarce in the physical sense (supply problem) or it is available but not managed effectively (demand problem) [48]. Many countries in the arid areas of the world, particularly Central and West Asia and North Africa, are definitely water scarce in the physical sense, but in many other regions the problem comes from either inadequate water management (obsolete or absence of water supply systems) or population poverty [48].

In terms of freshwater uses, agricultural water consumption makes up 70%, industrial water 22%, and domestic water the rest 8% of global water consumption (Fig. 3.7) [51]. Thus, 30% of water ends up as either industrial or municipal wastewater, the latter

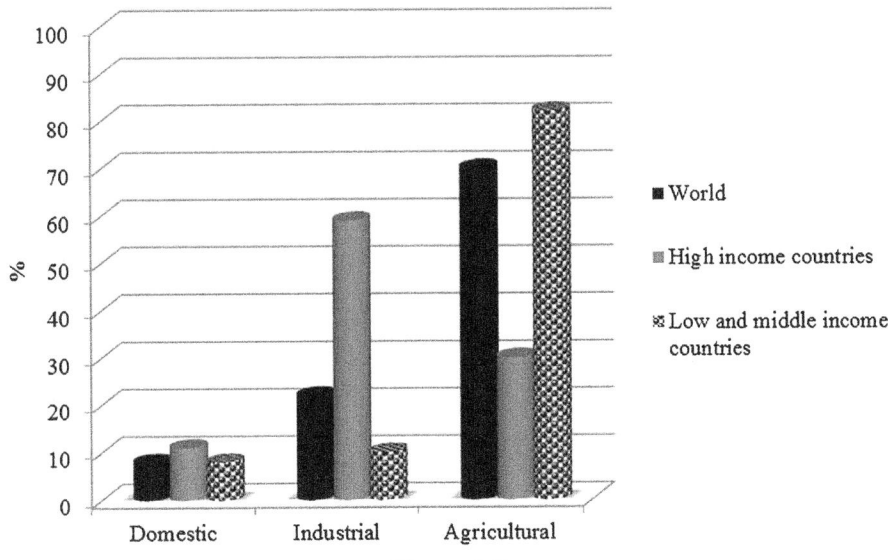

Figure 3.7
Competing water uses for main income groups of countries [51].

containing some quantities of treated or untreated industrial effluents. Wastewater reuse (direct use without treatment) and/or recycling/reclamation (use after treatment) offer an obvious solution and is integral to sustainable water management. Reuse and recycled wastewater can offer an alternative source of water that, depending on the water quality requirements, can be used for diverse purposes, as agricultural irrigation and industrial makeup water. Water recycling projects for nonpotable end uses have become a common practice, with more than 3300 projects registered worldwide.

Industry is increasingly implementing reuse and recycling of process (industrial) wastewater (IWW) for several purposes due to growing economic pressure and stricter regulatory requirements. For instance, in Japan industrial water recovery rates reached about 80% by the year 2000 [52]. Process water is produced by industrial operations, and it can be reused for purposes such as irrigation, washing, production line needs, fire protection, etc. For example, on a global scale two-thirds of industrial water is used for cooling. In power plants and refineries, cooling water represents the 90% of water used and reclaimed water is widely used for this purpose [52]. In Europe, industrial use of water accounts for about 32% of total water abstractions, and excluding cooling water, the main industrial water users are the chemical industry, the steel and metallurgy industries, and the pulp and paper industry [53].

In general, water quality requirements for industrial applications are industry- and process-specific. There are applications where process water can be directly reused and others that require high-quality water. Process water can be treated on-site or in decentralized treatment facilities and, depending on the quality, can be either used for industrial or domestic applications. Furthermore, process water can be reused/recycled within a facility itself, or between several facilities through industrial symbiosis, as for example, the utilization of high COD/BOD wastewater for biogas production through anaerobic digestion. Finally, treated and sometimes untreated industrial wastewater is directed to centralized municipal wastewater treatment plants for further processing.

Several countries have established goals for reclamation, and accordingly they implemented advanced municipal wastewater (MWW) treatment methods. According to AQUASTAT,[8] the United states reuses $\sim$7% of its treated wastewater, Australia 21%, Singapore 38%, Saudi Arabia 94%, and Israel 100% [54]. An interesting fact is that although in many developed countries most of the wastewater generated is treated, only a small portion is reused (Fig. 3.8). Thus, there is a great potential in MWW reuse, which could directly contribute to sustainable management of available water resources.

---

[8] AQUASTAT is FAO's global water information system, developed by the Land and Water Division. It is the most quoted source on global water statistics. http://www.fao.org/nr/water/aquastat/main/index.stm.

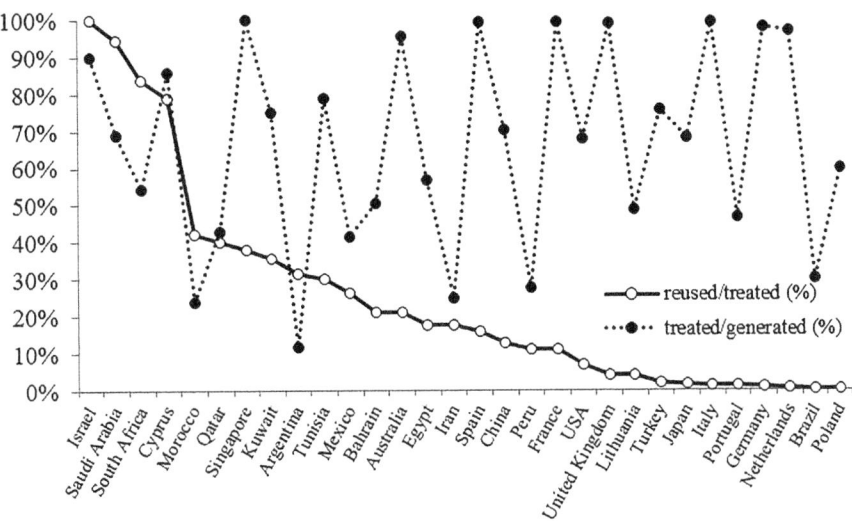

Figure 3.8
Treated (as % of generated) and reused (as % of treated) municipal wastewater.
Source: AQUASTAT.

With a few exceptions, treated wastewater applications are restricted to nonpotable uses, or at most to indirect potable uses, where recycled water is used for recharging groundwater aquifers and augmenting surface water reservoirs (EPA) [55]. Among several potential applications for treated wastewater, agricultural and industrial ones are the most common [54]. For example, cooling tower applications of reclaimed municipal wastewater represents one of the most successful water reuse applications [52]. But with about 70% of the world's freshwater currently used for irrigation, agriculture remains the largest user of reclaimed wastewater, and estimates indicate that about 20 million hectares of agricultural land are irrigated with treated and untreated wastewater worldwide (Figs. 3.9 and 3.10) [49,56−58]. Naturally, the use of reclaimed wastewater in the humid regions of developed countries, such as the eastern part of North America, northern Europe, and Japan, is not substantial, while in the arid and semiarid areas of developed countries, such as western North America, Australia, and southern Europe, the primary use of reclaimed wastewater is irrigation [49]. According to data published in 2006 for southern Europe, reclaimed water was reused predominantly for agricultural irrigation and for urban or environmental use, while in northern Europe the uses were mainly for urban or environmental applications or industrial use [50]. Furthermore, there is an escalating interest for artificial groundwater recharge with reclaimed wastewater to hold back saline intrusion in coastal aquifers in Europe.

Another facet of this subject is the water−energy nexus [56,59]. Water and energy are mutually dependent—energy production requires large volumes of water, and water

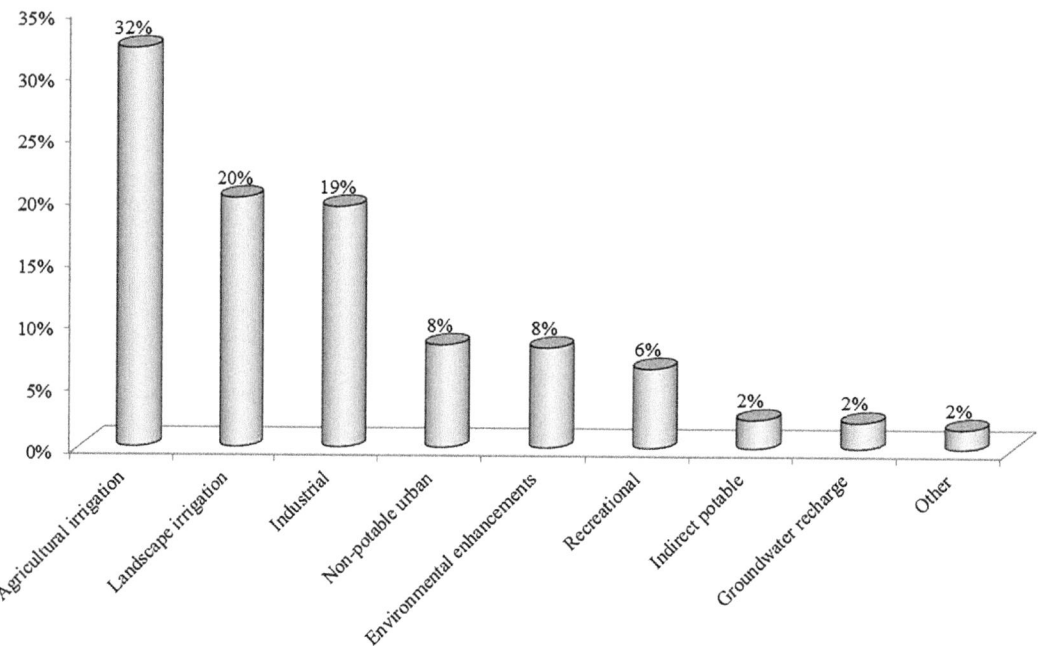

Figure 3.9
Global water reuse after advanced (tertiary) treatment [57].

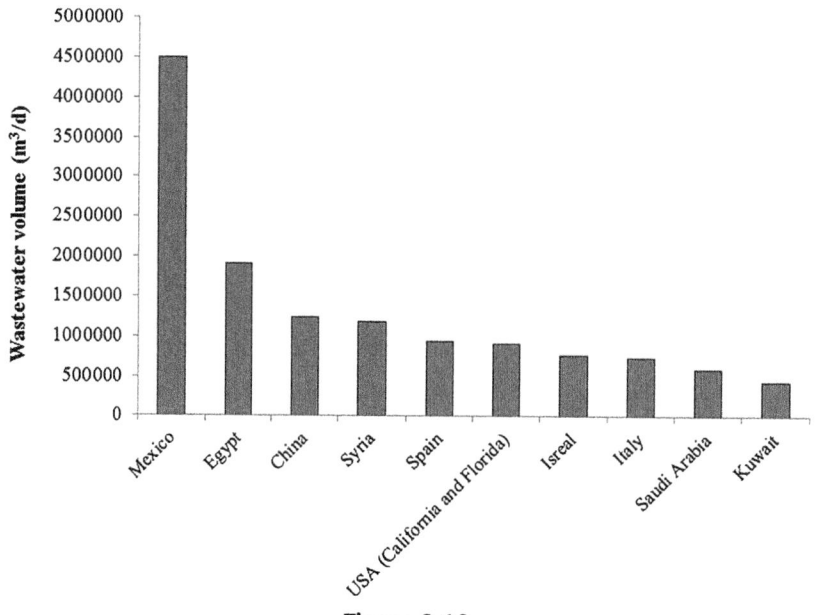

Figure 3.10
The 10 countries with the largest volume of wastewater used for irrigation [58].

abstraction, treatment, and distribution require large amounts of energy. Thus, sustainable management of either resource requires consideration of the other. Water reuse reduces energy use by eliminating additional wastewater treatment and in turn, reclaimed water can be used in energy- and water-intensive industries, as for example, in power plants and refineries.

Conventional biological wastewater treatment is designed to effectively remove organic matter, suspended solids and in some cases nutrients (nitrogen and phosphorous species), and pathogens. Advanced tertiary treatment technologies, such as membrane technologies, can be employed for further removal of dissolved solids and micropollutants. However, the challenges are growing steadily and new chemicals pose serious threats, as for instance, pharmaceuticals and hormones [60]. The number of anthropogenic chemicals is increasing fast, and they are posing serious challenges to wastewater treatment plants. Therefore, it is clear that new strategies and advanced technologies are needed to treat municipal wastewaters. Thus, while recycling and reuse of wastewater is a necessity, its utilization clearly requires adequate risk assessment. A good example is the use of reclaimed wastewater reuse for irrigation, one of the main uses of reclaimed wastewater in some parts of the world, where the risks are associated with food contamination, soil salinization, groundwater quality degradation, and others [60].

The major concerns for users of reclaimed wastewater are related to its quantity and quality [61]. Quantity should be assured otherwise potential users will probably not participate in relevant projects. Quality issues can be addressed by a range of treatment options, which are available such that any level of water quality can be achieved depending upon the use of the reclaimed water (Fig. 3.11) [56]. However, technology alone is not enough to ensure the safe use of reclaimed water and the setting of quality standards, as ELVs and EQSs, along with formulation of regulations and implementation guidelines, are of paramount importance. Moreover, legislation is the basis for permitting water reclamation facilities.

3.9.2 *Wastewater Reclamation in Legislation*

Wastewater treatment in developed countries is enforced by stringent effluent quality regulations (ELVs and EQSs) and increased public awareness on environmental issues. Furthermore, technical solutions are readily available and are further developed to deal with both municipal and industrial wastewaters, and produce effluents suitable for a range of applications. However, the same is not true for many developing countries, where treatment facilities are unavailable or inadequate, and much of the wastewater used, for example, by farmers, is not treated [49]. The following analysis concerns municipal wastewater reuse, since water quality requirements for industrial applications are process-specific and consequently no general guidelines exist.

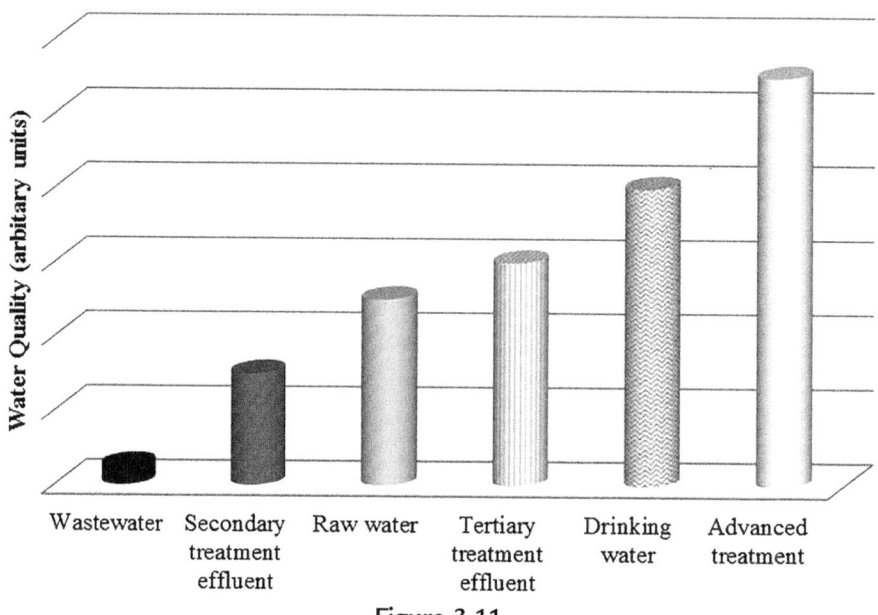

Figure 3.11
Treatment technologies and water quality [57].

In the European Union, it is expected that the implementation of the Water Framework Directive will favor the implementation of municipal wastewater reclamation and reuse at a larger scale, while Industrial Emissions Directive forces the industries to adopt strategies for wastewater management. Since 1991, the Urban Wastewater Treatment Directive already urged the Member States to reuse treated water "whenever appropriate" although a legal definition of the term "appropriateness" is still pending [50]. In comparison to the Urban Wastewater Treatment Directive, the Water Framework Directive introduced a holistic approach supporting reclamation of the municipal wastewater. However, the Water Framework Directive sets forth the principles to achieve sustainable water governance, but not the means [50]. Moreover, there are no guidelines or widely used standards addressing the issue of treated wastewater reuse in EU level [62], although several countries (eg, Greece, Italy, Cyprus) have developed national standards for treated wastewater reuse. The European Commission has recently initiated discussions in all levels on the possibility of forming a regulation establishing common standards for water reuse [62].

Internationally, there are a number of guidelines and fewer cases of laws on municipal wastewater reuse. Laws that are specific to water reuse in irrigation are in force in Tunisia (1975), Italy (1977), Israel (1978), and California (1978) [63]. In California, wastewater is collected and conveyed to one of the Sanitation Districts' 10 water reclamation plants (WRPs) for treatment through a three-stage (tertiary) process. According to the definitions given in the Regulations Related to Recycled Water, "Conventional treatment means a treatment chain that utilizes a sedimentation unit process between the coagulation and filtration processes and produces an effluent that meets the definition for disinfected tertiary recycled water" [64].

Tertiary recycled water can be used for a broad range of reuse applications, except for direct drinking water and the manufacturing of food and drink. There are several regulations either approved or under preparation considering the final use of the recycled water, such as Title 17 and Title 22 of the Regulations Related to Recycled Water. Chapter 3 of Title 22 concerns the water recycling criteria. Article 3 sets criteria for recycled water uses in irrigation, impoundments, cooling, and other purposes. Furthermore, Articles 5.1 and 5.2 concern the indirect potable reuse (IPR): groundwater replenishment—surface and subsurface application. California Recycled Water Regulations and Guidance were updated in 2015 [64].

Regarding guidelines, there are already several developed around the world. In 2006, WHO published a third edition of its Guidelines for the safe use of wastewater, excreta, and greywater in agriculture and aquaculture. These Guidelines propose a flexible approach of risk assessment and risk management, backed up by strict monitoring measures. In this context "wastewater" refers to domestic sewage and municipal wastewaters that do not contain substantial quantities of industrial effluent. WHO Guidelines are presented in four separate volumes:

- Guidelines for the safe use of wastewater, excreta, and greywater. Volume 1: Policy and regulatory aspects;
- Guidelines for the safe use of wastewater, excreta and greywater. Volume 2: Wastewater use in agriculture;
- Guidelines for the safe use of wastewater, excreta and greywater. Volume 3: Wastewater and excreta use in aquaculture;
- Guidelines for the safe use of wastewater, excreta and greywater. Volume 4: Excreta and greywater use in agriculture.

U.S. EPA Guidelines for Water Reuse is a comprehensive up-to-date document in support of regulations and guidelines developed in the United States [56]. The types of reuse applications are presented in Chapter 3 and are as follows: urban, industrial, agricultural, environmental, groundwater recharge (nonpotable), and potable. Treatment requirements for water reuse are presented in "Chapter on Energy and the Environment" and include disinfection and advanced treatment methods. Table 3.3 shows the types of treatment processes and suggested uses at each level of treatment [54,56].

In Australia the first guidelines issued in 1999 were for sewerage systems. Following that, in the period 2004—09 the Australian Guidelines for Water Recycling were published.

Table 3.3: Types of Wastewater Treatment Processes and Uses [54,56].

Use	Indicative Treatment Process
• No uses recommended at this level • Surface irrigation of orchards and vineyards • Nonfood crop irrigation • Restricted landscape impoundments • Groundwater recharge of nonpotable aquifer • Wetlands, wildlife habitat, stream augmentation • Industrial cooling processes	Primary treatment: sedimentation Secondary treatment: biological oxidation, disinfection
• Landscape and golf course irrigation • Toilet flushing • Vehicle washing • Food crop irrigation • Unrestricted recreational impoundment • Indirect potable reuse: groundwater recharge of potable aquifer and surface water reservoir augmentation	Tertiary/advanced treatment: • chemical coagulation, filtration (including membrane technologies and activated carbon adsorption), disinfection (chlorination, UV, ozonation, pasteurization), advanced oxidation (eg, UV/H_2O_2) • Advanced treatment, environmental buffer (surface water supply reservoirs or groundwater aquifers), and potable water treatment
• Direct potable reuse	Advanced treatment and potable water treatment

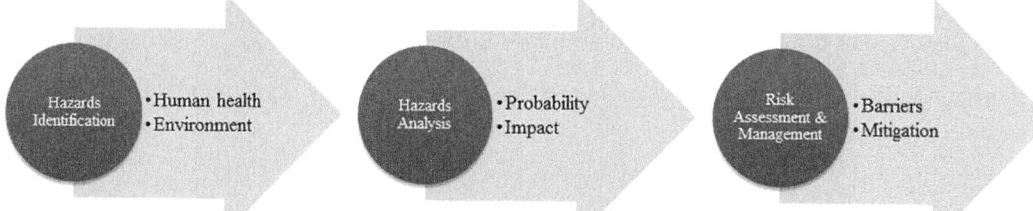

Figure 3.12
Typical risk assessment procedure.

These guidelines follow a risk management approach [65]. Every recycled water system should have a risk management plan based on the framework shown in Fig. 3.12. Treatment processes for designated uses of recycled water from treated sewage according to Australian Guidelines are presented in Table 3.4.

3.10 Wastewater Treatment Technologies

3.10.1 Historical Background and Overview

Wastewater treatment is the process of removing contaminants from wastewater; it includes mechanical, physical, chemical, and biological processes, and the objective is to produce environmentally safe treated effluent.

Table 3.4: Treatment Processes for Designated Uses of Recycled Water From Treated Sewage According to Australian Guidelines [65].

Use	Indicative Treatment Process
Dual reticulation, toilet flushing, washing machines, garden use Dual reticulation—outdoor use only or indoor use only Municipal use—open spaces, sports grounds, golf courses, dust suppression, etc. or unrestricted access and application	Advanced treatment required, such as: • secondary, coagulation, filtration, and disinfection • secondary, membrane filtration, UV light
Municipal use, with restricted access and application	Secondary treatment with disinfection
Municipal use, with enhanced restrictions on access and application	• Secondary treatment with >25 days lagoon detention or primary treatment with >50 days lagoon detention • Secondary treatment
Landscape irrigation—trees, shrubs, public gardens, etc.	Secondary treatment or primary treatment with lagoon detention
Commercial food crops consumed raw or unprocessed	Advanced treatment to achieve total pathogen removal required (eg, secondary, filtration, and disinfection)
Commercial food crops with limited or no ground contact and eaten raw (eg, tomatoes, capsicums) and crops with ground contact with skins removed before consumption (eg, watermelons)	Secondary treatment with >25 days lagoon detention and disinfection
Commercial food crops; aboveground crops with subsurface irrigation and crops with no ground contact and skins removed before consumption (eg, citrus, nuts)	Secondary treatment with disinfection
Commercial food crops with no ground contact and heavily processed (eg, grapes for wine production, cereals) and crops cooked/processed before consumption (eg, potatoes, beetroot)	Secondary treatment or primary treatment with lagoon detention
Nonfood crops—trees, turf, woodlots, flowers	Secondary treatment or primary treatment with lagoon detention

Industrialization and urbanization in the 19th century in combination with poor sanitation led to severe outbreaks of cholera causing heavy loss of life in large European cities. A major reason was the contamination of drinking water resources with pathogens originating from untreated municipal (urban) wastewater. Thus, municipal wastewaters drew the attention, while industrial wastewaters had to wait for several decades. From the first comprehensive sewer network in Europe in Hamburg in 1848, it took 68 years to develop the activated sludge treatment in 1916, and since then, 74 additional years to the development of advanced wastewater treatment technologies as membranes in the 1990s (Table 3.5). The technologies developed for municipal wastewaters are also used for the

Table 3.5: Timeline for Modern Wastewater Treatment [52].

Year	Development
1740	Chemical treatment of sewage discharges in Paris using lime as the precipitant.
1853	First comprehensive sewerage system completed in Hamburg, Germany.
1850—1910	Many patents applied for in the United Kingdom and the United States for chemical treatment of sewage.
1868—70	Frankland's tests on filtration of sewage through soil and gravel.
1890	First true biological filter at Lawrence Experimental Station, Massachusetts State Board of Health, the United States.
1906	Imhoff tank designed in Germany.
1916	First full-scale activated sludge plant at Worcester. First full-scale AS plant in the United States at Houston, Texas.
1936	Denitrification used in Sheffield.
1964	Development of basis for consistent nitrification by Downing, Painter and Knowles, WPRL, Stevenage, the United Kingdom.
1972	Biological phosphorous removal described by Barnard in South Africa.
1990s	Membrane biological reactors developed in Japan.

treatment of industrial wastewaters. Similar to municipal wastewaters, industrial wastewaters treatment has been evolved following a series of phases, from no treatment and direct discharge to reuse and recycling.

So, the development of wastewater treatment was driven by these outbreaks of waterborne diseases (Fig. 3.13). Since then, the evolution of wastewater treatment was driven by the need for environmental (water quality) protection; in particular, nutrient removal in the 1960s and 1970s after recognizing wastewater as the major cause for the eutrophication of surface waters, and removal of persistent organic pollutants such as PCBs, PAHs, and heavy metals until the beginning of the 1990s [52,66]. This evolution was supported by the emergence and development of the environmental legislation in the 1970s, which—apart from municipal—also prompted industrial wastewater treatment. Since the beginning of 2000, the attention was drawn to the so-called contaminants of emerging concern (CEC), such as pharmaceuticals, personal care products, and perfluorinated compounds. These chemicals were not previously detected in municipal water or are being currently detected at levels that may be significantly different than expected [66]. The situation is similar in industry where new raw chemicals are used or produced, and industrial wastewaters now contain fractions of these chemicals.

The wastewater treatment can be conducted in centralized or decentralized systems. Following treatment, the effluents are usually discharged to surface waters. Industrial wastewaters are usually treated on-site, although limited amounts are also sent to

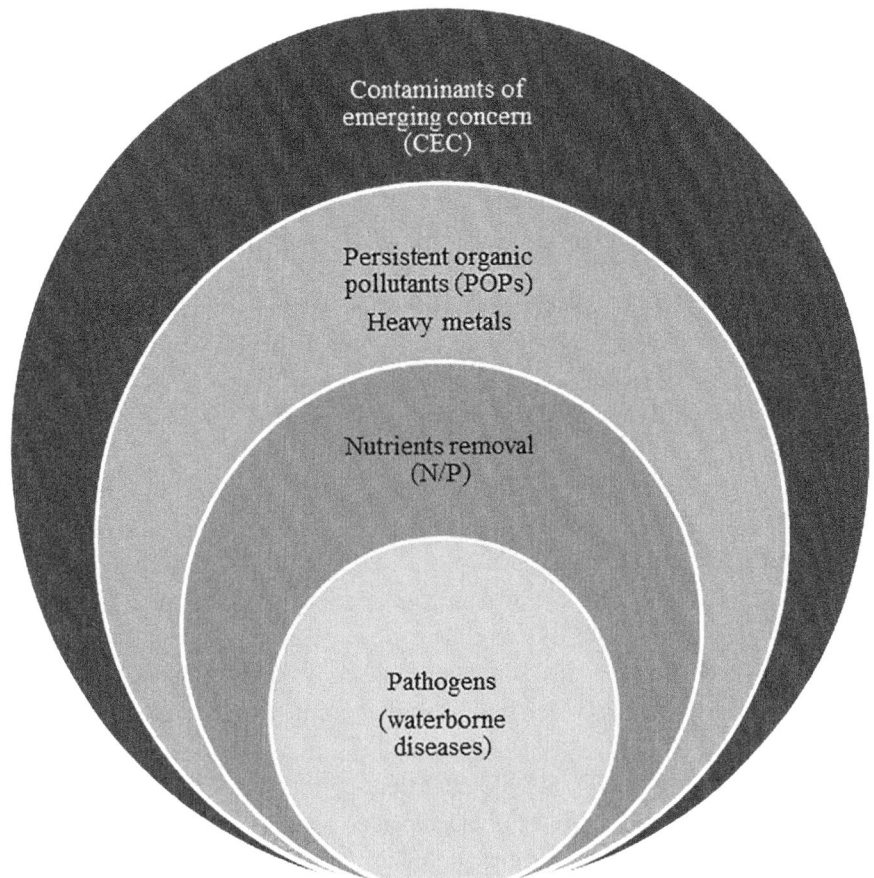

Figure 3.13
Drivers for wastewater treatment plants development and upgrade.

centralized municipal systems [32]. The typical wastewater treatment plant is divided into the following units (Fig. 3.14) [67–71]:

- *Preliminary treatment*—removal of large solids (rags, sticks, floatables, grease) by screening and removal of grit (sand, gravel, cinders, etc.). The processes employed are purely mechanical.
- *Primary treatment*—removal of a portion of suspended solids and organic matter by settling. While preliminary treatment is purely mechanical, primary treatment may employ physicochemical methods as coagulation/flocculation in order to enhance settling. Coagulation is a process of destabilization of the charge of colloidal particles in wastewater so that to build up to larger particles, which are easily settled.

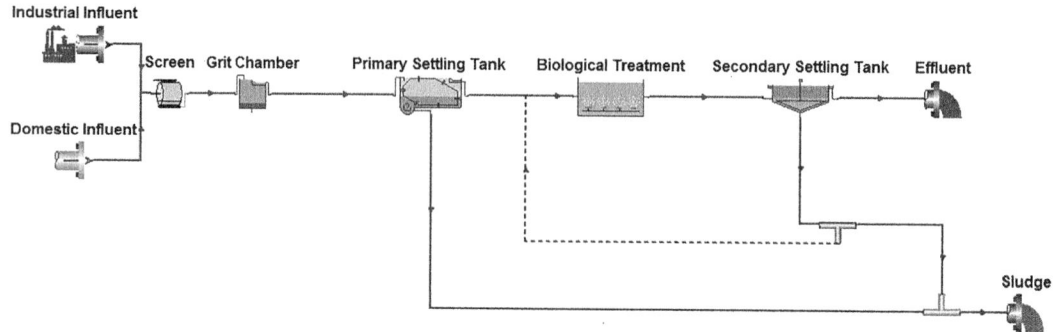

Figure 3.14

Unit operations in a typical wastewater treatment plant [67,68,70]. *Treatment process diagram generated using PetWin version 4.1, EnviroSim Associates Ltd.*

Flocculation produces larger particles from small colloidal particles, which then could be easily removed by gravitational settlement or filtration. More information on coagulation/flocculation process is provided in Section 3.11. Flotation is a mechanical treatment unit that removes solid or liquid particles from wastewaters with an aid of air. Air bubbles attach to the particles that need to be removed and under the effect of buoyant forces rise to the surface. Then, skimmers remove the waste at the top of the flotation tank.

- *Secondary treatment*—typically a biological treatment step for the removal of biodegradable organic matter (in solution or suspension) and suspended solids, also combined with nutrients removal processes (often included in tertiary step definition). The contaminants of major concern in the wastewater are significantly reduced during this stage in terms of biochemical oxygen demand (BOD) and chemical oxygen demand (COD). The objectives of the biological treatment are the following: oxidation of particulate and dissolved biodegradable constituents; capturing and converting suspended and colloidal solids into a biological floc; removal of nutrients, eg, nitrogen, phosphorous.
- *Tertiary treatment*—removal of residual suspended solids by employing filters. Disinfection is also typical in this step. Moreover, an advanced treatment step can be included in this definition where additional methods are used for further purification of the wastewater, such as adsorption, ion exchange, membrane separation, advanced oxidation, etc., for the removal of dissolved and suspended materials. This step is crucial when the purpose is the reuse of the treated wastewater.
- *Solids (sludge) treatment*—collection, stabilization, and subsequent disposal. This step includes the processes of thickening to increase the sludge solids before treatment, anaerobic digestion, dewatering of the digestate, and final treatment of the dewatered sludge by composting or drying (Fig. 3.15).

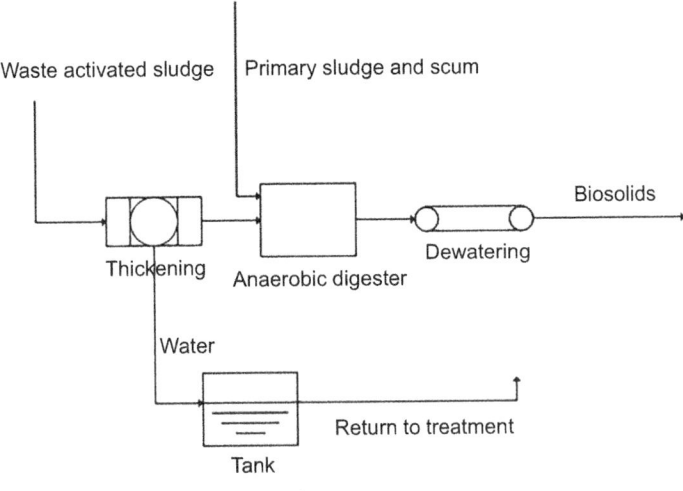

Figure 3.15
Sludge treatment steps.

In the following paragraphs, the secondary and tertiary treatment steps will be further analyzed.

3.10.2 Secondary Treatment

The heart of the secondary treatment is the biological process. Biological processes employed in wastewater treatments can be divided into two main categories: suspended growth and attached growth (or biofilm) processes [70]. In suspended growth processes the microorganisms responsible for treatment are maintained in liquid suspension by appropriate mixing methods. Activated sludge process is the most common suspended growth process. The term "activated sludge" refers to both the treatment process and the microorganisms that exist in the biological reactor. In attached growth processes, microorganisms are attached to an inert material and typical processes are trickling filters and rotating biological contactors.

Biological processes can be operated under aerobic (with dissolved oxygen supplied by an air stream), anoxic (absence of free oxygen, presence of nitrate), or anaerobic (absence of oxygen) conditions. While oxidation of the organic matter and in general nitrification, which is ammonia oxidation to nitrite and nitrate, take place in aerobic conditions, denitrification, which is the reduction of nitrate to nitrogen, takes place in anoxic conditions (Fig. 3.16) [72]. Denitrification requires absence of oxygen and presence of nitrate and biodegradable organic matter. In industrial wastewater treatment plants, where wastewater is poor in organic matter, methanol can be added to the anoxic tank to accelerate the process and increase its efficiency. Anaerobic ammonia oxidation is also

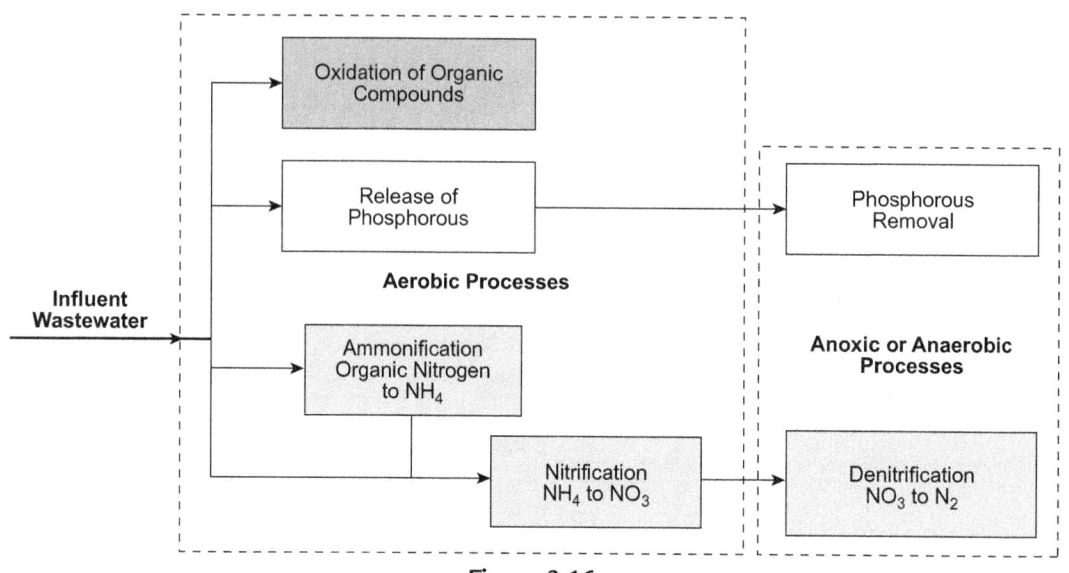

Figure 3.16
Biological processes in wastewater treatment plant [72].

possible (deammonification) and involves two steps: partial nitrification of ammonia and anaerobic oxidation of ammonia and nitrite to nitrogen. Obviously, the process requires that aerobic nitrification is accomplished before the anaerobic step and provides a means for nitrogen removal without organic matter consumption in contrast to the classic aerobic nitrification/anoxic denitrification process [70]. Furthermore, anaerobic—aerobic sequence is utilized for enhancing phosphorous removal.

In many occasions water reclamation schemes rely on secondary treatment. However, this level of treatment is suitable only for restricted agricultural irrigation applications and for some industrial applications such as cooling. A separate reference ought to be provided to Membrane Bioreactors (MBR), a technology which replaces conventional secondary treatment processes when stricter effluent standards are required [50]. MBRs are utilized in activated sludge processes and are increasingly adopted to treat both municipal and industrial wastewater [73—75]. MBRs combine biological processes and membrane filtration, and as a consequence they do not employ a secondary sedimentation tank for effluent clarification, avoiding in this way the problems of poor sludge settling. MBR achieves superior effluent quality compared to the conventional activated sludge system in terms of suspended solids and requires smaller footprint. Its main operational deficiency limiting its wider adoption is that of membrane fouling. Furthermore, the adoption of MBR is a necessity in many reuse projects, either as a pretreatment step for other membrane processes or with the effluent directly reused for demanding applications as unrestricted irrigation [50].

Several combinations and sequences of aerobic, anaerobic, and anoxic reactors are possible for both municipal and industrial wastewater. The simplest configuration is the one depicted in Fig. 3.14, where the biological treatment is replaced by a simple aerobic step. More efficient configurations are, eg, Ludzack-Ettinger, Modified Ludzack-Ettinger, Modified Bardenpho, Johannesburg and a Modified Bardenpho with an MBR. These configurations are basically biological nutrient removal (BNR) processes, used for the removal of total nitrogen and total phosphorous from wastewater through the use of microorganisms under different operational conditions [56].

In the Ludzack-Ettinger process, readily biodegradable substrate is used for denitrification, as the anoxic reactor is placed before the aerobic one. The recycled activated sludge flow rate is 20−100% of the influent flow rate [76]. The process allows high nitrogen removal rates; however, its main drawback is that its efficiency depends on the flow rate of the return of activated sludge. The general layout of Ludzack-Ettinger configuration is presented in Fig. 3.17.

A modification to the process described above can be presented in terms of providing an internal recycle of the activated sludge to the anoxic bioreactor in order to bring greater quantities of nitrate back for denitrification. Recycled activated sludge and nitrified recycle flow rates are 50−100% and 100−400% of the influent flow rate, respectively [76]. The process accelerates the rate of the denitrification and significantly increases total nitrogen removal. Better performance occurs due to the fact that more nitrates are provided to the anoxic zone. The modified Ludzack-Ettinger process shown in Fig. 3.18 outperforms Ludzack-Ettinger process in total nitrogen removal.

In the three-stage modified Bardenpho process, also known as the Westbank process or the A^2/O process, the influent is mixed with return activated sludge under anaerobic conditions for phosphorous release. The denitrification of internal nitrates recycle flow takes place in the anoxic tank. The phosphorous uptake occurs in the aerobic reactor along with the nitrification process. Recycled activated sludge and nitrified recycle flow rates are 30−50% and 100−300% of the influent flow rate, respectively [76]. The advantage of the

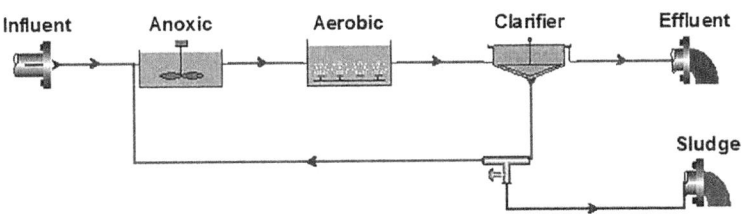

Figure 3.17
The flow diagram of Ludzack-Ettinger process. *Treatment process diagram generated using PetWin version 4.1, EnviroSim Associates Ltd.*

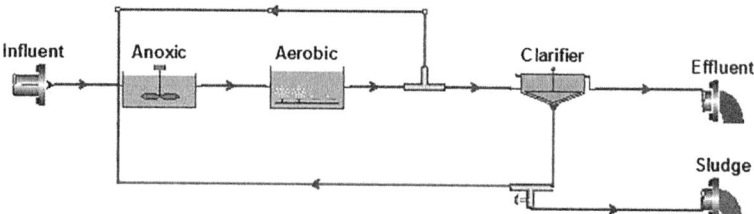

Figure 3.18
The flow diagram of the modified Ludzack-Ettinger process. *Treatment process diagram generated using PetWin version 4.1, EnviroSim Associates Ltd.*

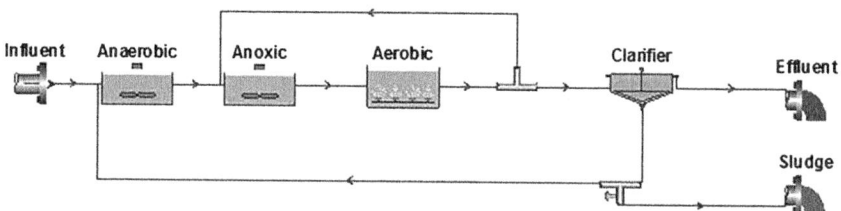

Figure 3.19
The general layout of the modified Bardenpho process. *Treatment process diagram generated using PetWin version 4.1, EnviroSim Associates Ltd.*

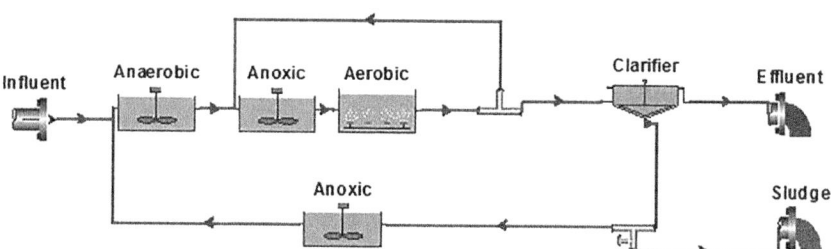

Figure 3.20
The flow diagram of the Johannesburg process. *Treatment process diagram generated using PetWin version 4.1, EnviroSim Associates Ltd.*

modified Bardenpho process is that it allows reducing phosphorous, ammonia, and nitrates, thus making it one of the most efficient biological nutrient removal processes. The general layout of the configuration in presented in Fig. 3.19. The modified Bardenpho configuration has superior total nitrogen removal performance among the configurations considered so far.

The Johannesburg process is pretty similar to the modified Bardenpho process except for one anoxic tank that is added before the anaerobic tank in order not to send the nitrates into the anaerobic zone. The general layout of the process is presented in Fig. 3.20.

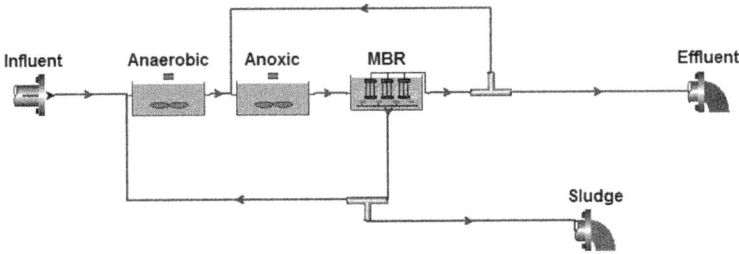

Figure 3.21

The flow diagram of the modified Bardenpho configuration with MBR. *Treatment process diagram generated using PetWin version 4.1, EnviroSim Associates Ltd.*

The Johannesburg configuration also has superior total nitrogen removal capacity. It also outperforms the modified Bardenpho process in total phosphorous removal.

The last configuration is a three-stage modified Bardenpho process with an MBR and is presented in Fig. 3.21. The configuration with the MBR outperforms the simple three-stage modified Bardenpho process in total suspended solids (TSS) removal.

3.10.3 Tertiary Treatment

Secondary effluents contain several residual constituents [70]:

- organic and inorganic suspended and colloidal particulate matter;
- dissolved organic substances, eg, volatile organic compounds (VOCs), surfactants, pharmaceuticals, and inorganic substances, eg, ammonia, nitrate, phosphorous, heavy metals; and
- biological constituents, bacteria, protozoa, and viruses.

These residual constituents may have several potential impacts on the environment and human health. Thus, in many occasions further treatment of wastewater is required in order to conform to regulations and requirements for the protection of the environment (ELVs and EQSs) or to satisfy certain quality criteria derived from the final use of the wastewater (recycling/reuse). In these cases, tertiary treatment employing advanced technologies is required. A conceptual diagram of treatment and uses of secondary effluent is given in Fig. 3.22. Nonpotable uses are discussed in Section 3.9.1, and potable uses are discussed in Section 3.11.3.

The most common tertiary processes are [54,70,77]:

- *Filtration*: Depth filtration (the most common method), surface filtration. Depth filters consist of a deep bed of solid media such as sand and anthracite, and the driving force is gravity. Depending on the type of filter, the effective size of the media varies between 0.4 and 2.0 mm in average diameter. Surface filters are shallow in the range of

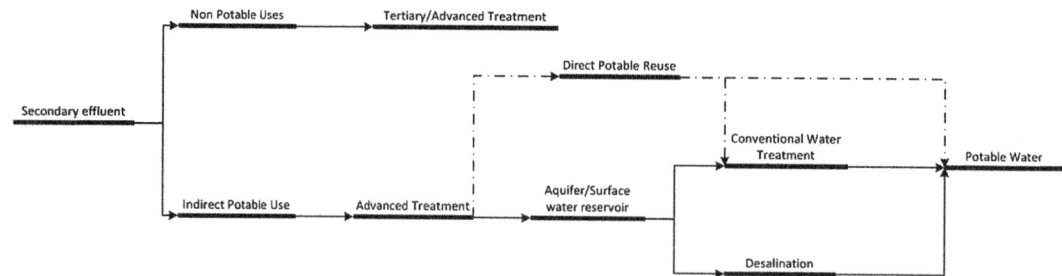

Figure 3.22
Treatment and uses of secondary effluents.

millimeters, while the materials used are synthetic fabrics from nylon, stainless steel, and other fibers.

- *Membrane separations*: Microfiltration (MF), ultrafiltration (UF), nanofiltration (NF), and reverse osmosis (RO). The driving force is pressure difference. These processes are described in more detail in Section 3.11.

- *Disinfection*: Disinfection is designed to inactivate microorganisms (pathogens) by utilizing chlorination (oxidant, the most common method), UV (commonly used alterative to chlorination), ozone (oxidant), and pasteurization. More information on disinfection process is provided in Section 3.11.

- *Advance oxidation*: It is a class of different oxidation technologies such as UV/H_2O_2, $ozone/H_2O_2$, ozone/UV, UV/TiO_2, and Fenton reactions (Fe/H_2O_2, Fe/ozone, $Fe/H_2O_2/UV$) that are typically added to the end of a treatment train. They are utilized in several applications in industrial wastewater treatment, but they are also valuable for reclaimed water treatment for the removal of residual organic compounds as pharmaceuticals.

Furthermore, several other processes may be utilized and the most common are [4,68,70,77]:

- *Chemical precipitation*: It is the utilization of chemicals for removing substances by precipitation. Typical applications involve phosphorous (lime precipitation) and heavy metals removal.

- *Absorption and stripping*: Absorption is the transport of a gas from the gas phase into the liquid phase and in water treatment is used in aeration (O_2 transfer) and is utilized in deoxygenation of the effluent, typically the last step in the treatment process.
  In stripping/aeration process air—water contacting devices, as packed columns, they are used to bring in contact gas and liquid phases and remove substances, like VOCs, from the latter.

- *Adsorption and ion exchange*: In adsorption, dissolved substances are transported from the liquid phase into a porous solid adsorbent granule by diffusion and then sorbed on the surface (physisorption) or react with it (chemisorption). Typical adsorbents are several types of activated carbon and zeolites. Ion exchange is similar except that it involves ions (cations, anions, or both) and is a stoichiometric process driven by both diffusion and electrostatic forces. Typical ion exchangers are ion exchange resins, zeolites, clays, and other materials. Ion exchange is typically used for water softening and demineralization.

- *Evaporation*: It is a distillation process where the components are separated by vaporization and condensation, and can be used for the removal of salts. Evaporators are used in this process, which are expensive processes utilized only when absolutely necessary.
- *Electrodialysis*: It is an electrochemical process in which ionic species are transported through ion-selective membranes from one liquid to another under the force of an electric potential. Ion exchange membranes are essentially ion exchange resins cast in sheet form. The electrical potential (DC voltage) applied is the driving force for ions to move with the membranes forming a barrier to the ions of the opposite charge.

The uses of these processes are shown in Table 3.6 [70].

There are several alternative wastewater tertiary treatment trains; a generic configuration is shown in Fig. 3.23. The configuration shown does not mean that all processes are used simultaneously; it just indicates the most common sequence of individual processes.

3.10.4 Case Study: Refinery Wastewater Treatment and Reclamation

Petroleum refining is a water-intensive industry as many of the processes in refineries use water and steam. Cooling is the primary use of water, consuming more than 90% of the

Table 3.6: Uses of Tertiary/Advanced Processes.

Residual Constituent	Process
Organic and inorganic suspended and colloidal particulate matter	Filtration, membrane separations, adsorption, ion exchange, distillation, chemical precipitation
Dissolved organic matter	Reverse osmosis, adsorption, ion exchange, stripping (VOCs), distillation, advanced oxidation processes, chemical precipitation (total organic carbon), electrodialysis
Dissolved inorganic matter	Depth filtration (phosphorous), reverse osmosis, adsorption, ion exchange, stripping (ammonia), distillation, chemical precipitation (phosphorous), electrodialysis
Biological species	Depth filtration (protozoa), micro/ultrafiltration (bacteria, protozoa), reverse osmosis, adsorption (bacteria), ion exchange (bacteria), distillation, advanced oxidation processes (viruses), chemical precipitation (protozoa), electrodialysis, disinfection

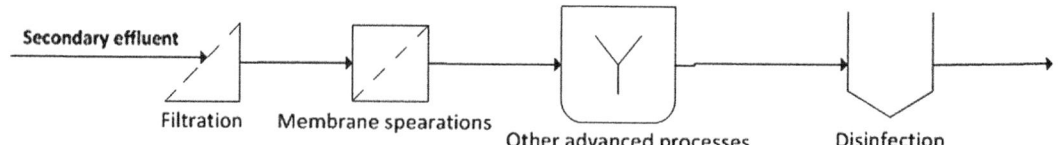

Figure 3.23
Generic configuration for tertiary/advanced wastewater treatment.

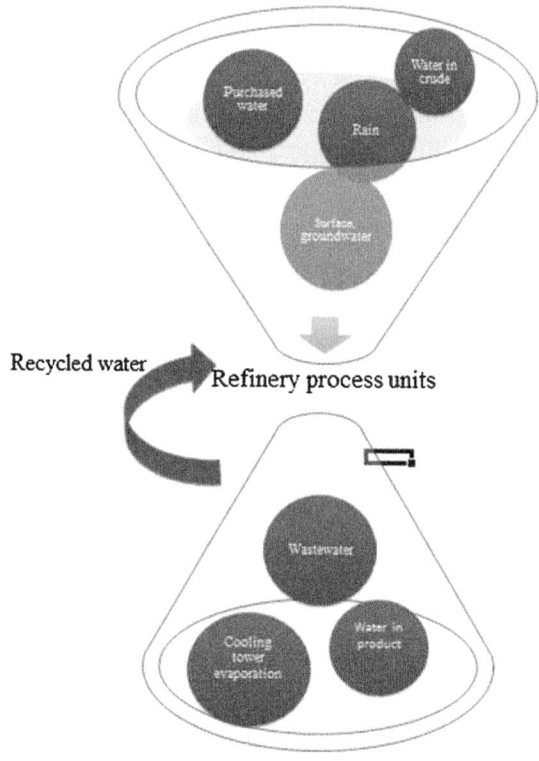

Figure 3.24
Refinery water balance [78].

total water, while about 40% of the water used is lost due to evaporation. Fig. 3.24 shows a qualitative refinery water balance [78].

The need to recycle and reuse water in refineries is increasing in a fast pace. The decision whether to use raw, recycled (treated), or reused water depends on the end use, since each use has its own specifications. The contaminants present must be compatible with the process where the water is going to be used. The reclaimed refinery wastewater potential uses are shown in Table 3.7 [78,79].

Table 3.7: Reclaimed Water Uses in Refineries [78,79].

Use	Water Source
Process water: desalter makeup, coker quench water, coker cutting water, flare seal drum, FCC scrubbers, hydrotreaters	Sour water treated in strippers for removal of H_2S and NH_3
Boiler feedwater and cooling water makeup	Treated and upgraded (tertiary treatment) refinery wastewater, pretreated stormwater, treated and upgraded municipal wastewater
Fire water and utility water	Treated refinery wastewater, stormwater

Refinery wastewater can be defined as the wastewater generated from industrial processes such as desalting, distillation, thermal cracking, and production of different by-products. Depending on the source of wastewater, it can be classified into four major categories [78]:

- *Desalter effluent* is produced in desalter units, which remove inorganic salts that are present in crude oil as an emulsified solution of salt. This water contains free hydrocarbons, phenols, benzene, sulfides, and ammonia.
- *Sour water* comes from condensed steam, which is used as a stripping medium in distillation and as a diluent to reduce the hydrocarbon partial pressure in catalytic cracking and other applications, and contains sulfide and ammonia as well as free hydrocarbons, phenols, and benzene.
- *Tank bottom water* is coming from crude tanks, gasoline tanks, and slop tanks. This water contains free hydrocarbons and sulfides.
- *Spent caustic* is formed due to the extraction of acidic components from hydrocarbon streams and includes residual hydrogen sulfide, phenols, organic acids, hydrogen cyanide, and carbon dioxide.

It is important to mention that segregation and pretreatment of different classes of wastewaters produced in refineries is an efficient method for a viable wastewater management and allows recovery of several components [79]. For example, sour water is pretreated for the removal of sulfides and ammonia by stripping with steam, flue gas, or inert gas in trays or packed columns. Stripping can also remove a part of phenols and cyanides. Other processes employed are air oxidation and vaporization followed by incineration.

As it is evident, refinery wastewaters contain relatively high concentrations of pollutants such as oil and grease, sulfides, ammonia, cyanide, phenols, several free hydrocarbons, benzene, toluene, ethylbenzene, xylene, polycyclic aromatic hydrocarbons (PAHs), and heavy metals. This is due to the fact that water is used in many different processes and frequently contacts a variety of organic and inorganic chemicals [78]. Many of these

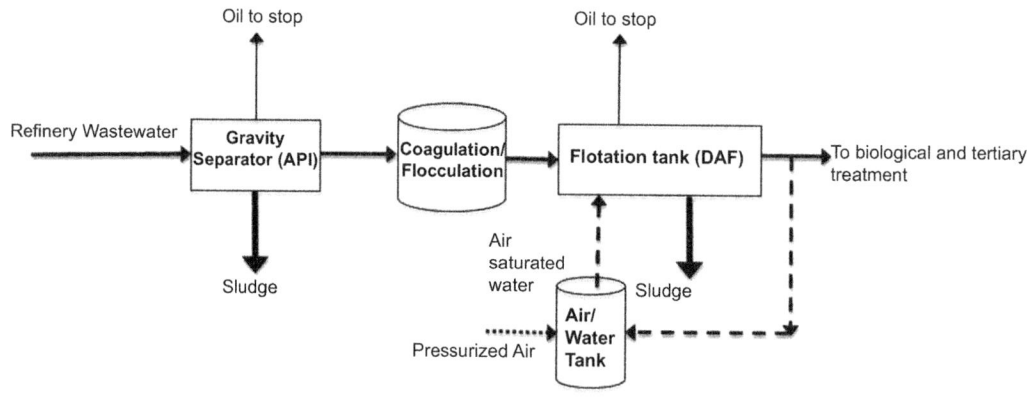

Figure 3.25
Typical refinery wastewater treatment plant [78].

compounds are extremely toxic and difficult to degrade in wastewater treatment facilities. Furthermore, certain compounds, as phenols and cyanides, inhibit activated sludge activity reducing the efficiency of the biological treatment [80].

Typical refinery wastewater treatment plants are very similar to municipal wastewater treatment plans, the major difference being the primary and secondary oil/water separation (Fig. 3.25). These two steps of oil removal are typically required to achieve the necessary removal of free oil from the collected wastewater prior to feeding it to a biological system. These specialized systems are API separator, named after the American Petroleum Institute, which provides the relevant design standards, followed by a dissolved air flotation (DAF) unit [78]. An API separator separates three phases (oil, solids, and water) that are usually present in refinery wastewater by employing the difference in specific gravity to allow hydrocarbons to float on the surface and be skimmed off and the sludge to settle. The effluent from API is treated in DAF units, where first coagulation/flocculation takes place followed by flotation, where the material is skimmed off, and settling of sludge. Flotation is achieved by dissolving air in the wastewater under pressure and then releasing the air at atmospheric pressure in a tank, where bubbles are formed, which adhere to the suspended matter causing it to float. Sludge is further treated by decanters and centrifugation.

The conventional biological treatment processes that are applied for the treatment of petroleum and petrochemical wastewater can only partially remove the contaminants involved. Often, existing regulations/specifications governing the discharge/reuse of petroleum wastewater necessitate the adoption of advanced treatment techniques including membrane processes [81]. The application of MBRs is an effective process that can achieve advanced wastewater treatment [75]. Furthermore, membrane technologies are

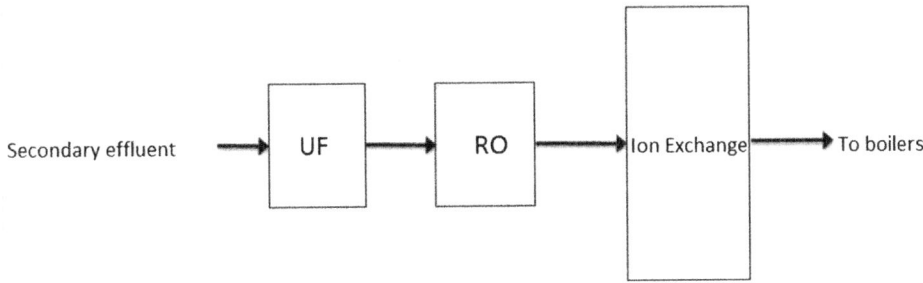

Figure 3.26
Tertiary refinery wastewater treatment for boiler feed [82].

used in tertiary treatment and have become popular since the last decade in the petrochemical industry. For example, RO is used as pretreatment for ion exchange demineralizers to produce pure water for boiler feed and process uses as shown in Fig. 3.26 [82]. During the decade 2000–10, a total of more than 20 UF or UF/RO systems were installed at petrochemical facilities worldwide for water reuse. More of the UF systems are used in MBRs than in tertiary filtration [82]. However, it should be mentioned that there are several problems identified, such as fouling by oil and organics and scaling by metals, which led to rapid and expensive membrane replacement.

3.11 Drinking Water Treatment Technologies
3.11.1 Treatment of Surface Water and Groundwater

Similar configurations, with some differences in comparison to tertiary wastewater treatment, are used for the treatment of surface water and groundwater intended for human consumption. The relevant processes are applied in somewhat different sequence in the case of drinking water treatment. Conventional water treatment process includes three main stages: clarification, filtration, and disinfection as shown in Fig. 3.27.

Prior to clarification, untreated water is pumped to screeners, which mechanically remove sticks, aquatic plants, hair, leaves, rags, and other floating debris. Screeners have a wide variety of forms and range from microscreeners to trash racks. This treatment prevents

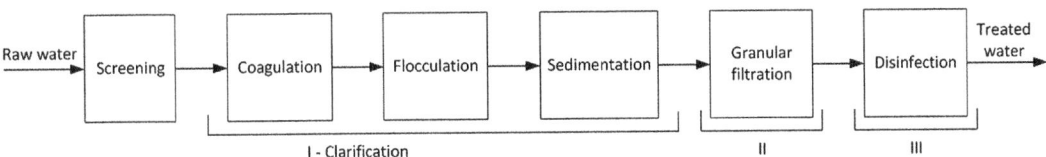

Figure 3.27
Conventional drinking water treatment plant.

downstream equipment from damaging and further treatment interference. After screening, water is clarified by a coagulation—flocculation—sedimentation sequence treatment. At the coagulation stage, charge of impurities residing in water is decreased by the added coagulant. The majority of water pollutants possess negative surface potential. To reduce their charges and cancel electrostatic interactions, positively charged coagulant should be added. Aluminum- and iron-based chemical compounds are common positively charged coagulants. Introduction of them to treated water results in their dissociation and release of positively charged ions (ie, Al^{3+} or Fe^{3+}). Negatively charged pollutants react with positive ions of the coagulant and form microflocs. The coagulation process is implemented at rapid mixing of water that ensures thorough and even distribution of the coagulant. Flocculation is a subsequent treatment, where long-chain low surface charge polymer (flocculant) is introduced. Microflocs formed at the coagulation stage are adsorbed on the flocculant chain, which results in the aggregation of destabilized particles

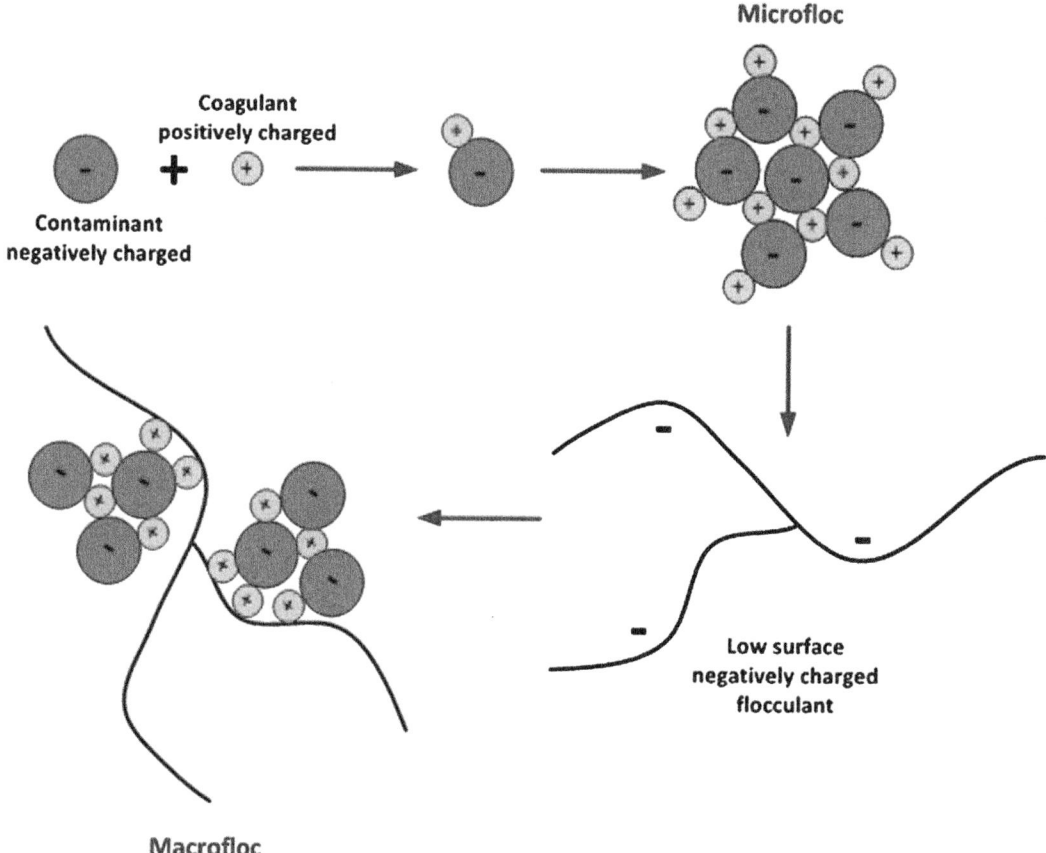

Figure 3.28
The coagulation—flocculation process.

and the formation of macroflocs. Such flocculants as starch, sodium alginate, polyacrylamide are widely used nowadays. Fig. 3.28 schematically presents the coagulation—flocculation process. At the last stage of clarification, gravitational sedimentation of the macroflocs occurs.

The water leaving the sedimentation tank still contains floc particles. To remove them, treated water passes through granular media, usually a bed of sand. When polluted waters are applied to porous media, water goes through open spaces between the sand particles, and the contaminants are either trapped in the pores or attached to the filtration media itself (Fig. 3.29). There are three main filtration mechanisms by which granular media retain pollutants, ie, mechanical screening, collision, and flocculation.

The last step of drinking water treatment process, disinfection, is utilized to decrease concentration of disease-producing microorganisms to an acceptable level. Conventional disinfection is a chlorine-based technique. Chlorine gas, chloramines, and chlorine dioxide are able to inactivate three categories of human pathogens: bacteria, viruses, and cysts. Disinfection by-products formed during this stage may also constitute health concerns. For example, reaction of chlorine gas with dissolved organic matter already existing in the treated water may lead to the formation of carcinogenic trihalomethanes. Modern water treatment techniques allow the minimization of the concentration of disinfection

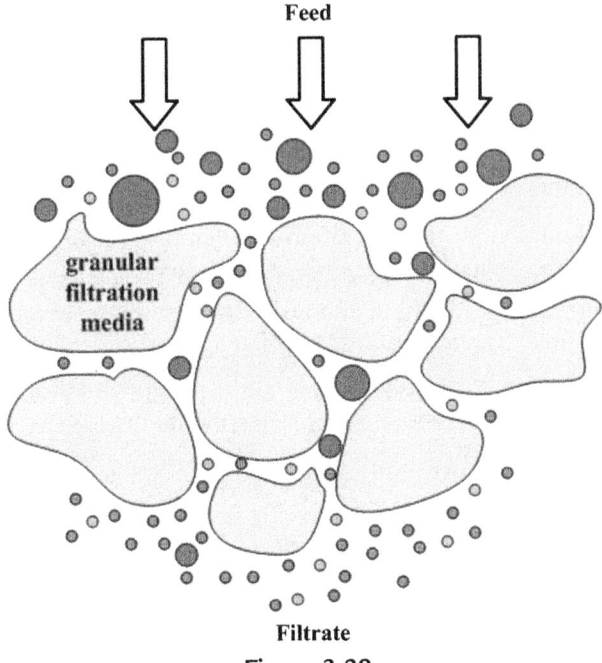

Figure 3.29
Schematic representation of the granular filtration process.

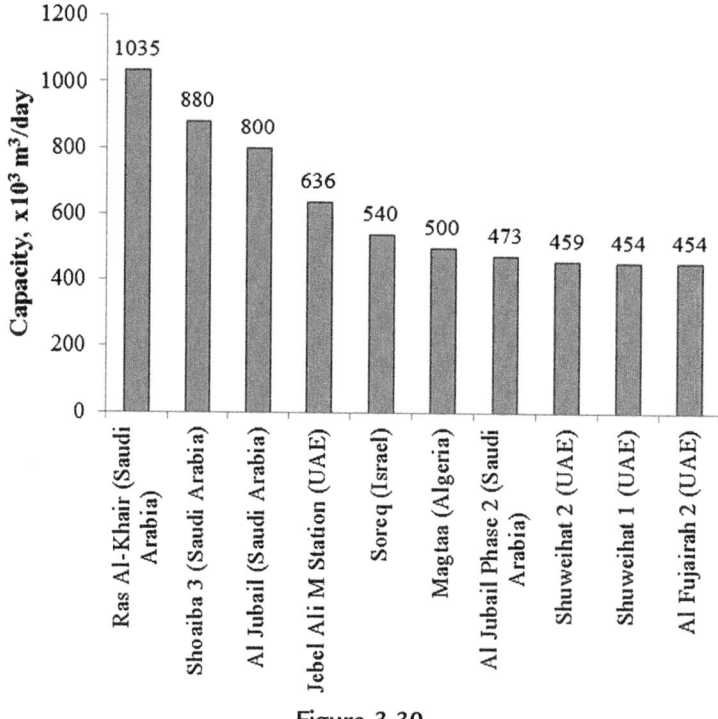

Figure 3.30
Capacity of the 10 biggest desalination plants in the world [84].

by-products or even totally avoid them. Principles and examples of advanced technologies currently utilized will be described in the next section.

3.11.2 Desalination of Salty Waters

The first recorded desalination project occurred when God thought Moses to turn bitter water into the potable one. Thousands years later, desalination became an emerging technology in drinking water production, and nowadays use of desalinated waters has become a common practice. The number of desalination plants worldwide as of 2013 is more than 17,000 in 150 countries, producing 80 million cubic meters per day, providing water for 300 million people [83]. Fig. 3.30 presents the daily capacity of the 10 biggest desalination plants globally [84].

The most commonly used techniques until mid-50s involved evaporation and distillation. It is in this period when RO emerged as a technology for salty water desalination. At present, RO accounts for ~60% of installed capacity [83]. A typical desalination train is shown in Fig. 3.31. As in the case of IPR of wastewater and other specialized treatment plants, membrane processes are the heart of the plant. Salty water pretreatment using

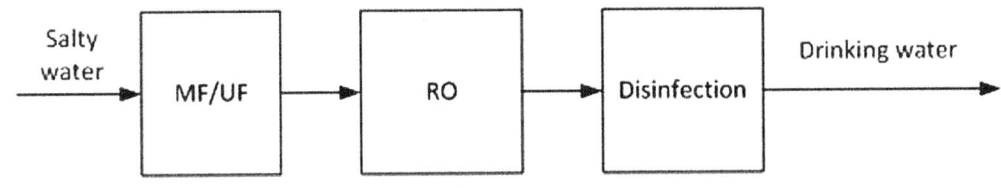

Figure 3.31
A typical desalination plant for the production of drinking water.

micro and ultrafiltration membrane systems prior to RO desalination further extended the membrane's useful life, thereby reducing the operational costs.

During the past decades, membrane separation processes have grown from a simple laboratory tool to an industrial technology with significant technical and commercial impact. In many cases, membranes are faster and more efficient than conventional separation methods. Membrane is a selective barrier that allows some constituents to pass through, while blocking the passage of others. The movement of material across a membrane requires a driving force, which could be pressure, concentration, temperature, or electrical potential. Industrial membrane processes are mainly driven by hydraulic pressure. There are four categories of hydraulic pressure—driven membranes: MF, UF, NF, and RO. Magnitude of hydraulic pressure applied to membrane depends on membrane pore size, ie, higher pressure required for smaller pores and vice versa. RO and NF processes operate at pressures significantly higher than those employed for UF and MF. For instance, RO and NF membranes reject solutes as small as 0.0001 and 0.001 μm, respectively, which are both in the ionic or molecular size range. Typical operating pressures for NF systems range from 0.5 to 1.5 MPa, and for RO systems they are from 5 to 8 MPa. Low-pressure membrane processes, on the other hand, are typically used to remove particulate and microbial contaminants. Their minimum solute rejection sizes are 0.01 and 0.1 μm, respectively. MF and UF systems generally operate at pressures of 50—500 kPa. Fig. 3.32 summarizes target contaminants, pore size, and operating pressures for each process.

In drinking water treatment, RO technology is used for desalination of salty waters. The technique allows the removal of all the pollutants including ions. It is at the same time a great benefit of RO technology and a disadvantage that accompanies the process. Since for normal human body functioning minerals are required, desalinated water has to be subjected to remineralization, ie, ions reinjection. UF and MF processes, in their turn, usually employed as pretreatment for RO, substitute for granular filtration in the conventional water treatment process and for disinfection. As pretreatment technology, UF/MF is used for retention of relatively big pollutants, such as natural organic matter. It allows decreased fouling rates of RO membranes and results in less energy consumption. UF/MF is also efficient as disinfection method when combined with a low

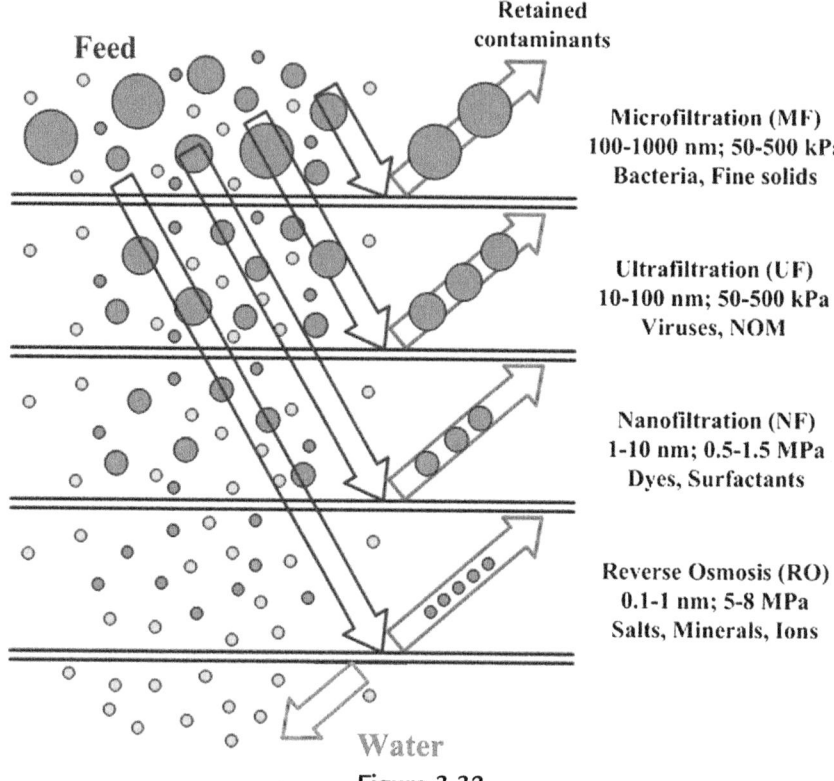

Feed

Retained
contaminants

Microfiltration (MF)
100-1000 nm; 50-500 kPa
Bacteria, Fine solids

Ultrafiltration (UF)
10-100 nm; 50-500 kPa
Viruses, NOM

Nanofiltration (NF)
1-10 nm; 0.5-1.5 MPa
Dyes, Surfactants

Reverse Osmosis (RO)
0.1-1 nm; 5-8 MPa
Salts, Minerals, Ions

Water

Figure 3.32
Filtration spectrum of pressure-driven membrane processes.

dose of chlorine or its derivatives. Such approach removes harmful microorganisms, provides residual effect, and reduces concentration of harmful disinfection by-products to the minimum. The residual effect is maintained by the chemical disinfectant, which prevents pathogenic microorganisms from growing, on the way from the water treatment plant to the consumer.

Despite the fact that the pressure-driven membranes produce high-quality product with no addition of additives or chemicals, they also have a number of limitations. For instance, they require hydraulic pressure that leads to high energy consumption and severe membrane fouling. For the minimization of productivity losses (water flux reduction), the membranes undergo cleaning by chemical reagents. Such an aggressive treatment results in membrane degradation and penetration of pollutants into the filtrate. The increase in operating pressure—another common practice to overcome mass transport problems—also leads to a decrease in the contaminants removal. The concerns on these critical issues have stimulated and facilitated the development of alternative nonconventional membrane technologies. Forward osmosis (FO) technology has gained vast popularity in recent years.

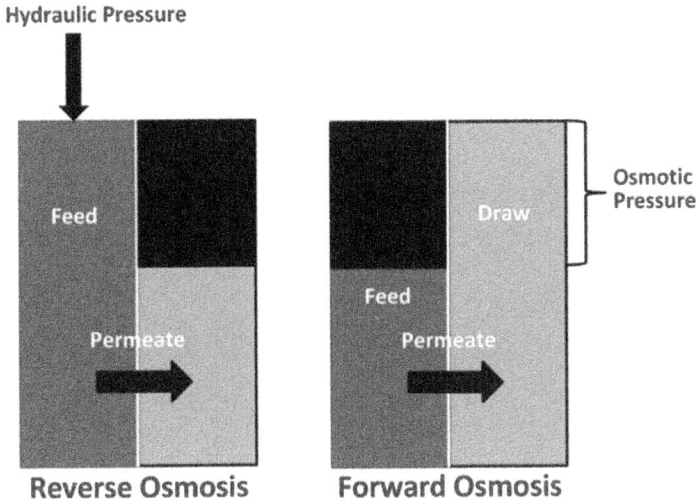

Figure 3.33
Solvent flow in the reverse osmosis and forward osmosis processes.

Unlike conventional membrane processes, FO is driven by osmotic pressure gradient. Fig. 3.33 demonstrates the difference between RO and the direct osmosis process.

From the feed side, treated water penetrates the membrane into the draw solution, which possesses a high osmotic pressure. Such substances as sugars, salts, nanoparticles, volatile solutes, etc. can be utilized as draw solutions. Depending on the further application, treated water would be directly sent for consumption or extracted from the draw solution by an assisting technology. FO has the potential advantages of low energy demand, insignificant membrane fouling, and rare membrane cleaning. This is why FO is a potential alternative technology to pressure-driven membrane processes, attracting thus much attention worldwide nowadays. The FO process has a wide range of applications, including water treatment. For instance, the Al Khaluf treatment plant in Oman with a capacity of 200 m³/day is using FO as part of the desalination process [9]. The plant utilizes direct osmosis to dilute the draw solution before it is desalinated by RO to produce potable water with 120 mg/l total dissolved solids. The chlorine-tolerant and fouling-resistant FO membranes produce a virtually particulate-free feed to the RO process. The diluted RO feed is reported to decrease desalination energy requirements by more than 20% [85].

3.11.3 From Toilet to Tap: Reclaimed Wastewater for Potable Use

From toilet to tap: the idea of producing drinking water from wastewater is controversial, but it gains momentum in some places like California, the United States, where severe droughts pose serious challenges. But it is not only droughts; even in Europe, which is rich in freshwaters, the consumption of drinking water is alarming. For example, in the United Kingdom the consumption of domestic (tap) water is ~150 L/inhabitant per day, a figure that has been steadily growing every year by 1% since 1930 [86]. It is interesting to mention that only 4% of this water is used for drinking purposes and 30% is lost in toilet flushing. In the United States and Australia, this figure is higher than 300 L/inhabitant per day. Thus, the discussion on alternative drinking water sources is timely.

The relevant methodologies are:

- indirect potable use through groundwater recharge, which requires suitable aquifer;
- indirect potable use through surface water augmentation, which depends on the availability of reservoir sites; and
- direct potable use for which public perception is main barrier.

For indirect potable reuse (IPR), the secondary effluent is first treated by using advanced treatment methods and then it is discharged to water aquifers and/or surface water reservoirs (environmental buffers), from where it is abstracted and further treated in *Drinking Water Treatment* facilities (*see* Fig. 3.22). In environmental buffers, naturalization and dilution occur. It should be noted that "unplanned" IPR occurs in many places around the world, where recycled water enters the source water supply upstream of the offtake point for a drinking water supply [59]. An example of a typical advanced treatment train used in IPR is shown in Fig. 3.34.

IPR has been in practice for over 40 years, and the associated epidemiological studies that have been conducted show that there is no conclusive evidence of increased risk for communities using IPR water [87]. There are several successful IPR projects, as for example, the Orange County Water District's Groundwater Replenishment System, in California, which is the world's largest wastewater purification system for IPR and can produce up to 265,000 m^3/day of highly purified recycled water that serves the water demands of nearly 600,000 residents [88,89]. A nonexhaustive list of IPR projects is provided in Table 3.8 [87,89]. The disadvantages of IPR are primarily related to capital and operational costs, energy consumption, and associated greenhouse gas footprint [89].

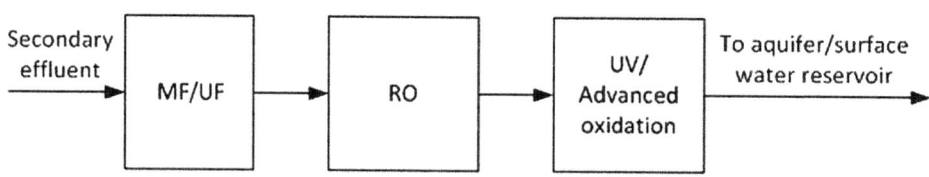

Figure 3.34
An example of a typical advanced wastewater treatment train used in indirect potable reuse.

Table 3.8: Operational IPR Projects Worldwide [87,89].

Place	Year	Treatment	Buffer
Montebello Forebay, California, the United States	1962	Secondary treatment, chloramination and inert media filtration (added in 1977)	Aquifer
Orange County Water District, California, the United States	1975–2004	Lime clarification, recarbonation, multimedia filtration, granular activated carbon, filtration and chlorination, RO (added in 1977), advanced oxidation with hydrogen peroxide and UV (added in 2001)	Aquifer
	2007 (upgraded plant)	MF/RO and advanced oxidation (UV and hydrogen peroxide)	
Upper Occoquan, Virginia, the United States	1978	Lime clarification, two-stage recarbonation, sand filtration, granular activated carbon, ion exchange, post carbon filtration, chlorination	Reservoir
San Diego, California, the United States	1981	Several treatments tested including RO and granular activated carbon (added in 1985). MF/RO, and advanced oxidation using UV light and hydrogen peroxide (added in 2002)	Reservoir
Hueco Bolson, Texas, the United States	1985	Two-stage powdered activated carbon treatment, lime treatment, two-stage recarbonation, sand filtration, ozonation, GAC filtration and chlorination	Aquifer
West Basin Municipality, California, the United States	1995	MF/RO UV and advanced oxidation processes	Aquifer
Chelmer, Essex, England	1997	MF and UV	Reservoir
Singapore	2000	Ultrafiltration, RO, UV, stability control and chlorination	Reservoir
Wulpen, Belgium	2002	MF/RO and UV disinfection	Aquifer

In the direct potable reuse (DPR) process, the treated wastewater is directly treated in drinking water plants. Unlike IPR, DPR does not involve the use of an environmental buffer, although engineered buffers can be used. Most DPR projects involve mixing of the recycled water with conventionally sourced water prior to delivery to consumers [89]. The first DPR project was implemented in the town of Chanute, Kansas (the United States), during a drought period between 1952 and 1957 and operated for a total of 5 months [61]. Five DPR projects are currently in operation [89]:

- Namibia (Windhoek), started in 1968 and upgraded in 2002 and its capacity is 21,000 m^3/day;
- South Africa (Beaufort West Municipality), started in 2011 and its capacity is 1000 m^3/day;
- USA, New Mexico (Cloudcroft), started in 2011 and its capacity is 100 m^3/day; and
- USA, Texas (Big Spring), started in 2013 and its capacity is 7000 m^3/day

The benefits of DPR are presented in a White Paper prepared by experts for the National Water Research Institute Fountain Valley, California [90]. According to this study, existing technical ability can reliably produce purified water that meets all drinking water standards, and DPR can secure stable source of water supply with parallel benefits for agriculture, environmental preservation and enhancement, and energy conservation. The main alternative, which is desalination, is energy intensive, and brine disposal is a serious environmental issue [90]. However, regulatory considerations, public health concerns, public acceptance barriers, and other political issues hold back DPR development.

3.11.4 Case Study: Municipal Wastewater Reclamation

Ashkelon desalination plant in Israel

In 1999, the Israeli government initiated a long-term, large-scale Sea Water Reverse Osmosis desalination program. The program has been designed to provide for the growing demands on Israel's scarce water resources and to mitigate the drought conditions experienced since the mid-1990s. Construction of the first Israeli desalination plant—Ashkelon desalination plant—has been initiated in 2003 by private companies that won the government's public tenders [91]. The capital cost of the plant was 167 million EUR. The water supply was initiated in August 2005 and the full operation started in December 2005. The 70,000 m^2 desalination plant takes water directly from the Mediterranean Sea and passes it through three main treatment stages, ie, pretreatment, RO, and posttreatment. At the pretreatment stage, seawater undergoes granular and 20 μm cartridge filtration. Next, ions are removed (<20 mg/L for Cl^- and <40 mg/L for Na^+) by spiral wound RO membranes at 70 bars. Since water leaving this stage does not possess any ions, it is required for normal body functioning, to accomplish a remineralization step. After the mineralization, water possesses pH 7.5—8.5, total dissolved solids <300 mg/L, and turbidity <0.5 NTU.[9] The plant desalinates 396,000 m^3 of seawater per day and utilizes pressure center design, triple line intake, energy recovery, and unique boron removal (<0.4 mg/L) system to increase the process efficiency and reduce water costs. Cost of the desalinated water produced at the Ashkelon plant is one of the world's lowest ever *prices* for *desalinated water*: 0.53 USD/m^3.

NEWater reclamation plant in Singapore

To overcome a water shortage in 1972, the Singapore's government drew the first water master plan, and started reclaiming used water for nonpotable use in order to conserve potable water and close its water loop. In 2002, the Public Utilities Board embarked on the NEWater initiative.[10] NEWater reclaims secondary effluents using MF followed by RO

---

[9] www.veoliawaterstna.com.
[10] Public Utility Board, Singapore. www.pub.gov.sg.

and ultraviolet disinfection. Alkaline chemicals are then added to restore the pH balance, after which the NEWater is ready for use. The treated water is used for both potable (indirect) and nonpotable use. Indirect potable use refers to the use of reclaimed water to augment drinking water supplies by discharging it to reservoirs, which are subsequently treated for potable consumption. For instance, Ulu Pandan—one of four NEWater plants—was officially opened at 2007. Its area, capacity, and capital cost is 26,000 m^2, 148,000 m^3/day, and 130 million SGD, respectively.[11] NEWater is audited twice a year by the National Environment Agency panel. A panel of international experts in the industry also provides ongoing peer advice to the scheme. NEWater's permit is based on WHO and USEPA guidelines. Unregulated parameters are derived from the Contaminant Candidate Lists. The final NEWater product pH ranges in 7.0—8.5, turbidity is <2 NTU, and conductivity <200 μs/cm. Price of 1 m^3 of NEWater water is $1.3054.

3.12 Water Economics and Management

3.12.1 Introduction

The particular significance of water can be realized from the fact that most Ancient civilizations have been developed close to water (eg, the Minoan, Egyptian, Arab, and Roman civilizations). For all these civilizations, water has been vital for life, for commerce, and cultural exchange. For some of them, water has also been worshipped as a deity.

Water, as it is found in nature, is a renewable public good, but at the same time exhaustible, rival, and mobile, with a lot of competitive uses: domestic, industrial, and agricultural. Because the degree with which it gets renewed is site- and time-specific, namely it is related to the weather, time, and geomorphological conditions applying in a geographic area, it can also be regarded as an exhaustible source. However, when water is used at home, it is regarded as a private good. Besides water being a necessity for direct human life, water is also essential for production in agriculture and industry. Namely, production is impossible without the water input.

Hence, water is a scarce resource, which demands careful management in order not to affect human sustainable development negatively. Water is characterized by an extremely uneven distribution, since only six countries in the world (eg, Brazil, Canada, China, Indonesia, and Russia) possess half of the world water supply. According to the World Business Council for Sustainable Development—WBCSD [92], from the 3% available freshwater on our planet, only 0.5% can be used and is stored in underground aquifers, or comes from rainfall, natural lakes, rivers, and man-made storage facilities. The rest 2.5% stays frozen in Antarctica and thus is not available for the satisfaction of human needs.

---

[11] Ulu Pandan NEWater plant. www.keppelseghers.com.

Given the human population explosion, the needs for consumption of drinking water, food, energy, and other goods as well as sanitation services etc., result in the demand for water exceeding its supply. On top of that, the experienced climate change and the subsequent change in precipitation patterns, together with the fact that water is a locally produced and consumed good and cannot be transferred, cause water supply shortages to be particularly felt in certain geographic areas. Therefore, water economics and management must be applied separately in each catchment area or at a river basin level, and problems must be solved first at a local level and then nationally and internationally. Since countries already suffering from water shortages are those countries with the highest population growth, strict measures must apply to conserve water, recycle, or desalinate.

3.12.2 What Do Water Economics Do?

Economics are here to work hand-in-hand with technology in order to save the situation. Economics inform us that water resources suffer from the widely known problems of public goods. The "tragedy of the commons," stemming from the absence of property rights, has led to water resource freeride, abuse, overconsumption, and pollution with repercussions for everybody. Economics can define prices, fees, taxes, and other motives that will cause demand to equal supply at all times at a local level. Moreover, economics can design the motives to prevent withdrawal from surface waters and underground water at an excessive rate, the latter causing also the problem of desertification.

It is true that some basic water quantity must be available to everybody at reasonable prices. Water prices should, but they do not always, transmit signals for the protection and rational consumption of this good, its opportunity cost, and economic value. Economic measures can rectify and prevent water pollution and the squandering of water to inefficient uses. Economics can also contribute to the design and assignment of sufficient property rights for the water.

While public goods such as water are considered to be for free, because they are free in nature, one needs expensive technology and infrastructure to have it available when and where needed. If water price is all inclusive and well informed, markets will not fail and can work toward sustainability. Applying all economic tools locally will also solve the problem globally. What earned Elinor Ostrom the Nobel Prize in Economics is her analysis on the governance of the commons, basically supporting that small communities can find the ways to organize themselves in a sustainable way and that sustainability starts from them [93].

Water economics nowadays are mostly concerned with the valuation of water services, originating from the preservation of wetlands, forests, and recreation. Valuation is the foundation for proper and all-inclusive pricing and financing in the water sector. Water economists are concerned with the full costing of water, the value identification from reduced load of contaminants such as nitrogen [94,95], economic valuation of climate change impacts on water [96], and other relevant topics. Since water demand has long exceeded our basic survival needs, policy makers are now turning to the study of behavioral patterns and their correspondence to water demand. This approach goes beyond what people need and concentrates on how much are consumers really ready to pay for.

3.12.3 Valuing Versus Pricing

While nonspecialists would consider these two concepts as being the same, they are in fact quite different. According to Adam Smith (Wealth of Nations), things with the greatest value in use, usually have little value in exchange and vice versa. Hence, value is a broader concept than price, which encompasses both the utility satisfaction derived from using the good and the purchasing power the good brings with it. Price is determined by the powers of demand and supply, while value is not so easily affected by them and has deeper and stable roots.

For example, water has both a *use value* and a *nonuse value* (Fig. 3.35), i.e., value from actually using the water for drinking, washing, etc. and a value from not using it, eg, the knowledge that water is there if you need it. The use value is further divided into the *direct value* (value from direct usages such as drinking), *indirect value* (value from indirect usages such as ecosystem support services), and an *option use value* (value from knowing that water will be there if you need it). Also the nonuse value is further divided into the *bequest* (value from knowing you can leave water intact for generations to come) and *existence value* of a good (the intrinsic value of water). The use value is more tangible, objective, and easy to understand. The nonuse value is more subjective and relies on ethical beliefs, individual perceptions, and future expectations. Specifically, examples of direct use values are:

- drinking,
- cooking,
- washing and showering,
- waterborne sanitary waste disposal,
- landscape and garden watering,
- swimming pool filling,
- car washing, etc.

Examples of indirect use values are:

- ecosystem support services,
- climate change mitigation,
- erosion control, etc.

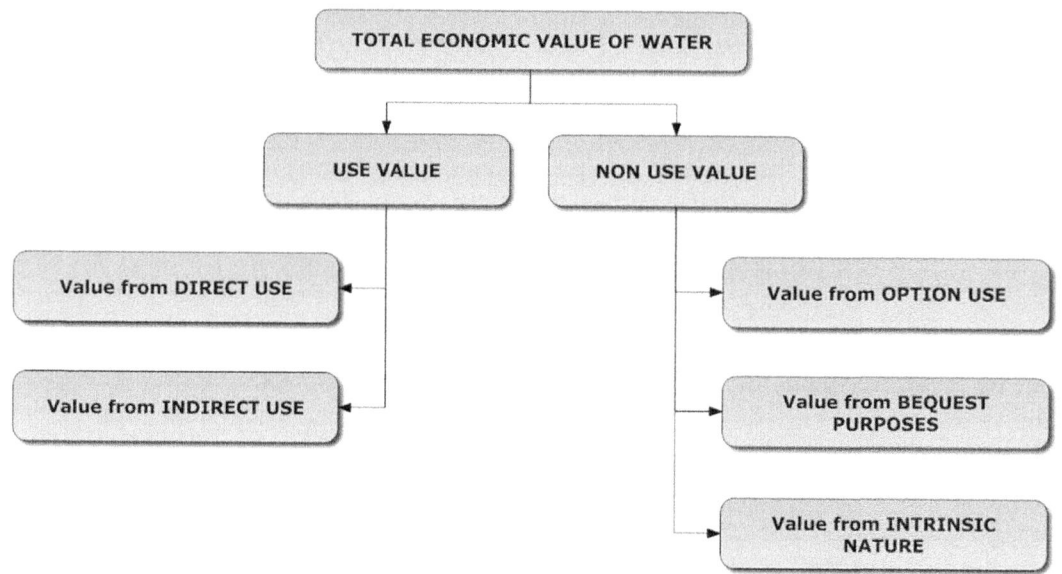

Figure 3.35
The water value typology. *Adapted from Ref. [113].*

As far as the nonuse value components are concerned, the use option value of water is related to the uncertainty felt about whether, where, and in what quantities and qualities water will be needed in the future. The bequest value of water stems from our desire to leave water resources intact for future generations, while the existence or intrinsic value of water stems from the inner need to preserve water ecosystems and from having the knowledge that water is there for us. In whatever place of the world we may live, we are all interested and affected from news about water pollution or overexploitation. The option value is usually not related to the derivation of satisfaction from using the good, but to the satisfaction from resting assured that under no circumstances will you be left with no water in your life. Due to uncertainty, consumers do not know whether they will need to use the good. It is also quite legitimate for consumers to derive satisfaction simply from knowing that the good is there for them, if and when they need it. For example, suppose there is a lake at quite a distance from your living place. You do not use it and you do not have any direct benefit from the preservation of that lake. However, if the water of that lake is preserved, you know that there is a possibility that one day you would fancy going swimming or fishing there.

Usually consumers buy a good for the certain price defined in the market. Nevertheless, sometimes they know that they would be willing to pay a higher price for the good purchased, because it is valuable to them. This extra amount, which consumers would be

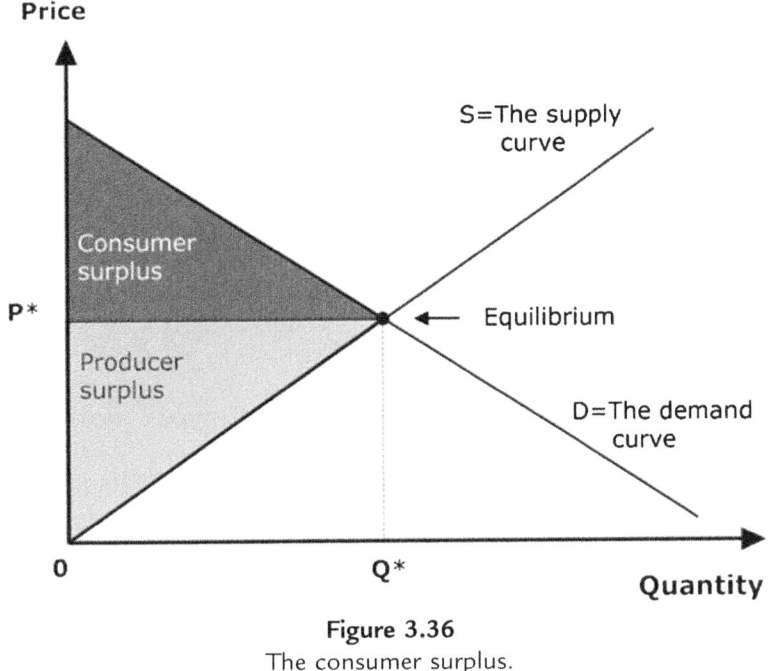

Figure 3.36
The consumer surplus.

willing to pay above the market price and this is an amount they only know for themselves, is called the "consumer surplus." Fig. 3.36 illustrates a typical demand curve, D and a supply curve, S for a good. Their intersection produces the market equilibrium. The area below the demand curve and above the dotted line representing the equilibrium price represents the consumer surplus, while the area above the supply curve and below the dotted line standing for the equilibrium price is the producer surplus, an analog of the consumer surplus from the point of view of the producer.

If policy makers were informed about the consumers' surplus about water, they would have charged price more efficiently and fairly. The identification of the full economic value of water is performed through sophisticated methods of valuation, some of them of revealed preference (revealed by the actual purchasing actions of consumers) and some of them classified as stated preference (intention to buy as stated by consumers). The former correspond to real market situations, where consumers are confronted with the choice of a real purchase. This is done with hedonic pricing and travel cost methods. Prices for water and other environmental services or disservices can be derived from real estate prices and travel cost decisions. The latter are estimated through the cost and time consumers incur for enjoying a water-related good.

Stated preference methods correspond to the adoption of contingent valuation and discrete choice experiments in order to derive the utility underlying the consumption of a certain good. Through hypothetical scenarios in surveys, consumers are asked to state directly their *willingness to pay* (WTP) or *willingness to accept* (WTA) under the contingent valuation method.[12] Those measures can also be indirectly observed under a discrete choice experiment[13] framework, where consumers are asked to select among various configurations of a good. The choices are bundles of characteristics among which, one of them is price. Under this valuation method the part-worth values of a good can be derived, namely a separate value can be derived for each attribute characterizing the good. The second case study peruses a discrete choice experiment.

3.12.4 The Equivalence of Willingness to Pay and Willingness to Accept

The consumer surplus is an approximation of the WTP and/or WTA values. Efthimoglou and Xepapadeas [97] accept that if the income elasticity of demand is very small, then the consumer surplus is a satisfactory approximation of the money equivalent of welfare variation, which is caused by a change in the price of the good. However, in many empirical studies WTP does not equal WTA. There are five main explanations in literature for this. The first is due to reference-dependent preferences [98]. According to this, consumers' preferences are a function of an initial endowment or a status-quo situation, which has an effect on the marginal rates of substitution between goods. A second explanation [99] attributes the above disparity to the transaction costs that one has to incur to replace an item that he/she has lost. Third, Kolstad and Guzman [100] support that the disparity can arise from the design of the experiment. Fourth, it is the uncertainty, irreversibility, and limited learning opportunities [101] that make the consumer delay the decision until he/she becomes more informed. Fifth, the diversity is due to the substitution effect between other market goods and a public good. Hanemann [102,103] shows that if there is only a small income effect, there can still be a substantial disparity between WTP and WTA, if the aforementioned elasticity of substitution is sufficiently low.

The knowledge of WTP or WTA prices is very important for environmental cost—benefit analysis. Herein, the cost—benefit analysis for water projects also falls. The history of nonmarket valuation is closely intertwined with water projects in the United States, because they have been one of the most important reasons for the development of

---

[12] The contingent valuation method (CVM) is used in surveys to inquire on the WTP/WTA for a good. See also *Case study 1*.

[13] The discrete choice experiment (DCE) can estimate a separate WTP for each attribute characterizing a good, since each good is perceived as a bundle of characteristics. WTP under this framework is not inquired directly, but indirectly within a certain configuration of the characteristics of the good. See also *Case study 2*.

cost—benefit analysis around the 1940s [102,103]. Actually, according to this source, nonmarket valuation was first applied in 1957 for the recreation benefits from California State Water project.

Given that water is a necessity, nobody can or should be prohibited from using a certain quantity of that. Water economics concentrates on valuations for its deterioration or the improved supply. Water economics also must identify the incentives and trade-offs from the different allocations of water. Furthermore, water economics are concerned with the study of costs and benefits from the distribution of water to different social groups.

Water consumption is interlinked with growth. With population and economic growth, water consumption becomes gradually more competitive. Thus, the rules for its management become more complicated too [104]. For example, poor African countries may still be experiencing a stage of simultaneous economic growth and water consumption growth. This is not the case for the more developed and industrialized European countries.

3.12.5 Average Pricing Versus Marginal Pricing

It is common knowledge that water is generally underpriced. Up-to-date water prices do not reflect its full economic value, but rather its physical supply cost: cost of network, pipes, energy consumed for pumping, cleaning, staff, various administrative fees, etc. Often, the consumer is not confronted with the physical supply cost either, because this is subsidized. Foremost, water pricing is "locked" on the recovery of historic costs and not future replacement costs.

Although volumetric pricing is mandatory for many places and uses around the world, agriculture still applies flat rates depending on acreage and crop type internationally. *Average pricing* is more or less a nonvolumetric method, however, a bit arbitrary, since the type of units to which it is calculated might be an ad-hoc construction and hence difficult to compare worldwide. Through nonvolumetric pricing, the full water supply costs are not fairly and equitably distributed across users.

To estimate the *marginal value* of water, we need to know the process through which water generates business or social value, for example, by building its profit function in a production process (as a difference of cost function from revenue function) in a water industry and then equate its marginal cost to marginal revenue. The latter are first derivatives of the cost and revenue functions, respectively. The average value is most often provided by the total value divided by the number of the produced water units. Nowadays, water prices have not yet internalized the environmental costs incurred in the water production and consumption process, because not all of them have been measured and there are sometimes political will hindrances. Therefore, any type of marginal pricing applied is not based on full costing. Therefore, it is incomplete.

Usually, water utilities use a blocks pricing method for water. These can be either increasing or decreasing blocks of pricing, uniform or seasonal pricing schemes:

- *Increasing block rates*: As the amount of water used increases, so does the price in each block of water consumption. This type of charging encourages conservation and efficient use, since consumers are penalized for their high consumption.
- *Uniform rate structures*: This structure applies the same charge per unit of consumed water for all sectors of economy. Uniform or flat rates do not encourage efficient use of water and do injustice to low water consumers to whom are also borne the negative consequences of water resources exhaustion.
- *Decreasing block rates*: This is the opposite to increasing block tariffs and hence, provides discounts to high-water-volume consumers. This method treats water as an input for production, but ignores its environmental dimension that concerns everybody.

Furthermore, various other combinations of the above pricing schemes can be possible, depending on the water supply situation prevalent. Prices applying in each block of consumption can also be of average or marginal type. Utilities might be charging a certain quantity at a low base price and then, from that quantity upward, they apply other pricing combinations. Since everybody is entitled to a basic water provision, any water quantity higher than the subsistence level should be charged with higher rates to reflect the additional value reaped by high-volume consumers and the cost incurred to the rest of the society. According to Randall [105], when a water sector is characterized by increasing marginal costs and interdependence among its users (which are evidence for a water sector being in a mature phase), this causes conflicts and external costs to increase too. Water utilities currently apply charging with the aim to collect enough revenue to pay for inflation, operating costs, chemicals, equipment, and construction costs. They do not seem to adopt a full cost approach. For example, they do not take into account the cost incurred from damages to water ecosystems, etc.

3.12.6 Water Price Regulation and Water Utilities

Water pricing is a means of exercising a public policy about water. Water prices have been steadily rising in the past years due to the high demand and the fact that water quality has deteriorated and demands additional processing before it is provided to consumers. Today's water consumers have been bequeathed (from their ancestors) water sources containing a lot of accumulated pollutants, which have not been there as a result of the current consumers' activities, but of the past generations. Less progress toward efficient water pricing has taken place in agriculture compared to domestic and industrial sectors

[106]. Agriculture water is still heavily subsidized, transmitting distorted messages to farmers.

According to the Water Framework Directive [107], the regulatory framework in Europe dictates the establishment of a binding revenue requirement so that the reasonable costs of providing water and sewerage services can be covered by the utilities. This market in the United States is more mature: economic regulation of the water sector in the United States follows the rate-base/rate-of-return methodology that prevails for distribution-oriented electricity and natural gas utilities [108]. The rate-of-return method is supposed to protect consumers from unfair prices while at the same time allowing utilities to earn a fair rate of return on their investment. The rate of base, on the other hand, means a specified rate of return, following the rules from a regulatory authority. Also, examples of regulation perspectives in South Australia dictate recovery of the costs of water planning and management activities as well as recovery of capital expenditure and the determination of a revenue cap control [109]. The latter is a system of setting prices by limiting the revenue for a certain period.

To safeguard a sustainable water supply, besides the pricing considerations for water, communities must finance the establishment of water and sanitation infrastructures that will guarantee uninterrupted supply and an acceptable water quality. Water utilities are natural monopolies. Unless they are regulated, they reduce production and charge excessive prices. What is peculiar with these businesses is that they encourage or they should encourage their customers to buy less of their product. This is not the case for other businesses, which encourage customers to buy as much as possible and then they increase their profits by economies of scale.

The economic crisis starting in Europe early in 2007 has caused a decrease in the consumption of most products sold by utilities, such as electricity, energy, water, etc. This does not mean that people have adopted a more conservative stance in front of water resources, but rather this has happened due to their lower income. This may sound fair enough, but there is always the risk that if consumption drops by a large amount, water utilities will not be able to materialize their planned investments. Therefore, water prices must rise at least until there is a sufficient match of the costs and revenues of water utilities, without at the same time leaving low and poor consumers deprived of water.

3.12.7 Water Management

Given the peculiar aforementioned explained nature of water, water management caters for the spatial and temporal matching between water demand and supply, and copes with its intermittent nature. Water management can take many forms worldwide, depending on the local needs and circumstances as well as culture. Since water problems are site specific, no uniform way of management can be adopted.

There are two principal ways to face sudden need for water: *storage* and *transport*. The latter is very expensive given the low price for water. Water supply is also capital intensive, because it involves a long-lived infrastructure. This entails that water supply is faced with huge fixed costs and uncertainty since it takes a lot of time before it is fully materialized.

Water resources management demands nowadays a holistic approach, which handles simultaneously the water quality and quantity, the groundwater and surface water with respect to both the urban and natural landscapes. It concerns modeling of dose–response for every possible contaminant on every possible living organism. Overall, water management, taking the tools suggested from water economics, applies water-use charges, water pollution charges, and tradable permits for water withdrawals and release of specific pollutants, all these with the aim of efficient use and conservation of water resources. However, for appropriate water management, one needs to know the exact quantities demanded at each time, at each place. This cannot be applied for all countries worldwide, since metering is not there for all of them.

Transboundary water management

Transboundary water management requires regional cooperation so that the riparian sides both participate in decisions, rights, and obligations stemming from the common propriety of water. According to United Nations sources,[14] there are 148 countries which include some territory (river or lake) falling inside areas characterized as transboundary river basins or catchment areas. Transboundary water management bears an additional caveat over the BAU (Business-as-Usual) case. This is related to geopolitical balances that involve complicated diplomatic relations. Transboundary water management is a more complicated task since it may involve stakeholders with different cultures and economic growth, hence different abilities to handle water quality problems and ecosystem services provided from water. Consequently, it is possible that one country may free ride on other countries' infrastructure and water technology developments.

Urban water management

Water sources can originate from surface sources, groundwater, recycled water, and desalinated water. There are different management priorities between the urban water and the agricultural water sectors. However, there is one common goal: to reduce water consumption or to make more efficient use of it, through reuse whenever possible. All management targets are usually set by catchment area or river basin area.

---

[14] www.un.org/waterforlifedecade/transboundary_waters.shtml.

As far as industries are concerned, they must be motivated to further reduce their freshwater use through recycling processes and efficiency technologies. Local suppliers must renew their infrastructure to prevent leaking. The application of wastewater treatment services should be expanded. The water utility must clean the water to the level of drinking quality, appropriate for consumers to use. Then, the used water ends in the wastewater treatment plant, where it must be sufficiently cleaned and returned to the environment.

One way to reduce water consumption is by increasing prices. It is a challenging task to achieve freshwater reduction without hampering economic growth.

Rural water management

In order to abide by the aims of water use reduction and efficiency increase, agriculture needs to embark on less-water-intensive cultivations and the use of recycled water where appropriate. If no adequate water resources are available, agriculture must move to areas where water is abundant and be practiced only up to the point of securing national food security. Countries should take into consideration their comparative advantage in national production. Water must be spent on profitable economy sectors. If agriculture is not one of them, it should be abandoned in the long term [110]. Overall, water management in the agricultural sector must go hand-in-hand with the improvement of agricultural practices.

3.12.8 Case Study 1: Recycled Water Economic Valuation With the Contingent Valuation Method

Crete is a Greek island characterized by prolonged, arid summer periods. Agriculture absorbs 81% of its freshwater resources. This situation is further aggravated in summer periods through tourism. Menegaki et al. [111] launched a CVM survey (through a detailed questionnaire) in July to October 2003 for the valuation of recycled water for irrigation in agriculture. The study defined four qualities of recycled water, designated for use in different irrigation applications. These are shown in Table 3.9.

The survey was split into two versions: One was focused on farmers' WTP for various types of recycled water for the irrigation of representative Cretan crops, such as the olive tree and the tomato plant. The other version of the survey was focused on consumers' WTP for the goods that would be irrigated with recycled water, namely olive oil and tomatoes.

Overall, farmers responded that they would be willing to pay for recycled water for irrigation at a price that would be less than that of freshwater (about 55% of its price), while consumers stated that they would pay for olive oil produced from olive trees irrigated with recycled water for a price equal to 88% of the one currently sold in market.

Table 3.9: Water Qualities Originating From Recycled Water [111].

Water Quality	Explanation	Appropriate Uses	
0	Untreated wastewater: polluted wastewater in the form it is found in sewerage.	None!	
1	Primary treatment: wastewater that has been subject to a first cleaning stage. The pollution level has been reduced by 30–40%.	Appropriate for the irrigation of forested land in a controlled way (on land with limited access), landscapes and flowers by surface or subsurface application.	
2	Secondary treatment: wastewater has been subject to a second cleaning stage. The pollution level has been reduced by 95%.	It is appropriate for surface tree irrigation, such as olive trees, vineyards, industrial trees, and other trees where water does not come into contact with the crops.	
3	Tertiary treatment: wastewater has been subject to a third cleaning stage. The pollution level is reduced by 99%.	Appropriate for the irrigation of cultivations which are consumed by humans on the condition that the edible parts do not come in contact with this water.	
4	Potable water.	Appropriate for the irrigation of any crop and human use.	

Since the research has confirmed the generally positive attitudes of both the rural and urban consumers toward recycled water, it recommends to policy makers a "pay-as-you do not use recycled water irrigation" pricing scheme for those users who continue using freshwater in applications for which one could use recycled water instead.

3.12.9 Case Study 2: The Economic Valuation of a Wastewater Treatment Plant With a Discrete Choice Experiment

In 2007, Genius et al. [112] launched a survey on Cretan farmers' preferences about wastewater treatment. Although the construction of wastewater treatment plants has been implemented in the urban areas, the construction in rural areas is still pending. Taking into account that irrigation water in Crete is of potable quality and only 36% of agricultural land is irrigated, it is obvious that additional irrigation water sources are welcome. The survey used a choice experiment and derived WTP prices for each individual attribute characterizing wastewater treatment plants. The examined attributes were as follows:

- Reduction of environmental pollution (depending on the level of treatment, pollution can be reduced by 50%, 90%, and 99%).
- Increase on irrigated agricultural land size.
- Recycled water can be applied for the irrigation of forests, olive trees, or tomato plants.
- Impact on landscape from the construction of the plant (it can be considered as small, medium, or high).
- Possibility for odor (it can be small, medium, and high).
- Generation of jobs (the wastewater treatment plant can offer 3, 12, or 20 working days per month).
- Increased water bills (to cover the cost of the wastewater treatment plant).

With the above attributes, a fractional factorial design[15] was used with two blocks and nine pairs of choices. The sample was 597 farmers from villages with 200–1000 inhabitants. WTP results for the attributes found significant are shown in Table 3.10 [24].

All attributes were significant except for landscape effects. This reveals that farmers' utility is not affected by landscape effects and that farmers are interested more for the profit-making aspects of the wastewater treatment plants, such as the provision of additional irrigation water. Moreover, younger and more educated farmers appeared as more likely to be in favor of a wastewater treatment plant project.

3.12.10 Conclusion

Sustainable development is impossible to substantiate unless sustainable water resources management takes place. Water economics is here to make us change our attitude to water: to comprehend its value, to use it efficiently, and price it properly. We can reduce its contamination by bringing agriculture into line and use recycled water derivatives when

---

[15] This is a design that reduces the huge and difficult to handle number of available choices by using only a fraction of the full factorial design eg, suppose we have six attributes, three with four levels and three with five levels, this generates $4^3 * 5^3$ choices for the full factorial design.

Table 3.10: The Attributes of a Wastewater Treatment Plant [112].

Attribute	WTP Price in €	
Value of decrease of odor probability from medium to small.	0.18—0.45	
Value of irrigated land increase.	0.32—0.48	
Value of water application in tomato plants.	0.12—0.23	
Value of water application in olive trees.	0.21—0.39	
Value of one-day job increase.	0.03—0.04	

applicable. Water economics is a valuable tool for a fully informed price for water and the design of relevant mechanisms to manage and conserve it.

References

[1] G.T. Miller, Living in the Environment: Principles, Connections and Solutions, Wadsworth Publishing Company, California, 1996.

[2] EPA, The Importance of Water to the U.S. Economy, U.S. Environmental Protection Agency, Office of Water, Washington, 2013.

[3] EC, Directive 2000/60/EC of the European Parliament and of the Council of 23 October 2000 Establishing a Framework for Community Action in the Field of Water Policy, O.J. L 327, 22.12.2000, 2006, pp. 1–73.

[4] V. Inglezakis, S. Poulopoulos, Adsorption, Ion Exchange and Catalysis, Design of Operations and Environmental Applications, first ed., Elsevier, Amsterdam, 2006.

[5] I. Shiklomanov, World fresh water resources, in: P.H. Gleick (Ed.), Water in Crisis: A Guide to the World's Fresh Water Resources, Oxford University Press, New York, 1993.

[6] S.C. Nixon, D.T.E. Hunt, C. Lallana, A.F. Boschet, Sustainable Use of Europe's Water?, State, Prospects and Issues, Environmental Assessment Series, No 7, European Environment Agency, Copenhagen, 2004.

[7] EC, Comparative Study of Pressures and Measures in the Major River Basin Management Plans, Final Report, European Commission, Brussels, 2012.

[8] WWAP (World Water Assessment Programme), The United Nations World Water Development Report 4: Managing Water under Uncertainty and Risk, UNESCO, Paris, 2012. http://unesdoc.unesco.org/images/0021/002156/215644e.pdf.

[9] K. Lutchmiah, A.R.D. Verliefde, K. Roest, L.C. Rietveld, E.R. Cornelissen, Forward osmosis for application in wastewater treatment: a review, Water Res. 58 (2014) 179–197.

[10] IPCC, Revision of the "Revised 1996 IPCC Guidelines for National Greenhouse Gas Inventories IPCC Expert Group Scoping Meeting Report, Intergovernmental Panel on Climate Change, Geneva, 2003.

[11] J. Alcamo, J.M. Moreno, B. Novaky, M. Bindi, R. Corobov, R.J.N. Devoy, C. Giannakopoulos, E. Martin, J.E. Olesen, A. Shvidenko, Europe, in: M.L. Parry, J.P. Palutikof, P.J. Van der Linden, C.E. Hanson (Eds.), Climate Change 2007: Impacts, Adaptation and Vulnerability, Contribution of Working Group II to the Fourth Assessment Report of the Intergovernmental Panel on Climate Change, Cambridge University Press, Cambridge, 2007, pp. 541–580.

[12] UN, UN Human Settlements Programme. Water and Sanitation in the World's Cities: Local Action for Global Goals, Earthscan Publications, London, 2003.

[13] J.B. Zimmerman, J.R. Mihelcic, J. Smith, Global stressors on water quality and quantity, Environ. Sci. Technol. 42 (12) (2008) 4247–4254.

[14] G. Fischer, F.N. Tubiello, H. van Velthuizen, D. Wiberg, Climate change impacts on irrigation water requirements: global and regional effects of mitigation, 1990–2080, Technol. Forecast. Soc. Change 74 (2006) 1083–1107.

[15] J. Bruinsma, World Agriculture: Towards 2015/2030. An FAO Perspective, Earthscan, London, 2003, 444 pp.

[16] IPCC, Climate Change 2007: Synthesis Report, Intergovernmental Panel on Climate Change, Geneva, 2007.

[17] G.A. Weyhenmeyer, T. Blenckner, K. Pettersson, Changes of the plankton spring outburst related to the North Atlantic Oscillation, Limnol. Oceanogr. 44 (1999) 1788–1792.

[18] I. Durance, S.J. Ormerod, Climate change effects on upland stream macroinvertbrates over a 25-year period, Global Change Biol. 13 (2007) 942–957.

[19] M. Davidson, P. Boet, Climate change impacts on structure and diversity of fish communities in rivers, Global Change Biol. 13 (2007) 2467—2478.

[20] N.W. Arnell, Climate change impacts on river flows in Britain: the UKCIP02 scenarios, Water Environ. J. 18 (2004) 112—117.

[21] P.S. Lake, Flow-generated disturbance and ecological responses: floods and droughts, in: P.S. Wood, D.M. Hannah, J.P. Sadler (Eds.), Hydro Ecology and Eco Hydrology: Past, Present and Future, John Wiley and Sons, Chichester, 2007, pp. 75—92.

[22] U. Ulbrich, W. May, J.G. Pinto, P. Lionello, The Mediterranean climate change under global warming, in: P. Lionello, P. Malanotte-Rizzoli, R. Boscolo (Eds.), Mediterranean Climate Variability, Elsevier, Amsterdam, 2006, pp. 398—415.

[23] J. Luterbacher, et al., Mediterranean climate variability over the last centuries. A review, in: P. Lionello, P. Malanotte-Rizzoli, R. Boscolo (Eds.), Mediterranean Climate Variability, Elsevier, Amsterdam, 2006, pp. 27—48.

[24] F. Giorgi, Climate change hot-spots, Geophys. Res. Lett. 33 (2006) L08707.

[25] J.W. Hurrell, Decadal trends in the North Atlantic Oscillation: regional temperature and precipitation, Science 269 (1995) 676—679.

[26] R. Trigo, E. Xoplaki, E. Zorita, J. Luterbacher, S. Krichak, P. Alpert, J. Jacobeit, J. Saenz, J. Fernandez, F. Gonzalez-Rouco, R. Garcia-Herrera, X. Rodo, M. Brunetti, T. Nanni, M. Maugeri, M. Turkes, L. Gimeno, P. Ribera, M. Brunet, I. Trigo, M. Crepon, A. Mariotti, Relations between variability in the Mediterranean region and mid-latitude variability, in: P. Lionello, P. Malanotte-Rizzoli, R. Boscolo (Eds.), Mediterranean Climate Variability, Elsevier, Amsterdam, 2006, pp. 179—226.

[27] P. Alpert, M. Baldi, R. Ilani, S.O. Krichak, C. Price, X. Rodo, H. Saaroni, B. Ziv, P. Kishcha, J. Barkan, A. Mariotti, E. Xoplaki, Relations between climate variability in the Mediterranean region and the Tropics: ENSO, South Asian and African monsoons, hurricanes and Saharan dust, in: P. Lionello, P. Malanotte-Rizzoli, R. Boscolo (Eds.), Mediterranean Climate Variability, Elsevier, Amsterdam, 2006, pp. 149—177.

[28] P. Lionello, J. Bhend, A. Buzzi, P.M. Della-Marta, S. Krichak, A. Jansà, P. Maheras, A. Sanna, I.F. Trigo, R. Trigo, Cyclones in the Mediterranean region: climatology and effects on the environment, in: P. Lionello, P. Malanotte-Rizzoli, R. Boscolo (Eds.), Mediterranean Climate Variability, Elsevier, Amsterdam, 2006, pp. 325—372.

[29] F. Giorgi, P. Lionello, Climate change projections for the Mediterranean region, Global Planet. Change 63 (2008) 90—104.

[30] J.G. Lammers, Pollution of International Watercourses: A Search for Substantive Rules and Principles of Law, Martinus Nijhoff Publishers, The Hague, 1984.

[31] UNEP, The Greening of Water Law: Managing Freshwater Resources for People and the Environment, United Nations Environment Programme, Nairobi, 2010.

[32] UNEP, Clearing the Waters — A Focus on Water Quality Solutions, United Nations Environment Programme, Nairobi, 2010.

[33] K.,S. Lanz, S. Scheuer, EEB Handbook on EU Water Policy under the Water Framework Directive, European Environmental Bureau, Brussels, 2001.

[34] European Parliament, Simplification of European Water Policies, Policy Department, Economy and Science, Brussels, 2007.

[35] J.P. Deason, T.M. Schad, G.W. Sherk, Water policy in the United States: a perspective, Water Policy 3 (2001) 175—192.

[36] U.S Congress, Federal Water Pollution Control Act to Provide for Water Pollution Control Activities in the Public Health Service of the Federal Security Agency and in the Federal Works Agency, and for Other Purposes, As amended through P. L., 2002, pp. 107—303 http://www.epw.senate.gov/water.pdf.

[37] Water Law of the People's Republic of China, Decree No 74, 2002. http://www.mwr.gov.cn/english/01.pdf.

[38] Water Law of the People's Republic of China on Prevention and Control of Water Pollution, 2008. http://www.mwr.gov.cn/english/02.doc.

[39] NWC, Water Policy and Climate Change in Australia, Australian Government, National Water Commission, Canberra, 2012.

[40] NWC, Australian Environmental Water Management: Framework Criteria, National Water Commission, Canberra, 2012. http://archive.nwc.gov.au/__data/assets/pdf_file/0020/22169/Australian-environmental-water-management-framework-criteria.pdf.

[41] P.H. Gleick, et al., The World's Water, vol. 7, Island Press, Washington, 2011.

[42] A.J.B. Greenwood, The first stages of Australian forest water regulation: national reform and regional implementation, Environ. Sci. Policy 29 (2013) 124–136.

[43] M. Seliverstova, Current russian water legislation, in: M. Sengupta, R. Dalwani (Eds.), Proceedings of Taal 2007: The 12th World Lake Conference, 2008, pp. 1137–1141.

[44] UNEP, Freshwater Law and Governance: Global and Regional Perspectives for Sustainability, United Nations Environment Programme, Nairobi, 2015.

[45] P. Whitehouse, Measures for protecting water quality: current approaches and future developments, Ecotoxicol. Environ. Saf. 50 (2) (2001) 115–126.

[46] G. Grimvall, Å. Holmgren, P. Jacobsson, T. Thedéen, Risks in Technological Systems, Springer, New York, 2010.

[47] W. Czernuszenko, P.M. Rowiński, Water Quality Standards and Dispersion of Pollutants, Springer, New York, 2005.

[48] F.R. Rijsberman, Water scarcity: fact or fiction? Agr. Water Manage. 80 (2006) 5–22.

[49] T. Sato, M. Qadir, S. Yamamoto, T. Endoe, A. Zahoor, Global, regional, and country level need for data on wastewater generation, treatment, and use, Agri. Water Manage. 130 (2013) 1–13.

[50] D. Bixio, C. Thoeye, J. De Koning, D. Joksimovic, D. Savic, T. Wintgens, T. Melin, Wastewater reuse in Europe, Desalination 187 (2006) 89–101.

[51] WBSDC, Water: Facts and Trends, World Business Council for Sustainable Development, Geneva, 2005.

[52] P. Lens, L.H. Pol, P. Wilderer, T. Asano, Water Recycling and Resource Recovery in Industry: Analysis, Technologies and Implementation, International Water Association, London, 2002.

[53] EC, Updated Report on Wastewater Reuse in the European Union, Task A-03 Update of the Final Report on Wastewater Reuse in the European Union, European Commission, Brussels, 2013, http://ec.europa.eu/environment/water/blueprint/pdf/Final%20Report_Water%20Reuse_April%202013.pdf.

[54] EPA, Guidelines for Water Reuse, EPA/600/R-12/618, U.S. Environmental Protection Agency, Washington, 2012, http://nepis.epa.gov/Adobe/PDF/P100FS7K.pdf.

[55] WRF, Framework for Direct Potable Reuse, WateReuse Research Foundation, Alexandria, 2015. https://www.watereuse.org/wp-content/uploads/2015/09/14-20.pdf.

[56] EPA, Water Recycling and Reuse: The Environmental Benefits, U.S. Environmental Protection Agency, Washington, 1998. http://www3.epa.gov/region9/water/recycling/pdf/brochure.pdf.

[57] S.M. Scheierling, C. Bartone, D.D. Mara, P. Drechsel, Improving Wastewater Use in Agriculture. An Emerging Priority, Policy Research Working Paper 5412, The World Bank, Washington, 2010.

[58] B. Jiménez, T. Asano, Water Reuse: An International Survey of Current Practice, Issues and Needs, IWA Publishing, London, 2008.

[59] NWC, Using Recycled Water for Drinking, an Introduction, Waterlines Occasional Paper No 2, National Water Commission, Canberra, 2007.

[60] I.K. Kalavrouziotis, P. Kokkinos, G. Oron, et al., Current status in wastewater treatment, reuse and research in some Mediterranean countries, Desalin. Water Treat. 53 (8) (2015) 2015–2030.

[61] D.R. Rowe, I.M. Abdel-Magid, Handbook of Wastewater Reclamation and Reuse, CRC Press, New York, 1995.

[62] EC, A Blueprint to Safeguard Europe's Water Resources, COM (2012) 673, European Commission, Brussels, 2012, http://eur-lex.europa.eu/legal-content/EN/TXT/PDF/?uri=CELEX:52012DC0673&from=EN.

[63] IIT, Review of Wastewater Reuse Projects Worldwide. Collation of Selected International Case Studies and Experiences, Indian Institutes of Technology, Mumbai, 2011. http://52.7.188.233/sites/default/files/012_EQP.pdf.

[64] CDPH, California Code of Regulations. Regulations Related to Recycled Water, California Department of Public Health, Sacramento, 2014. http://www.waterboards.ca.gov/drinking_water/certlic/drinkingwater/documents/lawbook/RWregulations_20140618.pdf.

[65] NWQMS, Overview of the Australian Guidelines for Water Recycling: Managing Health and Environmental Risks 2006, Environment Protection and Heritage Council, the Natural Resource Management Ministerial Council, Canberra, 2008. https://www.environment.gov.au/system/files/resources/5a40b1bc-1928-4dee-a8d2-82104e06b456/files/overview-water-recycling-guide-21a.pdf.

[66] C. Prasse, D. Stalter, U. Schulte-Oehlmann, J. Oehlmann, T.A. Ternes, Spoilt for choice: a critical review on the chemical and biological assessment of current wastewater treatment technologies, Water Research 87 (2015) 237–270.

[67] F.R. Spellman, Spellman's Standard Handbook for Wastewater Operators, vol. 1, Technomic Publishing Co., Lancaster, 1999.

[68] P.A. Vesilind, Wastewater Treatment Plant Design, IWA Publishing, London, 2003.

[69] C. Forster, Wastewater Treatment and Technology, Thomas Telford Ltd, London, 2003.

[70] Metcalf and Eddy Inc, Wastewater Engineering. Treatment and Reuse, fourth ed., McGraw-Hill, Singapore, 2004.

[71] A.A. Zorpas, V.J. Inglezakis, Sludge Management – From the Past to Our Century, Nova Science Publishers, New York, 2012.

[72] S. Jeyanayagam, True confessions of the biological nutrient removal process, Fla. Water Res. J. (2005) 37–46.

[73] S. Malamis, Biological Treatment of Wastewater with the Use of Membranes (Ph.D. thesis), National Technical University of Athens, Greece, 2009 (in Greek).

[74] B. Lesjean, A. Tazi-Pain, D. Thaure, Moeslang, H. Buisson, Ten persistent myths and the realities of membrane bioreactor technology for municipal applications, Water Sci. Technol. 63 (1) (2011) 32–39.

[75] S. Di Fabio, S. Malamis, E. Katsou, G. Vecchiato, F. Cecchi, F. Fatone, Are centralized MBRs coping with the current transition of large petrochemical areas? A pilot study in Porto-Marghera (Venice), Chem. Eng. J. 214 (2013) 68–77.

[76] L.K. Wang, N.K. Shammas, Y.T. Hung, Advanced Biological Treatment Processes, Humana Press, New York, 2009.

[77] J.C. Crittenden, R.R. Trussell, D.W. Hand, K.J. Howe, G. Tchobanoglous, MWH's Water Treatment Principles and Design, third ed., Wiley & Sons, New Jersey, 2012.

[78] IPIECA, Petroleum Refining Water/wastewater Use and Management, Operations Good Practice Series, London, 2010. http://www.ipieca.org/sites/default/files/publications/Refining_Water_0.pdf.

[79] L.K. Wang, Y.T. Hung, H.H. Lo, C. Yapijakis, Handbook of Industrial and Hazardous Wastes Treatment, CRC Press, New York, 2004.

[80] V.J. Inglezakis, S. Malamis, A. Omirkhan, J. Nauruzbayeva, Z. Makhtayeva, T. Seidakhmetov, A. Kudarova, Investigating the inhibitory effect of 4-nitrophenol, phenol and cyanide in the activated sludge process employed for the treatment of petroleum wastewater, in: International Conference on Industrial Waste & Wastewater Treatment & Valorization, IWWATV 2015, Athens, Greece, 2015.

[81] M.T. Ravanchi, T. Kaghazchi, A. Kargari, Application of membrane separation processes in petrochemical industry: a review, Desalination 235 (2009) 199–244.

[82] J.M. Wong, Water reuse for petroleum oil, product processing industries, Ind. Waterworld 165 (2012) 18–22.

[83] K. Zotalis, E.G. Dialynas, N. Mamassis, A.N. Angelakis, Desalination technologies: Hellenic experience, Water 6 (2014) 1134–1150.

[84] A. Bennet, Developments in desalination and water reuse, Filtr. Sep. 52 (4) (2015) 28–33.

[85] A. Bennet, Desalination and water reuse: what's the future for forward osmosis? Filtr. Sep. 50 (5) (2013) 28–30, 32–34.

[86] Waterwise, Water − The Facts, Why Do We Need to Think about Water?, Waterwise, London, 2012. http://www.waterwise.org.uk/data/resources/25/Water_factsheet_2012.pdf.

[87] C. Rodriguez, P. Van Buynder, R. Lugg, P. Blair, B. Devine, A. Cook, P. Weinstein, Indirect potable reuse: a sustainable water supply alternative, Int. J. Environ. Res. Public Health 6 (3) (2009) 1174−1209.

[88] Brown, Caldwell, Recycled Water Study, Prepared for City of San Diego, San Diego, 2012, http://www.sandiego.gov/water/pdf/purewater/2012/recycledfinaldraft120510.pdf.

[89] ATSE, Drinking Water through Recycling. The Benefits and Costs of Supplying Direct to the Distribution System, Australian Academy of Technological Sciences and Engineering, Melbourne, 2013.

[90] E. Schroeder, G. Tchobanoglous, H.L. Leverenz, T. Asano, NWRI White Paper, Direct Potable Reuse: Benefits for Public Water Supplies, Agriculture, the Environment, and Energy Conservation, Prepared for the National Water Research Institute, Fountain Valley, California, 2012. http://www.nwri-usa.org/documents/NWRIWhitePaperDPRBenefitsJan2012.pdf.

[91] A. Tenne, Seawater Desalination in Israel: Planning, Coping with Difficulties, and Economic Aspects of Long-term Risks, Water Authority, State of Israel, Desalination Division, 2010. http://www.water.gov.il/hebrew/planning-and-development/desalination/documents/desalination-in-israel.pdf.

[92] WBCSD, Water. Facts and Trends, World Business Council for Sustainable Development, 2005. Geneva, http://www.unwater.org/downloads/Water_facts_and_trends.pdf.

[93] WI, WorldWatch Institute, "State of the World 2014. Governing for Sustainability", Worldwatch Institute, Island Press, Washington, 2014.

[94] N.M. Nelson, et al., Linking ecological data and economics to estimate the total economic value of improving water quality by reducing nutrients, Ecol. Econ. 118 (2015) 1−9.

[95] J. Ramajo-Hernández, S. del Saz-Salazar, Estimating the non-market benefits of water quality improvement for a case study in Spain: a contingent valuation approach, Environ. Sci. Policy 22 (2012) 47−59.

[96] D. Andreopoulos, et al., Estimating the non-market benefits of climate change adaptation of river ecosystem services: a choice experiment application in the Aoos basin, Greece, Environ. Sci. Policy 45 (2015) 92−103.

[97] P. Efthimoglou, A. Xepapadeas, Public Enterprises, Theoretical and Applied Approach, Stamoulis Publications, Athens, 1990 (in Greek).

[98] A. Tversky, D. Kahneman, Loss aversion in riskless choice: a reference-dependent model, Quarterly J. Econ. 106 (4) (1991) 1039−1061.

[99] A. Randall, J.R. Stoll, Consumer's surplus in commodity space, The American Economic Review 70 (3) (1980) 449−455.

[100] C.D. Kolstad, R.M. Guzman, Information and the divergence between WTA and WTP, J. Environ. Econ. Manage. 38 (1999) 66−80.

[101] J. Zhao, C.L. Kling, A new explanation for the WTP/WTA disparity, Econ. Lett. 73 (2001) 293−300.

[102] W.M. Hanemann, Willingness to pay and willingness to accept: how much can they differ? Am. Econ. Rev. 81 (3) (1991) 635−647.

[103] M.W. Hanemann, The economic conception of water, in: P.P. Rogers, M.R. Llamas, L. Martinez-Cortina (Eds.), Water Crisis: Myth or Reality?, Taylor & Francis, London, 2005, pp. 61−91.

[104] V. Yevjevich, Water and civilization, Water Int. 17 (4) (1992) 163−171.

[105] A. Randall, Property entitlements and pricing policies for a maturing water economy, Aust. J. Agric. Econ. 25 (1981) 195−212.

[106] EEA, Water Prices, Indicator Fact Sheet, European Environment Agency, Copenhagen, 2003, http://www.eea.europa.eu/data-and-maps/indicators/water-prices/water-prices/at_download/file.

[107] EC, Directive 2000/60/EC of the European Parliament and of the Council of 23 October 2000 Establishing a Framework Action in the Field of Water Policy, O. J., L327, 22/12/2000, 2000, pp. 0001−0073.

[108] J.A. Beecher, J.A. Kalmbach, Structure, regulation, and pricing of water in the United States: a study of the Great Lakes region, Util. Policy 24 (2013) 32–47.

[109] ESCOSA, Economic Regulation of SA Water's Revenues. Statement of Approach, Essential Services Commission of South Australia, Adelaide, 2012. http://www.escosa.sa.gov.au/library/ 120713-EconomicRegulationOfSAWatersRevenue-StatementOfApproach.pdf.

[110] W.L. Nieuwoudt, G.R. Backeberg, A review of the modelling of water values in different use sectors in South Africa, Water SA 37 (5) (2011) 703–710.

[111] A.N. Menegaki, N. Hanley, K.P. Tsagarakis, The social acceptability and valuation of recycled water in Crete: a study of consumers' and farmers' attitudes, Ecol. Econ. 62 (1) (2007) 7–18.

[112] M. Genius, A.N. Menegaki, K.P. Tsagarakis, Assessing preferences for wastewater treatment in a rural area using choice experiments, Water Resour. Res. 48 (4) (2012) 1–11.

[113] I.J. Bateman, R.T. Carson, B. Day, M. Hanemann, N. Hanley, T. Hett, M. Jones-Lee, G. Loomes, S. Mourato, E. Ozdemiroglu, D.W. Pearce OBE, R. Sugden, J. Swanson, Economic Valuation with Stated Preference Techniques, A MANUAL, Edward Elgar, UK, USA, 2002, p. 458.

CHAPTER 4

Soil Environment

M.K. Doula
Benaki Phytopathological Institute, Kifisia, Greece

A. Sarris
Institute for Mediterranean Studies, Rethymno, Greece

Chapter Outline

Environment and Development. http://dx.doi.org/10.1016/B978-0-444-62733-9.00004-6

4.1 Introduction: Basic Definitions
4.1.1 Soil: The Skin of the Earth

Soil is generally defined as the top layer of the earth's crust, formed by mineral particles, organic matter, water, air, and living organisms. It is the interface between earth, air, and water and hosts most of the biosphere [1]. Soil, however, is not merely the sum of these constituents, but a product of their interactions. Soil is an extremely complex and variable medium; a typical sample of mineral soil comprises 45% minerals, 25% water, 25% air, and 5% organic matter; however, these proportions may vary [2].

Weathering is the driving process of soil development and describes the means by which soil, rocks, and minerals are changed by physical and chemical processes into other soil components. Therefore, weathering is an integral part of soil development. Depending on the soil-forming factors in an area, weathering may proceed rapidly over a decade or slowly over millions of years. Because it develops over very long timescales, soil can be considered a nonrenewable natural resource.

Soil formation is a dynamic rather than a static process [3] while five major factors influence the kinds of soil that develop. Wherever these five factors have been the same on the landscape, the soil will be the same. However, if one or more of the factors differ, the soils will be different. The factors are [4]:

- Climate (mainly temperature and precipitation). Climate determines the nature of the weathering that occurs. Temperature and precipitation, for example, affect the rates of physical, chemical, and biological processes that define the profile development.

- Living organisms. Native vegetation, microbes, soil animals, and human beings are factors that influence organic matter accumulation, profile mixing, nutrient cycling, and soil structural stability.
- Nature of parent material. Geological processes have brought to the Earth's surface numerous parent materials in which soils form. The nature of parent materials influence mainly soil texture[1] and thus many physical properties of soil such as downward movement of water, composition, natural vegetation, and the quantity and type of clay minerals present in the soil profile.
- Topography of the site, which relates to the configuration of the land surface and is described in terms of difference in elevation, slope, etc. The topography of the land can hasten or delay the processes of climate forces and therefore can modify their effects as well as the vegetative effects, having a major direct effect on soil formation and on the type of soil that forms.
- The length of time that parent material have been subjected to weathering. The time required for the development of a horizon, however, is influenced by the parent material, the climate, and the vegetation, emphasizing the interaction of time with the other soil-forming factors.

A soil is distinguished from weathered parent material by the vertical differentiation it exhibits due to biological activity, so that the properties that are singled out in most systems of soil classification must be displayed in the soil profile [5]. Soil only develops where there is a dynamic interaction between the air, water, living organisms, and geology. It is these dynamic interactions, which contribute to the multiple functions that soils perform.

Despite the above theoretical terms, there are different concepts as to what soils are, depending on the purpose for which a soil is used. For example, to a mining engineer, soil is the debris covering the rocks or minerals that must be quarried. It is, therefore, a nuisance and must be removed. To a highway engineer, soil may be the material on which a roadbed is to be placed and if its properties are unsuitable, it will need to be removed and replaced with rocks and gravels. To an average homeowner a good soil is rich, dark, and crumbly as opposed to "hard clay," which resists being spaded into a seedbed for a flower or vegetable garden. The farmer, along with the homeowner, looks upon the soil as

---

[1] Soil texture is determined by the size and type of particles that make up the soil (including the organic but mostly referring to the inorganic material). The size of the ex-rock pieces varies substantially, from large bits of gravel to much, much smaller clay pieces [4]:

- Gravel—particles greater than 2 mm in diameter.
- Coarse sand—particles less than 2 mm and greater than 0.2 mm in diameter.
- Fine sand—particles between 0.2 and 0.02 mm in diameter.
- Silt—particles between 0.02 and 0.002 mm in diameter.
- Clay—particles less than 0.002 mm in diameter.

a habitat for plants. However, the farmer earns a living from the soil and is therefore forced to pay more attention to its characteristics. For the farmer, soil is more than useful—it is indispensable. All these different perspectives of what this medium is, have led to a misunderstanding and devaluation of soil importance for our life and made soil the poor relevant of the other two major life components, ie, water and air. One more reason for this devaluation is that although water and air degradation are fast-seen processes with obvious consequences on human health and the quality of the environment, the degradation of soil is a very slow process that may occur for many years without giving obvious consequences or with consequences that may be easily underestimated (eg, reduced fertility, need of more intense fertilization), but when at the last stage degradation is nonreversible.

4.1.2 Soil Functions

Soil is a dynamic and living resource, which needs minimal and suitable conditions to carry out its indispensable functions for its conservation, to produce food, and for supporting the environment quality [6].

The soil can be thought of as a medium providing society with various benefits or services. Life on earth would be impossible without the soil and all the things that it performs for the humanity. Whatever people eat, drink or breathe, or wear, almost all comes from the soil or are dependent upon it [7].

Owners of soil or land might appreciate its value in supporting the foundations of their house and in providing a garden. Less obvious are the many other benefits the soil is providing, both to them and their descendants. Soils regulate neatly all of the water and biogeochemical cycles that are critical for maintaining critical elements of both the climate and biodiversity. Interference with these processes and ecological health of the soil is a major factor that underlies climate change and biodiversity loss.

Soil provides people food, biomass, and raw materials. It serves as a platform for human activities and landscape and as an archive of heritage and plays a central role as a habitat and gene pool. It stores, filters, and transforms many substances, including water, nutrients, and carbon. In fact, it is the biggest carbon store in the world (1500 gigatons) [1,8].

Drinking water quality is directly linked with soils; soil is a filtering and buffering medium for contaminants. A number of physical and chemical properties of soils result in clean groundwater for people and animals. Overloading a soil with contaminants as well as limiting its permeable surface by sealing and compaction can severely affect its functioning as a filtering (actually the biggest filter on Earth) and buffering medium for water. Drastic changes in soil pH can dramatically affect the retention capacity for contaminants, eventually triggering the sudden release of contaminants into the groundwater.

Soils are the home of one of the largest pools of *biodiversity* on earth. Soil biodiversity is defined by the variation in soil life, from genes to communities, and the variation in soil habitats, from microaggregates to entire landscapes [9]. There are an enormous number of organisms living in soil, mostly belonging to species yet to be fully described and studied. There is more biomass inside the soil than on it. Only little is known about this ecosystem, mostly due to the lack of methods for effectively isolating the different organisms present in the soils of the world. Only few species have been fully described and isolated, often leading to the discovery of new sources of pharmacologically active natural substances, for example penicillin. Soil biodiversity reflects the mix of living organisms in the soil. These organisms interact with one another and with plants and small animals forming a web of biological activity. Soil is by far the most biologically diverse part of Earth. Soil biota play many fundamental roles in delivering key ecosystem goods and services, such as releasing nutrients from soil organic matter (SOM), forming and maintaining soil structure, and contributing to soil water entry, storage, and transfer [10].

Specific attention is given nowadays to the role of the soil in greenhouse gas (GHG) emissions and to the concern about soils being "chemical time bombs," which is associated to the nonlinear behavior of the soil system. It is only in recent times that the full importance of soils for *global climate change* has been recognized. More research is now devoted to this in order to understand better its role in relation to the increase of GHGs in the atmosphere. Actually, soils in the world contain an estimated amount of c. 1.500 Gt of carbon (650 Gt in vegetation), and it is therefore the most important compartment of carbon in the terrestrial biosphere. Maintaining and eventually even increasing this large organic carbon pool is of crucial importance for limiting the increase of CO_2 in the atmosphere. A number of agricultural practices have been recognized as having a substantial beneficial effect on soil organic carbon (SOC) content. Promoting the adoption of such practices would help reverse the current trend of SOM depletion in agricultural land [11].

Concern about threats to human health is often a top human preoccupation so that any aspect of air and water that is related to human health gains immediate political attention. On the contrary, knowledge about health-related effects of soil degradation is limited. Usually these are perceived in relation to soil pollution by chemicals and the possible role in contaminating the food chain. "Healthy food from healthy soils" is an immediately understandable slogan for anybody. Although the scientific links between the soil and food quality are very complex and in some cases there is no real evidence to help distinguish between real and imaginary risks, thanks to many international awareness-raising campaigns, there is nowadays an increasing public interest in organic farming and sustainable agriculture. However, this is only one piece of the puzzle. Other off-site effects of soil degradation with immediate implications for our daily life should be also highlighted and included in these awareness-raising campaigns.

4.1.3 Soil Quality, Health, Fertility, and Resilience

The concept of soil quality emerged in the early 1990s, and the first official definition of this term was proposed by the Soil Science Society of America Ad Hoc Committee on Soil Quality (S-581) in 1997 [12].

Soil quality was defined as "the capacity of a specific kind of soil to function, within natural or managed ecosystem boundaries, to sustain plant and animal productivity, maintain or enhance water and air quality, and support human health and habitation." For the committee proposing this definition, the term soil quality is not synonymous with soil health, and they should not be used interchangeably. Soil quality is related to soil functions, whereas soil health presents the soil as a finite and dynamic living resource [13].

Therefore, *soil health* is defined as "the continued capacity of soil to function as a vital living system, within ecosystem and land-use boundaries, to sustain biological productivity, maintain the quality of air and water environments, and promote plant, animal, and human health" [14].

These two definitions may appear similar, but soil health concept directly mentions plant health, which is not the case in the definition of soil quality of Karlen et al. [12]. In a simple manner, the Natural Resources Conservation Service of the United States Department of Agriculture [15] proposes the following definition: "soil quality is how well soil does what we want it to do"; because of the numerous possible uses of soil, the meaning of the term soil quality heavily depends on the ecosystem considered. In agricultural soils, plant and animal productivity and health would be of the greatest importance, whereas it would not be the same in urban soils. Even in a given ecosystem, eg, cultivated soils, their multifunctionality makes it difficult to define a healthy soil without first defining the targeted goal or aim. Such goals could be plant health, atmospheric balance, avoidance of erosion, etc.

While *soil fertility* is defined as the quality of a soil that enables it to provide essential chemical elements in quantities and proportions for the growth of specified plants, *productivity* is the capacity of a soil for producing a specified plant or sequence of plants under a specified system of management. Productivity, therefore, emphasizes the capacity of soil to produce crops and is expressed in terms of yields [4].

Resilience refers to the capacity of a system to absorb change without significantly altering the relationship between the relative importance and numbers of individuals and species of which the community is composed [16]. Therefore, soil resilience refers to the ability of soil to resist or recover from an anthropogenic or natural perturbation. There is a close interdependence between soil quality and soil resilience since resilient soils have a high quality and vice versa [17].

4.1.4 Soil Composition and Types—Classification

Characteristics of the soil vary widely from place to place, eg, the soil on steep slopes is generally not as deep and productive as soil on gentle slopes; soil that has developed from sandstone is more sandy and less inherently productive than soil formed from rocks such as limestone; soil that has developed in tropical climates is quite different than a soil found in temperate or arctic areas. To describe soil variations, soil scientists have set up classification systems that recognize a large number of individual soils, each having distinguishing characteristics.

Examination of a vertical section of a soil reveals the presence of more or less distinct horizontal layers. Such a section is called *profile*, and the individual layers are known as *horizons*. Every well-developed, undisturbed soil has its own distinctive profile characteristics. These are useful in classifying and surveying soils but are of greatest importance in determining how the soils can best be used. The uppermost layers or horizons of a soil profile are darker in color than the lower horizons (A horizons). This difference is due to the accumulation of organic matter that results from the decay of plant roots and of other organic residues incorporated into the upper soil layers. Also weathering tends to be more intense in the upper horizon than in the lower layers [4].

Surface soil is the major zone of root development for crop plants. It contains many of the nutrients available to plants and supplies much of the water necessary for their growth. Through proper cultivation and the incorporation of organic residues, the topsoil can be kept loose and open to assure balanced air and water supplies for plant roots. It can be treated easily with commercial fertilizers and limestone, permitting the soil's fertility, and to a lesser degree its productivity, to be raised or stabilized at levels consistent with economic crop production.

The underlying layers contain comparatively less organic matter than those nearer the surface. They are characterized by an accumulation of varying amounts of substances such as silicate clays, iron and aluminum oxides, gypsum, and calcium carbonate. These underlying layers are referred to as B horizons. Therefore, the subsoil is comprised of those soil layers underneath the topsoil. It is not seen from the surface and is not commonly disturbed by soil tillage; however, its characteristic may affect land uses, eg, crop production is affected by root penetration into the subsoil and by the reservoir of moisture and nutrients it represents, or downward movement of drainage water is sometimes impeded or enhanced by subsoil due to its specific properties [18].

In order the wide variation of soil types to be understood, some definitions and terms of soil classification are given in the following.

Soil Taxonomy comprises six categories of classification and can be compared with those used for the classification of plants: (1) order (the broadest category), (2) suborder, (3) great group, (4) subgroup, (5) family, and (6) series (the most specific category) [4].

As regards soil orders there are 12 soil orders, ie, Alfisols, Andisols, Aridisols, Entisols, Gelisols, Histosols, Inceptisols, Mollisols, Oxisols, Spodosols, Ultisols, and Vertisols. Photo 4.1 presents these different soil orders while the global distribution of the 12 soil orders can be seen in Map 4.1 available from the US Department of Agriculture [19].

The major characteristics of the soil orders are summarized in Table 4.1.

Further classification results in a wide variety of different soil types. For instance, over 320 major soil types have been identified in Europe and within each there are enormous variations in physical, chemical, and biological properties. Almost 23,000 soil series in various combinations with different slopes and surface textures have been identified in the United States [20]. The major soil types have been recorded in maps, as for example, within the European Soil Information System (EUSIS). In particular, the Soil Geographical Database of Europe at scale 1:1,000,000 can be used to summarize the distribution of the major soils of Europe [21].

4.2 Soil Threats and Degradation

Soil is subject to a series of degradation processes or threats. Its structure plays a major role in determining its ability to perform its functions, and any damage to soil's structure also damages other environmental media and ecosystems.

Soil degradation is a serious threat for an increasing number of areas all over the world. It is defined as "a process that causes deterioration of soil productivity and low soil utility as a result of natural or anthropogenic factors which namely are displacement of soil material, and internal soil deterioration" [22,23].

Nine major soil threats are recognized and are subject of specific actions taken worldwide. These are erosion, decline in organic matter, local and diffuse contamination, sealing, compaction, decline in biodiversity, salinization and sodification, floods, and landslides. A combination of some of these threats can ultimately lead arid or subarid climatic

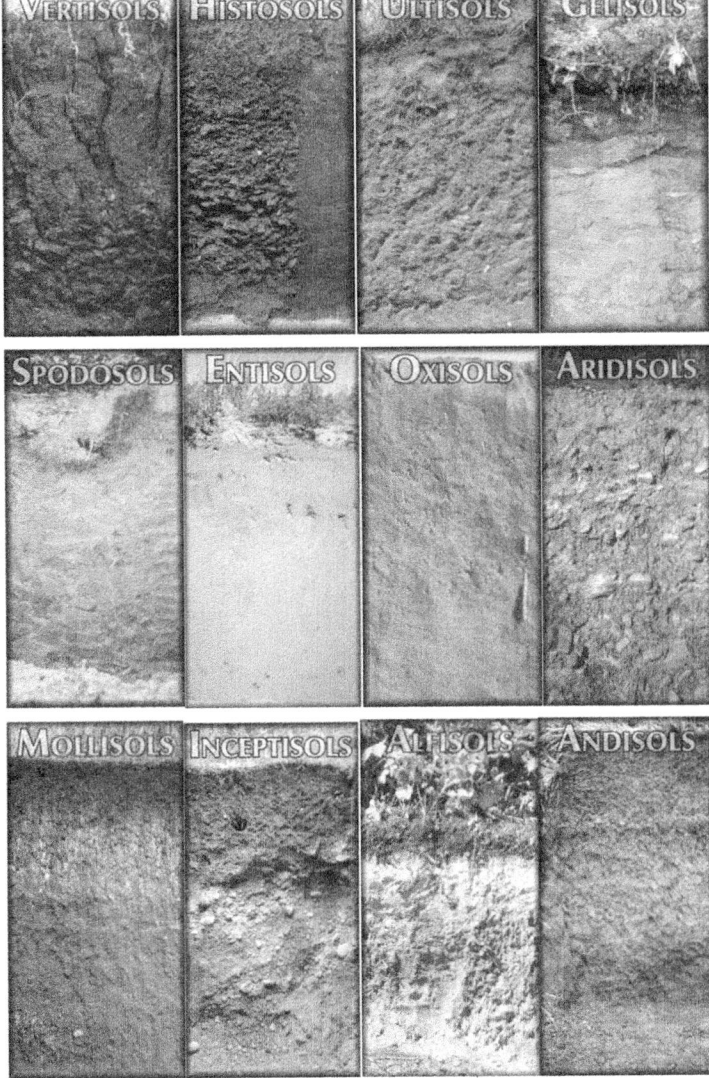

Photos 4.1
The 12 soil orders [19].

conditions to desertification [1]. However, desertification is not a primary threat to soil, but it caused a combination of threats and conditions.

The phenomenon of soil degradation generated by human activities is a very ancient feature. Historic soil erosion was known to occur in cycles during which periods of erosion were followed by periods of stability or soil formation. Soil loss was not only experienced as negative and in many cases it may have deliberately been induced by the

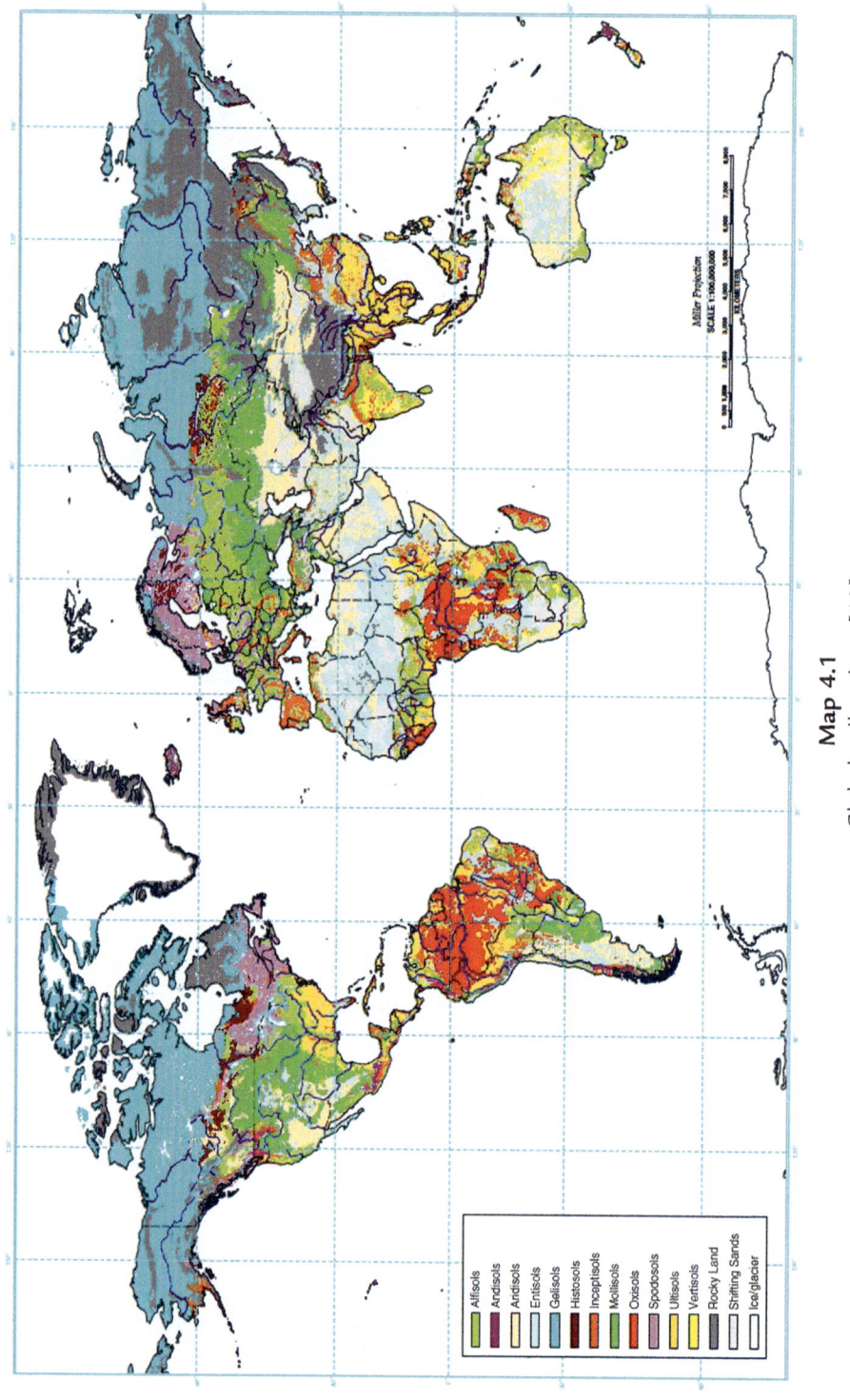

Map 4.1
Global soil regions [19].

Table 4.1: Major Characteristics of the Soil Orders [4,19].

Soil Order	Description	% of World Surface	Major Land Uses	Fertility
Alfisols	Found in semiarid to moist areas. Result from weathering processes that leach clay minerals and other constituents out of the surface layer and into the subsoil, where they can hold and supply moisture and nutrients to plants. Formed primarily under forest or mixed vegetative cover. Productive for most crops.	About 10%	Cropland, forests	High
Andisols	Common in cool areas with moderate to high precipitation, especially areas associated with volcanic materials. Formed by weathering processes that generate minerals with little orderly crystalline structure. Can result in an unusually high water- and nutrient-holding capacity. They include weakly weathered soils with much volcanic glass as well as more strongly weathered soils.	About 1%	Cropland, forests	Moderate
Aridisols	Common in deserts. Too dry for mesophytic plant growth. The lack of moisture greatly restricts the intensity of weathering processes and limits most soil development processes to the upper soil parts. Often accumulate gypsum, salt, calcium carbonate, and other materials that are easily leached from soils in more humid environments.	About 12%	Rangeland	Low
Entisols	Occur in many environments and mainly in areas of recently deposited parent materials or in areas where erosion or deposition rates are faster than the rate of soil development, such as dunes, steep slopes, and floodplains. Soils that show little or no evidence of pedogenic horizon development.	About 16%	Wetland, rangeland, forest, cropland	Low to moderate
Gelisols	Common in the higher latitudes or at high elevations. Soils that have permafrost near the soil surface and/or have evidence of cryoturbation (frost churning) and/or ice segregation.	About 9%	Best left wild	Low
Histosols	Commonly called bogs, moors, peats, or mucks; have a high content of organic matter and no permafrost. Most are saturated year round, but a few are freely drained. They form in decomposed plant remains that accumulate in water, forest litter, or moss faster than they decay. If drained and exposed to air, microbial decomposition is accelerated and the soils may subside dramatically.	About 1%	Wetland, cropland	Moderate

Continued

Table 4.1: Major Characteristics of the Soil Orders [4,19].—cont'd

Soil Order	Description	% of World Surface	Major Land Uses	Fertility
Inceptisols	Commonly found in semiarid to humid environments that generally exhibit only moderate degrees of soil weathering and development. Have a high range in characteristics and occur in a wide variety of climates.	About 17%	Cropland, forests	Low to moderate
Mollisols	They are extensive soils on the steppes of Europe, Asia, North America, and South Africa. Soils with dark-colored surface horizon relatively high in content of organic matter; they are base rich throughout and therefore are quite fertile.	About 7%	Cropland, rangeland	High
Oxisols	Highly weathered soils of tropical and subtropical regions, dominated by low-activity minerals, such as quartz, kaolinite, and iron oxides. They tend to have indistinct horizons. They characteristically occur on land surfaces that have been stable for a long time and have low fertility and low capacity to retain additions of lime and fertilizer.	About 8%	Cropland, forests	Low
Spodosols	Commonly occur in areas of coarse-textured deposits under coniferous forests of humid regions and tend to be acid and infertile. They are formed from weathering processes that strip organic matter combined with aluminum (with or without iron) from the surface layer and deposit them in the subsoil.	About 4%	Forests	Low
Ultisols	Commonly occur in humid areas and formed from fairly intense weathering and leaching processes that result in a clay-enriched subsoil dominated by minerals, such as quartz, kaolinite, and iron oxides. They are typically acid soils in which most nutrients are concentrated in the upper few centimeters. They also have a moderately low capacity to retain additions of lime and fertilizer.	About 8%	Forests, cropland	Low to moderate
Vertisols	Have high content of expanding clay minerals and undergo pronounced changes in volume with changes in moisture. They have cracks that open and close periodically, and that show evidence of soil movement in the profile. Because they swell when wet, vertisols transmit water very slowly and have undergone little leaching. They tend to be fairly high in natural fertility.	About 2%	Cropland, forests	High

local communities to produce fertile sediments. Soil erosion has been at the basis of the creation of large alluvial plains that were extensively cultivated in ancient times. Ancient societies adapted to the redistribution of resources (soil and water) and exploited the environment as how it had become. Historic erosion had many different complex causes ranging from the destruction of forests for fuel or timber to the neglect of common lands and extremes in climate. However, large efforts had been made at the end of the 19th century to reestablish forests, and there had been considerable success in reducing soil degradation.

Soil is constantly changing and evolving, and while some degradation processes are natural, human activity can accelerate these processes, and introduce others, and thereby impair the soil's capacity to carry out the functions we require from it. In the last 50 years, pressure on soil has been considerably accelerated in many places mainly due to intensification and expansion of grassland and cropland which has led to a number of soil degradation processes with different severity and area coverage, among which soil erosion is considered one of the most dangerous processes [24].

At present, soil is a threatened natural resource, and even though there is a wealth of know-how related to land management, improvement of soil fertility, and protection of soil resources, some 17% of the land surface has already been strongly degraded and the affected area is still growing. Overexploitation, overgrazing, inappropriate clearing techniques, and unsuitable land use practices have resulted in severe nutrient decline, water and wind erosion, compaction, and salinization. The resulting decrease in productivity has especially affected marginally suitable lands that were taken into cultivation due to population pressure, or that were not given the opportunity to recuperate for a sufficient long time after prolonged cultivation [25].

The degree of soil degradation depends on soil's susceptibility to degradative processes, land use, and the duration of degradative land use [26]. Human activities contributing to land degradation include unsuitable agricultural land use, poor soil and water management practices, deforestation, removal of natural vegetation, frequent use of heavy machinery, overgrazing, improper crop rotation, and poor irrigation practices. Natural disasters, including droughts, floods, and landslides, also contribute.

As an effort to estimate the problem of soil degradation at the global level, a global assessment of soil degradation (GLASOD) was undertaken in the early 1990s [24,27] and a land degradation assessment of drylands (LADA) was initiated by GEF (Global Environment Facility) and UNEP (United Nations Environment Programme) in 2000 and is now being developed with FAO (Food and Agriculture Organization of the United States). In 1990, Oldeman et al. [24] published a world map known as GLASOD, showing the extent of human-induced soil degradation. According to this study, 1964 million hectares, that is, 15.1% of the total land surface, or about one-third of the land used for agriculture, was affected by all forms of soil degradation. Of the affected area, 55.6% was reported as damaged by water erosion, 27.9% by wind erosion, 12.2% by chemical, and 4.2% by physical degradation [24].

Damage to soils from human activities is increasing, and despite the general recognition that soil degradation is a serious and widespread problem worldwide, it has not been quantified, at least accurately, and its geographical distribution and real extent are unknown. Nowadays some studies are beginning to question the data, arguing that degradation estimates are overstated. A major reason suggested for the overestimation of land degradation has been underestimation of the abilities of local farmers [28]. These authors argue, "experts need to discriminate more carefully between a naturally bad state, a temporary bad state and a degraded state of land."

4.2.1 Erosion

Soil erosion is recognized as one of the most important soil degradation processes worldwide. In many places, soil erosion is the most severe consequence of soil degradation with respect to restoration of soil quality, and controlling erosion is a prerequisite for a healthy soil. Erosion has severe effects on soil functions—such as the soil's ability to act as a buffer and filter for pollutants, its role in the hydrological and nitrogen cycle, and its ability to provide habitat and support biodiversity. From the almost 2000 million hectares of degraded soils worldwide, water-eroded land is estimated to be almost 56% of land, wind-eroded almost 28%, while only the 4% of the eroded land is due to physical degradation processes [29].

Soil erosion is a natural process that can be exacerbated by human activities and is the wearing away of the land surface by physical forces such as rainfall, flowing water, wind, ice, temperature change, gravity, or other natural or anthropogenic agents that abrade, detach, and remove soil or geological material from one point on the earth's surface to be deposited elsewhere [30].

Soil erosion rates over much of Europe were very low under the mixed farming that occurred before and just after the World War II. In the 1960s and 1970s several scientists who had worked on soil conservation schemes in Africa and Asia returned to Europe where they could see that erosion rates were becoming high as well. They began to measure the long-term impacts of slow processes, not only erosion by rain-wash but also those caused by plowing, grazing, and the loss of soil with root crops. In most of Europe, soil erosion and soil contamination began to be serious issues with the advent of modern agriculture after the World War II. At that time, agricultural policy focused on increasing agricultural production to sustain food security [31]. To make farming more effective and raise farm income, mechanization and intensification took place. Pesticides were developed to control plant diseases. Small fields got consolidated to enlarge fields. Furthermore, fields were leveled to make tillage, plant treatment, and harvesting operations more effective. Soil quality was gradually becoming less important as a deciding factor for the agricultural system. As a consequence, organic matter and soil biodiversity decreased and soils that were susceptible became more compact and sensitive to erosion.

In the beginning of the 19th century agriculture could be seen to be having a great impact on land degradation and erosion in the United States of America. This impact resulted in the abandonment and afforestation of huge areas in New England and the Appalachians.

The worse event of soil erosion with extreme social, economic, and environmental consequences, due to severe drought and the failure to apply dryland-farming methods to prevent wind erosion, is known as *Dust Bowl era*; it is characterized as a natural disaster and took place during the period of drought from 1931 to 1939. The Dust Bowl, also known as the *Dirty Thirties*, was a period of severe dust storms that greatly damaged the ecology and agriculture of the US and Canadian prairies. The drought came in three waves, 1934, 1936, and 1939—40, but some regions of the high plains experienced drought conditions for as many as 8 years. The eroding soil from once productive range and croplands filled the air with billowing clouds of dust that subsequently buried farm equipment, buildings, and even barbed-wire fences [32], thus, making the living conditions of many Great Plains inhabitants unbearable [33].

Soil erosion is increasing worldwide; however, precise erosion estimates are not possible due to the lack of comparable data, and therefore it is difficult to assess the total earth's area affected by erosion.

In Europe, an estimated 115 million hectares or 12% of Europe's total land area are subject to water erosion, and 42 million hectares are affected by wind erosion [1]. At present, it is estimated that in the Mediterranean region [34], which is considered a vulnerable area due to the specific climatic conditions which favor erodible phenomena, water erosion could affect the loss of 20/40 tons per hectare of soil after a single cloudburst, and in extreme cases the erosion could be even more than 100 tons per hectare [34—36].

The major drivers for water erosion are intense rainfall (particularly pronounced in clay soils after long droughts), topography, low SOM content, percentage and type of vegetation cover, and land marginalization or abandonment. Following the geographical distribution of these major drivers, several areas with a high risk of erosion (including some hots pots) are located in the Mediterranean regions. Erosion risk is also observed across western and central Europe. Even though the risk is relatively limited in, eg, France, Germany, and Poland, water erosion can still be a substantial problem in these countries. On the other hand, the analysis shows hilly to mountainous areas (Pyrenees,

Apennines, and the Alps) with very low or no erosion risk, mainly because of the existence of large forest areas with soils stabilized through tree roots [34].

Similarly, in the United States soil is eroding at a rate that is 10 times faster than the rate at which it is being replenished; however, the rate of soil erosion is much faster in other parts of the world such as Africa, India, and China, where erosion rates are 30—40 times faster than the rate of replenishment. In areas of Africa, the combination of soil depletion and soil erosion has led to the prediction of plummeting crop yields [37]. It is estimated that the total annual cost of erosion from agriculture in the United States is about US$44 billion per year, ie, about US$247 per hectare of cropland and pasture [20].

In Africa, many countries have already lost a significant quantity of their soils to various forms of degradation. Many areas in the continent are said to be losing over 50 tons of soil per hectare per year. This is roughly equivalent to a loss of about 20 billion tons of nitrogen, 2 billion tons of phosphorus, and 41 billion tons of potassium per year. Serious erosion areas in the continent can be found in Sierra Leone, Liberia, Guinea, Ghana, Nigeria, Zaire, the Central African Republic, Ethiopia, Senegal, Mauritania, Niger, the Sudan, and Somalia. Consequently, yield reduction in Africa due to past soil erosion may range from 2% to 40%, with a mean loss of 8.2% for the continent [20]. In South Asia, annual loss in productivity is estimated at 36 million tons of cereal equivalent valued at US$5400 million by water erosion and US$1800 million due to wind erosion. On a global scale the annual loss of 75 billion tons of soil costs the world about US$400 billion per year, or approximately US$70 per person per year [20].

Both wind and water erosion involve three common steps, ie, (1) detachment of soil particles, (2) transport of particles, and (3) deposition of particles in a new location. A soil's susceptibility to wind or water erosion is determined by its erodibility (mainly soil texture and organic matter content) and the climate's erosivity.

Wind erosion is a process that detaches soil particles from the surface. Once detached these particles are transported by either suspension into air and/or rolling along the soil surface. Fine sand, silt, or clay-size particles can be transported to great distances by strong winds. While larger particles rolling along the soil surface move shorter distance, they also shatter other soil particles along the way. As the wind speed decreases, soil particles are deposited on the surface in a new place. Wind erosion most commonly occurs in arid and semiarid regions, because of the frequent occurrence of dry and windy conditions [38,39].

Wind erosion removes mainly the finest soil particles and therefore, results in an ongoing decrease in soil fertility. The effects, however, of wind erosion on agricultural productivity are detectable only after years or decades. Wind erosion is additionally influenced by the interactions of various components (such as land use) resulting in a high temporal variability in the actual wind erosion risk of a particular site [38].

In the case of water erosion, raindrop impact is the primary cause of particle detachment, moving them away. Some of the detached particles float into soil pore spaces. This can clog and seal soil pores and result in reduced water entry (infiltration) into the soil. If the rainfall rate exceeds the rate at which water can infiltrate the soil, the excess water runs off and often carries the detached soil particles with it. Detached particles (sediment) are carried with flowing water down the slope. How many particles and how far they are transported depend on the velocity and volume of the running water. As the water velocity slows down, it loses the energy needed to continue carrying the detached suspended soil particles, and the soil particles are then deposited in their new location [40,41].

Another type of erosion exclusively owed to human activities is tillage erosion. Tillage erosion is the net redistribution of soil within the landscape as a result of farming activities and is one of the most important soil degradation processes on sloping croplands worldwide [42]. Tillage contributes to the denudation of upper portions of hill slopes and causes accumulation of soil on lower portions of the hill slopes. It therefore plays an important role in soil redistribution across the landscape, along with the traditionally recognized processes of wind and water erosion [43].

Both on-site and off-site effects of soil erosion can be recognized as seen in Table 4.2, the impact extent of which is depended on local conditions and environmental, economic, and societal priorities of the area. On-site damages are most frequently measurable, and soil management plans can be developed to address this threat [44]. On the contrary, the off-site effects, although well-known and frequently reviewed, may cause damages to areas far away than the eroded ones, however, the exact extent, causes, and impacts need a more

Table 4.2: On-Site and Off-Site Damage due to Water Erosion [44].

On-Site Damages	Off-Site Damages
Loss of organic matter	Floods
Soil structure degradation	Water pollution
Soil surface compaction	Infrastructure burial
Reduction of water penetration	Obstruction of drainage networks
Supply reduction at water table	Changes in watercourse shape
Surface erosion	Water eutrophication
Nutrient removal	
Increase of coarse elements	
Rill and gully generation	
Plant uprooting	
Reduction of soil productivity	

broad and multidiscipline study. For instance, the silting of artificial water reservoirs implies enormous costs to hydroelectric power plants and water authorities.

Less frequently appreciated is also the way that sediments from eroded soil can accumulate in river channels, reducing channel capacity, blocking culverts, increasing bank erosion, all of which cause flooding and inundation. Sediments accumulating in channels as a result of erosion have an impact that lasts for decades and centuries. Managing a river to prevent flooding through a loss of channel sediment and discharge capacity needs to be planned over many decades and centuries. Erosion can be also a major threat to human health, particularly in densely populated urban areas. Recent examples exist of massive wind erosion problems all over the world, for instance, in China, Australia, and Iceland [45]. Moreover, sediments are often associated with a number of contaminants and nutrients that are major causes of the degradation of bathing water quality in coastal areas with severe economic implications on tourism. A combination of the above two effects is the case of mercury contamination of rivers in the Amazon where soils of the area are naturally rich in mercury [46]. Erosion incidents deposited mercury in the Tapajos River causing adverse health problems to local people through the fish they consume.

4.2.2 Organic Matter Decline—Biodiversity Loss

SOM is a key component of soil, controlling many vital functions, is a source of food for soil fauna, and contributes to soil biodiversity. SOM provides the physical environment for roots to penetrate the soil and for excess water to drain freely from the soil. Organic matter can hold up to 20 times its weight in water, contributing to the water retention capacity of soils.

Although soil scientists would expect to find different behavior in different soils at different "critical" concentrations of SOM, it seems widely believed that a major threshold is 2% SOC (c. 3.4% SOM), below which potentially serious decline in soil quality will occur [47].

A global map [48] of estimated soil carbon stocks to 1 m depth at a nominal spatial resolution of 1 km [48] was generated based on the SOC and bulk density values included in the HWSD [49] which provides a tool for visualizing the distribution of carbon stocks worldwide.

SOC accounts for more than 95% of the total carbon accumulated in pastures and perennial crops, and nearly 100% of the total carbon accumulated in cropland ecosystems. It contributes to the resilience of agricultural ecosystems and increases sustainability of rural livelihoods, which in turn enhances socioeconomic development [50]. In addition, SOC is among the mandatory items to be reported for agricultural land use under the

Kyoto Protocol, and it is one with the highest potentials both in terms of enhancement of carbon sink and reduction of carbon emission.

The amount of organic matter in a soil depends on a range of factors and reflects the balance between accumulation and breakdown of organic substances. The main factors that control this balance are climate, soil type, vegetative growth, topography, and tillage [51].

SOC is a characteristic that is mostly affected by bioclimatic conditions and land use. Some 45% of soils in Europe have low or very low organic matter content (0−2% organic carbon). However, in recent years both land use and climate have undergone dramatic changes that in turn cause changes in SOC. With regard to the European Union, the changes are particularly driven by numerous land use regulations (eg, Nitrate Directive, Water Framework Directive, Biodiversity, Climate Change, Natura 2000, etc.). In addition, many regions of the European Union are experiencing climate evolution, such as temperature rise and changes in patterns of precipitation [52]. As a result of combined land use and climate changes in the European Union, the loss of SOC is substantial and is estimated at the rate equivalent to 10% of the total fossil fuel emissions at the pan-European scale [53].

As practices for SOM increase, the followings could be mentioned:

- Incorporation of plant residues (after harvest)
- Addition of organic materials such as manure, compost, or other organic substances
- Mulching, thus the coverage of the soil with raw plant residues
- Crop rotation that includes pasture and fodder species
- Less or minimal soil cultivation, especially during dry and hot periods (no-till farming may also be adopted)
- Maintaining high soil humidity levels

Therefore, to keep the soil productive, its nutrient levels need to be replenished on a regular basis. Crop residues should be returned to soil to maintain or even increase their organic matter status. A sufficiently high organic matter level is important to increase soil stability, soil's water-holding capacity, and the nutrient-holding capacity and supply. However, additional organic and/or inorganic fertilization is inevitable to restore and maintain the optimal productivity of the soil. Loss of organic matter and soil biodiversity and consequently reducing soil fertility are often driven by unsustainable agricultural practices such as overgrazing of pasturelands, overintensive annual cropping, deep plugging on fragile soils, cultivation of erosion-facilitating crops (eg, maize), continuous use of heavy machinery destroying soil structure through compaction, unsustainable irrigation systems contributing to the salinization, and erosion of cultivated lands [25].

4.2.3 Salinization and Sodification

Salinization is the result of the accumulation of salts and other substances from irrigation water and fertilizers and is regarded as one of the major causes of desertification and therefore is a serious form of soil degradation. Sodification is the process by which the exchangeable sodium (Na) content of the soil is increased. Na ions accumulate in the solid and/or liquid phases of the soil as crystallized $NaHCO_3$ or Na_2CO_3 salts. High levels of salts will eventually make soils unsuitable for plant growth [54].

Salt-affected soil often exhibits a white or gray salt crust on the ground [55]. The pH of the soil is around 8.5, and the salt interferes with the growth of all but the most specially adapted plants.

Salinity is one of the most widespread soil degradation processes on the Earth. According to some estimates, the total area of salt-affected soil is about 1 billion hectares. They occur mainly in the arid–semiarid regions of Asia, Australia, and South America [56].

In Europe, salt-affected soil occurs in the Caspian Basin, the Ukraine, the Carpathian Basin, and the Iberian Peninsula. Soil salinity affects an estimated 1 million hectares in the European Union, mainly in the Mediterranean countries [56]. Another study estimates that 3.8 million hectares in Europe are affected [57]. However, while several studies show that salinization levels in soils in countries such as Spain, Greece, Romania, and Hungary are increased [58], systematic data on trends across Europe are not available.

When alkalinity takes place, the high pH level does not, in most cases, permit plant life. Excess sodium on the exchange complex results in the destruction of the soil structure that due to a lack of oxygen cannot sustain either plant growth or animal life. Alkaline soils are easily eroded by water and wind.

Saline soils are usually categorized into three types, ie, saline, sodic, and alkaline sodic soil [54].

- *Saline soil* contains a lower amount of Na adsorbed onto soil particles. This type of soil is often seen in sandy soil containing lower amounts of clay and organic matter. Saline soil in deserts is usually of this type.
- *Sodic soil* contains a large amount of Na adsorbed onto soil particles. This type of soil contains large amounts of clay.
- *Alkaline sodic soil* (or alkaline soil) is a type of sodic soil that is highly alkaline with the pH value more than 8.5. This type of soil contains higher amounts of carbonate and bicarbonate which can be hydrolyzed to alkaline products.

Salt-affected soil can be divided into five main groups [56]:

- saline soil (Solonchak) with high amount of water soluble soils;
- alkaline soil (Solonetz), high alkalinity and high exchangeable sodium percentage (ESP);
- magnesium soil: high magnesium content in the soil solution;
- gypsiferous soil: strong gypsum or calcium sulfate ($CaSO_4$) accumulation; and
- acid sulfate soil: highly acidic iron or aluminum sulfate accumulation.

The cause of soil salinity and sodification is multifactorial, and any process that affects the soil–water balance may affect the movement and accumulation of salts in the soil. These processes may be hydrology, climate, irrigation, drainage, plant cover and rooting characteristics, and farming practices. In semiarid areas, salinization often occurs on the rims of depressions and edges of drainage ways, at the base of hill slopes, and in flat, low-lying areas surrounding sloughs and shallow bodies of water. These areas receive additional water from below the surface, which evaporates, and the salts are left behind on the soil surface. Summer fallow management practices may cause increased salinization by increasing the soil moisture content to the point that water moves to seeps on hill slopes. Salts accumulate as the water evaporates from these seeps [59].

According to USDA-NRCS [59], some early signs of soil salinity that may assist in identifying future substantial problems are:

- increased soil wetness in semiarid and arid areas to the point that the soil does not support equipment;
- the growth of salt-tolerant weeds; and
- irregular patterns of crop growth and lack of plant vigor.

On the other hand, advanced signs that require emerge action are:

- white crusting on the surface;
- a broken ring pattern of salts adjacent to a body of water;
- white spots and streaks in the soil, even where no surface crusting is visible; and
- the presence of naturally growing, salt-tolerant vegetation.

Drainage is essential for reclaiming such degraded soils since soluble salts can be leached (washed) and moved through the soil (rather than run off the surface). The purpose is to leach salts below the plant root zone. If the soil is poorly drained because of compaction, drainage potential should be ensured first, for example, by adding an amendment like organic matter.

4.2.4 Sealing

Soil sealing is the covering of the ground by an impermeable material or, in other words, the removal of topsoil layers [60]. It leads to the loss of important soil functions, such as food production, water storage, or temperature regulation, while it affects fertile agricultural land, puts biodiversity at risk, increases the risk of flooding and water scarcity, and contributes to global warming [57].

The increase in sealed surfaces due to changes in land use together with a decrease in forest cover has increased the frequency and size of storm runoff, causing flooding, mudflows, and landslides [26]. Increases in damage from flooding have also resulted from the development of floodplains for industry and habitation.

As a result of sealing, the natural water cycle is altered, producing larger volumes of runoff and higher peak flows [61]. It also precludes rain from infiltrating the ground and recharge aquifers. As water can neither infiltrate nor evaporate, water runoff increases, sometimes leading to catastrophic floods. Cities are increasingly affected by heat waves, because of the lack of evaporation in summer. Landscapes are fragmented and habitats become too small or too isolated to support certain species. In addition, the food production potential of land is lost forever. It is estimated that 4 million tons of wheat are potentially lost every year due to soil sealing [62].

Sealing address is without doubt a big challenge for the modern communities. In the context of the Soil Thematic Strategy [1], the European Commission points out the need to develop best practices to mitigate the negative effects of sealing on soil functions. Furthermore, the Roadmap to a Resource Efficient Europe [63] proposes that by 2020, EU policies take into account their impacts on land use with the aim to achieve no net land take by 2050.

Given the irreversibility of already exploited and sealed soils, the best way to mitigate and address sealing is the development and implementation of best practices but also the enhancement of their adoption by specific and targeted legislative frameworks, plans, and tools that will assist the implementers. Limiting soil sealing should have a priority over mitigation or compensation measures, since soil sealing is an almost irreversible process. Therefore, smarter spatial planning can limit urban sprawl as, for example, development of potential inside urban areas, through the regeneration of abandoned industrial areas (brownfields). Mitigating measures include using permeable materials, supporting "green infrastructure," and making wider use of natural water harvesting systems. Only where on-site mitigation measures are insufficient, compensation measures that enhance soil functions elsewhere should be considered [62].

Three-tiered approach has been proposed [62]:

- *Limiting* the progression of soil sealing with improved spatial planning or by reassessing "negative" subsidies that indirectly encourage soil sealing;
- *Mitigating* damage when soil sealing cannot be avoided, through measures such as the use of permeable surfaces instead of conventional asphalt or cement and building green roofs; and
- *Compensating* valuable soil losses by action in other areas to offset drawbacks in ecofunction. Measures may take the form of payments, as in the Czech Republic and Slovakia, or the restoration of already sealed soil. Good practices have been identified notably in the cities of Dresden (Soil Compensation Account) and Vienna.

4.2.5 Landslides

Landslides are the gravitational movement of a mass of rock, earth, or debris down a slope. Landslides occur when the stability of a slope changes from a stable to an unstable condition. Such changes can be caused by a number of factors, acting together or alone. Natural causes of landslides include groundwater pressure, loss of vegetation cover (eg, after a fire), erosion of the toe of a slope by rivers or ocean waves, saturation by snowmelt or heavy rains, and earthquakes. Human causes include deforestation and removal of vegetation cover, cultivation, construction, and changes to the shape of a slope. Landslides can be very slow moving or very rapid and, in general, any area composed of very weak or fractured materials resting on a steep slope can and will likely experience landslides [64].

As regards the United States, landslides occur in every state, ie, the Appalachian Mountains, the Rocky Mountains, and the Pacific Coastal Ranges, and some parts of Alaska and Hawaii have severe landslide problems.

Currently, there are no data on the total area affected in Europe, although estimates have been made for Italy (7%), Portugal (1%), Slovakia (5%), and Switzerland (8%). The main landslide-prone regions include mountain ranges such as Alps, Apennines, Carpathians, Balkans; hilly areas on landslide-sensitive geological formations (eg, in Belgium, Portugal, and Ireland); coastal cliffs and steep slopes (eg, in the United Kingdom, France, Bulgaria, Norway, and Denmark); and gentle slopes on quick clay in Scandinavia. Landslides are possibly the most serious environmental issue in Italy [57].

Globally, landslides cause hundreds of billions of dollars in damages and hundreds of thousands of deaths and injuries each year [65].

Although the physical cause of many landslides cannot be removed, geologic investigations, good engineering practices, and effective enforcement of land-use management regulations can reduce landslide hazards.

4.2.6 Compaction

Compaction can detrimentally affect a number of soil functions by reducing the pore space between soil particles, increasing bulk density, and reducing or totally destroying the soil's absorptive capacity. Reduced infiltration increases surface runoff and leads to more erosion while decreasing groundwater recharge [66,67].

Heavy loads on the soil surface that cause compaction in the subsoil are cumulative and cause the bulk soil of the subsoil to increase significantly. Compaction results in a greatly reduced crop rootability and permeability for water and oxygen. The worst effects of surface compaction can be rectified relatively easily by cultivation, and hence it is perceived to be a less serious problem in the medium to long term. However, subsoil compaction can be extremely difficult and expensive to alleviate, and remedial treatments usually need to be repeated. Indeed, once the threshold of the preconsolidation stress is reached, compaction is virtually irreversible [68].

A direct impact of compaction and associated decrease of soil porosity (Photo 4.2) is the reduction in the available habitats for soil organisms, in particular, soil organisms living in surface areas, such as earthworms.

Photo 4.2
Destroyed soil structure with decreased porosity-compacted soil layer. *Authors' photos collection.*

Compaction damages earthworm tunnel structures and kills many of them. Alteration of soil aeration and humidity status due to soil compaction can also seriously impact the activity of soil organisms. Oxygen limitation can modify microbial activity, favoring microbes that can withstand anaerobic conditions. This alters the types and distribution of organisms found in the rest of the soil food web. In addition, compaction can significantly reduce the number of microarthropods involved in biological regulation. The degree of impact varies with both the type of microarthropod and soil. Although microarthropod populations may recover, this can take several months [68].

4.2.7 Contamination

Both terms, contamination and pollution, are used synonymously. However, according to the definition given by Knox et al. [69,70] trace element—contaminated soils are not considered to be polluted unless a threshold concentration exists that begins to affect biochemical and biological processes. Soil pollution is as old as man's ability to smelt and process ores, and goes back as far as the Bronze Age (2500 BC).

Pollution cases may be, according to their spatial dimensions, classified into the following two main types:

Diffuse Sources (*nonpoint sources*). Nonpoint sources are related to diffuse processes or human activities that cover large areas. Diffuse soil contamination is in general associated with atmospheric deposition, certain agricultural practices (soil amendment with sewage sludge, application of manure, mineral fertilizers, pesticides, fumigation), and inadequate waste and wastewater recycling and treatment. Pollutants can be washed by rainfall both into the soil and from soil into surface and groundwater. Currently, the most important soil contamination problems from diffuse sources are atmospheric deposition of acidifying and eutrophying compounds or potentially harmful chemicals, deposition of contaminants from flowing water or eroded soil itself, and the direct application of substances such as pesticides, sewage sludge, fertilizers, and manure which may contain heavy metals [71].

Heavy metals together with excessive nitrogen inputs are regarded as the main sources of contamination in agricultural soils and may be caused by human activities, such as fertilization and amendment practices, used to increase soil productivity. Metals like Hg, Cd, As, and Pb can contaminate the soil gradually and damage soil and ecosystem functioning. These contaminating elements will become part of the nutrient cycling resulting in biodiversity decline, water pollution, and consequently a potential danger for human health [72]. The excessive application of fertilizers or manures usually exceeds the functional soil ability to retain and transform nutrients and influences the soil capability to provide nutrients for plant growth and also its buffering and filtering capacity [73].

The saturation of soil with nitrogen or phosphate has led to losses of nitrates and phosphate, which move into groundwater waterways and coastal systems, causing eutrophication [74].

Effects of emissions from nonpoint sources in Europe and the United States have been detected even in remote areas such as Antarctica. At a European level, the atmospheric transport of heavy metals is a significant process: 30–90% of the metals emitted from each European country are deposited in other countries. Because this type of pollution may cover very large areas, even countries, the characterization, mapping and remediation, needs more detailed planning and technical installations than localized cases.

Localized sources (point sources). Point sources refer to discrete and localized contamination processes. Point source contamination is often linked to no operational industrial plants; power generation; industrial accidents; uncontrolled industrial, municipal, and agricultural waste (AW) disposals; and mining activities [71]. Contaminated sites can pose serious threats to health and to the local environment as a result of release of harmful substances to water resources, uptake by plants, and direct contact by people. Major pollutants include heavy metals, organic contaminants such as chlorinated hydrocarbons, and mineral oil. Point sources are generally responsible for high concentrations of pollutants in small areas. In such cases, pollution would be spreading from the source in a flow pattern, which is more or less localized and showing concentrations that decrease with increasing distance from the source of pollution [71].

As regards the European countries, it is difficult to quantify the real extent of local soil contamination as many European countries lack comprehensive inventories and there is a lack of EU legislation obliging Member States to identify contaminated sites. Estimates show that the number of sites in Europe where potentially polluting activities are occurring, or have taken place in the past, now stands at about 3 million [75]. Some locations, depending on their use and the nature of the contaminant, may only require limited measures to stabilize the dispersion of pollution or to protect vulnerable organisms from pollution. However, it should be noted that around 250,000 sites might need urgent remediation [76]. The largest and probably most heavily affected areas are concentrated around the most industrialized regions in northwest Europe, from Nord-Pas de Calais in France to the Rhein-Ruhr region in Germany, across Belgium and the Netherlands and the south of the United Kingdom. There are ∼3000 problem areas including former military sites, abandoned industrial facilities, and storage sites which may still be releasing pollutants to the environment leading to groundwater contamination and related health problems [77]. The contaminated sites in Ukraine are about 5 million hectares, mostly in human settlements and around the industrial factories, and in Lithuania nearly 3 million hectares. In the mining industry, which is a major driver of soil degradation in central and eastern European countries, the risk of contamination is associated with sulfur- and heavy-metal-bearing tailings stored on mining sites and the use of certain chemical reagents such as cyanide in the refining process. Throughout southeastern Europe, land which was already under stress from poor land management practices has been further damaged by military and refugee settlements, land mines (as much as 27% of Bosnia's plowed land is still mined), and other unexploded devices. In eastern Europe huge irrigation and hydroelectric projects coupled with poor water management have resulted in salinization and waterlogging of large areas, especially in Azerbaijan, Belarus, the Russian Federation, and Ukraine [77].

Worldwide, and according to the Environment Toxin Report 2013, published by the Green Cross Switzerland [78], the 10 most polluted areas in the world are:

*   Matanza-Riachuelo River, Argentina (VOC—volatile organic compounds, specially toluene, and heavy metals from improper management of industrial wastes and sewage)—Water
*   Hazaribagh, Bangladesh (chromium in wastes generated by the plenty of tanneries of the area)—Soil
*   Agbogbloshie Dumpsite, Ghana (lead, cadmium, and mercury from electronic wastes)—Soil
*   Citarum River, Indonesia (chemicals such as lead, cadmium, and chromium and pesticides discharged by local industries)—Soil and Water
*   Kalimantan, Indonesia (cadmium and mercury as by-products from gold mines of the area)—Soil and Water
*   Niger River Delta, Nigeria (oil by accidents or oil robbery)—Soil and Water
*   Dzerzhinsk, Russia (chemicals including sarin, lead, and phenols as well as toxic by-products generated from local industry)—Soil and Groundwater
*   Norilsk, Russia (heavy metals)—Soil and Groundwater
*   Kabwe, Zambia (lead from lead mines)—Soil and Water
*   Chernobyl, Ukraine (radionuclides due to the nuclear accident of 1986)—Soil and Water

Waste landfilling is also an important potentially contaminating activity. Application of farm manures, sewage sludge, and composted green wastes lead to air pollution (odor and ammonia) and to diffuse water (nitrate and phosphate) pollution. Moreover, the potential of soil contamination is greatly increased in landfills that do not comply with the minimum requirements set by the legislative framework, eg, the European landfill directive [79].

Threshold values for soils are difficult to evaluate since heavy metals toxicity and metal bioavailability are not only dependent on the total content in soils but also in other environmental factors [80]. At European level only threshold values related to the application of sewage sludge in agricultural soils have been defined [81].

The determination of natural background values is very difficult since the geochemistry of most of the European ecosystems is greatly influenced by human activities [82].

The U.S. Environmental Protection Agency (U.S. EPA) and many Public Health agencies in the United States have published what they consider to be "safe levels" of certain metals in soil based on their own research or that of other organizations. However, these organizations do not always agree on what is safe, and "safe levels" specifically for gardens where food is grown can be higher than what is considered safe by other agencies (Table 4.3).

Soil remediation

All remediation options have advantages and limitations that make them more or less applicable in any particular case, and a wide range of site-specific technical

Table 4.3: Recommended Limits of Heavy Metals for Soil in mg/kg According to U.S. EPA.

	Land Application of Biosolids for Home Vegetable Gardens	Human Health Screening Level (HHSL)	Regional Screening Level for Superfund Sites, Residential Soil	Total Threshold Limit Concentration for Hazardous Toxic Wastes
Agency and Online Access	U.S. EPA Clean Water Act Title 40, Section 503.13 http://ecfr.gpoaccess.gov	California Office of Environmental Health Hazard Assessment http://oehha.ca.gov/risk/chhltable.html	U.S. EPA Region 9 www.epa.gov/region9/superfund/prg	California Department of Toxic Substances Control CA Code of Regulations, Title 22, Section 66261.24 www.dtsc.ca.gov/lawsregspolicies/title22/index.cfm
Antimony		30	31	500
Arsenic	41	0.07	Use HHSL[a]	500
Barium	—	5200	15000	10,000
Beryllium	—	16	160	75
Cadmium	39	1.7	Use HHSL[a]	100
Chromium (III)	—	100,000	120,000	2500
Cobalt	—	660	23	8000
Copper	1500	3000	3100	2500
Lead	300	80	Use HHSL[a]	1000
Mercury	17	18	23	20
Molybdenum	—	380	390	3500
Nickel	420	1600	1500	2000
Selenium	100	380	390	100
Silver	—	380	390	500
Thallium	—	5	—	700
Vanadium	—	530	390	2400
Zinc	2800	23000	23000	5000

[a]If the U.S. EPA Regional Screening Level for a metal results in a risk for cancer that is four times greater than the Human Health Screening level (HHSL), the HHSL value is used instead.

factors determine which remediation options are most appropriate. Some of these factors relate to the nature of the relevant pollutant linkages, such as the type, amount, lateral and vertical distribution of pollutants and affected media, and the properties of pathways. Others relate to the general characteristics of the site, such as its size, location, accessibility, topography and wider environmental setting, and the existence (or proposed construction) of buildings and other structures. The current or intended use of the site also needs to be taken into account to ensure that remediation does not compromise soil functions, including geotechnical properties.

Other factors also affect the choice of the most appropriate option such as the legal and commercial context within which the site is being handled, the views of key stakeholders (eg, site owners, purchasers, funders, regulators, and the local community), and the costs and benefits of using any particular option.

The key question

For any individual site, two questions should be answered:

- Does the contamination matter? And, if so,
- What needs to be done about it?

The answers to both the questions above depend to some extent on when the contamination happened. For "new" contamination, the accepted principle is that deterioration of the environment needs to be avoided. This principle underlies the approach in regimes aimed at controlling potentially polluting activities, such as Pollution Prevention and Control (PPC).

In deciding whether contamination matters, the amount, or concentration, of any contaminants present is always going to be a significant factor, but it does not provide the whole answer. It is also necessary to consider to what extent the substances present may harm human health or the wider environment, including damage to property. In short, what risk, if any, is caused by contaminants, and is that risk unacceptable? This need to make judgments about the degree of risk also applies to deciding what to do about the contamination. Technical obstacles as well as potentially large costs mean that it is often neither feasible nor realistic to think in terms of total cleanup of past damage. Instead, the goal is to find solutions that identify and deal with risks from contamination in a sustainable way [83].

In order to facilitate the evaluation of the potential benefits of a remedial action as compared to the impacts that may be caused by it, the following question should be asked prior to initiating the work:

Will the environmental work result in a net positive benefit to the environment?

In most cases the evaluation of the potential impacts and benefits of conducting a project will show that its merit will be conditional. And it should be recognized that when environmental work is performed for the benefit of a specific location, it is often at the expense of another.

Sustainable development in contaminated land management (CLM) and more specifically *sustainable remediation* is a growing field of knowledge. At a strategic level, remediation of contaminated sites supports the goal of sustainable development through:

- the act of conserving land as a resource;
- prevention of spreading of pollutants to the air, the soil, and the water; and
- reducing the pressure on development.

Although all these positive effects occur due to remediation, some negative effects also arise on the environment, economy, and society. These negative impacts should not exceed the benefits of a remediation [84].

There are no united guidelines or common methodology for sustainable remediation assessments used by all nations internationally. According to Woodward et al. [85], this is a possible barrier for implementing sustainable remediation. Another possible barrier is the difficulty to equate results in a consistent metric since many of the factors influencing the outcome need a qualitative assessment. There are a variety of views and no uniform picture of what sustainable remediation is and how it should be assessed. Lesage and Zoller [86] have the following view on sustainable remediation:

> *Sustainable remediation is developing methods that do not require extraordinary resources, or resources better used elsewhere. It is working with nature, by using supporting natural processes technologies, rather than against it. It is achieving balance between risk mitigation and the expenditures required to achieve it, through optimization based on well-defined criteria.*

Once relevant pollutant linkages have been identified as a result of risk assessment, an important task is the definition of the boundary within which remediation options are considered so that potential conflicts between different objectives can be addressed and the most appropriate overall decision can be made. One way for the definition of this boundary is to specify at the outset of options appraisal a series of objectives that the remediation strategy has to achieve to be considered acceptable to all those involved.

Objectives will be linked to the [87]:

- degree to which risks need to be reduced or controlled;
- time within which the remediation strategy is required to take effect;
- practicability of implementing and, where appropriate, maintaining the strategy;
- technical effectiveness of the strategy in reducing or controlling risks;

- durability of the strategy (ie, will it provide a robust solution over the design life?);
- sustainability of the strategy (ie, how well it meets other environmental objectives, for example, on the use of energy and other material resources, and avoids or minimizes adverse environmental impacts on off-site locations, such as a landfill, or on other environmental compartments, such as air and water);
- cost of the strategy (bearing in mind that the person who makes the decision about remediation may not be the person who has to pay);
- benefits of the strategy—all remediation strategies should deliver direct benefits (the reduction or control of unacceptable risks)—but many have merits that extend well beyond the boundaries of the site; for example, remediation may enhance the amenity or ecological value of an area or contribute toward improved economic activity by removing blight or encouraging regeneration; and
- legal, financial, and commercial context within which the site is being handled including the specific legal requirements that remediation has to comply with, and the views of stakeholders on how unacceptable risks should be managed.

It is important to mention that after the implementation of any soil remedial action, an appropriate soil monitoring and protection system/strategy should be developed and implemented for almost 10 years, unless differently required. The strategy should include the establishment of appropriate soil indicators that will be monitored periodically to continually assess the risk level of the area. At the same time, a code of good practices which identifies pressures and impacts on soils and outlines best practice in relation to sustainable soil management under different land uses should be developed and distributed to the stakeholders of the area as well to the local and regional authorities [87].

4.2.8 Biodiversity

Rapidly advancing soil degradation is severely threatening soil biodiversity, eventually leading to the extinction of species yet to be discovered and fully studied. Implications for human health of the degradation of the soil ecosystem need still to be fully understood.

Hence, soil degradation by erosion, contamination, salinization, and sealing all threaten soil biodiversity by compromising or destroying the habitat of the soil biota. Management practices that reduce the deposition or persistence of organic matter in soils, or bypass biologically mediated nutrient cycling, also tend to reduce the size and complexity of soil communities. It is, however, notable that even polluted or severely disturbed soils still support some level of microbial diversity.

Little is known about how soil life reacts to human activities, but there is evidence that soil organisms are affected by SOM content, the chemical characteristics of soils (eg, pH, the amount of soil contaminants or salts), and the physical properties of soils such as porosity and bulk density, both of which are affected by compaction and sealing.

A limited number of data concerning the dynamics of soil biodiversity are available and these generally refer to a few groups of soil organisms. Mushrooms, for instance, are a group of soil organisms for which a relatively long history of records exists. From this type of data set, it has been possible to show mushroom species decline in some European countries. For example, a 65% decrease in mushroom species over a 20-year period has been reported in the Netherlands, and the Swiss Federal Environment Office has published the first-ever "Red-List" of mushrooms, detailing 937 known species that face possible extinction in Switzerland [88].

4.3 Desertification

Desertification is defined by the United Nations Convention to Combat Desertification (UNCCD) [89] as "land degradation in arid, semi-arid and dry sub-humid areas resulting from various factors, including climatic variations and human activities." The most recent terminology adopted by the UNCCD includes areas suffering from "desertification, land degradation, and drought" and reflects the wider endorsement of the convention by countries that do not have drylands within their national territories.

Desertification causes a progressive loss of soil fertility, through the destruction of the structure and composition of the soil, which does not permit good agricultural productions or the existence of vegetation with varied natural species [90]. The desertification has been wrongly confused with depopulation. However, these two phenomena can in fact be related. The loss of soil fertility ends up leading to a decline in agriculture, to land abandonment, and ultimately to emigration.

There are several factors that contribute to desertification; some are natural (intense rainfall events, drought) while others are directly related to human activities (agriculture, industry).

Desertification is exacerbated by climate change and vice versa. As severe weather events increase in frequency and severity due to climate change, dryland degradation tends to increase. Worse still, desertification and climate can form a "feedback loop" with the loss of vegetation caused by desertification reducing carbon sinks and increasing emissions from rotting plants. The result is more GHGs in the atmosphere and a continuation of the vicious cycle involving climate change and desertification.

Desertification is experienced on 33% of the global land surface and affects more than 1 billion people (Map 4.2), half of whom live in Africa [20]. In all, more than 110 countries have drylands that are potentially at risk. In Africa, a billion hectares or 73% of

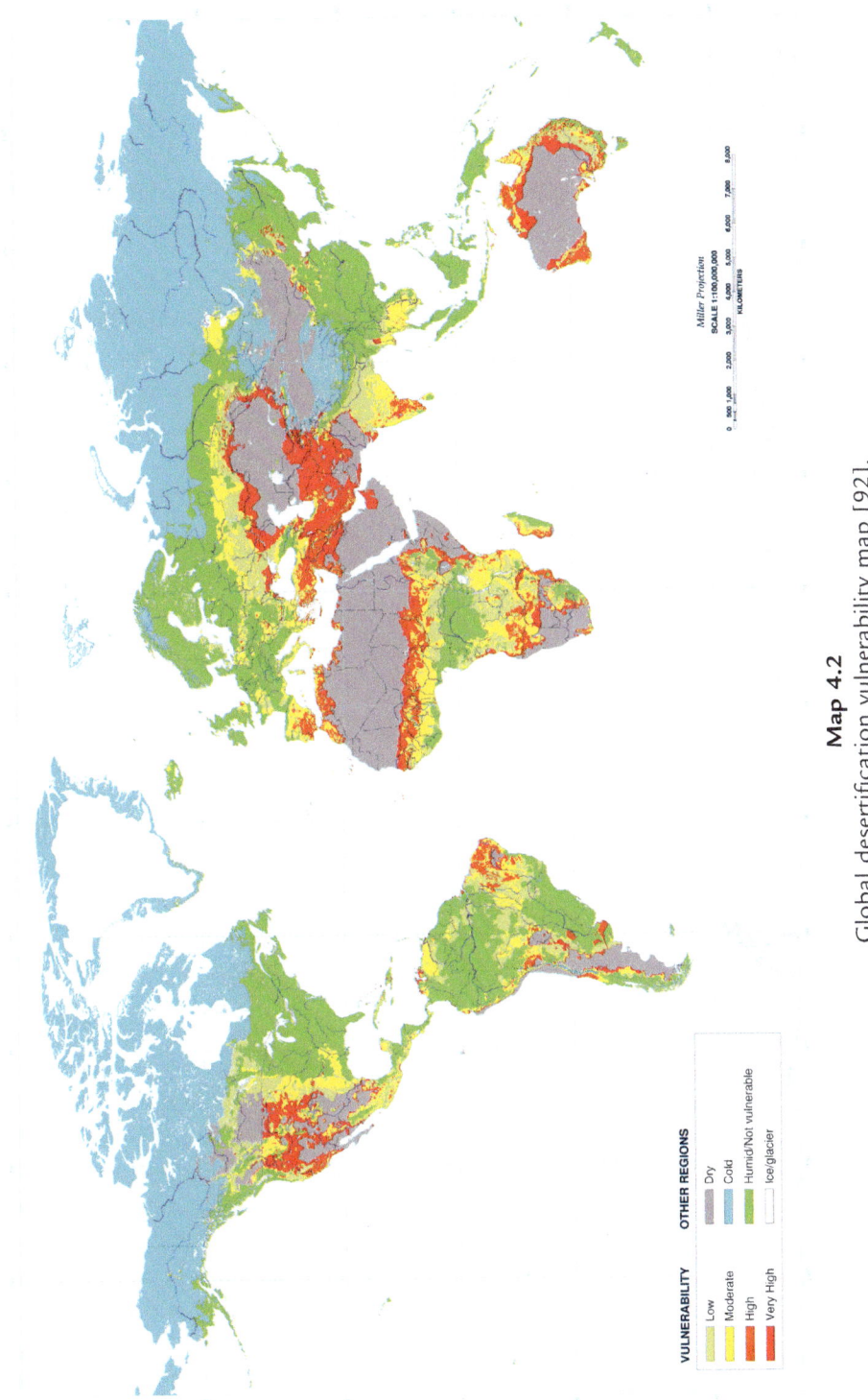

Map 4.2
Global desertification vulnerability map [92].

its drylands are affected by desertification with another 1.4 billion hectares affected in Asia. But it is not just a problem for developing countries; in fact the continent that has the highest proportion of drylands subject to desertification is North America with 74% affected. Five countries of the European Union are also affected while many of the most affected areas are in the former Soviet Union [91]. In 1998, the Soil Science Division of the World Science Resources [92] produced the Global Desertification Vulnerability Map (Map 4.2) by categorizing four vulnerability classes based on soil climate and soil classification.

Desertification is a serious problem in the African continent. It has been estimated that 319 million hectares of Africa are vulnerable to desertification hazards due to sand movement. An FAO/UNEP assessment of land degradation in Africa suggests that large areas of countries north of the equator suffer from serious desertification problems. For example, the desert is said to be moving at an annual rate of 5 km in the semiarid areas of West Africa. Desertification is not, however, a new degradative phenomenon in Africa. Archaeological records suggest that Africa's arid areas have been getting progressively drier over the past 5000 years What is new is the coincidence of drought with the increasing pressures put on fragile arid and semiarid lands by mounting numbers of people and livestock. This is basically what is accelerating land degradation throughout much of Africa. In the wetter areas, however, there is a better chance that degradation can be halted and the land restored [93].

Within the European Union, Bulgaria, Cyprus, Greece, Hungary, Italy, Latvia, Malta, Portugal, Romania, Slovakia, Slovenia, and Spain consider themselves affected by desertification and are included in UNCCD list [94]. The situation is most serious in southern Portugal, much of Spain, Sicily, southeastern Greece, and the areas bordering the Black Sea in Bulgaria and Romania. In southern, central, and eastern Europe, 8% of the territory current shows very high or high sensitivity to desertification, corresponding to about 14 million hectares, and more than 40 million moderate sensitivities are included [76].

According to China's State Forestry Administration (SFA) and Ministry of Land and Resources, deserts now cover almost one-fifth of the country's territory, while the area threatened by desertification amounts to more than one-quarter of China's landmass. The impact of this is felt most acutely in the driest areas in western China, which are also among the poorest. The government estimates that the livelihoods of 400 million people are either threatened or affected by desertification, land degradation, the encroachment of the Gobi, Taklamakan, and Kumtag deserts as well as other deserts and sandy lands in western China [93].

Agricultural activities, because they are based on the use of soil, contribute decisively to desertification. Some activities are as follows.

4.3.1 Arable Lands

*   Removal of the vegetation cover
*   Unsustainable agricultural practices (deep plowing and destruction of soil structure)
*   Loss of SOM (eg, soil wash due to rainfall)
*   Compaction—heavy machinery
*   Nutrients loss—monocultures (eg, wheat, vineyards), nonnative plants cultivation. Intensive agriculture exhausts the soil's nutrients and minerals needed to sustain plant life
*   Pollution/contamination

4.3.2 Irrigated Areas

*   Excessive use of water—water erosion
*   Insufficient irrigation system
*   Salinization—formation of salt layers on soil surface

4.3.3 Pastures

*   Overgrazing—excessive use of the same pastures. Overgrazing removes the grass and other vegetation that protects the soil from erosion
*   High density of animals per area

4.3.4 Forests

*   Deforestation
*   Use of fast-growing exotic species (eucalyptus)
*   Fires

Desertification has substantial economic consequences. The World Bank estimates that at the global level, the annual income lost in the areas affected by desertification amounts to 42 billion dollars each year, while the annual cost of mitigating desertification would cost only 2.4 billion dollars [95].

Economic pressures can lead to the overexploitation of land and usually hit the poor the hardest. Forced to extract as much as they can from the land for food, energy, housing, and source of income, they are both the causes and the victims of the desertification. Desertification brings hunger and poverty. People living in areas threatened by desertification are forced to move elsewhere to find other means of livelihood. Usually they migrate toward urban areas or go abroad. Mass migration is a major consequence of desertification. From 1997 to 2020, some 60 million people are expected to move from the desertified areas in sub-Saharan Africa toward Northern Africa and Europe [95].

4.3.5 Actions Against Desertification

- Land restoration and fertilization. Use of organic amendments, like composts to increase SOM
- Combating wind effects by constructing barriers and stabilizing sand dunes with local plants
- Reforestation. Trees play several roles, ie, fix soil, act as wind breakers, enrich soil in nutrients, adsorb water during rainfall, etc.
- Adoption of sustainable agricultural practices (ie, development of Codes of "Good Farming Practices"). Agriculture diversity must be preserved. Soil "breathing" during certain time periods (no cultivation, no grazing) should be ensured
- Development of integrated scenarios for changing societal behaviors (modern and traditional) of everyday life (social, commercial, professional, etc.) that affect and intensify desertification
- Education and training of local communities

Many parties to the convention have now prepared national action programs to strengthen activities to combat desertification and drought [96,97]. However, there is no indication that governments are developing structures through which bottom-up action programs could be implemented at the local level [98]. In addition, inadequate resource mobilization is hampering the affected developing countries' efforts to fulfill their commitments under the convention. A recent analysis of the CCD [99] argues that the convention model was ill-advised as "it has tied people into a series of COP (Conference of the Parties) performances which demonstrate no linkage with real problems on the ground." The desertification problem remains poorly understood as the available data show: estimates of areas affected range from one-third of the world's surface area to about 50%, and people affected from 1 in 6 to 1 in 3 [99].

4.4 International Policy

Actions to protect soil quality necessitate tackling collectively the different threats. Therefore, policies need to be adaptive and make allowance for the reality of complex evolving systems, avoiding the pitfalls of a command and response strategy. Unlike other media, no specific objectives and targets have been set for soil conservation, and it is rarely considered in sectorial planning activities such as transnational transport corridors. At the national level, some countries have produced legislation, policies, and guidelines to ameliorate or prevent further soil degradation, but policy measures are primarily aimed at combating pollution in other areas and affect soils only indirectly. Statutory soil monitoring is carried out in a number of countries but rarely specifically for soil protection.

The Earth Summit in Rio de Janeiro in 1992 took a step forward in bringing problems associated with land resources to a wider attention. In *Agenda 21* [100], Chapters 10, 12−14 relate to land, covering the integrated approach to management of land resources, desertification and drought, mountain region development, and sustainable agriculture. In the discussions of deforestation, biological diversity, and freshwater resources (Chapters 11, 15, and 18), significant emphasis is placed on land as a productive resource, the importance of sustainable land use, and environmental pollution and conservation. *Agenda 21* has remained a primary basis for land resources policy although a further landmark of awareness of land at the highest policy level is found in the review prepared for the UN Millennium Summit [101]. This review identifies the threats to future global food security arising from problems of land resources.

The *United Nations Convention to Combat Desertification in Those Countries Experiencing Serious Drought and/or Desertification, Particularly in Africa* (UNCCD) is a Convention to combat desertification and mitigate the effects of drought through national action programs that incorporate long-term strategies supported by international cooperation and partnership arrangements. The Convention, the only convention stemming from a direct recommendation of the Rio Conference's Agenda 21, was adopted in Paris, France, on June 17, 1994 and entered into force in December 1996. It is the first and only internationally legally binding framework setup to address the problem of desertification. The Convention is based on the principles of participation, partnership, and decentralization—the backbone of good governance and sustainable development. It has 196 parties, making it truly global in reach. To help publicize the Convention, 2006 was declared "International Year of Deserts and Desertification," but debates have ensued regarding how effective the International Year was in practice [102]. Once significant effort has been made to design a framework program, international solidarity might facilitate the launching of specific projects and activities under the agreed policies, in an effective manner and without creating excessive transactional burden. Because programs need to be adapted to particular regional circumstances, most of the specific requirements are described in the five regional implementation annexes for Africa, Asia, Latin America and the Caribbean, the northern Mediterranean, and Central and Eastern Europe.

The UNCCD developed a 10-year strategic plan and framework to enhance the implementation of the Convention (*The Strategy*) between 2008 and 2018.

With the adoption of The Strategy, affected Parties are expected to align their National Action Plans (NAPs) and other relevant implementation activities relating to the Convention with The Strategy. The Strategy heralds the UNCCD's mission as "to provide a global framework to support the development and implementation of national and regional policies, programmes and measures to prevent, control and reverse desertification/land degradation and mitigate the effects of drought through scientific and technological excellence, raising public awareness, standard setting, advocacy and resource mobilization, thereby contributing to poverty reduction."

The Strategy supports the development and implementation of national and regional policies, programs, and measures to prevent, control, and reverse desertification/land degradation and mitigates the effects of drought through scientific and technological excellence, raising public awareness, standard setting, advocacy, and resource mobilization.

Four strategic objectives with their own long-term impacts will guide the actions of all UNCCD stakeholders and partners in seeking to achieve the global vision. These four strategic objectives are:

1. to improve the living conditions of affected populations;
2. to improve the condition of affected ecosystems;
3. to generate global benefits through effective implementation of the UNCCD; and
4. to mobilize resources to support the implementation of the Convention through building effective partnerships between national and international actors.

Additionally, the UNCCD works closely with the other two "Rio Conventions," the Convention on Biodiversity (CBD) and the United Nations Framework Convention on Climate Change (UNFCCC) (each of the three conventions derives from the 1992 Earth Summit in Rio de Janeiro) [102].

Approaches to soil conservation have been greatly modified since the 1970s. Work used to concentrate on mechanical protection, such as bunds and terraces, largely to control surface runoff. This has been supplemented by a new approach [103,104] which calls for greater attention to biological methods of conservation and the integration of water conservation with soil protection, through improved management of soil–plant–water relationships, including reduced disturbance by tillage [105]. Within the international agricultural research system, the Consultative Group on International Agricultural Research, there is now a commitment to natural resource management, and explicit recognition of degraded land and desertification as environmental problems [106].

Despite these developments, there is no clear indication that the rate of land degradation has decreased. As yet, there are no continuously monitored indicators of soil condition that would permit quantitatively based assessments of changes over time, comparable to the monitoring of deforestation.

It has been suggested that soil monitoring should become a basic task of national soil survey organizations [107], but this proposal has yet to be widely adopted.

An international program was set up to develop a set of land quality indicators [108], comparable to those used to monitor economic and social conditions. The program continues on a modest scale under the Global Terrestrial Observation System.

4.4.1 Concern in Europe

Worrying trends emerged from research findings and monitoring programs about the status of European soils. This made the European Union decide to analyze and describe the threats being faced by the soils of Europe and suggest a foundation for their protection.

Different community policies contribute to soil protection, particularly environment (eg, air and water) and agricultural (agri-environment and cross-compliance) policy. However, even if exploited to the full, existing policies are far from covering all soils and all soil threats identified. For these reasons, the Commission adopted a *Soil Thematic Strategy* [1] and a proposal for a *Soil Framework Directive* on September 22, 2006 with the objective to protect soils across the European Union, but this directive has not been endorsed yet by the European Council. There remains opposition to the proposals in several Member States who say that soil protection solely should be up to Member States, with emphasis on sharing best practice examples and further development of (voluntary) guidelines.

The Soil Thematic Strategy [1] sets the frame, and the proposal for a framework Directive [109] sets out common principles for protecting soils across the European Union. Within this common framework, the EU Member States will be in a position to decide how to best protect soil and how to use it in a sustainable way on their own territory. The overall objective is protection and sustainable use of soil, based on the following guiding principles:

- Preventing further soil degradation and preserving its functions:
  - when soil is used and its functions are exploited, action has to be taken on soil use and management patterns and
  - when soil acts as a sink/receptor of the effects of human activities or environmental phenomena, action has to be taken at source.
- Restoring degraded soils to a level of functionality consistent at least with current and intended use, thus also considering the cost implications of the restoration of soil.

The main aspect of the Soil Thematic Strategy was the proposal, by the European Commission, for a Soil Framework Directive. This would require Member States to systematically identify damaged soils, combat soil degradation, and to identify areas where there is a risk of erosion, landslides, loss of organic matter in soils, or compaction or salinization of soils. Member States would then adopt risk reduction and remediation plans for affected areas, within national remediation strategies [110].

In 2012, the European Commission released a new communication [111], which provides an overview of the implementation of the Thematic Strategy for Soil Protection since its adoption in September 2006. According to this report, the European Parliament adopted its first reading on the proposal for the Soil Framework Directive in November 2007 by a majority of about two-thirds. At the March 2010 Environment Council, a minority of Member States blocked further progress on grounds of subsidiarity, excessive cost, and administrative burden, and up to date no further progress has since been made by the Council. Nevertheless, some countries are already adopting aspects of the EU Soil Thematic Strategy in their national legislation [76].

The "Roadmap to a Resource Efficient Europe" [63] proposes that "by 2020 EU policies take into account their direct and indirect impact on land use in the EU and globally, and the rate of land take is on track with an aim to achieve no net land take by 2050; soil erosion is reduced and the soil organic matter increased, with remedial work on contaminated sites well underway." Within the Thematic Strategy for Soil Protection [1] proposals have been prepared for a dedicated "Soil Framework Directive."

The European Commission has founded numerous soil-related projects which have produced very useful results for soil threats as they are defined in the Soil Thematic Strategy, for soil quality monitoring, remediation/rehabilitation, protection against natural hazards, and people use. For example, LIFE program has funded about 147 soil-related projects since its launch in 1992, and there has been an increasing focus on soil protection since the publication of the Thematic Strategy in 2006. In specific, since 2014, LIFE has cofinanced 21 projects related to soil sealing, 13 projects related to soil biodiversity, 24 projects related to soil carbon capture, 11 projects related to soil monitoring, 12 projects related to water and soil, 43 projects related to sustainable agriculture, and 23 projects related to land contamination. Similarly, other EU-funding instruments have funded many soil-related projects (CASCADE, DIGISOIL, Geoland2, iSoil, LUCAS, ENVASSO, ESVA, Ramsoil, SOTER, and many others).

Thus, technologies/methodologies/practices have been developed for urban and rural areas, for agricultural areas, for industrial and brownfield areas, soils that accept different types of wastes, recycling of wastes/nutrients/water on land, etc. In the framework of these projects, many Web applications (eg, SoilPro, Prosodol, sigAGROasesor, Agrolca Manager, AgroStrat, AgriClimateChange and other), databases and platforms (iSOIL, PanGeo e-SOTER, Soilection, EUGRIS, and other), and networks (EURODEMO, NICOLE, GS Soil, agriXchange, EUGRIS and others) have been developed as outcomes.

However, despite the very useful results of these projects and the well-designed and demonstrated actions, there is a big gap between the projects' outcomes and their applicability/adoption by the target end users (ie, authorities, policy makers, land owners), which, in turn, does not enhance the implementation of the European soil policy at local, regional, national, and European level. One of the reasons for this is that the available soil data and projects' data are not translated into problem-solving technology, and the language of delivery of soil information and technologies is complex so that local and regional authorities (LRAs) and other nonexperts (land users, other stakeholders), who need it, find it difficult to avail themselves of the information.

However, bridging the gap between the already produced results and the adoption by the stakeholders is not the only issue of concern when trying to identify the reasons of the low applicability of the EU projects' results. Another one is the limited success of the awareness-raising campaigns implemented by the projects. Awareness raising has been carried out for environmental issues around water and air, but less so for soil. Some projects (eg, SOILCONS-WEB, VOLANTE) began making an effort in this direction, but much more need to be done considering that much effort has gone into awareness raising among key stakeholder groups, such as farmers; however, awareness-raising targeted to activities of the LRAs, which is a significant gap with regard to soil conservation, is limited. More information campaigns and decision-support tools, designed especially for LRAs, will help bridge the gap between lessons learnt and those implementing land use policy.

Low applicability is owed also to the fact that the projects focus mainly on one subject (eg, remediation of metals-contaminated land) and develop a technology/methodology for this specific problem without providing to the end users the technologies/ methodologies that should be applied before and after the proposed technology. Therefore, another important aspect that has not been considered when considering projects' applicability is that results from different projects could be combined and applied in sequential steps in many different cases, which, however, requires a decision-making tool. For example, and considering the two LIFE projects BioReGen and PROSODOL:

1. By using BioReGen an end user may apply remediation of metals-contaminated land by growing high-productivity plants that act as bioaccumulators of certain metals in soil, and thus offering cost-effective options for remediation.
2. By using the monitoring system of PROSODOL the user could implement an initial characterization of the contaminated area, set up a list of appropriate soil quality indicators, evaluate the risk level of the area, and finally develop a periodical monitoring system of the soil quality after the implementation of BioReGen results and by using the Web GIS application developed during PROSODOL (http://www.prosodol.gr).

Many such combinations could be done between the results of the EU-funded projects to provide a holistic approach of soil protection that will assist the implementation of EU soil policy at local/regional level.

One more reason, apart from those mentioned, is the fact that this knowledge is dispersed in the Internet, meaning that there is a lack of an overall tool, which will assist not the knowledge searching, but it will provide the information to implement the knowledge.

It is obvious therefore that although many of the EU projects have produced positive results that could feed into soil policy, in practice this has rarely happened. The Athens Soil Platform Meeting (in 2013)—a thematic seminar for LIFE projects from across the EU—identified the need for projects to develop strategies for building contacts and fruitful working relationships with legislators at regional, national, or EU level [112]. Another proposal suggested establishing a Common European Platform to transfer knowledge from the scientific community to public authorities and policy makers, by making a range of decision-support tools widely available to members of such a pan-European network [113]; in other words, a Pan-European Soil Platform in which all available tools/technologies/methodologies developed so far by the EU soil-related projects will be included in a multilanguage Web platform which will target not primarily the scientists or the researchers but the LRAs, the policy makers as well as the land owners, the farmers, and the public, in general.

4.4.2 Societal Challenges

There has become a consensus that the soil is one of the great challenges that need to be addressed, if Europe is to meet its aim of achieving sustainable development.

Soil in its broadest sense is part of our human habitat; we depend ultimately on it for almost everything. Our dependence is not reflected in the way we organize or manage it. Inadvertently, it is all too easy to focus on one thing that the soil is doing and then to neglect the others, especially those things that are of long term. In order to achieve a land use that is in all senses sustainable, it is necessary for people to have a better, more holistic understanding of what is going on in the soil.

Why are seemingly ignorant decisions often made?

If the soil is so critical to life and human well-being why is it neglected? Why is it threatened? Many people may be ignorant about the long-term consequences of land use. Soil might be perceived as a limitless resource. Land is often owned by people with legal rights that entitle the owners to manage the land as they see fit. Soil degradation is often slow and complex and it is not always easy to demonstrate. Fortunately, neglect is not always a disaster and the soil does have a great capacity to survive and recover from ill-treatment (soil resilience). The problem is often that if the soil is used exclusively for one thing, then often it cannot adequately do all of the other functions that are vital for the maintenance of fertile soils or ecological processes such as regulating the water cycle. The problem then is an organizational and institutional one.

As a first step for the adoption of soil-protective measures, the society should realize what ecosystem health is, how an ecosystem's components interact with each other, and that people are also an ecosystem component. Then, scientists, politicians, policy makers, and other relevant stakeholders should provide to the members of the society clear and understandable answers to the following questions:

- What are the key functions or services provided by the soil and how should these be measured and monitored?
- How can sustainable soil conservation and protection be achieved?
- What is known about protecting and conserving soil functions from good practice and from case studies?
- How can strategies be identified for conserving and protecting soils, within the context of current environmental and agricultural policies?
- How can combating soil degradation be incorporated into soil conservation and protection strategies?

During the last decade, many people expressed concern about the way land use and pollution were reducing the resilience of the soil and its ability to withstand all of the threats that it is facing. There is no doubt that soil degradation has resulted in soils becoming both less fertile and less able to regulate water and cycle nutrients. On the other hand, it is clear that not all areas are affected and that there are many places that serve as examples of good practice.

The majority of people, except from people working on the land, are completely detached from the soil. Nevertheless, there is an enormously enthusiastic minority passionate about gardening and horticulture, and there is a great interest in nature and landscape. Still, preoccupations with air and water are incorporated to the daily lives of the urban population, while soils are generally considered as completely irrelevant. This makes air and water protection easy to justify to the general public, who might not understand the need for soil protection.

Raising consciousness about the importance of soil conservation and protection for the welfare of our modern societies has to be a major policy goal.

Given the legal framework developed by the European Union or internationally, Local and Regional Authorities (LRAs) have a crucial role to play in the protection of European soils. Everything should be done in order to alert LRAs to the importance of their role and to help community members and LRA officials take a leadership role in ensuring that future development reflects environmental protection as well as social and economic community goals.

Regions and Municipalities have major powers in key sectors such as education, the environment, transport, and economic development. Local and regional authorities are vital to the democratic life and are key actors in the conception and implementation of common policies. These units of self-government have the capacity to support development projects directly in their territories and on the ground, and to establish a full cooperation with national governments and international institutions in order to create optimal conditions for sustainable and inclusive growth. Over the years, local and regional governments have proven and continue to prove that by interacting, working together, and exchanging best practices, local and regional leaders are better able to tackle challenges and pave the way to a better future.

Soil is an integral part of our environmental, social, and economic systems, providing food, biomass, and raw materials, serving as a habitat and gene pool, controlling the quality and quantity of water flow, climate change mitigation and adaptation, and biodiversity; soil also performs storing, filtering, and transformation, as well as social, cultural, and religious, functions. In this way, soil plays an integral part in the regulation of natural and socioeconomic processes that are necessary for human survival, such as the water cycle and the climate system. Because soil forms the basis of many different human activities, it has a significant economic value, which, however, is barely recognized. Ensuring that soil is in a good state to deliver its essential functions is vital for the sustainability of Europe's environment and economy. Therefore, all initiatives and actions that aim for soil conservation and sustainable management bear economic benefits for local population.

Thus, the LRAs, as key players for soil sustainability, should be provided with the appropriate tools to mitigate soil threats as well as to improve soil quality by a sustainable way. The LRAs should adopt soil conservation measures and integrate soil conservation into regional and town planning policies as well as into other public policies (agriculture, energy, waste disposal, transport, energy, infrastructure, etc.). This, in turn, will bring significant economic benefits to local communities, as local authorities will be able to plan and develop targeted actions to protect soil and to implement innovative ideas/technologies in all production or other sectors (eg, cultural, aesthetic), where soil is involved.

However, the LRAs rarely know what exactly to implement and which could be the most appropriate practices, methodologies, technologies, strategies that are most suitable for each case of soil degradation, that are, of course, affected by local conditions and peculiarities. There is also a lack of knowledge on how to set targets for soil properties and how to monitor them in the short term and in the long term. This is why it is crucial that the scientific community provide the LRAs with simple and easy understandable tools, which will include all the produced knowledge to date as well as specific implementation instructions.

The LRAs should be urgently provided with and trained in such decision-making tools. It is true, however, that there are many European soil and environmental platforms and portals; however, none of them targets the immediate involved stakeholders (eg, the LRAs), since these platforms/portals target mainly the scientific community. Therefore, they are basically little understandable by the key-holders, ie, those who can adopt and implement specific measures and finally solve the problems, namely the RLAs and the policy makers.

It has to be well understood that the local authorities and the decision makers require functional summaries of the environmental problems highlighting major issues (eg, straightforward identification of sensitive areas irrespective of source; ability to identify remedial/protective actions, to determine the effect of remedial actions; and appropriate monitoring actions) and not long and nonunderstandable scientific reports. Thus, the development of practical and simple decision-making tools that will encourage local policy makers, authorities, and individuals to take a more holistic approach to soil conservation and improvement and at the same time will provide rapid and reliable data/information and Web-monitoring tools will be very useful for the stakeholders.

Knowing exactly what to implement and how, which benefits to expect, and also the cost of these actions, LRAs will have significant economic benefits, since

1. they could perform proper financial planning actions to seek the necessary funds as well as rational use of available funds and achieve best value for money;
2. they could identify those practices which will lead to the desired result, depending on the goals they have set, so that national funds to be utilized in the best possible way. It is thus crucial for the authorities to be able to identify the most fit-for-purpose and cost-effective solutions;
3. addressing degraded soils' impacts would economically benefit the citizens and through them the community as a whole. For example, addressing loss of soil productivity, degradation of the plant cover, land contamination, soil loss/sealing/compaction, intensive and mainly nonsustainable agriculture, desertification risk, and also remediating brownfields and waste disposal areas, improving urban areas by introducing greening practices, etc., would bring net economic benefits to the agricultural sector; improve food quality and safety; increase product competitiveness; increase available clean areas; decrease inputs like water, fertilizers, and pesticides; reduce health risks for human and animals, aesthetically improve degraded areas with subsequent increase in human well-being and tourism; and many others; and
4. instead of using expensive technologies (eg, dig-and-dump) they can identify alternative methodologies/technologies derived from many research projects that are of low cost and have proven environmental benefits, however, still unknown to the wider public (LRAs included) due to the limited dissemination and awareness-raising campaigns.

The anticipated social impacts are also significant considering that regional and local authorities are those authorities which are closest to the citizens, and through many participatory processes the voice of citizens can be heard while they can express their views on issues that are concerned with the local community and influence its socioeconomic development. The undertaking of initiatives and the design of projects/ activities for soil protection and improvement with immediate effect on the quality of local environmental and the prosperity of the civil society will strengthen the relationship of trust between citizens and LRAs and allow their smooth and effective cooperation. This, in turn, will bring substantial benefits for the progress and the success of the actions planned and undertaken and also for the society due to the strengthening of the relationships between its members, leading to an overall improvement of citizens' social life.

Furthermore, the awareness raising and also the knowledge on the actual scientific possibilities/solutions for soil protection and improvement given to the LRAs and to the citizens in a way that they can read and understand will make them active actors in all activities and will significantly increase the acceptance of the designed plans by the civil society.

As far as citizens are concerned, an overall improvement of their social life may be seen due to improvement of their economic status and also of the local environment.

It should be also mentioned that the impacts of soil degradation are more severe for small and poor farmers who are marginalized for lack of credit, spirit of initiative, or know-how. For them the loss of marginal production, but vital for their survival, is extremely penalizing. The improvement in this situation by adopting low-cost but effective practices as derived from many EU projects (however, unknown to the farmers or other potential users) will narrow societal disparities and will further assist the development of the society.

Considering the growing social demand regarding the quality of the everyday environment and of the productive sectors, the establishment (or the consolidation) of such a participatory (in other words democratic) circle at territorial level would help to improve the environment, will boost development, and strengthen the social network.

4.5 Building a Strategy for Soil Protection at Local Scale—The Case of Agricultural Wastes Disposal on Soil

4.5.1 Agricultural Wastes and Soil Disposal

Organic wastes originated from agricultural sector are produced in huge quantities worldwide and include materials that are traditionally used by farmers, such as different types of manures and composts, but also wastes that their landspreading or reuse in agriculture either prohibited or restricted by law (eg, olive mill wastes, OMW).

AW is a general term used to describe waste produced on a farm through various farming activities. These activities can include, but are not limited to, dairy farming, horticulture, seed growing, livestock breeding, grazing land, agroindustrial activities (transformation and preservation of agricultural products), market gardens, nursery plots, and even woodlands. Agricultural and food industry residues, refuse, and wastes constitute a significant proportion, over 30%, of worldwide agricultural production, and can be in the form of solid, liquid, or slurries depending on the nature of agricultural activities [114].

Given that AWs are not restricted to a particular location, but are rather distributed widely, their disposal may affect natural resources, such as surface water and groundwater, soil and crops, as well as human health [115].

Although the quantity of wastes produced by the agricultural sector is significantly lower than the quantity generated by other industries, their pollution potential is high on a long-term basis. For instance, the landspreading of manures and slurries can cause nutrient and organic pollution of soils and waters. Given also that animal excreta also contains a plethora of organic chemicals and pathogens, the risk for surface water and groundwater contamination can be high [114].

The characteristics and composition of AW are variable depending on many factors, which are mainly local (yield, water quality and availability, soil properties, agricultural practices used, type of livestock species, breed and age, diet, type of production, waste-handling practices, etc.) or regional (eg, climatic conditions). Thus, the composition and properties of the wastes produced from a specific activity in one area differ from those produced in another one, or for the same area and the same activity, the waste's composition and properties differ from one year to another. All these make the problem of sustainable AW management more complex.

Generally, AWs have variable water content and high C/N ratio and are rich in organic substances and also inorganics, which, however, are different for the different waste categories [116].

For example, the wastes produced by olives processing are very rich in potassium, phosphorous, and iron, while the wastes originated from pistachios processing have very low potassium content and are very rich in iron, manganese, copper, and chlorides.

Since the sources of AWs are diverse, they can often be potentially hazardous and detrimental to the terrestrial and aquatic ecosystems. Uncontrolled and improper handling can often lead to many environmental adverse effects. Overapplication of AW to cropland and pasture can result in decrease in crop production due to inhibitory amounts of nitrite nitrogen (NO_2-N) or salts added to soil. Application of dairy effluents or feedlot manure to soils can also reduce their permeability and thus adversely affect crop growth. Excess loadings of nitrogen and phosphorous from AW applied to land may cause eutrophication of water bodies or contamination of drinking water [117].

The options of AW management, offering more advantages from an environmental point of view, would be those permitting recovery and recycling of the available contained resources. The key tenet in support of landspreading of wastes is that it recycles nutrients and organic matter to the land, which would otherwise be lost in disposal to landfill or thermal destruction.

Farmers may spread treated as well as untreated AW on soil. The treated wastes include mainly composted materials originated from various sources (eg, manures, wastes from food processing), while the untreated wastes are mainly raw materials and are applied on soil at the time of the their production or after a preliminary treatment (mainly sedimentation or evaporation). In each case, it should be always kept in mind that these materials are or were wastes. In this sense, the measures, which should be taken to protect the environment, and particularly soil and water bodies, are dependent on waste types, characteristics, hazard risk, and quantities produced.

Depending on wastes' properties and their hazard potential, two approaches for safe and sustainable reuse of organic wastes on soil may be proposed, ie, *a bottom-up* approach (ie, farmers to authorities), appropriate for traditionally used wastes (manures and composts), and a *top-down* approach (ie, authorities to farmers) for wastes that are hazardous and should be reused under specific restrictions. Both approaches are consistent of concrete steps and require close cooperation between farmers and local/regional authorities. Monitoring and decision-support tools have been developed for these two approaches during two LIFE projects (PROSODOL [87] and AgroStrat [118]) that are anticipated to assist authorities and farmers to reuse organic wastes on soil by a sustainable way, in accordance with the Section 4.4.2.

4.5.2 Traditionally Used Agricultural Waste-Monitoring Strategy

This group of AW includes mainly manures (after stabilization) and composts, which are traditionally used by farmers for thousands of years. Although farmers are familiar in their use, landspreading of this group should be also monitored and applied on soil following rules that consider soil quality, waste characteristics and nutritional potential, and also cultivation characteristics. For this type of AW, the *bottom-up* approach seems to be more appropriate. As Scheme 4.1 shows, the strategy includes eight steps:

1. Physical, chemical, and biological characterization of the wastes,
2. Soil characterization and analysis,
3. Establishment of soil/water quality criteria,
4. Definition of quantified cultivation targets,

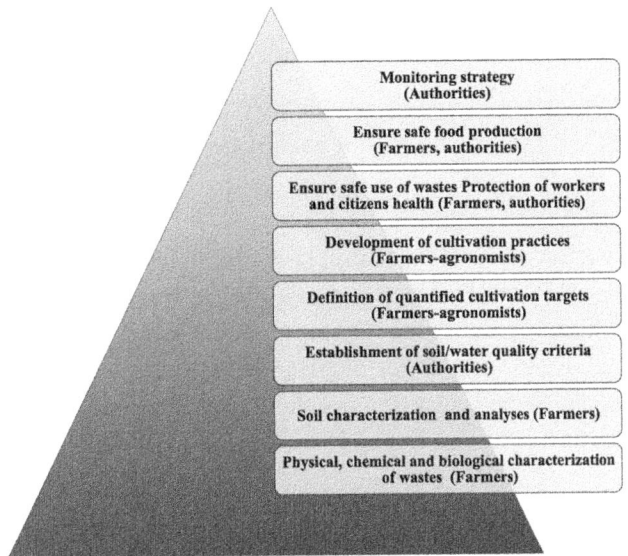

Scheme 4.1

The strategy for safe and sustainable reuse of agricultural waste (AW) that are traditionally used for agricultural purposes.

5. Development of cultivation practices,
6. Ensure safe use of wastes and protection of workers and citizens,
7. Ensure safe food production, and
8. Design of a regular monitoring strategy.

Physical, chemical, and biological characterization of wastes

Prior reuse, the suitability of wastes should be proved. Suitable wastes need to be identified through physical and chemical characterization. Toxicity to humans, soil, and water bodies are recommended to be also defined in order to ensure public and workers' health. Soil toxicity could be assessed by using the standard methods for the determination of (1) nitrogen mineralization and nitrification in soils and the influence of chemicals on these processes (ISO 14238); (2) the effects on earthworms (ISO 11268-1); (3) the chronic toxicity in higher plants (ISO 22030); and (4) soil biomass or soil respiration (ISO 14240). The selection among the available methods should be based on several factors, such as current soil quality, present and future use of the area, amounts of produced waste and treatment level, and others. Many other toxicity tests are also available, eg, extraction procedure; toxicity characteristics leaching procedure; synthetic precipitation leaching procedure; and others.

Wastes' nutritional status is also an important factor that should be identified in order the appropriate fertilization plan to be determined. Nutritional status can be defined by evaluating the results of the chemical analyses and comparing them with generally accepted values for composts and AW, which can be found in the literature.

The farmers or the AW users can undertake this first step, which is considered mandatory, by collecting samples and sending them to a chemical laboratory for analysis.

Soil characterization analyses

Without doubt, the choice among cultivation practices, available waste types, and fertilization schemes depends upon the type of cultivated crops and soil properties. For this, prior the selection and application of any cultivation practice or waste type, soil to be cultivated should be analyzed for a series of parameters that determine its quality and fertility. Soil analysis should be repeated annually, not only to assist farmers to identify the most appropriate cultivation practice but also in order to define any potential adverse effects caused to soil health due to previous practices or waste use. Soil analysis is therefore recommended to determine its level of available nutrients in order to establish the level of baseline of micronutrients. When wastes are applied on soil, the main concern of the scientists and the authorities is the organic load and the toxic substances (eg, polyphenols) of the wastes. However, these should not be the only issue of concern. Specific care should be taken also for inorganic constituents, eg, K, Cl^-, NO_3^-, SO_4^{2-}, P, Mg, Fe, Zn, and others, since the very high amounts disposed on soil change its quality properties drastically, while the concentrations of the inorganic soil constituents and the electrical conductivity steadily increase over the years [119].

If waste landspreading is planned, then soil heavy metals content should be defined and compared to national thresholds in order to ensure that wastes landspreading will not overload soil with excess concentrations of heavy metals. If soil heavy metals are above the established thresholds, then waste landspreading must be avoided. Moreover, other soil properties as well as hydro- and geomorphological characteristics of the area should be considered, eg, infiltration rate, depth to water table, soil texture, slope, etc. [87].

Collection and analyses of soil samples from the neighboring area (ie, not cultivated) is also recommended in order to define soil properties of undisturbed areas to be used during the monitoring stage of soil quality after wastes landspreading.

Same as the first step, the second one is also mandatory and can be undertaken by the farmers or the AW users; however, assistance from experts (eg, agronomists) will ensure correct sampling.

Establishment of soil/water quality criteria

In order to maintain environment quality at waste reuse areas, it is important to ensure, through this third mandatory step, that the reuse will not cause any adverse effect on soil and water quality and will not negatively affect established standards for the surrounding area (eg, aesthetic, touristic, etc.). These standards and preconditions should be carefully studied before waste landspreading, considered or developed, if there are not existed. General and specific area properties and regional and local development plans as these are or will be defined by the local/regional priorities should be considered as well.

In order the farmers and the responsible authorities to be able to identify changes in soil quality, an initial soil survey at field, region, or larger level should be performed and be available for the future monitoring of the area. This will assist authorities to establish soil and water quality criteria that must be met under all circumstances of AW reuse.

Definition of quantified cultivation targets

Having identified the properties of the land to be cultivated and wastes properties, farmers should proceed to the next step, which is the setting and the quantification of their targets. The most acceptable strategy for maximizing the agronomic and economic benefits is to specifically quantify the anticipated benefits, economic and environmental ones. Generally speaking, the main goals of farmers are:

*   high yield and subsequent profit,
*   good-quality crops that satisfy market demands,
*   low cultivation and operational costs, and
*   conservation of soil health to ensure future fertility and productivity.

In order farmers to benefit the most by the reuse of AW, they should exactly determine what they are trying to achieve, eg, restoration of the productivity of an eroded soil, provide supplemental nutrients to a high value crop, etc., and to determine what practical and workable combinations of organic materials and mineral fertilizers are most appropriate to accomplish the proposed task. Having completed this mandatory step, farmers could proceed to the next one and define the most appropriate cultivation practices for their case and special area's conditions.

Development of cultivation practices

Soil must always maintain all its functions and its absorption capacity to ensure a sustainable system and for this, the ultimate goal when applying AW on land should be to apply them in such a way that the soil either filters the potential toxic elements effectively, or electrochemically absorbs them or decomposes them in order that a clean solution infiltrates through the soil body.

When considering of using organic wastes in crop production or field application, application rates must be carefully estimated and based upon soil fertility, crop requirements, and waste-specific characteristics. Therefore, this step, which is considered also mandatory, will assist end users to precisely estimate the AW amounts to be distributed. The concentration of available soil nutrients depends on soil properties, cropping and fertilization history, land management, and climatic conditions. Plant nutrient requirements depend on soil fertility, crop type, and the target yield.

After the definition of all the appropriate soil and waste parameters, the rate of nutrients to be applied, meaning the doses of AW and supplementary inorganic fertilizers, can be estimated, considering that the applied nutrients should be equal to or greater than the nutrients removed by the crop over time so that soil fertility can be maintained.

Irrigation water quality and composition should be also taken into account. Irrigation water contains soluble salts, some of which are considered nutrients (eg, potassium, sulfur) or pollutants (eg, heavy metals, nitrates). Therefore, the chemical analysis of the water to be used for irrigation can provide valuable data and sometimes may be a restrictive factor to a chosen practice (eg, in case of high heavy metals and nitrates content, or high salinity). If the nutrient content of the irrigation water is considerable, then the respective nutrient amounts should be extracted from the total estimated nutrient supplement. An example of doses estimation considering soil, wastes, and irrigation water parameters is given by the "Cultivation Management Software" developed by the LIFE AgroStrat Project [118].

Three more parameters should be determined [87]:

1. *Maximum permitted AW amount.* This is the maximum amount of AW that a soil can afford based on its physicochemical properties. To estimate the maximum amount one should consider the concentrations of all waste's elements/substances and define the one that is the restrictive factor for the application (taking also into account local soil properties). The concentration of this restrictive element/substance is then used for the estimation of the maximum waste amount to be distributed. For instance, this element/substance could be the one with the highest concentration or with the lowest threshold.
2. *Annual permitted application of AW.* The annual rate of application could be determined by taking into account the maximum permitted amount and the general rules of soil fertilization.
3. *Time of AW application for different crops.* The time of application has to be defined considering the annual rainfall rate, intensity and distribution throughout the year, and the temperature, in relation to water balance, soil properties and processes, microbial activity, and AW decomposition.

Although farmers have, in general, experience in developing cultivation practices, however, this experience is based mainly on traditional habits. Therefore, the cooperation with experts (agronomists) is recommended in order to ensure that the cultivation practices will be the most appropriate given the local environmental, social, and economic circumstances.

Ensure safe use of wastes and protection of workers and citizens' health

A vital priority when considering reuse of AW on soil is the protection of workers' and citizens' health during and also after landspreading. Waste poses a threat to the environment and to human health if it is not managed properly and recovered or disposed of safely. There are safe ways of dealing with any waste, while any waste can be hazardous to human health or the environment if it is wrongly managed. Deciding whether any waste poses a problem requires consideration not only of its composition but also of what will happen to it. Even everyday items may cause problems in handling or treatment. Anything unusual in waste can pose a problem, and what should be identified as potential problems in a consignment of waste are significant quantities of an unexpected substance, or unusual amounts of an expected substance. For this, the users should handle wastes by following the specific instructions given for the specific type of waste, while it is also important that the responsible local, regional, or governmental services undertake or supervise the monitoring of all appropriate actions that ensure safe reuse.

Ensure safe food production

Apart from the ordinary tests for the harvested crops (eg, nutritional status, pesticides residues, etc.), other constituents, typical of the wastes used during the cultivation, should be also measured by following a well-designed sampling and laboratory analysis plan while the analyses results should be compared to standards for safe food production. This step is recommended to be performed in cooperation with the special authorities' agencies.

Design of a regular monitoring strategy to assess potential risks
for soil/water bodies and safe reuse

Apart from assisting farmers and AW users during the different steps of this hierarchy, local and regional authorities should act as absolute monitoring authorities. However, in this case of AW types, this final step could be voluntary.

Nevertheless, all information should be available to the authorities at any time so as to be able to monitor, implement monitoring strategies, and assess the collected data. Therefore, farmers must inform the responsible authorities and provide them with detailed plans regarding data collected during the previous steps. Period and duration of landspreading should be taken into account and a time plan should be submitted also, so as the authorities to be able to design the appropriate monitoring plan.

Local, regional, or governmental authorities perform the monitoring of the AW reuse areas; however, farmers could also play a significant role in monitoring and maintenance of soil and water quality. An effective monitoring system has to consider the geomorphology, the hydrology, soil types of the application area, and the peculiarities and the characteristics of the AW as well as local meteorological conditions.

It is, therefore, recommended that a monitoring strategy fully suited to AW reuse in agricultural sector should include:

1. An optimized set of soil quality indicators

    In order the cultivated or disposal area to be monitored, the establishment of a set of soil and water indicators is required. This requires a scientific work to be done, and a strategy should be designed and implemented by experts. It includes soil and water sampling in order to identify background levels of key soil and water parameters, as well as the definition of the parameters that are most likely to be influenced by the reuse of waste on land. The latter are also depended on the properties and characteristics of the waste type to be used. For example, it was defined that when OMWs are applied on soil, the parameters that are mainly affected are soil pH, electrical conductivity, organic matter, total nitrogen, polyphenols, available phosphorous, exchangeable potassium, and available iron, and therefore these parameters can be used as indicators to assess safe use of OMWs [87]. On the contrary, electrical conductivity, organic matter, total nitrogen, polyphenols, available phosphorous, exchangeable potassium, available copper, and available zinc were defined as the most appropriate parameters when pistachio wastes are distributed on land [118].

    If such a methodological study could not be performed, then it is recommended to identify the most appropriate soil and water parameters by assessing quality parameters of the surrounding area and start monitor them over time. Some common and sensitive soil parameters can be also used, as for example, soil pH, electrical conductivity, polyphenols, total organic carbon, nitrogen, and phosphorous. For water, biochemical oxygen demand (BOD), electrical conductivity, nitrates, phosphates, pH and some bio-indicators for water life could be used.

    Additionally, it is important to ensure that legislative restrictions as regards mainly heavy metals in soil and water are met.

    In general, the monitoring of quality indicators within a defined ecological zone requires [120]:
    *   Direction of change—positive or negative, increase or decrease, etc.
    *   Magnitude of percent change over the baseline values
    *   Rate of change—duration: months, years
    *   Extent of change—percentage of the area being monitored, ie, what percentage of the area has changed with respect to the selected indicator during a specified period

    Monitoring of soil and water indicators needs to set up sampling strategies allowing assessment of changes in systems' quality.

2. Threshold values for the quality indicators

   In general, changes in soil and water quality can be assessed by measuring appropriate indicators and comparing them with critical limits or thresholds at different time intervals, for a specific use in a selected area-system. A critical limit or threshold level is the desirable range of values for a selected indicator that must be maintained for normal functioning of the ecosystem health. Within this critical range, the system performs its specific functions in natural ecosystems [120].

   Thus, when a set of indicators are proposed, this list should be accompanied by thresholds level for each one of the indicators in order to assist evaluation of the collected data and of the chemical analyses results. The thresholds could be identified based on EU directives, on national and international laws, but also on the international literature mainly as far as soil is concerned.

   The peculiarity of soil indicators (appropriate for waste reuse or disposal areas) is that they mainly correspond to soil properties associated with fertility and not to pollutants in the classical sense, such as heavy metals, and therefore are not included in national laws. Nevertheless, international literature can provide general limits as these properties have been extensively studied for many years. Given the complexities of setting limits and the uniqueness of each targeted area/region, it may be more efficient to develop guidelines that can help in setting up limits under certain land and environment conditions [121].

   Thus, although a general definition of soil indicators' thresholds could be performed after searching in international literature and national legislative frameworks, it should be highlighted that the definition of indicators' thresholds would be more effective and representative of each target area if they would be determined after evaluation of data collected from the areas of interest and by taking into account local characteristics and values of the indicators of representative control samples.

3. Periodical soil and water quality monitoring

   The next step is monitoring the impact of AW reuse on soil, water bodies, and the environment through a systematically planned sampling scheme combined with different eco-bio-toxicological tests.

   As regards soil, an initial georeferenced grid or free (based on the main soil types of each target area) soil sampling should take place at depth increments in order to define the current situation in representative, benchmark soils of the area. Emphasis should be given to identify control soils, ie, soils that have never accepted AW or other wastes in the area as well as soils in which AW have been applied intensively. It is recommended that for the initial characterization of the area, the collected soil samples should be analyzed for all soil physicochemical properties. After the initial characterization of the area, soil samples should be collected annually from hot spots, which would have been identified during the initial characterization of the area, and be analyzed only for the soil quality indicators. For the annual monitoring of the area, a georeferenced soil-sampling scheme should be planned and implemented by authorities or through them by areas' owners [87,118].

Software application tools for soil monitoring could facilitate adoption of the monitoring system by authorities and individuals and enhance continuous and effective monitoring. Such tools have been developed by EU-funded projects, eg, from LIFE PROSODOL project, and are described in the following section as it is considered an obligatory tool for potential hazardous wastes.

Neighboring water bodies are also recommended to be monitored periodically. Water samples should be collected from all first, second, and third catchment order discharge and in the groundwater both before and after rainfalls. This way a database could be developed where the ground and surface water will be recorded spatially and temporarily.

In addition, it is recommended to record and monitor areas of waste reuse using GIS Web tools which allow easy and visualized evaluation by authorities, scientists, and other stakeholders [87,118].

4.5.3 Potentially Hazardous Wastes-Monitoring Strategy

Despite the fact that they are originated from natural products, these types of AW may contain a plethora of potential hazardous constituents, as for example, polyphenols, pesticide residues, heavy metals, and also pathogens. Therefore, a mandatory management and monitoring strategy must be developed by the responsible authorities in order to ensure that soil distribution will not harm human beings, animals, and the environment. An example of these waste types is OMW produced in huge amounts worldwide but especially in the Mediterranean region, as Spain, Italy, and Greece are the three world leader countries in olive oil production [87].

In this case the strategy includes seven steps (Scheme 4.2):

1. Physical, chemical, and biological characterization of the wastes,
2. Identification of potential and current waste disposal areas and recording them in a GIS geodatabase,
3. Characterization of disposal areas—risk assessment,
4. Evaluation of risk level,
5. Adoption of soil quality indicators and thresholds,
6. Defining the conditions of landspreading, and
7. Periodical monitoring—evaluation of the results.

As regards soil quality protection from landspreading, these measures are considered as being efficient for maintaining soil quality and sustainability in long term.

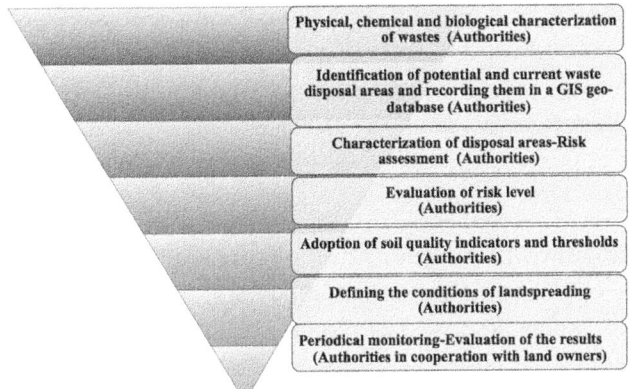

Scheme 4.2
The strategy for safe and sustainable reuse of potential hazardous agricultural waste (AW) on soil.

Physical, chemical, and biological characterization of the wastes

Exactly as it is described in paragraph "Physical, chemical, and biological characterization of the wastes" under Section 4.5.2, wastes must be fully characterized; however, in this case the monitoring should be more detailed and integrated, considering the variability of the wastes composition. Therefore, a well-designed sampling wastes strategy must be developed and performed by experts.

The appropriateness of wastes for landspreading must be proved through appropriate chemical/physical/biological characterization, and waste must be characterized for their hazardous potential. According to the results of this assessment and the national or transnational legislative restrictions, the competent local/regional/governmental authority may permit (or not) the landspreading. Given that the wastes could be distributed on land (with or without preconditions), the authorities must proceed to the next steps and define the preconditions and the monitoring measures.

Identification of potential and current waste disposal areas and recording them in a GIS geodatabase

Authorities must identify the current disposal areas in its territory and record them in an inventory. The inventory will contain all licensed disposal areas and as many as possible nonlicensed ones. Local inventories should be created as a first step under the responsibility of local or regional authorities, which afterward will be integrated into a national inventory under the responsibility of governmental agencies. GIS mapping of the disposal areas and the establishment of a digital database is strongly recommended.

GIS has been extensively used in applications related to environmental risk assessment. An example of a Web-based map application tool developed during LIFE PROSODOL project is presented in this paragraph for evaluating the location suitability of the olive oil production facilities and further, the waste disposal areas depending on several anthropogenic (residential areas, road network), environmental (slope, archaeological sites, lake and rivers area, Natura areas, land use—Corine), and geological (hydrolithology, geology, faults) criteria factors. The pilot area of this study is Rethymnon prefecture, Crete, Greece.

Different scenarios are tested by weighting accordingly all these features giving evaluation suitability maps for the above risk assessment modeling techniques. For the development of the Web-based tool, the geological, hydrological, and land use features in the vicinity of all OMW disposal sites of the pilot prefecture of Rethymno, Crete, Greece, were used.

In order to assess the appropriateness of the location of these facilities, two of the most used approaches to solve multicriteria problems, such as risk assessment analysis and suitability modeling, the Weighted Sum Model (WSM) and the Analytic Hierarchy Process (AHP) methods were applied in the wider area of the municipality of Rethymnon in Crete.

In multicriteria problems such as suitability or risk assessment modeling, one must define the problem, break the model into submodels, and identify the appropriate input data. Since the input criteria data (which are actually in most cases raster layers) will be in different numbering systems with different ranges, to combine them in a single analysis, each cell for each criterion must be reclassified into a common preference scale such as 1—10, with 10 being the most favorable. An assigned preference on the common scale indicates the relative confidence that we have in the influence of a criterion compared to another. The preference values are on a relative scale.

The preference or suitability values not only should be assigned relative to each other within the respective criterion layer but should also have the same meaning between other criteria layers. On the other hand, each of the criteria in risk assessment analysis may not be equal in importance. Depending on the technique used, like the WSM or the AHP method, various weighting schemes are introduced and different suitability location areas are exposed in the final Web map application. ArcGIS Desktop was selected for creating the risk assessment maps using the ArcGIS ModelBuilder Tool [87].

Thirteen criteria were gathered and selected for the area of interest. Three of them compose the anthropocentric main factor, seven of them compose the environmental factor, and the rest compose the geological main factor in the underlying model. So, there are three main groups of criteria and thirteen extended subcriteria to take into account. The final process that needs to be done for estimating the risk of OMW disposal areas in respect with the three main criteria, anthropogenic, environmental, and geological (or 13 subcriteria), takes place in the tool (coded in Python) where two different approaches for the multicriteria analysis may take place, the WSM or the AHP method. A user may define the desire percentage of importance for each factor or subcriteria, while the above approaches give the user the ability to decide the analysis process. According to the method the user chooses, different risk assessment result maps may be emerged.

Seven scenarios were created both for the WSM and AHP method in order to determine the risk assessment for the OMW disposal areas in terms of the various components of the anthropogenic, environmental, and geological criteria and moreover to fill the bulk of the diversity of the importance of these various criteria. Specific maps were created for each one of the scenarios (Fig. 4.1).

- *Scenario 1*—For the WSM this scenario is based only on the anthropogenic aspect of the risk assessment analysis. The importance is given only on anthropogenic subcriteria (100%) while the rest are not taken into account (0%). As for the AHP the anthropogenic criteria are more important (biggest priority) than the environmental and geological criteria, while the environmental criteria are also more important than the geological ones.
- *Scenario 2*—For the WSM only on the environmental aspect of the risk assessment analysis is taken into account. The importance is given only on environmental subcriteria (100%) while the rest are not taken into account (0%). As for the AHP the anthropogenic criteria are more important (biggest priority) than the environmental and geological criteria, while the geological criteria are more important than the environmental ones.
- *Scenario 3*—For the WSM this scenario is based only on the geological aspect of the risk assessment analysis. The importance is given only on geological subcriteria (100%) while the rest are not taken into account (0%). As for the AHP the environmental criteria are more important (biggest priority) than the anthropogenic and geological criteria, while the anthropogenic criteria are more important than the geological ones.
- *Scenario 4*—For the WSM the importance is given in all factors and subcriteria (100%), which actually are normalized to have an equal weight of importance. As for the AHP the environmental criteria are more important (biggest priority) than the anthropogenic and geological criteria, while the anthropogenic criteria are less important than the geological ones this time.
- *Scenario 5*—For the WSM this scenario has given an advance in importance on the anthropogenic aspect of the risk assessment analysis (50%) while the rest are sharing the rest percentage (25% and 25%). As for the AHP the geological criteria are more important (biggest priority) than the anthropogenic and geological criteria, while the anthropogenic criteria are more important than the environmental ones.
- *Scenario 6*—For the WSM this scenario has given an advance in importance on the environmental aspect of the risk assessment analysis (50%) while the rest are sharing the rest percentage (25%/25%). As for the AHP the geological criteria are more

WSM   AHP

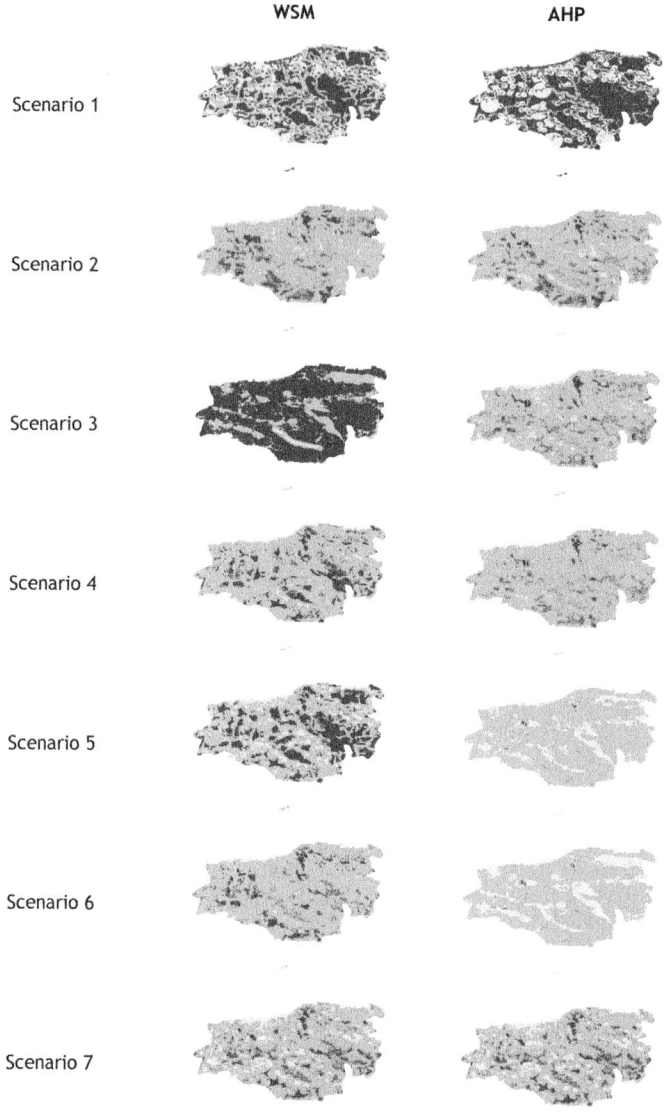

Scenario 1

Scenario 2

Scenario 3

Scenario 4

Scenario 5

Scenario 6

Scenario 7

Figure 4.1
Risk assessment scenarios result maps.

important (biggest priority) than the anthropogenic and geological criteria, while the anthropogenic criteria are less important than the environmental ones this time.

- *Scenario 7*—For the WSM the importance is given to all three main factors (100%) which actually are normalized to have an equal weight of importance, but this time giving an importance in residential area criteria a 70%, and a 30% for the road network criteria. In the environmental aspect of the criteria, full importance is given to slope,

Table 4.4: Suitability of the Known OMW Disposal Areas' Locations (Locations With Suitability Value <5), According to the Seven Mentioned Scenarios.

	Scenario 1	Scenario 2	Scenario 3	Scenario 4	Scenario 5	Scenario 6	Scenario 7
WSM	21%	100%	100%	92%	71%	98%	73%
AHP	22%	97%	95%	97%	85%	87%	73%

OMW, olive mill waste; *WSM*, weighted sum model; *AHP*, analytic hierarchy process.

aquifers, and coastline, while medium importance to the rest environmental subcriteria. For the geological aspect of the analysis only the hydrolithology subcriterion was given an importance of 80% having the rest subcriteria sharing the remainder percentage. As for the AHP the environmental criteria are more important (biggest priority) than the geological, the geological criteria are more important than the anthropogenic, while the anthropogenic criteria are more important than the environmental ones, giving in such a way a balanced importance to all main factors.

Below, in Table 4.4, the suitability of the known OMW disposal areas' locations (locations with suitability value <5), according to the seven mentioned scenarios for each of the two decision-making approaches, is shown.

Using the Google Maps API the user can load KML files, allowing the building of a 3D map application (Fig. 4.2). The implementation of the map application was handled with JavaScript over PHP pages of the PROSODOL Website. By selecting the user-defined

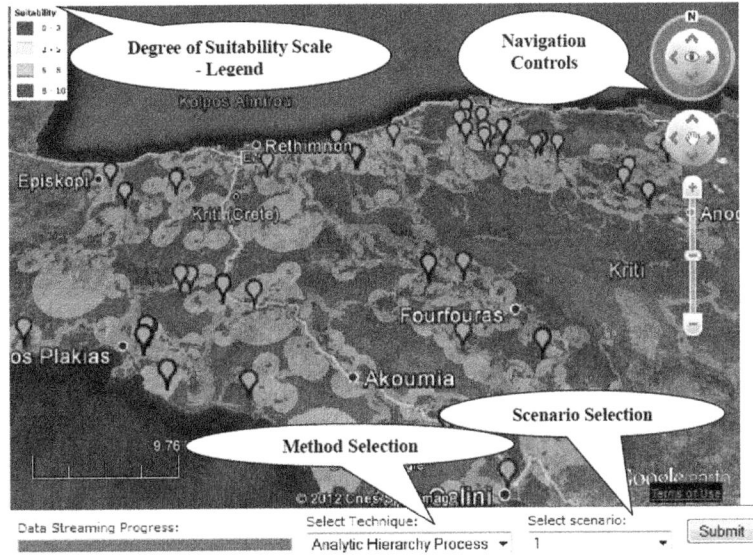

Figure 4.2
Risk assessment 3D web map application interface through the Website of LIFE PROSODOL [87].

method (WSM or AHP), the modeling scenario, and submitting the information to the application, the corresponding risk assessment map is loaded.

This tool is considered as highly significant and can provide authorities, ie, local, regional and national, with a powerful decision-making tool that would enhance the definition of the areas that could be appropriate for hazardous AW disposal, however, according to the criteria set and the priorities of each territory. This tool should be used in combination with the soil monitoring activities and specifically, as the initial stage of potential disposal areas' evaluation and selection.

Characterization of disposal areas—risk assessment

As a third step, authorities should proceed to complete and detailed characterization of the disposal areas and to the performance of risk assessment studies.

Recorded disposal areas should be characterized considering location, hydrogeology, physiography, geomorphology, land use, soil structure, texture, water permeability, coefficient of hydraulic conductivity (saturated or unsaturated), porosity, presence and depth of impermeable soil layers. Additionally, the collected data may include history of the site, extent and types of contaminants that may exist, hydrogeological and hydrological regime for the broader area, known/anticipated presence and behavior of receptors, sampling of soil and groundwater: comparison with generic guideline values or quality standards, sampling of soil and groundwater: site-specific modeling of fate, transport and exposure and comparison with toxicological values, and other parameters which may be considered necessary for the complete characterization of the area. Such a characterization will permit the performance of the risk assessment study of the area and the identification of the sites, which pose risk to human health and to the environment.

For a risk to exist, there must be a *source* (or hazard or pressure), a *pathway*, and a *receptor* (or target). This is the basis for the *Source—Pathway—Receptor (S-P-R) conceptual model* for environmental management. In addition, a conceptual model also provides information useful to the scoping of any investigation as it identifies the sites that pose the greatest risk to the environment and human beings and also identifies the S-P-R linkages that have the highest risk associated with them [122]. Thus, the detailed information obtained through the assessment will further assist the decision on the extent of measures, which are required to manage the risk, which may involve breaking the pathway or removal of the source or monitoring of the receptor.

Indicatively, a risk assessment study could comprise:

1. Preliminary investigation (desk study, site reconnaissance, and sometimes limited exploratory investigation). The goal of this preliminary stage is to assess whether potentially contaminating activities have taken place on the site, whether soil and/or water

pollution is suspected, and in some cases to confirm the existence of pollution. In short, this phase focuses on hazard identification.

2. Detailed investigation. The aims at the main site investigation stage are (1) to define the extent and degree of contamination, (2) to assess the risks associated with identified hazards and receptors, and (3) to determine the need for remediation in order to reduce or eliminate the risks to polluted or actual receptors.

3. Supplementary or feasibility investigations to better define the need for and type of remedial action or monitoring. The aim may be to assess the feasibility of various remediation techniques; this may include more detailed physical and chemical characterization of soils and laboratory studies on soil or groundwater treatability. Supplementary investigations may also be designed to improve understanding of the nature, extent, and behavior of contaminants.

The risk assessment, however, should not be limited to toxic constituents, like the polyphenols for the case of OMW, which may pose threat to human and animal health, but to consider also the potential progressive soil degradation due to the presence in wastes of other less hazardous or nonhazardous constituents, like nutrients and other inorganic wastes' constituents. This factor is often underestimated and the majority of risk assessment studies focus on the toxicity, which may be caused to soil and to humans from toxic substances/elements. Therefore, specific care should be taken also for inorganic constituents (eg, K, Cl^-, NO_3^-, SO_4^{2-}, P, Mg, Fe, Zn, and others), since the very high concentrations disposed on soil change drastically its quality properties, while their concentrations in soil as well as the soil electrical conductivity remain high even many years after the last disposal. For this, the performance of a complete soil physicochemical analysis and identification of the organic and the inorganic soil constituents are strongly recommended. Determination of phytotoxicity potential is also recommended.

The risk for each potential pathway is considered to be a combination of the probability that a hazard will reach the target (eg, high polyphenols concentration in soil due to OMW disposal) and the magnitude of harm if the target is exposed to the hazard (eg, phytotoxicity). The probability that a contaminant will reach a target in sufficient concentration to cause harm may be assessed qualitatively according to the scale: **high** (certain or near certain to occur), **medium** (reasonably likely to occur), **low** (seldom likely to occur), or **negligible** (never likely to occur). The magnitude of harm is assessed as **severe** (human fatality or irreparable damage to the ecosystem), **moderate** (eg, human illness or injury, negative effects on ecosystem function), **mild** (minor human illness or injury, minor changes to ecosystem), or **negligible** (nuisance rather than harm to humans and the ecosystem). The qualitative level of risk associated with each pollutant pathway is then assigned by the combination of the aforementioned probability with the

Table 4.5: Risk Assessment Rating.

Probability	Magnitude			
	Severe	Moderate	Mild	Negligible
High	High	High	Medium/low	Near zero
Medium	High	Medium	Low	Near zero
Low	High/medium	Medium/low	Low	Near zero
Negligible	High/medium/low	Medium/low	Low	Near zero

magnitude of harm. Thus, having identified all the crucial parameters the risk should be rated according to Table 4.5 [123].

Evaluation of risk level

The third step is the evaluation of the level of risk of the suspicious areas and exclude for further future disposal of all areas under *high risk*. For these areas a remediation plan should be developed and implemented immediately. For areas under *medium risk*, further assessment of the threat type and potential extent is strongly recommended in order to decide the conditions of waste disposal or the design and implementation of remediation actions. For these cases, decisions should be taken considering data collected during the risk assessment study is proposed. For areas under *low* or *near zero risk*, a management plan for the safe disposal wastes should be developed and implemented under the supervision of local authorities and the responsible governmental agencies.

Adoption of soil quality indicators and thresholds

Also for the case of potentially hazardous wastes, a set of appropriate soil indicators must be established and their respective threshold values must be defined, as described in paragraph "Design of a regular monitoring strategy to assess potential risks for soil/water bodies and safe reuse" of Section 4.5.2.

Defining the conditions of landspreading

It is very likely, some areas, although being of low or negligible pollution/degradation risk, to be inappropriate to accept wastes disposal due to their specific characteristics (eg, slope, soil infiltration rate). In order to ensure safe wastes disposal, soil and land data have to be considered in combination with bioclimatic conditions and management practices.

The decision of land distribution is proposed to be taken considering appropriate suitability criteria and after grouping land into suitability orders (ie, suitable or unsuitable) and then into suitability classes according to the degree of their limitations (eg, slight, moderate, severe limitations; currently not suitable and permanently not suitable for waste application) [124,125].

The following steps are proposed:

Step 1: Definition of suitable or unsuitable soils for wastes disposal
Soils with the potential to receive or soils that should be excluded from disposal/distribution/application are identified based on permanent physical and/or chemical characteristics.

Step 2: Estimation of the maximum permitted OMW amount
The soils that are suitable for waste application should be further studied in order to define the *maximum permitted waste amount* (or the maximum amount they can afford), based on their physicochemical properties and on wastes composition and considering legally applied thresholds for these properties.

Step 3: Estimation of annual permitted application of wastes
The annual rate and timing of wastes application could be determined by taking into account the maximum permitted levels of potentially toxic substances as defined by the valid legislative framework and also the thresholds as derived from the literature, especially for the nontoxic macronutrients (eg, P, K, N) and for the available forms of metals. The annual permitted application should be estimated after evaluation of the specific local environmental conditions and soil quality. Since the most of the AW constituents are nontoxic and are considered as important nutrients (N, P, K, organic matter, Fe, etc.), their application could be beneficial for soil quality and may improve fertility. However, due to the very high load of AW in these constituents, the disposal on soil should follow restrictions and rules and the annual application should be estimated by considering:

* The concentration of the specific elements/substances in soil,
* The concentration of the specific elements/substances in waste,
* The specific climatic, geomorphologic, and environmental conditions of the area that may affect the behavior of these elements/substances in soil (leaching, adsorption, decomposition, etc.), and
* The maximum permitted amount of each of these elements/substances that can be disposed on soil without changing its quality.

As regards the nutrients' content, the AW could be considered as a nutritional material (like fertilizers) and thus, annual dose estimation should follow the general rules of soil fertilization considering soil properties and purpose of use.

Step 4: Time of waste application
Finally, the time of application has to be defined considering the annual rainfall rate, intensity and distribution throughout the year, and the temperature, in relation to water balance, soil properties and processes, microbial activity, and wastes decomposition. The background philosophy is to apply the wastes at periods where rainfall-induced leaching of the soil water is not expected.

Periodical monitoring—evaluation of the results

Authorities should design an effective monitoring strategy and implement it in cooperation with scientists and local wastes users. Monitoring of soil quality indicators as defined in the previous fifth step should be performed once a year and preferably before wastes distribution. This requires annual soil sampling and chemical analysis in the framework of a defined monitoring strategy that land users or polluters must follow. Sampling and proceeding of soil samples to laboratory should be under the responsibility of the landowner. The results of the chemical analysis should be evaluated by an expert (eg, agronomist), and a technical report should be submitted to the responsible authorities. The report, apart from the evaluated results of soil and wastes analysis, should also include a detailed description of the wastes distribution plan (amount, timing, equipment use). The evaluation of the soil quality indicators within a defined ecological zone requires [120]:

* Direction of change—positive or negative, increase or decrease, etc.
* Magnitude of change—percent change over the Environmental Quality Standards (EQSs) or baselines values of the area
* Rate of change—duration: months, years
* Extent of change—percentage of the area being monitored, ie, what percentage of the area has changed with respect to the selected indicator during a specified period

Depending on the evaluation results, the responsible authorities may permit wastes disposal or not, while in case of continuation of waste disposal on soil, the maximum dose should be determined considering the risk level of the area. The responsible authorities should also establish a periodical monitoring strategy in order to be able to identify potential risks at any time. A specific inventory (a database) of each disposal site should be established and updated annually. This will facilitate the immediate identification of risky areas as well as will provide data regarding history of the site, specific local geomorphological characteristics, amounts of waste that have been disposed each year, results of waste and soil chemical analyses, and any other data that are considered useful and necessary for the effective protection of soil quality and function.

If, for any reason, observed that a disposal area is under risk of soil deterioration, then, after evaluation of risk level, changes in land distribution plans may be proposed or the authorities may require development and implementation of a remedial strategy.

To facilitate the implementation of the above measure the LIFE PROSODOL project developed a Web-based GIS tool, which presents soil constituents' distributions versus time and depth. Through this tool, local and regional authorities will have the opportunity to map and screen disposal areas rapidly, identify potential risky conditions, carry out systematic monitoring of the areas of interest, and facilitate decision making on the

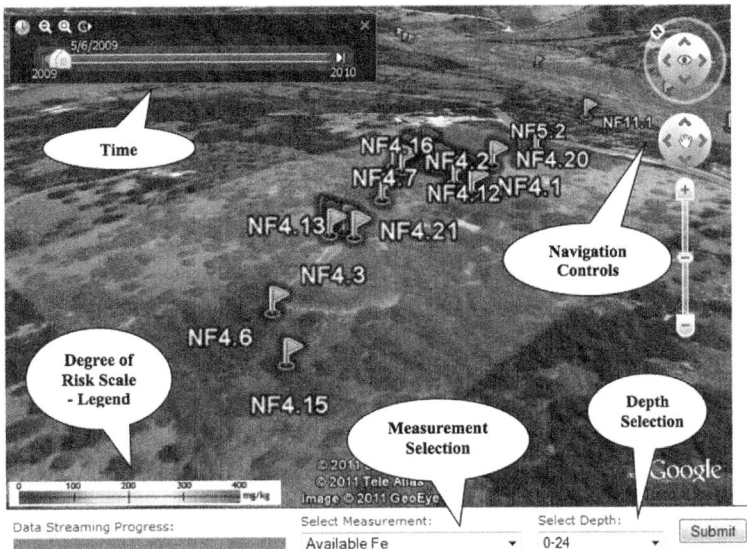

Figure 4.3
The interface of the map application for an olive mill waste (OMW) disposal area. *Yellow flags* (gray in print versions) represent sampling sites inside the area of interest [87].

appropriate measures to be taken at field or municipal scale. The tool was developed for the case of OMW; it monitors the eight soil indicators selected for this case (ie, soil pH, electrical conductivity, organic matter, total nitrogen, polyphenols, exchangeable potassium, available phosphorous, and available iron) and integrates the continuous monitoring of the OMW disposal areas into the regular activities of local/regional authorities and thus, allows the proper and continuous monitoring of such areas.

The adoption and application of such a tool, however, requires the cooperation of the disposal areas' owners, since repeated soil samplings at various sites are necessary for maps creation and update. The proposed application tool uses interpolation surfaces that indicate the distribution of the different physical and chemical parameters in the area of interest, so the user can rapidly obtain an idea of the possible diffusion of the chemical parameters and the degree of risk in the vicinity of the waste disposal areas. This, potentially, allows also the establishment of an Operational Center, which could be located, for instance, in cooperation with the Environmental Protection Office of the Local Government (District) in the premises of a Municipality, and can undertake the continuous monitoring of areas under risk and the scientific and consulting supporting of the owners.

The design of the particular software package needs to monitor a number of private fields that are spread around and make queries based on various spatial and chemical attributes.

Thus, it is proposed that, for each disposal area, one initial mapping should be carried out by performing soil sampling from various sites and for at least four times (eg, every 2 months). The sampling sites will be decided according to the generally accepted soil sampling rules, and a qualified person should be present and undertake the overall control. The collected soil samples should be analyzed for the parameters, proposed as suitable indicators. The maps that will be created should be used for no more than 5–8 years. After this period, the maps must be updated by repeating the sampling procedure. Fig. 4.3 presents the interface of the map application for OMW disposal areas; however, by making minor changes the tools can be implemented for other waste types.

The creation of the interpolation surfaces requires a number of data around the vicinity of wastes disposal areas in order to identify the distribution of the different chemical parameters in the area, so that to obtain an idea of the possible diffusion of the chemical parameters and the degree of risk in the vicinity of the waste disposal areas. Various interpolation algorithms could be used for mapping the specific parameters; however, the method of the Inverse Distance Weighting (IDW), which calculates cell values by averaging the values of sampling points in the vicinity of each cell based on distance, is proposed since it causes limited secondary effects (eg, bull's eye effect around isolated sample points, extreme trends away of the sample areas, etc.). Such an application allows:

- *Measurement selection*—The user selects the name of the chemical parameter in order to see the corresponding interpolated surface map.
- *Depth selection*—The user can choose the depth of soil for which the user wishes to see the value of the selected chemical parameter. Then the user can submit the information provided, and the map application starts to stream the data needed and present the corresponding interpolated surface map.
- *Navigation controls*—The user may navigate inside the map through the navigation controls, in any direction, angle, pan, and zoom, giving user the freedom of any view perspective.
- *Time slider*—An animation of the interpolated surface area map can be viewed through different time periods on the Google map.
- *Degree of risk scale—legend*—When the interpolated surface map is loaded, the corresponding scale of the risk degree of the selected chemical parameter is shown.

The 3D map application was designed in such a way that the end user can easily and effectively use and retrieve the surface interpolated information needed [87].

Finally, the competent authority should keep a register of permits issued and the record of each landspreading operation and make the necessary site visits and spot-checks to confirm that landspreading operations were in compliance with the permit conditions and the records would indicate where any pollution incident could be linked with a landspreading operation.

A record of each spreading operation will be kept (type of waste, quantity of waste applied, location of farm and field, date of spreading), including the results of monitoring and analysis, and supplied to the competent authority.

References

[1] COM, 231 Final. Communication from the Commission to the Council, the European Parliament, the European Economic and Social Committee and the Committee of the Regions, "Thematic Strategy for Soil Protection", 2006.

[2] W. Chesworth, Encyclopedia of Soil Science, Springer, Dordrecht, The Netherlands, 2008, ISBN 1-4020-3994-8.

[3] N.G. Juma, Major components of soil, in: The Pedosphere and Its Dynamics: What Is a Soil? 1999 (Online). Available, http//www.pedosphere.com/ (assessed in March 2015).

[4] C.N. Brady, The Nature and Properties of Soils, tenth ed., MacMillan Publ. Comp. NY, 1990.

[5] A. Wild, Russell's Soil Condition and Plant Growth, Longman Scientific & Technical, Harlow, Essex, UK, 1988.

[6] J.W. Doran, T.B. Parkin, Quantitative indicators of soil quality: a minimum data set, in: J.W. Doran, A.J. Jones (Eds.), Methods for Assessing Soil Quality, Soil Science Society of America, Madison, WI, 1996, pp. 25–38.

[7] USDA-NRCS, U.S. Department of Agriculture, Natural Resources Conservation Service the Soil Food Web, 2015. http://www.nrcs.usda.gov/Internet/FSE_MEDIA/nrcs142p2_049822.jpg (assessed in May 2015).

[8] D.C. Adriano, A. Chlopecka, K.I. Kaplan, Role of Soil Chemistry in Soil Remediation and Ecosystem Conservation, Soil Sci. Soc. Am. Spec. Publ., Madison, WI, 1998.

[9] UN-United Nations, Defending the Soil (Chapter C). United Nations Secretary General's Report A/544/2000, 2000.

[10] P. Lavelle, A. Spain, Soil Ecology, Kluwer Scientific Publications, Amsterdam, The Netherlands, 2001, ISBN 0-7923-7123-2.

[11] C.S. Holling, G.K. Meffe, Command and control and the pathology of natural resource management, Conserv. Biol. 10 (2) (1996) 328–337.

[12] D.L. Karlen, M.J. Mausbach, J.W. Doran, R.G. Cline, R.F. Harris, G.E. Schuman, Soil quality: a concept, definition, and framework for evaluation, Soil Sci. Soc. Am. J. 61 (1997) 4–10.

[13] J.W. Doran, M.R. Zeiss, Soil health and sustainability: managing the biotic component of soil quality, Appl. Soil Ecol. 15 (2000) 3–11.

[14] J.W. Doran, M. Sarrantonio, M.A. Liebig, Soil health and sustainability, Adv. Agron. 56 (1996) 1–54.

[15] USDA-NRCS, Healthy Soil for Life, U.S. Department of Agriculture, Natural Resources Conservation Service, 2015. soils.usda.gov/sqi (assessed in May 2015).

[16] L.J. Douglas, L.A. Lewis, Land Degradation; Creation and Destruction, Blackwell, Oxford, UK, 1995, p. 335.

[17] R. Lal, Degradation and resilience of soils, Philos. Trans. R. Soc. Lond. B Biol. Sci. 352 (1997) 997–1010.

[18] USDA-NRCS, U.S. Department of Agriculture, Natural Resources Conservation Service Soil Profile Gallery, 2015. http://www.nrcs.usda.gov/wps/portal/nrcs/detail/soils/survey/office/ssr7/tr/?cid=nrcs142p2_047970 (assessed in May 2015).

[19] USDA-NRCS, The Twelve Orders of Soil Taxonomy, U.S. Department of Agriculture, Natural Resources Conservation Service, Soil Survey Division, World Soil Resources, 1998. http://nrcspad.sc.egov.usda.gov/DistributionCenter/pdf.aspx?productID=587 (assessed in March 2015).

[20] H. Eswaran, R. Lal, F. Reich, Land degradation: an overview, in: E.M. Bridges, I.D. Hannam, L.R. Oldeman, F.W.T. Pening de Vries, S.J. Scherr, S. Sompatpanit (Eds.), Responses to Land Degradation, Proceedings of 2nd International Conference on Land Degradation and Desertification, Khon Kaen, Thailand, Oxford Press, New Delhi, India, 2001.

[21] European Soil Bureau, The European Soil Database. Version 1, EUR 19945, CD-ROM Cat.Nr. LBNA19945ENZ, Office for Official Publications of the European Communities, Luxembourg, 2004, ISBN 92-894-1947-4.

[22] R.S. Dwivedi, Spatio-temporal characterization of soil degradation, Trop. Ecol. 43 (1) (2002) 75−90.

[23] G. Wim, M. El Hadji, Causes, General Extent and Physical Consequence of Land Degradation in Arid, Semi Arid and Dry Sub Humid Areas, Forest Conservation and Natural Resources, Forest Dept. FAO, Rome, Italy, 2002.

[24] L.R. Oldeman, R.T.A. Hakkeling, W.G. Sombroek, World Map of the Status of Human-induced Soil Degradation: An Explanatory Note, rev. ed., UNEP and ISRIC, Wageningen, 1991.

[25] ISRIC, Soil Threats, World Soil Information, 2015, http://www.isric.org/about-soils/soil-threats (assessed in May 2015).

[26] EEA-UNEP, Down to Earth: Soil Degradation and Sustainable Development in Europe. A Challenge for the 21st Century, in: Environmental Issues Series No 6, EEA, UNEP, Luxembourg, 2000.

[27] UNEP, World Atlas of desertification, in: N. Middleton, D.S.G. Thomas (Eds.), United Nations Environment Programme, Edward Arnold, London, 1992, ISBN 0-340-55512-2.

[28] V. Mazzucato, D. Niemeijer, Overestimating Land Degradation, Underestimating Farmers in the Sahel, Drylands Issues Paper, International Institute for Environment and Development, London, 2001, http://www.iied.org/pdf/dry_ip101eng.pdf [Geo-2-169], (assessed in March 2015).

[29] GACGC, World in Transition: The Threat to Soils, Annual Report. German Advisory Council on Global Change, Economica Verlag GmbH, Bonn, 1994.

[30] W. Schäfer, K. Severin, T. Mosimann, J. Brunotte, A. Thiermann, R. Bartelt, Bodenerosion durch Wasser und Wind, in: Bodenqualitätszielkonzept Niedersachsen Teil 1: Bodenerosion und Bodenversiegelung, Niedersächsisches Landesamt f. Ökologie [Hrsg.], Hildesheim, Germany, 2003.

[31] P. Bullock, Soil information: uses and needs in Europe, in: P. Bullock, R.J.A. Jones, L. Montanarella (Eds.), Soil Resources in Europe, Official Publications of the European Communities, Luxembourg, 1999, pp. 177−182.

[32] NOAA, The National Oceanic and Atmospheric Administration of the U.S.A. Image ID: theb1365, Historic C&GS Collection. Location: Stratford, Texas. Photo Date: 18 April 1935. Credit: NOAA George E. Marsh Album http://www.photolib.noaa.gov/htmls/theb1365.htm, 2015 (assessed in May 2015).

[33] R.L. Baumhardt, Dust Bowl era, in: Encyclopedia of Water Science, United States Department of Agriculture (USDA), Bushland, Texas, USA, 2003, pp. 187−191, http://dx.doi.org/10.1081/E-EWS 120010100.

[34] L. Franchis, F. Ibanez, T. Fox, Plan Blue Papers-2, Threats to Soil in Mediterranean Countries-Document Review, UNEP, 2003, ISBN 2-912081-14-9.

[35] SEC, 620, Commission Staff Working Document of 22 September 2006, 2006.

[36] COM, 232. Proposal for a Directive of the European Parliament and of the Council Establishing a Framework for the Protection of Soil and Amending Directive 2004/35/EC, 2006.

[37] J.B. Marler, J.R. Wallin, Human Health, the Nutritional Quality of Harvested Food and Sustainable Farming Systems, Nutrition Security Institute, WA 98004, 2006. http://www.nutritionsecurity.org/PDF/NSI_White%20Paper_Web.pdf (assessed in March 2015).

[38] E. Bergsma, P. Charman, F. Gibbons, H. Hurni, W.C. Moldenhauer, S. Panichapong, Terminology for Soil Erosion and Conservation, ISSS, 1996, 2000, ISBN 90-71556-15-8.

[39] M.A. Wilson, Department of Geology, The College of Wooster. http://commons.wikimedia.org/wiki/File:KelsoSand.JPG (assessed in May 2015). Public Domain.

[40] UNL.EDU, Plant & Soil Science e-Library. http://passel.unl.edu/pages/informationmodule.php?idinformationmodule=1086025423&topicorder=9&maxto=20&minto=1 (assessed in May 2015).

[41] Soil-Net Photo Library. http://www.fondriest.com/environmental-measurements/parameters/water-quality/turbidity-total-suspended-solids-water-clarity/ (assessed in May 2015).

[42] G. Govers, D.A. Lobb, T.A. Quine, Tillage erosion and translocation: emergence of a new paradigm in soil erosion research, Soil Till. Res. 51 (1999) 167−174.

[43] K. Van Oost, W. Van Muysen, G. Govers, J. Deckers, T.A. Quine, From water to tillage erosion dominated landform evolution, Geomorphology 72 (2005) 193−203.

[44] A. Giordano, Pedologia forestale e conservazione del suolo, UTET, 2002.

[45] Y. Youlin, V. Squires, L. Qi, Global Alarm: Dust and Sandstorms from the World's Drylands, United Nations, 2001.

[46] IDRC-CRDI, The International Development Research Centre-Science of Humanity, 2015. http://www.idrc.ca/EN/Pages/default.aspx (assessed in May 2015).

[47] P. Loveland, J. Webb, Is there a critical level of organic matter in the agricultural soils of temperate regions: a review, Soil Till. Res. 70 (2003) 1−18.

[48] J. Scharlemann, R. Hiederer, V. Kapos, C. Ravilious. Updated Global Carbon Map, UNEP-WCMC & EU-JRC, Brussels, Belgium. http://eusoils.jrc.ec.europa.eu/esdb_archive/octop/Resources/Global_OC_Poster.pdf (assessed in June 2015).

[49] FAO/IIASA/ISRIC/ISS-CAS/JRC, Harmonized World Soil Database (Version 1.1), FAO, Rome, Italy and IIASA, Laxenburg, Austria, 2009. http://www.iiasa.ac.at/Research/LUC/luc07/External-World-soil-database/HTML/index.html?sb=1 (assessed in May 2015).

[50] SCAPE, Strategies, science and law for the conservation of the world soil resources, in: International Workshop, Selfoss, Iceland, September 14−18, 2005, 2005. AUI Publication no. 4. ISSN: 1670-5785.

[51] LIFE-So.S. Soil Sustainable Management in a Mediterranean River Basin Based on the European Soil Thematic Strategy. http://www.lifesos.eu.

[52] IPCC, Summary for Policymakers and Technical Summary, 2007. Geneva, Switzerland, p. 89.

[53] I.A. Janssens, A. Freibaur, B. Schlamadinger, R. Ceulemans, P. Ciais, A. Dolman, M. Heimann, G.-J. Nabuurs, P. Smith, R. Valentini, E.-D. Schulze, The carbon budget of terrestrial ecosystems at the country-scale − a European case study, Biogeosci. Disc. (2004). http://www.biogeosciences.net/bgd/1/167/SRef-ID: 1810−6285/bgd/2004-1-167.

[54] H.L. Bohn, B.L. McNeal, G.A. O'Connor, Salt affected soils, in: Soil Chemistry, second ed., Wiley Interscience, NY, 1985, pp. 234−261.

[55] JRC, Joint Research Center − European Soil Portal, Soil Data and Information Systems, 2015. http://eusoils.jrc.ec.europa.eu/library/themes/Salinization/ (assessed in May 2015).

[56] G. Tóth, K. Adhikari, Gy. Várallyay, T. Tóth, K. Bódis, V. Stolbovoy, Updated map of salt affected soils in the European Union, in: G. Tóth, L. Montanarella, E. Rusco (Eds.), Threats to Soil Quality in Europe EUR 23438 EN 151 pp, Office for Official Publications of the European Communities, Luxembourg, 2008, pp. 65−77.

[57] A. Jones, C. Bosco, Y. Yigini, P. Panagos, L. Montanarella, Soil Erosion by Water: 2011 Update of IRENE Agri-Environmental Indicator 21, JRC Scientific Report JRC68729, European Commission, Office for Official Publications of the European Communities, Luxembourg, 2012.

[58] J.M. de Paz, F. Visconti, R. Zapata, J. Sanchez, Integration of two simple models in geographical information system to evaluate salinization risk in irrigated land of the Valencian Community, Spain, Soil Use Manage. 20 (3) (2004) 333−342.

[59] USDA-NRCS, Soil Quality Resource Concerns: Salinization-soil Quality Information Sheet, U.S. Department of Agriculture, Natural Resources Conservation Service, 1998. http://www.nrcs.usda.gov/Internet/FSE_DOCUMENTS/nrcs142p2_053151.pdf (assessed in May 2015).

[60] Copernicus, Newsletter No 9, February 2015, European Commission, 2015. http://newsletter.gmes.info/issue-09-february-2015/article/participation-copernicus-programme-business-development-asset.

[61] Hidrología Sostenible. http://www.hidrologiasostenible.com/sustainable-urban-drainage-systems-suds/, 2015 (assessed in May 2015).

[62] G. Prokop, H. Jobstmann, A. Schönbauer, Overview of Best Practices for Limiting Soil Sealing or Mitigating Its Effects in the EU-27, Report on Best Practices for Limiting Soil Sealing and Mitigating Its Effects. Technical Report-2011-050, European Commission, 2011, ISBN 978-92-79-20669-6, http://dx.doi.org/10.2779/15146.

[63] COM, 571 Final of September 2011. Roadmap to a Resource Efficient Europe, 2011.

[64] USGS. http://landslides.usgs.gov, 2015 (assessed in May 2015).

[65] STATISTA. http://www.statista.com/statistics/267837/economic-damage-caused-by-mudslides/, 2015 (assessed in May 2015).

[66] http://www.dpi.nsw.gov.au/agriculture/resources/soils/structure/compaction.

[67] Landscape Resources. http://www.landscaperesource.com/images/articles/soil_compaction_001.jpg, 2015 (assessed in May 2015).

[68] R. Ruser, H. Flessa, R. Russow, G. Schmidt, F. Buegger, J.C. Munch, Emission of N_2O, N_2, and CO_2 from soil fertilized with nitrate: effect of compaction, soil moisture and rewetting, Soil Biol. Biochem. 38 (2006) 263–274.

[69] A.S. Knox, J. Seaman, D.C. Adriano, G. Pierzynski, Chemophytostabilization of metals in contaminated soils, in: D.L. Wise, D.J. Trantolo, E.J. Cichon, H.I. Inyang, U. Stottmeister (Eds.), Bioremediation of Contaminated Soils, Marcel Dekker, NY, 2000, pp. 811–836.

[70] A.S. Knox, J.C. Seaman, M.J. Mench, J. Vangronsveld, Remediation of metal- and radionuclides-contaminated soils by in situ stabilization techniques, in: I.K. Iskandar (Ed.), Environmental Restoration of Metal-Contaminated Soils, Lewis Publishers, 2001, pp. 21–60.

[71] M.K. Doula, V.A. Kavvadias, K. Elaiopoulos, Zeolites in soil remediation processes, in: V. Inglezakis, A. Zorpas (Eds.), Natural Zeolites, Bentham Publisher, 2012, pp. 519–568.

[72] SCAPE, Soil Conservation and Protection in Europe: The Way Ahead, The SCAPE Advisory Board, EUR 22187 EN, Heiloo, The Netherlands, 2006, p. 139.

[73] B. Maréchal, P. Prosperi, E. Rusco, Implications of soil threats on agricultural areas in Europe, in: G. Tóth, L. Montanarella, E. Rusco (Eds.), Threats to Soil Quality in Europe. JRC Scientific and Technical Reports, EUR-Scientific and Technical Research series, Office for Official Publications of the European Communities, Luxembourg, 2008, pp. 129–137.

[74] L. Van-Camp, B. Bujarrabal, A.-R. Gentile, R.J.A. Jones, L. Montanarella, C. Olazabal, S.-K. Selvaradjou, Reports of the Technical Working Groups Established under the Thematic Strategy for Soil Protection, EUR 21319 EN/2, Office for Official Publications of the European Communities, Luxembourg, 2004, 872 pp.

[75] EEA-European Environment Agency, Progress in Management of Contaminated Sites (CSI 015), 2007.

[76] JRC-Joint Research Center, The State of Soil in Europe, Reference Report, European Soil Portal, 2012.

[77] DANCEE-Management of contaminated sites and land in central and Eastern Europe, Ad Hoc International Working Group on Contaminated Land, Ministry of Environment and Energy, Danish Environment Protection Agency, Danish Cooperation for Environment in Eastern Europe, Copenhagen, 2000.

[78] Green Cross Switzerland. http://www.greencross.ch (assessed in May 2015).

[79] Council Directive 1999/31/EC of April, 1999 on the Landfill of Waste.

[80] L. Rodriguez, H.I. Reuter, T. Hengl. A framework to estimate the distribution of heavy metals in European soils, in: G. Tóth, L. Montanarella, E. Rusco (Eds.), Threats to Soil Quality in Europe, JRC-Joint Research Center, Scientific and Technical Reports (EUR – Scientific and Technical Research series) Office for Official Publications of the European Communities, Luxembourg, 2008, pp. 79–85.

[81] Council Directive 86/278/EC on the Protection of the Environment and in Particular Soil when Sewage Sludge Is Used in Agriculture.

[82] C. Reimann, R. Garrett, Geochemical background-concept and reality, Sci. Total Environ. 350 (2005) 12–27.

[83] CLARINET, Sustainable Management of Contaminated Land: An Overview, Report from the Contaminated Land Rehabilitation Network for Environmental Technologies, Federal Environment Agency, Austria, 2003.

[84] P. Bardos, J. Nathanail, B. Pope, General principles for remedial approach selection, Land Contamin. Reclam. 10 (2002) 137–160.

[85] D.S. Woodward, P. Hadley, M.S. Heaston, D. Chiang, Green vs. sustainable remediation: where should we be going? In situ and on-site bioremediation, in: "Proceedings of the 10th International In Situ and On-Site Bioremediation Symposium, May 5–8, 2009, Baltimore, Maryland, USA, 2009.

[86] S. Lesage, U. Zoller, What is sustainable remediation, J. Environ. Sci. Health Part A Toxic/Hazard Subst. Environ. Eng. 36 (2001).

[87] LIFE PROSODOL, Strategies to improve and protect soil quality from the disposal of olive oil mill wastes in the Mediterranean region, in: M.K. Doula, V. Kavvadias, K. Komnitsas, F. Tinivella, J.L. Moreno Ortego, A. Sarris (Eds.), Results and Achievements of a 4-year Demonstration Project — What To Consider; What To Do, 2012. http://ec.europa.eu/environment/life/project/Projects/index.cfm?fuseaction=home. showFile&rep=file&fil=PRODOSOL_Results_Achievements.pdf (assessed in May 2015).

[88] Swissinfo. http://www.swissinfo.ch, 2007 (assessed in March 2015).

[89] UN-United Nations, United Nations Convention to Combat Desertification in Countries Experienced Serious Drought and/or Desertification, Particularly in Africa, 1994. www.unccd.int/convention/history/INCDresolution.php?noMenus=1 (assessed in September 2014).

[90] http://www.gse.mq.edu.au/units/gse813/CSIRO/environ/desertification.htm.

[91] UNESCO. http://www.unesco.org/mab/doc/ekocd/chapter2.html, 2015 (assessed in January 2015).

[92] USDA, U.S. Dept. of Agriculture, Natural Resources Conservation Service, Soil Survey Division, World Soil Resources. (Online), 1998. http://www.nrcs.usda.gov/wps/portal/nrcs/detail/soils/use/?cid=nrcs142p2_054003 (assessed in March 2015).

[93] FAO/UNEP. http://www.fao.org/docrep/x5318e/x5318e02.htm, 2015 (assessed in March 2015).

[94] UN-United Nations, United Nations Convention to Combat Desertification in Countries Experienced Serious Drought and/or Desertification: Annex V Regional Implementation Annex for Central and Eastern Europe, 2001. http://www.unccd.int/convention/text/pdf/annex5eng.pdf (assessed 25.09.15).

[95] UNCCD, Economic assessment of desertification, sustainable land management and resilience of arid, semi-arid and dry sub-humid areas, in: Background Document of the 2nd Scientific Conference, April 9—12, 2013, Bonn, Germany, 2013.

[96] UNCCD, Fact Sheet 4: Action Programmes for Combating Desertification. United Nations Secretariat of the Convention to Combat Desertification, 2000. http://www.unccd.int/publicinfo/factsheets/showFS.php?number=4 [Geo-2-172], (assessed in April 2015).

[97] UNCCD, Action Programmes on National (NAP), Sub-regional (SRAP) and Regional Level (RAP). United Nations Secretariat of the Convention to Combat Desertification, 2001. http://www.unccd.int/actionprogrammes/menu.php [Geo-2-173], (assessed in April 2015).

[98] CSE, Green Politics: Global Environmental Negotiations 1, Centre for Science and Environment, New Delhi, 1999.

[99] C. Toulmin, Lessons from the Theatre: Should This Be the Final Curtain Call for the Convention to Combat Desertification? in: WSSD Opinion Series, International Institute for Environment and Development, 2001. http://www.iied.org/pdf/wssd_02_drylands.pdf [Geo-2-170], (assessed in April 2015).

[100] UNCED, Agenda 21: Programme of Action for Sustainable Development, United Nations, Rio de Janeiro, 1992.

[101] UN-United Nations, We the Peoples — the Role of the United Nations in the 21st Century, New York, 2000, http://www.un.org/millennium/sg/report/key.htm [Geo-1-001], (assessed in April 2015).

[102] L.C. Stringer, Reviewing the international year of deserts and desertification 2006: what contribution towards combating global desertification and implementing the united nations convention to combat desertification? J. Arid Environ. 72 (11) (2008) 2065—2074.

[103] T.F. Shaxson, N.W. Hudson, D.W. Sanders, E. Roose, W.C. Moldenhauer, Land Husbandry: A Framework for Soil and Water Conservation, Soil and Water Conservation Society, Ankeny, IA, 1989.

[104] D.W. Sanders, P.C. Huszar, S. Sombatpanit, T. Enters, Incentives in Soil Conservation: From Theory to Practice, Science Publishers for World Association of Soil and Water Conservation, Enfield, NH, 1999.

[105] University of Bern, FAO, ISRIC, DLD and WASW, WOCAT World Overview of Conservation Approaches and Technologies, in: FAO Land and Water Digital Media Series No. 9. CD ROM, Food and Agriculture Organization, Rome, 2000.

[106] M. Shah, M. Strong, Food in the 21st Century: From Science to Sustainable Agriculture, CGIAR System Review Secretariat, World Bank, Washington, DC, 1999.

[107] A. Young, Soil monitoring: a basic task for soil survey organizations, Soil Use Manage. 7 (1991) 126–130.

[108] C. Pieri, J. Dumanski, A. Hamblin, A. Young, Land Quality Indicators, World Bank Discussion Paper 315, World Bank, Washington, DC, 1995.

[109] COM, 232. European Commission: Proposal for a Directive of the European Parliament and of the Council Establishing a Framework for the Protection of Soil and Amending Directive 2004/35/EC, 22/9/2006, 2006.

[110] LIFE Focus, LIFE Among the Olives, Good Practice in Improving Environmental Performance in the Olive Oil Sector, European Commission Environment, Directorate-General, 2010.

[111] COM, 46 Final of February 2012 on the Implementation of the Soil Thematic Strategy and Ongoing Activities, 2012.

[112] Athens Soil Platform Meeting. http://en.bpi.gr/newsdet.aspx?id=30, 2013 (assessed in May 2015).

[113] LIFE Publication, LIFE and Soil Protection, European Commission-Environment Directorate-General, 2014, p. 8.

[114] A.K. Sarmah, Potential risk and environmental benefits of waste derived from animal agriculture, in: G.S. Ashworth, P. Azevedo (Eds.), Agriculture Issues and Policies Series – Agricultural Wastes, Nova Science Publishers, Inc., NY, 2009, ISBN 978-1-60741-305-9.

[115] EEA-European Environment Agency, Web of European Environment Agency, Environmental Terminology and Discovery Service (ETDS), 2012. http://glosary.eea.europe.eu/terminology (assessed in February 2015).

[116] M.J. López, R. Boluda, Residuos agrícolas, in: J. Moreno, R. Moral (Eds.), Compostaje, Ediciones Mundi-Prensa, 2008, pp. 489–518.

[117] M.A. Bustamante, R. Moral, C. Paredes, A. Perez-Espinosa, J. Moreno-Caselles, M.D. Perez-Murcia, Agrochemical characterization of the solid byproducts and residues from the winery and distillery industry, Waste Manage. 28 (2008) 372–380.

[118] LIFE AgroStrat – Sustainable strategies for the improvement of seriously degraded agricultural areas: The example of *Pistacia vera* L. Cultivation Management Software. http://www.agrostrat.gr/?q=en/CultivationManagementSoftware, 2014 (assessed in April 2015).

[119] V. Kavvadias, M. Doula, K. Komnitsas, N. Liakopoulou, Disposal of olive oil mill wastes in evaporation ponds: effects on soil properties, J. Hazard Mater. 182 (2010) 144–155.

[120] M.A. Arshad, S. Martin, Identifying critical limits for soil quality indicators in agro-systems, Agric. Ecosyst. Environ. 88 (2002) 153–160.

[121] M.K. Doula, V. Kavvadias, K. Elaiopoulos, Proposed soil indicators for Olive Mill Waste (OMW) disposal areas, Water Air Soil Pollut. 224 (2013) 1621–1632.

[122] D. Daly, Groundwater at Risk in Ireland – Putting Geoscientific Information and Maps at the Core of Land Use and Environmental Decision-making, John Jackson Memorial Lecture, Royal Dublin Society, 2004.

[123] K. Modis, G. Papantonopoulos, K. Komnitsas, K. Papaodysseus, Mapping optimization based on sampling in earth related and environmental phenomena, Stoch Environ. Res. Risk Assess. 22 (2008) 83–93.

[124] Soil Science Society of America, Utilization, treatment and disposal of waste on land, in: Workshop Proceedings, Chicago, 6–7 December, USA, 1986.

[125] MAFF, Dept. of Environment, Code of Practice for Agricultural Use of Sewage Sludge, London, 1989.

CHAPTER 5

Urban Environment

S. Malamis
National Technical University of Athens, Athens, Greece

E. Katsou
Brunel University, Uxbridge, Middlesex, United Kingdom

V.J. Inglezakis
Nazarbayev University, Astana, Republic of Kazakhstan

S. Kershaw
Brunel University, Uxbridge, Middlesex, United Kingdom

D. Venetis
EPTA SA, Athens, Greece

S. Folini
University of Florence, Florence, Italy

Chapter Outline

Environment and Development. http://dx.doi.org/10.1016/B978-0-444-62733-9.00005-8
287

5.1 Introduction

With half of the world's population living in cities, the process of urbanization is a critical aspect of global environmental management that requires attention. In the 1800s less than 5% of the world's population was urban. In the 1950s ~70% of the world population was rural; this figure dropped to around 45% in 2015; on the contrary, the urban population increased from 30% to 55% over the same time period. In 2005, the world's urban population was 3.17 billion out of a total of 6.45 billion. The year 2007 was a milestone for human history, since for the first time, 50% of the world's population was living in cities. The massive increase of the population in cities has been the result of both population increase and the tendency of people to move from rural regions to urban cities in order to find better working and living conditions. During the last 200 years, the world's population has increased six times. Over that same time period, the urban population has increased 100 times, since more people concentrate at cities. Consequently, the movement of people from rural areas to urban ones has been much more intense than population growth. In Europe ~35% of the European citizens live in cities having a population higher than 100,000 and another 40% in smaller urban towns [1].

Prior to discussing about urbanization and its consequences, it is necessary to define it. Urbanization is the process of the growing development of cities that occurs due to two principal reasons:

* migration to cities from nonurban areas; this is particularly significant in emerging economies where nonurban poverty affecting individuals and groups and their wish for improved living standards and finances are major drivers for migration and
* population increase and the need for more homes in connected locations that promote mobility.

The United Nations projections show that in the time period 2015–30 almost all of the global population increase will be absorbed by cities [2]. Urbanization can be quantified either based on the urban population compared to total population or by considering the rate at which the urban population is increasing. The city density of population is becoming acute in most cities, for example, in London, the United Kingdom (Fig. 5.1A and B), and in Chinese cities (Fig. 5.1C and D). Urban areas have been recast as key arenas in which the concept of sustainable development should be interpreted and applied. Cities only encompass 2% of the world's land surface, yet they are responsible for consuming more than 75% of the planet's resources and produce 75% of the world's waste [3]. Many of the cities of tomorrow are more likely to be mega or super cities, with single

Figure 5.1
City space in modern cities. (A) and (B) shows London's density of buildings, with some green (gray in print versions) space around. (C) is Taiyuan City in Shanxi Province, China, highlighting the huge expansion of accommodation in high-rise blocks. (D) is Yulin City, Shaanxi Province, next to the Yellow River. Note that the air quality in (C) and (D) is notably poorer than in (A) and (B); this is due to differences in pollution controls, rather than natural atmospheric characteristics. *Photos copyright Steve Kershaw.*

megacities sprawling urban regions representing the largest, most complex man-made structures ever created. For all of these reasons, we view the urban environment as a pressing issue requiring prompt attention.

Large cities provide opportunities for employment and promote the division of labor. This means that the employees can specialize their labor activities and thus productivity increases. Around the city, agricultural activities can flourish which will support the needs of the city. Furthermore, good transportation networks allow the effective, everyday commuting of people from their house to their working place. In order for a city to be able to grow, it should be supported by surrounding rural areas. The opportunities in cities are immense, but the problems are acute and the time to deal with them is already upon us. A successful city must balance social, economic, and environmental needs: it has to respond to pressure from all sides. Such a city should offer investors security, infrastructure (including water and energy), and efficiency. It should also put the needs of its citizens at the forefront of all its planning activities. A successful city recognizes its natural assets, its citizens, and its environment and builds on these to ensure the best possible returns.

5.2 Key Urban Environmental Problems

It is the perception of older generations that urban problems have worsened from the 1980s onwards, and this is supported by the clear evidence of increasing populations and pressures on available resources. The key problems are the following:

- space, especially living space, and overcrowding,
- resources, including food and water supply, and energy,
- mobility, thus the transport system, considered in Section 5.5,
- health issues, particularly due to increased stress and environmental pollution, leading to pressure on health services that have limited resources with restricted budgets and appropriate capacity of people with suitable expertise (doctors, nurses, administrators). An important component is air pollution which is further discussed in Section 5.6,
- sanitation, water and wastewater management considered in Section 5.8, and
- in geologically sensitive areas, the threat from natural hazards, such as volcanoes (Vesuvius and Naples; mudslides in Indonesia), earthquake zones, including tsunamis, due to unstable tectonics (Western United States, SE Asia).

Associated problems that are becoming more significant are:

- distance from work: the need to live in cities, where most of the work is,
- rate of dwelling construction outpaced by migration to cities,
- rents and space premium, eg, London rents,

- environmental degradation: loss of green space and increase in vehicle emissions, and
- manufacturing and city usage of resources.

There are two broad categories of environmental issues relating to urban areas; these are known as "cumulative" and "systemic" (Table 5.1). Cumulative issues are those that can arise in any human settlement and place, but can be exacerbated in towns and cities by the agglomeration and density of population and activities. For example, carbon dioxide emissions are not a problem of urban areas per se. On the contrary, systemic issues arise from the unique characteristics (economic, social, environmental, and others) of urban settlements. Over the past 150 years, important improvements have been seen in urban areas. The improvements and innovations of past generations are in many cases responsible for improving the way of living in cities. Although the main challenges of the urban environment seem to be persistent, contemporary cities also face challenges that are less visible but equally or even more important; such challenges include climate change, health hazards from air quality, exposure to a wide variety of chemicals that are produced, and others. These problems are intimately bound up with modern lifestyles and at the same time require the attention and action of many different parts of society [4].

We observe that many of the same challenges and problems have been repeatedly diagnosed by specialists in the field and that a broadly similar range of solutions has been proposed by experts over decades. Overall, progress in managing the urban environment cannot be said to be proportionate to the analytic effort expended on the topic. Urban environmental management presents a classic case of what people in the field have described as a "wicked problem." Although the discipline of urban planning has been widely practiced throughout most of the 20th century, urban settlements are, for the most part, better understood as emergent rather than planned systems. They have evolved in particular ways as agglomerations of people; accretions of buildings and roads;

**Table 5.1: Typical Examples of "Cumulative" and "Systemic"
Urban Environmental Issues [4].**

Parameter	Cumulative	Systemic
Built urban environment	Waste collection Urban density	Transportation, connection of different city regions Sewerage system; runoff collection system Rain harvesting
Health issues	Noise nuisance Infections and diseases	Hot spots of air pollution
Natural urban environment	Contaminated sites	Floods Green space

infrastructures for water and energy supply and the removal of sewage and waste; public and private spaces; places of business and residence; locations for the production and consumption of goods and services; facilities for entertainment, education, and health; and so forth. The behavior of urban systems, of which urban environments are a vital part, is therefore shaped by the trial-and-error accumulation of factors and forces that survive because they fit into and reinforce other aspects of the system. Innovation and change depend on complex interactions driven by a variety of forces embedded in institutions, markets, regulations, and technologies [5]. The development of new technologies is unlikely to be sufficient on its own to shift our cities onto a more environmentally sustainable trajectory. Despite numerous prestigious and small-scale examples of success stories, there is clearly a lack of progress in terms of creating environmentally sustainable cities, largely because of a "web of constraints" related to the adequacy of institutional arrangements, existing infrastructure commitments, appropriateness of policy instruments, consumer preferences, and the availability of information and skills. In our view, many of the past failures to get a grip on the challenges of the urban environment arise from well-meaning, but misguided, attempts to impose simple solutions on complex problems.

Chinese cities (Figs. 5.1 and 5.2) are growing almost exponentially to house the huge migration into cities from the countryside. An almost universal pattern is identified: children of farmers go to work in cities for a better life and do not return, leaving an aging rural community that continues to farm and produce the nation's food. Parts of the Chinese countryside that had an abundance of farming communities in the latter half of the 20th century have been steadily emptied in the 30 years from 1980s to present times, having a knock-on effect on food production. Nevertheless, increased use of fertilizers and genetically modified crops have to some extent offset the loss of manpower. However, the increase in fertilizers means higher nitrates, with the wider impact that has as nitrates are fed to the oceans via river runoff and also infiltrate to the aquifer. London's increased population since the 1980s led to high pressure on housing, in particular, leading to one of the most problematic accommodation issues in the United Kingdom. London property prices are so high, as of 2015, that it is frequently reported that even professionals earning good wages in respected professions cannot afford to buy property. In addition, increasing numbers of people cannot afford the high rents, which lead to considerable problems for working-age members of the community who thus have to move outside London. This creates an increased commuter force and thus impacts the transport system almost to overload for entry to London.

The environmental problems related to urbanization are more intense in cities having population greater than 10 million people (Fig 5.3); these cities are known as megacities. Four megacities are located in Europe, six in America, two in Africa, and fourteen in Asia. Thus, more than 50% of the megacities are located in Asia, and particularly in the south and east parts. In some Asian countries such as China and India, high population

Figure 5.2
(A) and (B): Smog developed in the lower slopes of the mountains west of Beijing, 50 km from the city is due to the high concentration of vehicles and population; (C): Green (gray in print versions) space development on the banks of the Yellow River in Yulin, Shaanxi Province, partly provides an attractive amenity and also acts as a buffer against flooding; (D): An electric car in Beijing, illustration of determined approaches to solving the pollution issues in Chinese cities.
Photos copyright Steve Kershaw.

growth is taking place together with urbanization, strong industrial expansion, and intense vehicle use. This is resulting in acute atmospheric pollution. Studies of megacities is required on different spatial scales in order to be able to understand the local-to-regional-to-global impacts of megacities and their implications [6]. For example, a global model was applied to investigate the outflow characteristics of important contaminants in megacities, showing the accumulation of pollutants in the region surrounding the megacity against the transport of the pollutants to other, further away locations or to the upper troposphere [7–9].

In the literature several methodologies and indicators have been applied to evaluate and rank cities against environmental, social, economic, infrastructural, and other parameters. For example, the World Bank publishes the World Development Indicators (WDI), and the United Nations reports the City Development Index (CDI). Gurjar et al. [9] evaluated megacities based on their ambient air quality and their trace gas and particle emissions.

Figure 5.3

Population of the 26 megacities of the world (data of 2011) [7].

5.3 Stress Posed to Natural Sources by Urbanization

Building construction, transport, and thus access to natural resources are key stresses, leading to environmental degradation in emerging, urbanized economies. Natural resources are stressed because of the need for increased construction materials; cement is reported as being responsible for 9% of global CO_2 emissions, with significant increases in emerging economies and in the two large mainly stabilized economies of India and China, where population growth and migration to cities play a major role in natural resource stress. The consequence is increased service sectors required to sustain the swelled populations. An important concept developed as part of the GreenHouse Gas (GHG) protocol that began in 2001, in relation to the use of resources, is that of "scopes" of carbon emissions that were designed to compartmentalize the types of emissions. The idea is that by such compartmentalization, producers of carbon emissions (ie, the entire world population) will become more acutely aware of the need for energy conservation. Actually the choice of "scope" as a term is not intuitively understandable, and a better term would be "categories"; nevertheless, the literature has settled on scope as a term. The three scopes are:

- Scope 1: all the emissions related to a person's place of work,
- Scope 2: energy usage at the place or work; however, the energy was produced off-site, and this principally relates to electricity, and
- Scope 3: all other carbon emissions that impact a person's life; this is a very complex array of subcategories, that can be exemplified by the modern phenomenon of smartphones, which utilize a very wide array of materials.

A somewhat hidden aspect of Scope 3 emissions that touches the lives of nearly all city dwellers is the importance of rare-earth elements that underlie much of modern electronic technology. For example, touch-sensitive screens are largely derived from a single mine in China. A very interesting and disturbing aspect relates to the element tantalum, which is required to miniaturize electronic capacitors required in microelectronics. The key source of tantalum is Congo in Africa, where it is mined along with the strategic metal cobalt in a deposit often referred to as coltan. Coltan production is not under the control of an organized mining system and is strongly exploitative of workers. Affluent societies that rely on microelectronics and are fed by manufacturers to buy the latest products each year are to some considerable extent underpinned by such essential supplies, and therefore extend the concept of resource issues beyond national boundaries into the economies and individual lives and freedoms of inhabitants in other countries. Figs. 5.4–5.7 illustrate processes in Shanxi Province in China, indicating the continued reliance on coal for power generation, but also show the new initiatives in wind and solar energy in the drier areas, typified by northern China, and where wind levels are more sustained.

5.4 Measures for Restoration/Conservation of the Urban Environment

Environmental management lies at the heart of urban restoration and conservation. Environmental management is an intuitive process utilized by most careful-thinking householders for their domestic economy and at its center is the importance of resource conservation and waste management. In order to develop environmental management at a scale that can lead to conservation and restoration of urban environments, a formalized structure is required. Such a structure was developed historically by the International Standards Organization (ISO) as a series of standards. These standards are comprehensive documents that lay out the processes by which environmental resources can be brought under an umbrella of control that minimizes waste and maximizes resources and efficiency in use of resources. The key standard is called Environmental Management System (EMS) ISO14001, which has a comparable British Standard called BS8555. A companion ISO standard to ISO14001 is the family of Quality Standards based on ISO9001. A brief description of ISO14001 EMS is provided here so the reader can see how this concept works. ISO14001 is based on a four-category system often called the "Plan-Do-Check-Act" model which is cyclic in action, going from one to the next of those four categories:

1. Planning the management of an organization, such as a business, school, university etc., requires understanding of all the aspects of the business of the organization and making baseline measurements of the energy and materials usage of the organization. A plan is produced in relation to the activities of an organization and is then carefully scrutinized to determine to what extent it can apply to the organization's processes.

Figure 5.4

(A): Pingshuo open cast coal mine in northern Shanxi Province, one of the two major coal-producing areas in China (the other is in northeast China); the mine here is nearly 2 km across. (B): Coal lorries are a permanent feature of the highways in Shanxi, where thousands of lorries travel daily. The uncertainty about price of coal (it has fallen significantly in the last few years) causes fears that the coal market may diminish, which inhibits the expansion of rail transport for coal. (C): Existing rail transport carries some of the coal; this photo also shows part of the huge expansion in wind turbines in northern China, served by sustained winds. (D): A coal-fired power station in Shanxi Province. *Photos copyright Steve Kershaw.*

2. The plan is enacted.
3. Then, the plan is monitored to assemble data on how well the efficiency of management is working.
4. Finally, actions to improve the management are put in place in light of the experience of the management system, leading to improvements. Thus, the process is cyclic, with regular monitoring, so that improvements are fed back into the system's operation on a continual basis, to result in continuous improvement, with no prescribed conclusion; it simply continues and is fine-tuned. In the case of ISO14001, yearly reviews are required by external auditors and every 3 years certification has to be renewed, thereby

Figure 5.5
Insights into wind turbine expansion in northern China. *Photos copyright Steve Kershaw.*

ensuring continual improvement of the organization's operations to guarantee continued certification. Alternative EMS include EMAS (European Eco-Management and Audit Scheme), widely used in Europe; and BS8555 (British standard applied only in the United Kingdom). A specific EMS for educational organizations, called EcoCampus, was developed for EMS application in British universities. Nevertheless, the application of the different EMS varies.

Prominent examples for the restoration of the urban environment are given below for the city of London:

• *Sanitation*: Development of a supersewer to cope with increased sewage and storm flow production in London. The supersewer will be built below Thames River and will stretch for 25 km, with a total construction cost of £4.2 billion. According to Thames Water the current combined sewerage system overflows on a weekly basis, discharging 39 million metric tons of untreated sewage straight into Thames River each year. In 2013, 55 million metric tons of sewage polluted the river. The super-sewer will retain the mixed stormwater and sewage, acting as a buffer tank during periods of increased rainfall. Subsequently, the mixed sewage and stormwater will be sent for treatment.

Figure 5.6
Coalbed methane is extracted from Carboniferous and Permian age sedimentary rocks in northern China. (A) and (B): These "nodding donkey" pumps move much more slowly than those used to pump oil; they slowly move water in the subsurface, several hundred meters down, and allow the methane to rise. (C): Green pipe (gray in print versions) is for gas, yellow (light gray in print versions) pipe for water used to displace the gas. (D): Not all the gas can be captured and is flared. *Photos copyright Steve Kershaw.*

- *Transport*: The new Crossrail, largely underground, is designed to link key mainline stations with the underground, increasing efficiency to London's transport and add capacity to the system.
- *London railway and underground systems*: Lengthening trains and platforms. As the rail lines are at capacity, the time-tabling of new services becomes difficult; the solution is to increase the size of trains. However, there are limitations since double-decker trains would require considerable rebuilding because of the bridge heights in London and cities into which commuters come from as far away as Leeds.
- *Wider transport involving routes into London*: Planning and land issues continue to delay implementation of the rail system that was intended to increase connectivity from the Channel Tunnel through to London and onward to northern parts of the

Figure 5.7

One advantage of countries that have a lot of space is that there is ample area for solar power fields, as here in Shanxi province, northern China. Currently they have only a small input into the energy mix, but are funded by a serious attempt to reduce the important pollution issues. *Photos copyright Steve Kershaw.*

United Kingdom. In the United Kingdom, the disruption to areas where the rail line would run, plus a key lobby that argues for enhancement of existing rail provision, is causing significant delays to the implementation of High Speed Railway 2 (HS2), with some uncertainty as to whether it will go ahead.

- *Water*: major problem as modeling predicts water shortages as climate continues to change. There is thus significant pressure on the water resources, including the triple-A grade aquifer in the chalk that underlies London.

- *Waste management*: London's waste requires increased recycling. Currently, the central London waste management is enacted at Walbrook Wharf, which collects all waste from the "square mile," the business district of central London. Fig. 5.8 shows Walbrook Wharf where separated recyclables and waste are loaded onto barges to be transported at high tide out of the city to a processing plant downstream, thereby using river transport and avoiding adding to vehicle movements. Waste is separated in a converted warehouse complex on the north bank of the River Thames near St. Paul's Cathedral. Sorted waste is loaded onto barges and taken downstream to a processing plant where waste is processed.

- Energy-from-waste (efw) power production (Fig. 5.9): Due to the pressure on the limited space for landfill, particularly in the densely populated southeast England, some

Figure 5.8
East London and the River Thames. Inset: Walbrook Wharf waste collection and recycling scheme. Note the "green roofs" approach to improve city space appearance in some buildings.
Photos copyright Steve Kershaw.

Figure 5.9
Energy-from-waste (efw) electricity generator, owned by Grundon private waste management company, near Heathrow Airport, London. The Grundon efw plant has an ergonomically designed roof aimed to prevent air turbulence affecting airplanes arriving to and departing from the airport. The tall waste chimney is artistically adorned with a large spiral, to improve the aesthetic appearance of the building. Emissions are monitored live by the Environment Agency to maintain the low emissions status of this plant.

efw power stations have been constructed, for example, the Grundon plant close to Heathrow Airport. The plant takes local municipal waste and burns it to drive turbines feeding electricity into the national grid. However, planning permission for such power stations is a lengthy process, noting that these are large constructions, so there is limited scope for efw power stations in London.

- Urban soil management and allotments (Figs. 5.10 and 5.11): Rosendale Gardens Allotments is the largest allotment in southern Britain and is within sight of the city of London in the distance. Allotments provide local vegetable production minimizing transport for parts of essential food, yet are mostly used by individuals and families to supplement food bought mostly in supermarkets. A coordinated approach to food production nearer city centers would be a valuable component to reducing carbon emissions.

5.5 Transport and Land Use

This section assesses how urbanization impacts transport and land use. Both these issues are essential aspects of everyday life in the cities. Adequate planning is required in order to have an effective transport system and to allow for land use for all the activities which take place in a city. Efficient planning can upgrade the quality of life in cities.

5.5.1 How Urbanization Affects Transport in Cities

Urbanization leads to concentration of people into existing limited space, so that there is a steady increase in transport problems to the point of gridlock, unless planning is put in place to manage the urbanization process in tandem with transportation. Cities were

Figure 5.10
Rosendale Gardens Allotments site in south London. This is one of the largest allotments in southern England and illustrates the value of possibilities of locally-grown food that avoids transport over long distances. *Photo copyright Steve Kershaw.*

Figure 5.11
The ultimate green (gray in print versions) roof. This photo shows a reconstructed old farmhouse
in central Iceland, showing how green roof ideas were used hundreds of years ago; this is not a
new concept! Imagine how some parts of modern cities might benefit from such an approach.
Photo copyright Steve Kershaw.

constructed in historical time without the consideration of increase of population density
so that urbanization merely compounds whatever transportation problems exist. The
increase of city populations leads to stresses on all transport systems; road and rail (both
surface and underground) fill to capacity if planning is not enacted to maintain flow,
particularly at key times of the day (rush hours). The transport system includes the set of
transport infrastructures and modes that support the commuting of passengers and freight.
It generally expresses the level of accessibility. The transportation is a vital aspect of the
proper function of cities. Every day people need to commute from their place of living to
their place of work and vice versa. Understanding the urban entities involves the analysis
of patterns and processes of the transport and land use system. This system is a very
complex and multiparameter including the interrelations between transport, land use, and
special interactions.

5.5.2 Case Studies Concerning Transport

In the UK Victorian age, city expansion was a major change, but little improvement on
city design has been implemented since then. Thus, the transportation within cities is
simply increased in density without planning reorganization of cities, and the current cities
are overloaded by increased population onto a system that was developed 150 years ago.

The problems of managing and solving London's urbanization problems must therefore be done piecemeal. One could believe that the only solution is to redesign the cities from scratch, but this is unachievable. A key example of a city solution is Milton Keynes, a new city constructed essentially in the 1960s with transportation issues built into its design so that bicycle and pedestrian route ways are separated from cars, and its roads are built in a grid system to allow ready movement of vehicles between districts of the city.

In Beijing some solutions to road transport include:

- Cars allowed to move on only certain days, related to the car registration plate; car ownership is strictly controlled by a lottery system, where only a limited number of new cars are allowed each year, and those wishing to have cars must apply for a random lottery space. Each month the lottery is run and it is random each time, so that most people never get the chance to have a car.
- Introduction of a metro system, which was developed for the 2008 Olympic Games in Beijing. It is an efficient, well-organized system, but is of rather lower capacity than, for example, the New York Metro or the London Underground, because in Beijing the trains are smaller in size. Nevertheless, the road transport issues of the numerous concentric ring roads (there are six ring roads as of 2015) are significant, with long delays accompanied by significant pollution, so a public transport system is very important and indeed would benefit from expansion to reduce ground-surface transport.

In China generally, the large population now in cities exists in large centers that are separated by long distances. Thus a critical component of China's urbanization is the linkage between city population centers; this has been achieved by expansion in three transportation routes starting in the 1980s and by 2015 has become a mature system:

- Paved road links form a major route network across the country, named "high-speed roads," and effectively equal to motorways in the United Kingdom and interstates in the United States. They are prefixed "G" (meaning Gong, the pinyin for "public" in Chinese). The density of traffic varies, but in rural areas vehicle density is much less than in the United Kingdom, for example, and is an efficient long-haul means of traveling the country and served by frequent service stations. G7, for example, links Beijing to north-western China, and traveling these roads feels very much like interstate travel in the United States, yet its development has been compressed into a 20-year timescale, having learned from Western countries.
- High-speed rail links; most of the main cities are linked by 300 km/h custom-built trains served by major new stations that were constructed since 2000. Thus, it is now very easy to move between Chinese cities and might be considered an extension of urbanization. In London, the United Kingdom, a low-emission zone in the Greater London area excludes heavily polluting diesel vehicles from entering the capital, and congestion charging within the city center is designed to reduce traffic and also of course raise money for the city. Cambridge, the United Kingdom, has increasing car

density because of a buoyant economy. The Park and Ride scheme with a guided bus-way routing through the city to outlying areas has not proved effective so far, and the common experience of drivers is heavy rush hour and weekend traffic on key routes into the city. Solutions may include more managed planning and perhaps the removal of car parking in the city center. A multistory car park directly next to the main shopping center is a key feature of the town, together with available street parking spaces spread through the city center area. These bring in much-needed revenue to Cambridge in parking fees, but result in an unpleasant journey to the city and increased pollution levels. Although generally air pollution is low, the concentration of so many cars in the city area, exacerbated by the continuing practice of not switching off engines in standing vehicles results in some local air pollution (although there are initiatives in the United Kingdom to highlight this and encourage a countrywide practice of switching off engines after 60 s of idling while the car is standing). For Cambridge, and indeed other cities, a forward-looking approach would be to move parking to the city perimeter and make the entire city center pedestrianized, with access for bicycles and permitted vehicles for passengers with mobility difficulties. An electric tram system would be a sensible option, together with an efficient public bus service between peripheral parking and the city center region. Electric bicycles are increasingly seen in Cambridge and are a potential option for personal transport over the short distances. An underground rail system in Cambridge has been mooted, but is unlikely, given the short distances and small size of the city. Whatever developments take place, all these possibilities require capital investment and cannot be achieved in the short term.

In the city of Athens, Greece, the cars are allowed to enter the center only on certain days. Specifically, the private cars in which the last digit of the registration plate number is odd are allowed to enter the center only at odd dates (eg, 9th November), while the cars that have an even number as last digit of their registration plate number are allowed to enter only on the even number dates (eg, 16th September). This measure was temporarily applied in 1979 as a response to the oil crisis of the time in order to decrease fuel consumption and has been a permanent measure since 1982 in order to decrease traffic congestion and air pollution in the city of Athens. Fig. 5.12 shows the traffic sign that is used to signal to the drivers that they are entering the ring. To promote the use of low-emission vehicles the government allowed since September 2012 at all days the entrance inside the ring to electric vehicles and to vehicles which have CO_2 emissions less than 140 g/km.

5.5.3 Land Use

Land use can be defined as the management and alternation of the natural environment into built environment such as agglomerations/settlements and seminatural use such as

Figure 5.12
Sign showing the driver that he/she is entering inside the vehicle limitation ring of Athens, Greece. *Photo copyright Simos Malamis.*

arable field. It can also be defined as the actions that inhabitants undertake in a certain land in order to produce, alter, or maintain it [10,11]. In the use of land in urban areas, two issues need to be specified: (1) the type of use which means the type of activities that are taking place and (2) the intensity of use.

The most important drivers for land use are the intense urbanization and the need to have appropriate infrastructure to support the population in cities. Many of the forests and farm land that existed in the 19th century in Europe and in North America have been converted into urban settlements in order to house the needs of the increasingly expanding cities. The increasing requirements for food also impacts on rural land with an increase of the land that is used to cultivate. In developed countries, the increasing demand for food, animal feed, and bioenergy will most likely require the intensification and increase in productivity of the agricultural land and even the conversion of additional land to agricultural land [1].

Many contemporary cities face significant challenges related to land use, such as the loss of green space, limited space for parks and open space, and significant pressures to the remaining ecosystems, wetlands, and wildlife. Although these changes may seem to be relatively small within a timescale of 10–20 years, in the scale of 50–100 years the effect can be huge. The public perceives these changes and associated problems when it visualizes residential and commercial development replacing undeveloped land.

(A)   **(B)**

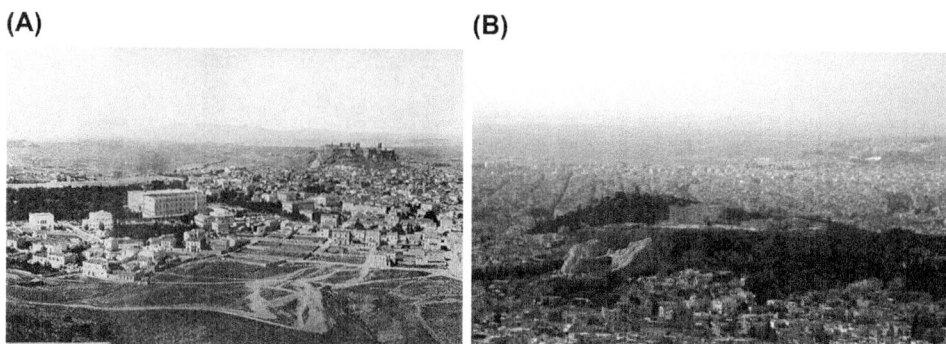

Figure 5.13
View of the city of Athens at the end of the 19th century (A) and now (B).

Fig. 5.13 shows the view of the city of Athens from Mount Lycabettus at the end of the 19th century, and now; a remarkable difference is seen in the occupation of free space, mainly from residential and office buildings. Urban growth rates are not slowing down, especially when viewed at the global scale, since this has been attributed to the increasing population and urbanization.

The analysis of the European Environment Agency (EEA) on the change of land cover in 36 European countries shows a change in the type of land for 1.3% of the total land available (68,353 km^2 from the available 5.42 million km^2) for the time period 2000–06. The rate of land change has slowed down compared to 15–20 years ago. As shown in Fig. 5.14 in Europe, 35% of the land is occupied by forests, 25% is arable land and permanent crops, while only 4% is artificial areas which include cities, out of which 80% are housing, services, and recreation. In Europe the demand for land, energy, and timber before the 19th century resulted in a decrease of the land covered by forest. However, during the two centuries a net increase in forest land has been observed. During the period 2000–06 forested land increased by ∼100,000 ha and the artificial areas by more than 600,000 ha [1].

The use of land for urban activities results in soil sealing. The latter can be defined as the loss of soil resources taking place due to the covering of land for housing, road, or other construction work, which is considered to be irreversible. In Fig. 5.15, the percent of soil seal is given for European countries. The European average is 1.81%; very small countries such as Malta, Luxembourg, and Liechtenstein have high soil seal coverage, while in countries where climatic conditions do not favor urban development the soil coverage is very low (ie, Iceland).

5.5.4 Modeling of Land Use

The planning for land use in cities is a complex multiparameter challenge. As land use decisions critically impact on environmental issues, it is important to control land use in

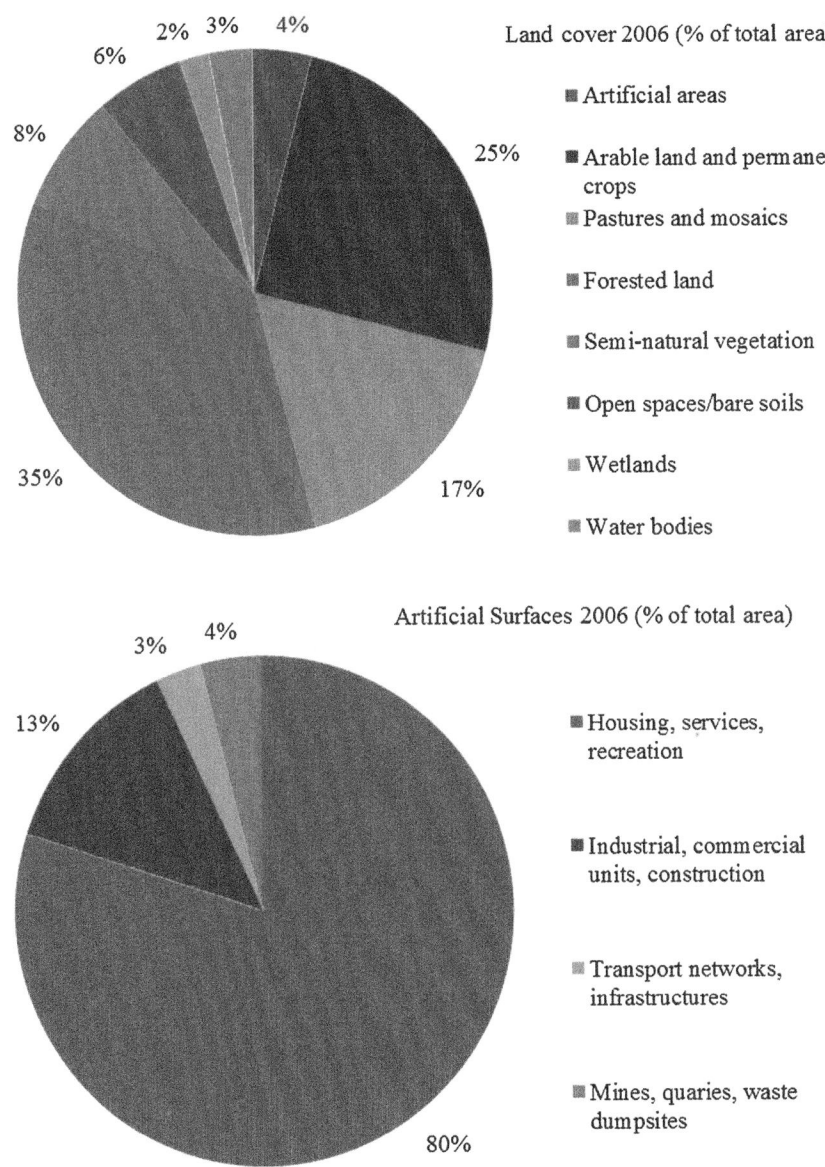

Figure 5.14

Types of land cover in 36 European countries (the 38 EEA member and collaborating countries, plus Kosovo, excluding Greece, Switzerland, and the United Kingdom) [1].

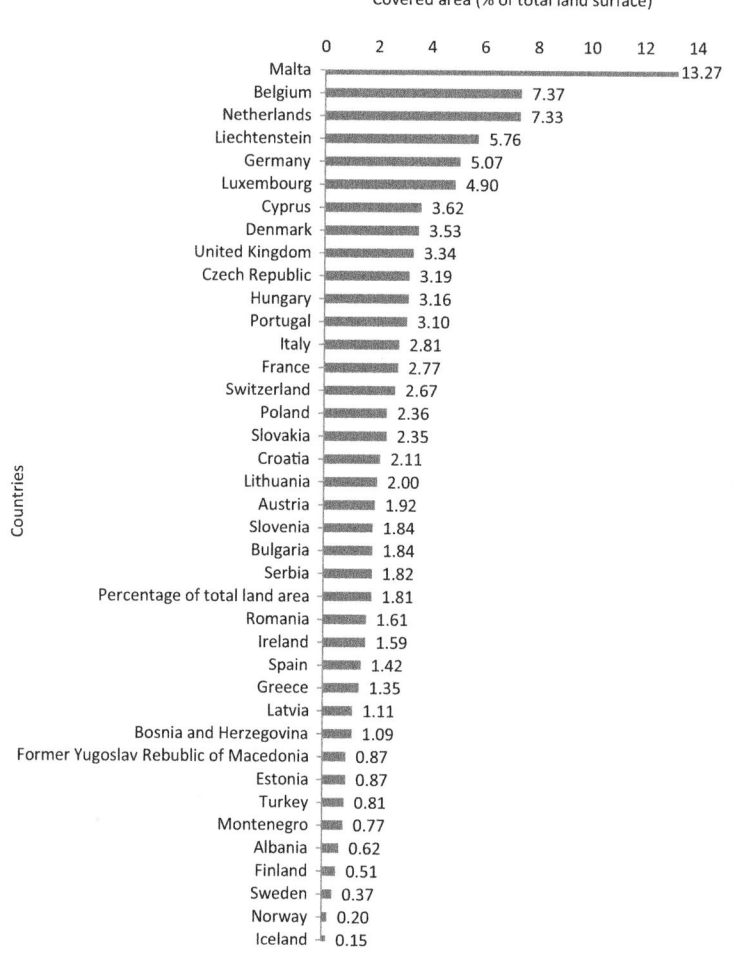

Figure 5.15
Level of soil sealing, in European countries, 2006 [1].

order to mitigate air, water, and land pollution, provide enough land for green and open spaces, and conserve/restore wetlands and coastal resources. Several models of urban land use have been developed, which are characterized by different level of complexity and include among others [12] (Fig. 5.16):

• *Von Thunen's model* (Fig. 5.16A): It is the oldest land use model and was developed by Von Thunen in 1826. It considers a central place, which is the central market, and its concentric impacts on surrounding land uses. The relative cost for the transport of agricultural products to the central market is the one that determines the use of agricultural land around the city. As a result, the most productive activities will compete for the land that is located closer to central market, while the activities which are not so

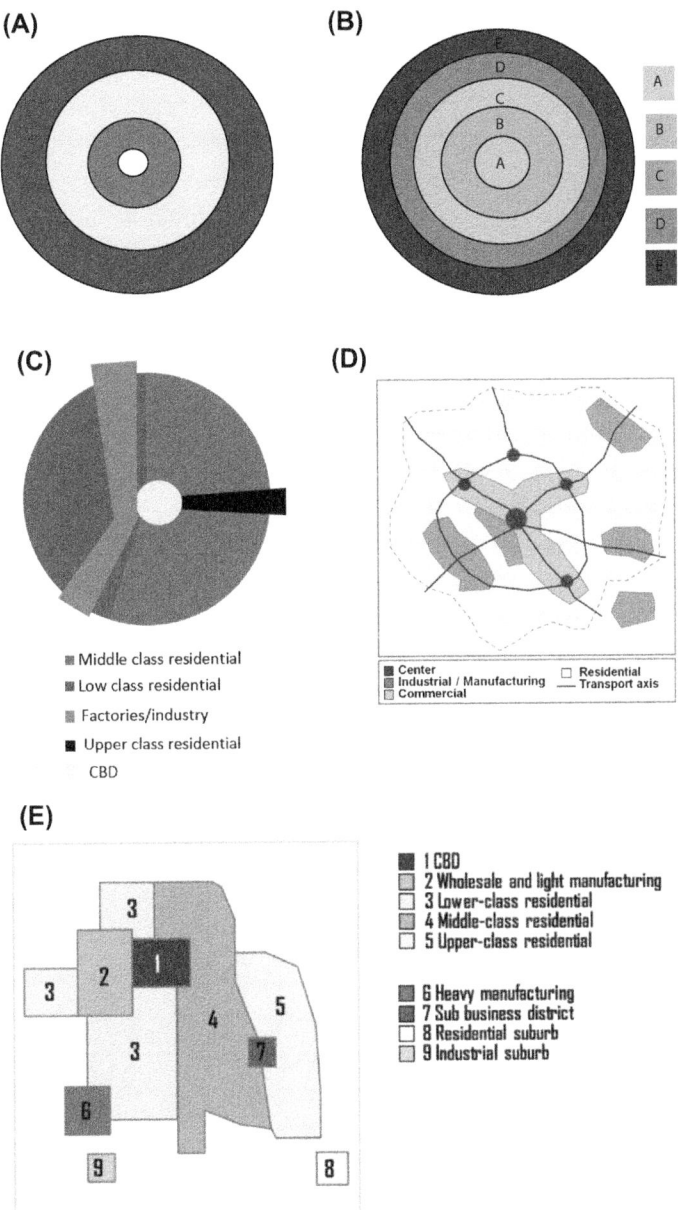

(A)

(B)

(C)

■ Middle class residential
■ Low class residential
■ Factories/industry
■ Upper class residential
  CBD

(D)

Center
Industrial / Manufacturing   Residential
Commercial   Transport axis

(E)

1 CBD
2 Wholesale and light manufacturing
3 Lower-class residential
4 Middle-class residential
5 Upper-class residential

6 Heavy manufacturing
7 Sub business district
8 Residential suburb
9 Industrial suburb

Figure 5.16
Visual representation of Von Thunen's model (A), Burgess model (B), Hoyt model (C), hybrid models (D), and the multiple nuclei model (E) [12]. *Copyright permission for (D) and (E) has been granted by Jean-Paul Rodrigue.*

productive will be located further away from the market center. Consequently, intensive farming will be located closer to the city center than forests. The main assumption of Von Thunen's model is that agricultural land use is formed as concentric circles around the central market; the latter consumes all the surplus production, which must be transported from the rural areas to the market. The model was developed in 1826 and was first applied to analyze the agricultural land use patterns in Germany of the 19th century. The main drawback of the model is that it does not consider differences in local, physical conditions since it has been developed in an isolated state.

- The *concentric zone model or Burgess model* (Fig. 5.16B): This model was one of the first attempts to explain the distribution of social groups in urban areas. The model considers that there is a correlation between the socioeconomic status (mainly income) of households and the distance from the Central Business District (CBD). The further away the household is from the CBD, the better the quality of housing, but the longer is the time taken to commute. This way concentric circles (zones) are developed around the CBD which is Zone A. Zone B is adjacent to the CBD, is known as zone of transition, and includes industrial activities that are located there in order to take advantage of nearby labor and markets as well as transport facilities such as ports and railways. Zone C is close to the industrial areas and includes the poorest housings which are usually occupied by first-generation immigrants. Zone D is a residential zone that is dominated by the middle class. It has the advantage that it is located relatively close to the industrial zones of employment (ie, zones A and B). Zone E consists of high class and expensive housing in a rural, suburbanized setting. This model was first applied to explain the land use of the city of Chicago in 1925 and can still be reflective of the land uses of a city. The main criticisms against this model is that it is too simplistic, reflecting the urban characteristics which were observed 50−100 years ago. It was developed at a time when the American cities were growing very fast. Furthermore, the model was developed for American cities and has limited transferability. It assumes a spatial separation of place of work and residence, which was not generalized until the 20th century.
- *The Hoyt model* (Fig. 5.16C): It was developed by the economist Homer Hoyt in 1939 and is a modification of the Burgess model. Although the Hoyt model accepts the notion of a CBD, it considers that all the zones expand outward from the CBD rather than in concentric zones; the zones are supposed to follow the routes of highways, railways, underground tube, and other transport means. Consequently, the effect of direction and time was also added to the impact of distance, which was overlooked by the concentric models. The land use often follow transport arteries, such as railway lines, resulting in the development of sectors. Thus, cities will grow along major transport axis. The model assumes that cities tend to grow in a pattern that resembles a wedge, starting from the CBD and growing outward. The low-income households will

be concentrated close to industrial zones, while the middle and high income will be located further away. This model recognizes the tight connection between transportation arteries and land uses and is applicable to several cities in the United Kingdom. Its main limitation is that it is does not consider private cars, which allow the transport from land that is further outside and is not well connected to the public transportation means.

- *The multiple nuclei model* (Fig. 5.16E): It was developed by Harris and Ullman in 1945, and, as its name implies, it assumes that cities grow through the gradual integration of several separate nuclei in the urban spatial structure. Although the city may have initially developed with a CBD, other smaller CBDs develop in the outskirts of the city. As a result, nuclei are formed in other parts of the city apart from the main CBD. The model is suitable for application in large cities and also takes into account the increased mobility given by private vehicles. The multiple nuclei may develop and expand for several reasons which include (1) the need to have suitable transport means (eg, ports, railways) for certain industrial activities, (2) the need to move to locations where land and house rents are cheaper, (3) the clustering of certain activities for mutual benefits and convenience for the inhabitants such as hospitals with pharmacies and Universities with bookstores, (4) the need for certain activities to be placed in different locations such as industrial activities and hospitals, and (5) benefits resulting from activities placed close to each other such as factories and residents. The model has some limitations since it makes assumptions of land being flat, even distribution of resources, transport costs, and distribution of people in residential areas which are simplifications.
- *Hybrid models* (Fig. 5.16D): These models attempt to consider concentric, sector, and nuclei arrangements of different processes in order to account for land use in cities. Consequently, hybrid models integrate the concentric arrangement for CBDs and sub-centers with the sector arrangement for transportation in order to develop the land use pattern. Using hybrid models it is possible to account for the change of the urban spatial structure; hybrid models integrate the various spatial effects of transportation on land use giving also the time dimension. The city can develop as nuclei in certain locations due to specific conditions, but also along the transportation routes.
- *The land market model*: It is a pure land economics-based approach, where the cost of land, supply, and demand are the drivers for land use. The higher the demand for the land, the higher will be its cost. The principle that is followed is that towards the center of the city, less land is available (ie, limited supply) and at a higher cost. As one moves toward the suburbs more land becomes available (ie, more supply) and the cost reduces. The accessibility and the ease of transportation are factors that significantly impact on the land rent and thus on its use. Within each city there are different categories of activities, and each one wants to occupy land and is willing to pay a specific price to acquire

it (or rent it). However, this market-based approach on land use is challenged by the structural modification of modern large cities.

- *Cellular automata models for land use*: They consist of dynamic models for land use applications. In these models space is divided into finite cell grids. Each cell grid represents a finite land unit. Certain land uses are defined for the different cells; when the use of one cell (ie, piece of land) is about to change, transition rules are applied. These models can dynamically and realistically simulate land use with a high level of detail. However, adequate and reliable data are required to apply them. The models can be integrated with Geographic Information Systems (GIS) for user friendly representation.

Regardless of the model that is used to guide decision making for land, it is recognized that this decision needs to be balanced and results from the trade-off of several, often conflicting, parameters. Land is finite but there are many different potential options for activities that can take place on each piece of land. Land is an open space that needs to integrate territorial cohesion, spatial planning, urban/rural diversity, housing, and transport. Land use needs to be productive, in the sense, that it must be used to extract minerals, to produce crops and feed the population, and produce crops that are used to produce bioenergy. Land has several functions including the protection of nature and ecosystems, for recreation, and flood mitigation. Proper land use can decrease greenhouse gas (GHG) emissions and can be used as a carbon storage tank mitigating the adverse impacts of climate change. It can also serve for adaptation measures to climate change [1].

5.6 Air Quality

The rapid urbanization has resulted in increasing air pollution emissions due to transport, energy production, industrial and commercial activities. All these activities are concentrated in densely populated areas.

5.6.1 Main Air Pollutants

The US Environmental Protection Agency (EPA) has specified six common air pollutants for which National Ambient Air Quality Standards have been set [13]: ozone (O_3), particulate matter (PM), carbon monoxide (CO), nitrogen oxides (NO_x), sulfur dioxide (SO_2), and lead (Pb). The description of air pollutants is covered in chapter "Atmospheric Environment". Ozone is formed close to ground through the chemical reactions taking place among nitrogen oxides and volatile organic compounds (VOCs) in the presence of sunlight. Sources of NO_x and VOCs include electric utilities, vehicle exhaust gas, petrol vapors, and chemical solvents. Urban emissions are important contributors to the problem of near-ground-level ozone formation [14—17]. The anthropogenic activities which lead to the generation of PM include combustion of fossil fuels, vehicle emissions [18], power plants, and several industrial processes. Currently, human activities result in 10% of the total mass of particulates in the atmosphere [19]. PM_{10} is defined as airborne particulate matter having a mean aerodynamic diameter less than 10 μm. $PM_{2.5}$ is defined as airborne particulate matter having a mean aerodynamic diameter less than 2.5 μm. Recently, the importance of $PM_{2.5}$ has attracted attention; there is also concern about superfine particles of PM_1 and even $PM_{0.1}$. $PM_{0.1}$ is so small that can pass through the lungs into the blood; therefore, these particles can affect the blood of humans and not just the lungs, creating even more health problems.

The incomplete oxidation of carbon results in the production of CO, instead of CO_2. Anthropogenic sources of CO production include the operation of internal combustion engines at an enclosed building, iron smelting, the exhaust from a home wood fire, industrial processes of incomplete combustion, and undiluted warm car exhaust without a catalytic converter [20]. As urban areas can be associated with such activities, CO is a temporary atmospheric pollutant in some urban areas. The NO_x are mainly emitted through combustion at high temperature from car engines and power plants. NO accounts for the majority of NO_x emissions. Usually, only a small part of NO_x emissions is directly emitted as NO_2 (5−10%). NO_2 is produced from the oxidation of NO. In diesel cars this is not the case, since direct NO_2 emissions can be up to 70% of the total NO_x [21]. The main producers of SO_2 emissions are the combustion of fossil fuels at power plants (73%) and other industrial facilities (20%). Other, smaller sources of SO_2 emissions include extracting metals from ore, the burning of high-sulfur-containing fuels by locomotives, large ships, and nonroad equipment. It is naturally formed during the eruption of volcanoes. The main source of atmospheric lead pollution has been from the vehicle fuel. To address this issue, the lead has been removed from petrol resulting in a dramatic decrease of lead in the atmosphere.

Other toxic heavy metals that can be found in air include arsenic (As), cadmium (Cd), and mercury (Hg). The human activities that result in arsenic emissions include fossil fuel combustion and smelters. In the past, pesticides were a major contributor to As emissions, but limitations have significantly reduced their contribution. The human activities resulting in the highest Hg emissions include the combustion of fossil fuels. Industrial activities such as metal (particularly gold) and cement production, and waste disposal emit Hg. The anthropogenic sources of cadmium (Cd) emissions to the atmosphere include industrial activities (nonferrous, iron and steel production, cement production), fossil fuel combustion, and waste incineration [21]. Organic air pollutants include benzene (C_6H_6) and benzo(a)pyrene (BaP). The greatest source of benzene emissions is from the incomplete combustion of fuels. In Europe, 80−85% of benzene emissions are generated from vehicle movement. Other sources of C_6H_6 are domestic heating, oil refining and storage, handling, and transportation of petrol. The BaP is an indicator of polycyclic aromatic hydrocarbons (PAHs) [21].

Pollutants of particular health concern in developed countries include particulate matter (PM), nitrogen dioxide (NO_2), and ozone [22]. In Europe, the most dangerous pollutants for human health are PM and O_3 followed by BaP and NO_2. Air pollution is related not only to health problems, but also to environmental problems including acid rain, eutrophication, and loss of vegetation [21]. It should be noted that air pollution is an environmental problem that both developing and developed countries face. There are several cities in developing countries that face significant air pollution challenges [23].

5.6.2 Impacts to Humans in Cities

Air pollution is one of the biggest environmental challenges in modern world causing serious health problems. According to the World Health Organization (WHO) air pollution in 2012 was linked to ~7 million mortalities [24]. The air pollution in cities can also cause respiratory diseases [25–27]. In any case, the urban air pollution is going to become the main environmental cause of premature mortality worldwide in 2050 [28]. A large number of toxicological and epidemiological studies have demonstrated that outdoor air pollution causes adverse effects on public health [22,29]. Air pollution is heavier in urban areas compared to rural areas due to activities such as heavy traffic congestion and industrial activities which lead to the emission of air pollutants into the atmosphere. Consequently, people living in the cities are exposed to more air pollution and face a greater danger of health problems compared to rural population.

All individuals can be potentially exposed to air pollution. Sensitive groups such as the elderly, children, and people with respiratory and heart problems can experience health problems even at moderate concentration of air pollutants. Several epidemiological studies have demonstrated that particulate matter and ozone have been connected to increased morbidity and mortality rates in several European cities [30,31]. There is sound scientific evidence that PM_{10} deteriorates airway disease as shown by the correlation of PM_{10} levels with reduced lung function [32], hospital admissions for asthma [33], and chronic obstructive pulmonary disease [33]. Once inhaled, particulate matter can cause heart and lung operation problems. Finer particles can penetrate deeper in the lung and cause more harm than coarser particulates. Thus, long-term exposure to $PM_{2.5}$ is associated with serious health outcomes. The scientific evidence from toxicological and epidemiological studies indicates that there is a causal relationship between long-term $PM_{2.5}$ exposure and cardiovascular effects, mortality, and probably for effects on the respiratory system. Evidence has shown that the long-term exposure of humans to $PM_{2.5}$ results in a mortality rate of 6% for every additional 10 µg/m^3 of exposure. At PM_{10} and $PM_{2.5}$ concentrations below 100 µg/m^3 the acute health effects are considered to be linearly correlated to exposure [31]. In a large study of 29 European cities, Katsouyanni et al. [34] estimated that mortality increases by 0.6% for an increase of PM_{10} by 10 µg/m^3 for acute exposure. As mentioned in Section 5.6.1, concerns have now been raised about the ultrafine particles

PM_1 and even $PM_{0.1}$. The effects can range from subtle biochemical and physiological changes to severe illness and death.

The inhalation of ozone can result in several health problems, particularly for frail groups such as people with lung diseases and asthma, the elderly and children. Ground-level ozone can adversely affect ecosystems and sensitive vegetation. Evidence shows that the increase of ground-level ozone by 10 µg/L, increases the mortality rate by 0.3%. Furthermore, the combined exposure to both O_3 and PM pollutants has shown to increase the mortality rate [31,35]. Carbon monoxide can cause significant harmful health problems by reducing the delivery of oxygen to the human organs and tissues. CO can cause death when it is inhaled at high levels. NO_2 can cause inflammation of the airways when it is found at high concentrations, while chronic exposure can adversely affect the lungs. Sulfur dioxide can result in the constriction of the airways and may cause mortality. Exposure to high concentrations of lead adversely affects the kidneys, the hemoglobin, the gastrointestinal tract, the reproductive system, and the nervous system [36]. The COMEAP has concluded that chronic exposure of the population to fine particulates, sulfur dioxide, and sulfate particles can adversely affect the cardiovascular system resulting in heart attacks and strokes [37].

Exposure to higher levels of air pollutants causes more harmful effects. However, for certain air pollutants (eg, fine particulate matter) there is no specified threshold value below which there is no effect on population [38]. According to the COMEAP report the $PM_{2.5}$ levels caused by anthropogenic activities in 2008 had an effect on mortality equivalent to nearly 29,000 deaths in the United Kingdom and a loss of total population survival of 340,000 years [37]. Based on these results, it can be inferred that air pollution decreases the average life by 11.5 years, if air pollution is solely responsible for the 29,000 deaths to 6 months if the timing of all deaths was influenced by air pollution. WHO has published a systematic review of the health effects of transport-related air pollution in Europe [39].

The adverse health effects caused by air pollution are a dramatic burden for the national health service of many countries. According to the British Department for Environment, Food and Rural Affairs (Defra), the total cost resulting from the health impacts of air pollution in the United Kingdom ranged from £9.1 to £21 billion in 2005 [40]. The health problems that required treatment among others included asthma, rhinitis, cancer, nonallergic respiratory morbidity, cardiovascular problems, male fertility, and pregnancy problems.

5.6.3 Air Pollution in Cities

In Europe, a significant proportion of the population is exposed to air pollution. In several cases the population is exposed to air pollution levels which exceed the EU air quality limits. During the period 1997–2008, it is estimated that 13–62% of the population may have been exposed to concentrations of PM, O_3, and nitrogen dioxide (NO_2) above the EU air quality limits. According to recent data, the EU28 percent population exposure above EU and WHO reference levels were as follows: 10–14% for $PM_{2.5}$, 21–30% for PM_{10}, 14–17% for O_3, 24–28% for BaP, 8–13% for NO_2, and less than 1% for SO_2, CO, Pb, and benzene [21]. However, it should be noted that the number of people who are affected varies from year to year as a result of variability in emissions, pollution buildup, and dispersion/deposition conditions.

Table 5.2 lists the 10 cities having the highest levels of N_2O concentrations and the greatest number of days that O_3 and PM_{10} concentrations in air exceeded the EU limit values for the year 2008. It can be seen that for PM_{10}, the cities having the highest exposure days above limit values are mainly east European cities located in Poland, Romania, and Bulgaria. In the case of the greatest days of O_3 exposure above the EU limits and the highest levels of N_2O average concentration several cities in North Italy are listed. These data are given for the urban background, which is representative of urban residential areas and therefore for the majority of the urban population. In streets with intense traffic the exposure of people to air pollution is usually much higher [21].

In China, air pollution is currently one of the main environmental concerns. The cities of Beijing, Shanghai, Guangzhou, Shenzhen, and Hong Kong are all cities where high economic development takes place, resulting in $\sim 20\%$ of the GDP in China. These cities face significant air pollution problems [41]. In Beijing the main sources of anthropogenic emissions include domestic heating, traffic, and industrial activity. In China the main air

Table 5.2: The Most Polluted Cities for Daily PM_{10}, O_3 Concentrations, and NO_2 Annual Mean Concentration in the Urban Background, 2008 [21].

Number of Days of PM_{10}, Exceedances of EU Limit Value of 50 $\mu g/m^3$ (Daily Mean)		Number of Days of O_3, Exceedances of EU Limit Value of 120 $\mu g/m^3$ (Maximum Daily 8 h Mean)		NO_2 Annual Mean Concentrations in $\mu g/m^3$ (the EU Limit Value is 40 $\mu g/m^3$)	
Plovdiv, Bulgaria	208	Turin, Italy	77	Brescia, Italy	62
Pleven, Bulgaria	185	Campobasso, Italy	74	Turin, Italy	60
Sofia, Bulgaria	176	Bologna, Italy	72	Brasov, Romania	58
Krakow, Poland	152	Bergamo, Italy	69	Modena, Italy	50
Timisoara, Romania	136	Athens, Greece	68	Milan, Italy	49
Rybnik, Poland	122	Novara, Italy	65	Trieste, Italy	48
Nowy Sacz, Poland	116	Cremona, Italy	64	Rome, Italy	43
Craiova, Romania	112	Brescia, Italy	64	Athens, Greece	42
Zabrze, Poland	108	Milan, Italy	62	Padua, Italy	41
Turin, Italy	106	Reggio nell Emilia, Italy	61	Genoa, Italy	41

Turkish PM_{10} data are not validated and therefore not part of this table reflecting the situation in 2008.
AirBase, 2010

pollutants are PM and SO_2 which occur due to intense industrial activity, the combustion of fossil fuels, power plants, and household heating [42].

5.6.4 Evolution and Improvement of Air Quality

In Europe, the implementation of EU policies to combat the air pollution has significantly contributed toward the improvement of the air quality in the member state countries. Specifically, the implementation of Directive 2008/50/EC on ambient air quality and cleaner air for Europe and of Directive 2004/107/EC relating to arsenic, cadmium, mercury, nickel, and PAHs in ambient air. The limits set for the air pollutants are for PM_{10} (50 µg/m^3 for 24 h and 40 µg/m^3 for 1 year), $PM_{2.5}$ (25 µg/m^3 for 1 year), sulfur dioxide (350 µg/m^3 for 1 h and 125 µg/m^3 for 24 h), nitrogen dioxide (200 µg/m^3 for 1 h, 40 µg/m^3 for 1 year), lead (0.5 µg/m^3 for 1 year), carbon monoxide (10 µg/m^3 maximum daily 8 h mean), benzene (5 µg/m^3 for 1 year), ozone (120 µg/m^3 maximum daily 8 h mean), arsenic (6 ng/m^3 for 1 year), cadmium (5 ng/m^3 for 1 year), nickel (20 ng/m^3 for 1 year), and PAHs (1 ng/m^3 for 1 year). Furthermore, several member states have set more strict limits at regional or national level [43,44].

During the last few years most of the air pollutants have decreased in Europe. Table 5.3 shows the change of the main air pollutants in Europe for the period 2003−12; the table also summarizes the sectors that are mainly responsible for these emissions and the reason

Table 5.3: Change of Air Pollutants Between Years 2003 and 2012 for the EU28 Countries and Reasons for This Change [21].

Pollutant	Change Between 2003 and 2012 (%)	Main Sources of Emission	Main Reason for Change
NO_x	−30	Transport (vehicle engines), energy (power plants), and industry sectors (combustion)	All sectors emitting NO_x have reduced their emissions
BaP	+21	Commercial, institutional, and household fuel combustion sector	Increase in solid fuels use (eg, wood) for domestic heating
SO_2	−54	Energy and industry sector, shipping activities	Decrease in energy sector due to gas-flue desulfurization and scrubbers
CO	−32	Commercial, institutional, and household fuel combustion	Decrease CO emissions from transport
Pb	−19	Fossil fuel combustion, waste incineration, and industry	Use of unleaded fuel
$PM_{2.5}$	−17	Commercial, institutional and household fuel combustion; industry; and transport	Use of biomass for fuel combustion
PM_{10}	−15	Commercial, institutional, and household fuel combustion; industry; and transport	Use of biomass for fuel combustion

for the change in the pollutant concentration. It seen that $PM_{2.5}$, PM_{10}, NO_x, CO, and Pb decreased significantly in the range of 15−32%, while remarkable decrease of SO_2 was observed (54%), which is mainly attributed to the desulfurization of flue gas and to the use of scrubbers in power plants. Some issue of concern has arisen from the increase (21%) of the polycyclic aromatic hydrocarbon, BaP. This is most likely attributed to the use of solid feed for energy production (mainly wood).

Despite the improvements in most of the air quality parameters, Europe has still a lot to do in order to accomplish air quality levels which do not pose unacceptable risks for human health and the environment. The decrease in air pollution has also decreased the severity of eutrophication, acidification, and ozone formation. However, such environmental problems still persist in several regions. Attention is now focused on fine and ultrafine particulate matter (ie, $PM_{2.5}$, PM_1, and even $PM_{0.1}$). The overall projections for Europe are positive since it is expected that the air quality in cities will improve. It is forecasted that more cities and countries will meet the air quality standards [28]. Within the next 5−10 years, improvements in housing and end-of-pipe solutions will have a positive impact on air quality. However, the continuous increase of the urban population in Europe will set new challenges for air quality in cities [28]. Despite the improvements of the air quality in European cities, still it is estimated that there are more than 400,000 premature mortalities for which air quality is responsible.

Several techniques have been developed in order evaluate the air pollution in cities. These include the application of exposure indicator variables, dispersion models, interpolation methods, and land-use regression models. The land-use regression models are increasingly being applied in the last years to evaluate outdoor air pollution. In their work, Hoek et al. [45] reviewed 25 land-use regression models which simulated the mean annual concentrations of selected air pollutants (ie, NO_2, NO_x, $PM_{2.5}$, soot content of $PM_{2.5}$, and VOCs). In the reviewed land-use regression models, air pollution was monitored in 20−100 different locations. These methods have been globally applied, including cities in Europe and North America. The main conclusion resulting from the application of such models is that they are able to provide better or equivalent simulations compared to geostatistical methods.

In the United Kingdom, the air pollutants in large cities have been decreasing in the last 50 years. The main improvements which have been implemented include the cleaner engine combustion and the obligatory use of catalytic converters for vehicles that are fueled with petrol. As a result, a decrease in air emissions has been observed, particularly for CO and VOCs. The particulate matter and NO_x emissions resulting from road transport have also decreased by 42% and 55%, respectively, from 1990 up to 2004 [46]. More recently, the rate of decrease of air pollutants has been lower, while in some cities the air pollutants have risen. Potential reasons for this increase may include the increase in the transport, the increased use of diesel vehicles, and higher background concentrations of pollutants from other sources. The main conclusion of Defra is that air pollution will continue to be an important environmental problem causing human health problems, unless major policy initiatives are undertaken to further reduce air pollution in cities [40].

Further measures need to be applied in urbanized areas in order to mitigate the effects of air pollution to human health and ecosystems. These could include:

*   More extended use of hybrid and electric cars in cities. This could be promoted by providing suitable incentives to people using such vehicles (eg, reduced vehicle taxation, subsidies).
*   Investment in new and greener public transportation, and gradual replacement of the older fleet.
*   Recovery of energy from biowaste produced in municipalities. Anaerobic digestion plants can be widely applied for the treatment of source-separated biowaste at municipality level.
*   Dense shrub and tree plantings in open spaces in cities can absorb significant amounts of dust and air pollutants and can also filter noise. Thus, green and quite areas can be developed within the city, improving the local climate by providing adequate ventilation and reducing the heat-island effect [47].

In North America, evidence from the megacities (Los Angeles and New York) shows that urban development can be implemented together with air quality improvement. This can serve as an example for other cities in Asia which are growing very fast. As air pollutants are transported by wind currents, the effect of air pollutant emitted in megacities may be felt in other locations far away from the megacities. The impact of pollutant emissions from megacities is felt long distances away from the local sources, but no policy mechanisms currently exist to mitigate air quality impacts resulting from such pollution transport [48].

Cities can be evaluated and ranked for their air quality based on indicators. To evaluate the performance of cities (and particularly megacities) against air pollution, the cities can be ranked based on the annual per capita per unit surface area emissions of selected air pollutants (eg, PM, CO, NO_x). For example, a multipollutant index (MPI) can be developed which takes into account the combined level of the different criteria pollutants. China has taken steps in order to improve the air quality of its cities. Significant efforts were undertaken to decrease air pollution in Beijing for the 2008 Olympic Games. Concentrations of fine particulate matter and ozone in Beijing often exceed healthful levels in the summertime. The air quality of the city is deteriorated by the neighboring Hebei and Shandong Provinces and the Tianjin Municipality. For example, the wind flow from the south, Hebei Province can contribute 50−70% of Beijing's $PM_{2.5}$ concentrations and 20–30% of ozone. Consequently, measures within Beijing are not sufficient to attain high air quality standards [49,50].

5.7 Urbanization and Climate Change

The cities are a major contributor of GHG emissions; it is estimated that cities contribute as much as 70% of the global GHG emissions [51]. Consequently, the urban activities are major drivers of the climate change. Urban areas are also among the most vulnerable areas to climate changes which can include problems due to temperature increase, rising of the sea level, or other more extreme weather conditions [52]. The very existence of coastal and near costal metropolitan cities can be at stake in the future due the increase in the sea level [53]. Other cities will have to invest heavily in infrastructure protecting against climate change (eg, major flood control works, stormwater capture systems, tree planting). Inhabitants will have to adapt to changes. Even cities which may not be directly affected by climate change may have to cope with issues such as increasing pressure to accommodate climate refugees from other cities [54]. In the CDP 2013 report on 110 global cities, it was concluded that 88 cities reported risks from temperature increase, 81 cities reported risks from intense/frequent rainfalls, 49 cities reported risks from droughts, 43 cities reported risks from floods, and 39 cities reported risks from sea level rise [55].

Although in terms of total GHG emissions, cities generate very large amounts, recent evidence shows that the per capita production of urban GHG emissions for individual cities is lower than the average respective ones in the countries where these cities are located [56]. Furthermore, recent evidence shows that the contributions of deforestation, agriculture, and combustion of fossil fuels for power generation in nonurban areas are significant. As it is noted by Satterthwaite [57], probably less than half of the global anthropogenic GHG emissions are generated within the boundaries of cities. Nevertheless, if the GHG emissions resulting from power plants are actually assigned to the place and individuals responsible for the consumption, rather than the generation, the urban contribution to GHG emissions rises [57]. This does not mean that actions are not necessary in cities to reduce GHG emissions. On the contrary, there is currently the need to develop effective adaptation strategies, particularly in the urban regions of low- and middle-income countries. Apart from the environmental problems caused to cities by climate change, it can also lead to economic and social inequalities [52].

5.8 Solid Waste Management

Urban and consumer-based lifestyle drives the economy, but also results in the production of waste materials. Municipal solid waste (MSW) is waste that is produced in urban areas.

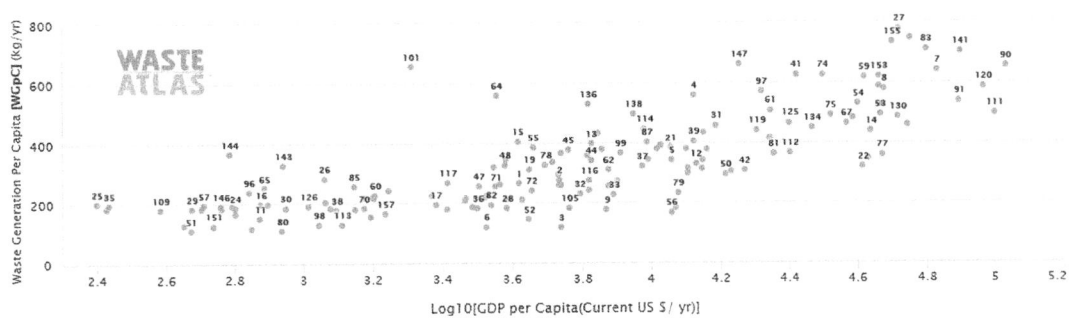

Figure 5.17

Generation rate of municipal solid waste (MSW) per capita versus GDP per capita. *Courtesy of Waste Atlas [59].*

The current, global MSW generation level is ∼1.3 billion metric tons per year, equivalent to 1.2 kg/capita/day [58]. The generation rate of MSW per capita is presented in Fig. 5.17 for several countries and ranges from 0.30 to 2.5 kg/capita/day [59].

The MSW generation is increasing in an alarming rate; between 2002 and 2012 the urban population increased by a factor of 1.04, but waste generation increased by a factor of 1.92. Annual total MSW generation growth rate (kg/year) varies in time and region, depending on the population growth and economic development, and ranges from negative to very high values [58–60]. The annual MSW growth rates in OECD countries do not exceed 2% and in some cases they even decreased after 1995 (Table 5.4 and Fig. 5.18) [61–63]. Negative values are observed, for example, in Canada, the United States, Australia, and the United Kingdom and may reflect reduced consumption and thus lower waste production due to the recent global financial crisis [63]. However, growth rates are high in countries such as China where the fast urbanization has led to remarkable increase of MSW. In China the total annual MSW generation growth rate from 1980 to 2006 was 23%, and the annual MSW generation rate per capita growth rate was 5.6% [64]. Such high MSW growth rates have been observed in other parts of the world as, for example, in

Table 5.4: Municipal Solid Waste (MSW) Annual Growth Rates [61–63].

Region	Based on Annual Waste Generation	Based on Annual Waste Generation per Capita
United States (1960–95)	0.9% to 3.8%	1.3% to 2.5%
United States (1995–2013)	−0.3% to 0.8%	−1.1% to 1.0%
European Union	−1.3% to 2.4%	−1.5% to 0.8%
Australia	2.2%	0.6%

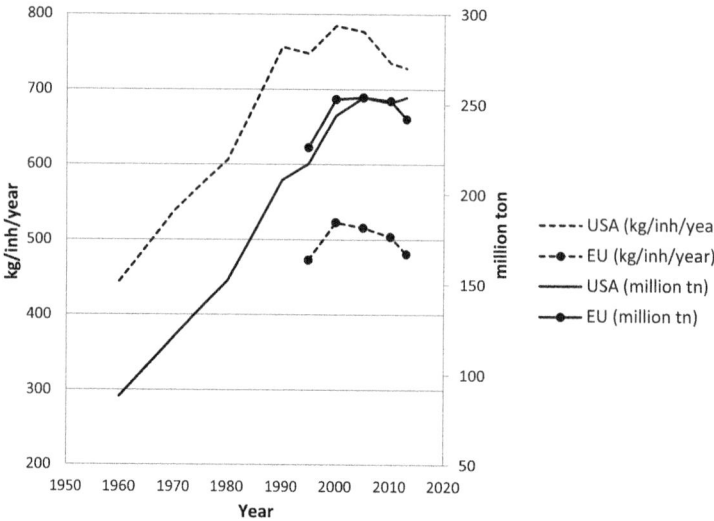

Figure 5.18
Municipal solid waste generated in the United States (1960–2013) and the European Union (1995–2013) [61,62].

Pakistan where the annual MSW generation growth rate per capita in Karachi from 2007 to 2012 was 12.8% [65]. Projected growth rates to the year 2025 are similar (Fig. 5.19).

Cities pose serious threats to the environment due to the increased amount of waste generation and their complexity. The impact of solid waste in both global and local

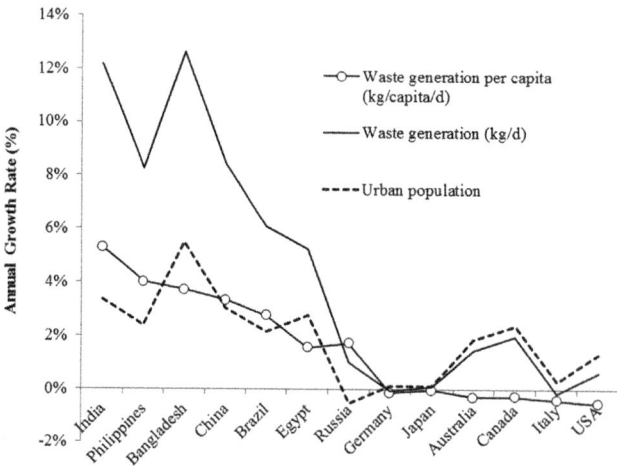

Figure 5.19
Current and projected municipal solid waste (MSW) and population annual growth rates in selected countries (2005–25) [58].

environment is growing fast. In low-income countries, waste collection rates are lower than 43% of the total waste generated; as much as 50% of the collected waste is often disposed in uncontrolled disposal sites and about 15% is processed through unsafe and informal recycling [58,66]. Uncollected solid waste contributes to flooding, air pollution, odor problems, and uncontrolled methane release, thus seriously threatening public health [58]. Furthermore, the organic content of MSW is a large producer of methane, contributing to global warming. Cities allocate a significant part of their budget for waste management. According to the latest data, as published by the World Bank [58], world cities spend more than $205 billion per year in waste management activities.

Inefficient waste management generates environmental pressures, leads to the consumption of more resources than required, and represents a significant cost to governments, businesses, and taxpayers. Moreover, waste materials have commercial value, a fact that used to be overlooked in the past. To achieve efficient waste management and a resource-efficient economy, waste disposal to landfills should be reduced, waste prevention should be promoted, and material reuse, recycling, and energy/materials recovery should be implemented [67]. The benefits of such a policy are not limited to more efficient resource use, but also offer a way to reduce GHG emissions from methane released in landfills. Furthermore, the use of recycled materials reduces the need for virgin materials and thus further reduces GHG emissions from primary production [67]. Experience in developed countries has demonstrated that by applying effective recycling schemes business opportunities are created and further jobs are provided [68]. Thus, by developing and implementing efficient waste management policies in cities costs can be reduced, economy can be stimulated, and natural resources and environment can be protected.

5.8.1 Waste Classification, Generation, and Composition

Solid waste can be classified into different types depending on their source as shown in Fig. 5.20. Definitions of MSW vary from region to region, and caution is needed when it comes to generation rates and relevant statistics. According to the EU's Landfill Directive (1999/31/EC) MSW is defined as "waste from households, as well as other waste which, because of its nature or composition, is similar to waste from households" [69]. According to OECD, MSW covers waste from households, including bulky waste, similar waste from commerce and trade, office buildings, institutions and small businesses, yard and garden, street sweepings, contents of litter containers, and market cleansing. Waste from municipal sewage networks and treatment works, as well as municipal construction and demolition, is excluded [58]. According to Eurostat and the EEA, MSW includes predominantly household waste (domestic waste) with the addition of commercial waste (business, sport, recreation, education, or entertainment), nonhazardous waste from industry, and waste from clinics and hospitals, which are similar in nature and composition to household

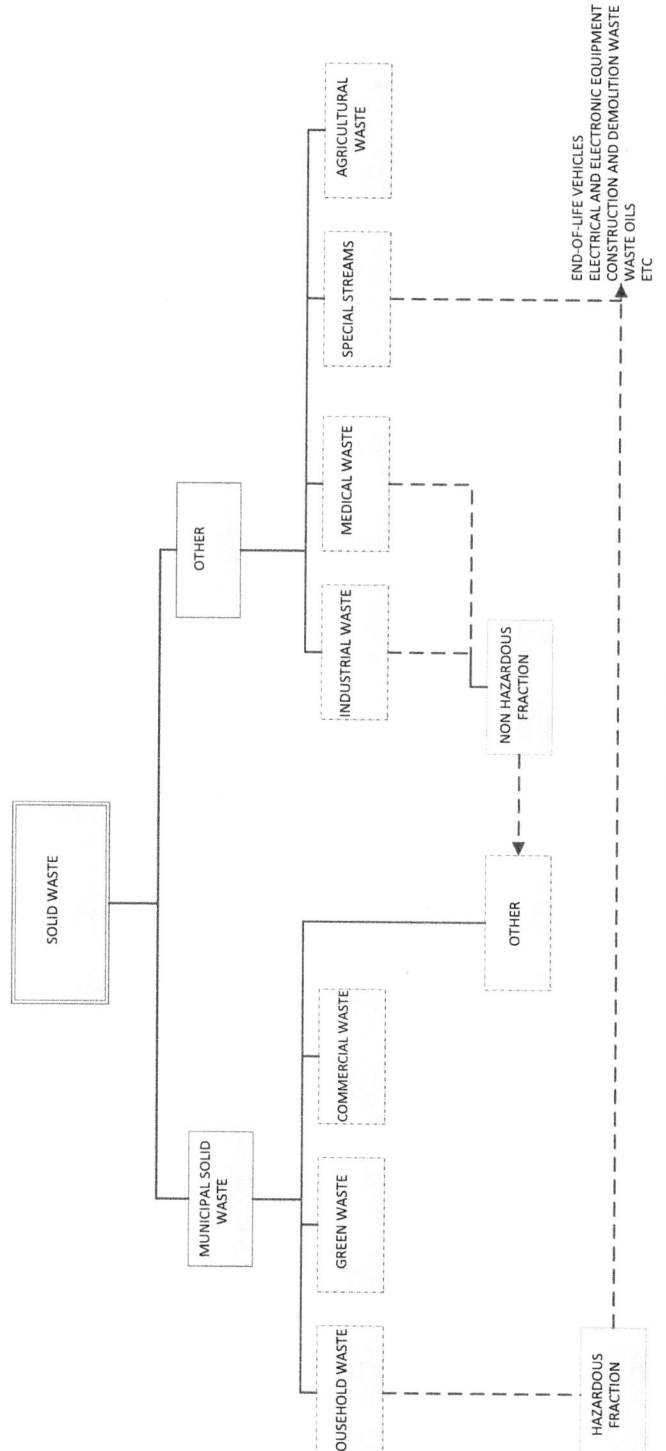

Figure 5.20
Solid waste categorization.

58.4 [image legend bar] 906.7

Figure 5.21
World map with municipal solid waste (MSW) generation per capita (kg/year). *Courtesy of Waste Atlas [59]*.

waste. In this context, municipal waste is understood as waste collected by or on behalf of municipalities [67].

The rapid growth of population and economic development inevitably leads to an increase of solid waste generation. The generation rate and composition of MSW varies greatly from region to region and changes significantly with time due to several factors, the economic wealth being the most important one. The annual MSW generation varies from about 900 kg/capita in North America to below 180 kg/capita in Africa (Fig. 5.21) [58]. In Europe in 2012, the annual MSW generation per capita was highest in Switzerland (694 kg/capita) and lowest in Albania (262 kg/capita) [28]. The United States generate annually a much higher amount of MSW per capita (802 kg/capita) than the European countries (336–602 kg/capita) which may in part be due to differences in data gathering and classification of waste; for instance the US data for MSW include some industrial wastes as well [63]. However, while MSW generation rates in developed countries stabilize, increase slowly, or even decrease, it is expected that waste generation will more than double over the next 20 years in developing countries. It is characteristic that China surpassed the United States as the world's largest waste generator in 2004 and will likely produce twice as much MSW as the United States by 2030 [58].

As it has already been mentioned, economic growth significantly affects waste generation; this can be clearly illustrated using the World Bank GNI (gross national income) per capita classifications (2012 data):

- Low income: $876 or less
- Lower middle income: $876 to $3465

- Upper middle income: $3466 to $10,725
- High income: $10,725 or more

As expected, urban solid waste generation per capita is higher in high-income countries and reduces as the income decreases. In terms of total urban waste generation, lower-middle-income countries figure is higher than that of upper-middle-income countries, probably a result of China's inclusion in the lower-middle-income group (Figs. 5.22 and 5.23).

In the preceding analysis, large-scale averages are used and it should be kept in mind that values vary considerably by region, country, and city. In Fig. 5.24 the waste generation per capita for some cities is presented. As is evident, waste generation in cities is generally higher than country or region averages since, depending on the region, urban residents produce about twice or even more as much waste compared to their rural counterparts [60].

Some representative values of MSW generation in cities of developing countries are as follows: 1.01 kg/capita/day in Kuwait city, 1.08 kg/capita/day in Chongqing, 1.11 kg/capita/day in Shanghai, 1.17 kg/capita/day in Hangzhou, 1.26 kg/capita/day in Bahrain, 1.3 kg/capita/day in Qatar, 1.33 kg/capita/day in Hong Kong, 1.39 kg/capita/day in Astana, 1.62 kg/

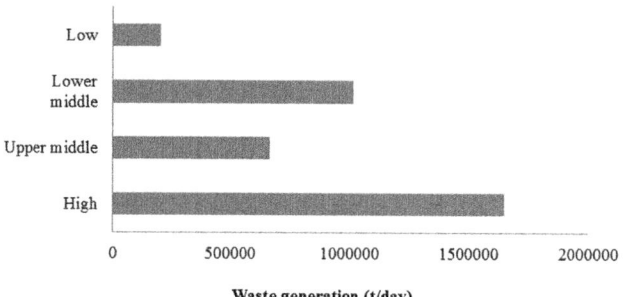

Figure 5.22
Total urban waste generation by income level [58].

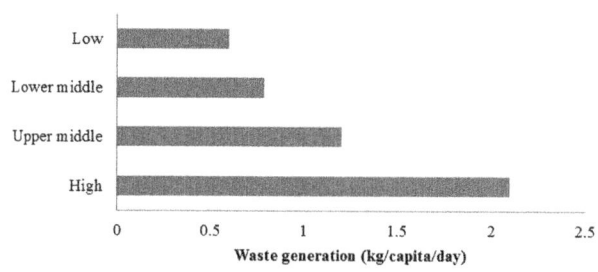

Figure 5.23
Urban waste generation per capita by income level [58].

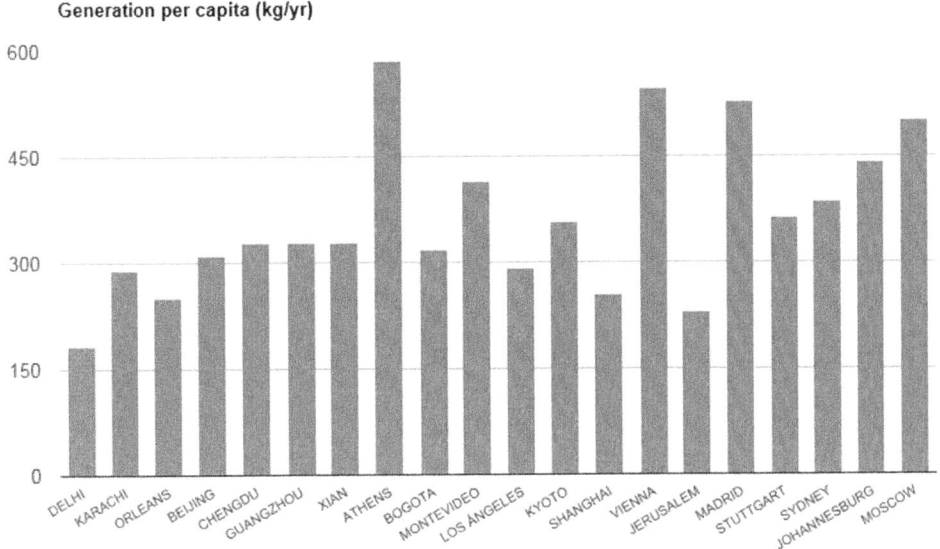

Figure 5.24

Municipal solid waste (MSW) generation per capita in selected cities. *Courtesy of Waste Atlas [59].*

capita/day in Kuala Lumpur, and 1.51 kg/capita/day in Lhasa [64,70–73]. The values are considerably lower in Indian cities, between 0.19 and 0.76 kg/capita/day [74].

The composition of MSW may vary from municipality to municipality, depending on the definition of MSW and the local waste management system. Based on global averages, MSW consists of organic waste (46%), paper (17%), plastics (10%), glass (5%), metals (4%), and other materials (18%). In Fig. 5.25, the MSW average composition in several parts of the globe is presented.

Low-income countries have the highest proportion of organic waste 64% compared to 28% in high-income countries. The opposite holds for recyclables (paper, plastics, glass, and

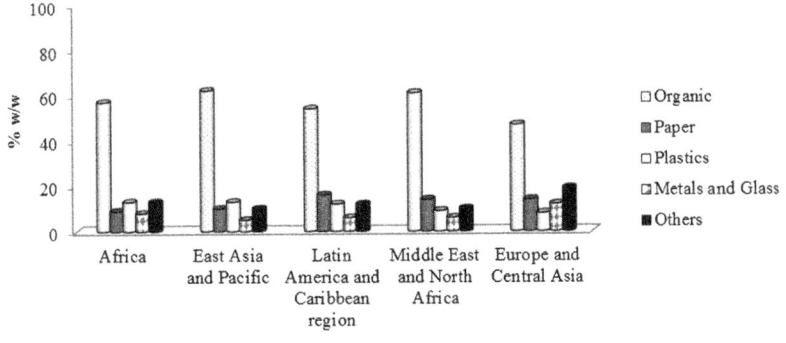

Figure 5.25

Average composition of municipal solid waste (MSW) per region [58].

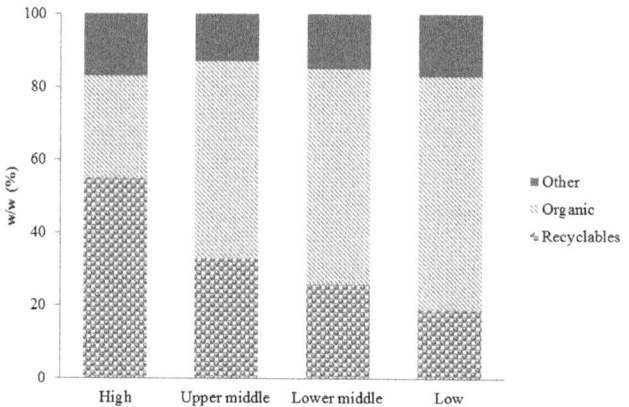

Figure 5.26
Average composition of municipal solid waste (MSW) by income level. Recyclables are paper, plastic, glass, and metals [58].

metals) which make up the highest proportion of MSW in high-income countries (Fig. 5.26). This is a result of urbanization and increase of purchase power of population which leads to the production of such materials. Paper makes up to 31% of MSW in high-income countries and only 5% in low-income countries. For example, data from European Union in 2005 show that the percent of biowaste in Ireland, Italy, Luxembourg, the Netherlands, Portugal, and Spain was 25%, 29%, 45%, 35%, 34%, and 49%, respectively, while paper and paperboard contributes 31%, 28%, 22%, 26%, 21%, and 21%, respectively, of the total MSW [60].

Diversion rate in high-income countries amounts to 55% and lower than 8% in the rest of countries, where 71−91% of waste is disposed in landfills and dumps sites (Fig. 5.27). With the exception of high-income countries, 13−49% of MSW is disposed in open uncontrolled dumps, posing a serious threat to both public health and the environment.

Diversion from landfills is seen as the panacea of waste management problems but should there be an upper diversion limit or 100% is the ultimate target? Integrated waste management systems and recycling schemes with high diversion rates consume a significant amount of resources. For example, separate collection schemes require an organized transportation network which means trucks and consumption of fuel. Furthermore, it is not always clear whether the benefits truly offset the environmental burden that may be caused by the additional waste-handling activities [75]. A study, supported by the EPA, developed models using life cycle assessment to estimate the cost of recycling in both monetary and environmental burden terms. The study concluded that there is a break point at about 20% diversion above which the cost starts to increase at an exponential rate (Fig. 5.28). The rapid cost increase is associated with the costs for separate collection of recyclables in the commercial and residential sectors [75,76]. Thus,

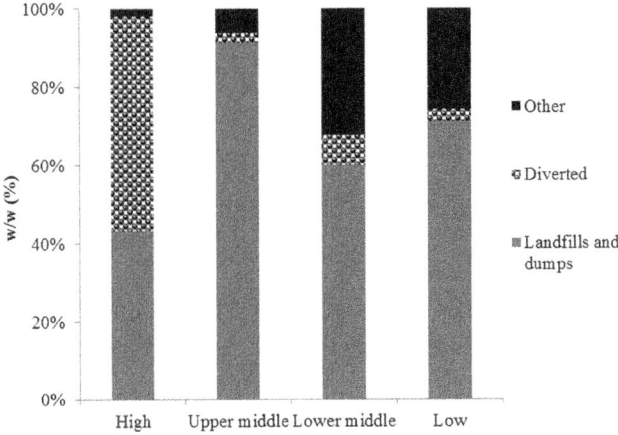

Figure 5.27

Municipal solid waste (MSW) disposal by income level. Diverted waste is the sum of the waste recycled, composted, and incinerated [58].

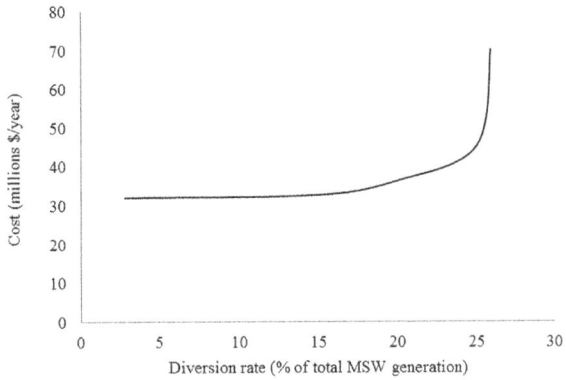

Figure 5.28

Variation of cost with diversion rate, based on a life cycle assessment. Approximate numerical values [75,76].

regardless of the environmental benefits, it might not be financially viable to target the highest possible diversion rates. Moreover, there are recycling schemes with high diversion rates which have proven to be successful. The conclusion is that waste management options should be evaluated through the application of tools such as life cycle assessment which can provide more detailed information than traditional evaluation tools.

5.8.2 Waste Management Planning

Waste management planning is the cornerstone of any national, regional, and local waste management policy and has a key role to play in achieving sustainable waste management. The main purpose of a waste management plan (WMP) is to provide an overview of all waste generated (status part) and to propose appropriate management options (planning part), including other aspects such as management policies [77]. In general, a WMP represents a road map for the future years and applies a combination of policy, legislation/ regulation development and enforcement, financial instruments, technical assistance, public awareness raising, and education. Traditional practices, still implemented in many parts of the planet, address the problems of solid waste in a partial way, dealing only with the management of the existing system, treatment plants, and final disposal; most efforts are concentrated on the end-of-life stage of materials. In contrast, Integrated Waste Management Plans, implemented in some developed countries, take into consideration the entire life cycle of waste from collection and transportation to rehabilitation of disposal sites. Waste reduction, a term used to describe waste prevention (source reduction) and diversion (reuse, recycling, recovery), is a key element of such a management plan [77]. These ideas are embodied in several policy documents, as Agenda 21 program run by the United Nations and the Waste Framework Directive issued by the European Commission in 2008.

The development of Integrated Solid Waste Management (ISWM) is based on:

- The integration of all elements of the solid waste life cycle based upon the 3R's processes (reduce, reuse, and recycle);
- The integration of technical, environmental, social, financial, institutional, and political aspects in order to guarantee system sustainability; and
- The active participation of all public, private, and community stakeholders in the conception and planning of processes and solutions.

The whole process of proper Solid Waste Management planning is based upon a generic policy, established by national authorities or international bodies as the European Commission in Europe (Fig. 5.29). In general, planning concerns the short-term and is very precise, strategy is about the middle-term and is less precise, and policy concerns the long-term and is the least precise.

Strategic planning can be defined as the process of determining needs and priorities, and necessary actions to be taken to develop waste management practices. In general, a national WMP is of a strategic nature, setting certain objectives, whereas regional or local master plans will be more action- or implementation-oriented [77]. A Strategic Plan should be prepared in the stages of the Strategy, Action Plan, Monitoring Plan, and Operational Plan [78]:

- The Action Plan specifies the objectives and activities that will be implemented to achieve them
- The Monitoring Plan defines what and how will be monitored and it includes indicators, time frames, etc.

Figure 5.29
Planning hierarchy.

- The Operational Plan defines how the action and monitoring plans will be implemented in practice defining the capacity needs in terms of human and other resources, the way these resources will be engaged, and how risks will be dealt with, etc.

Each stage has a distinct function which is taken into account in a step by step decision-making approach within the strategic planning process.

The IWMP is necessary to ensure that waste services keep pace with demand, are appropriate to needs, and are cost-effective. Furthermore, the planning process must ensure that the performance of the plan meets its objectives; the latter should ensure sustainable improvements to service coverage and standards. The main steps toward the development of a WMP include [70,79]:

1. Identification of the legislative requirements: The first step in the development of a WMP is to define the targets. A WMP has to consider at least the targets set in the legislation currently in force.
2. Definition of the waste streams to be included: An integrated waste management program shall not only handle household waste, but also other types of waste such as industrial and special types of waste, which may have different treatment and disposal requirements.
3. Definition of the planning period: A WMP can be intended either for immediate action or for long-term planning. In terms of local and regional planning this includes investments in waste treatment facilities; a minimum time frame of 20 years should be considered. Regardless of the total planning period, the WMP should be reevaluated in a shorter time. The need for revision is mainly caused by legislative changes and authorization problems but also due to political reasons.

4. Definition of the current status: The next step is to define the target area and to describe as clearly as possible the existing waste management situation. In this framework the following issues should be addressed: estimation of the waste quantities, waste composition, and existing infrastructure.
5. Identification of alternative waste treatment methods: The best available practices for the fulfillment of the targets have to be identified.
6. Formulation of alternative scenarios: Alternative waste management scenarios are formulated and compared.
7. Comparative assessment of the alternative scenarios: Having selected the alternative scenarios the selection of the final option has to be performed. This is usually done through the comparison of the developed scenarios against certain economic, financial, technical, and environmental criteria.
8. Financial analysis: The analysis includes the investment and operational costs.

5.8.3 Waste Management Options and Costs

Waste management options

A WMP has to be tailored according to the needs of the region where it will be implemented; the same applies to the waste management options that will be integrated into the plan. Competent authorities have to identify how these options can be combined within technically and economically feasible boundaries, always taking the legislative restrictions into consideration. The combination of different waste management options results into effective waste management schemes in order to achieve the targets of the WMP. An overview of the most common municipal waste management options that are used in IWMPs is provided in Fig. 5.30.

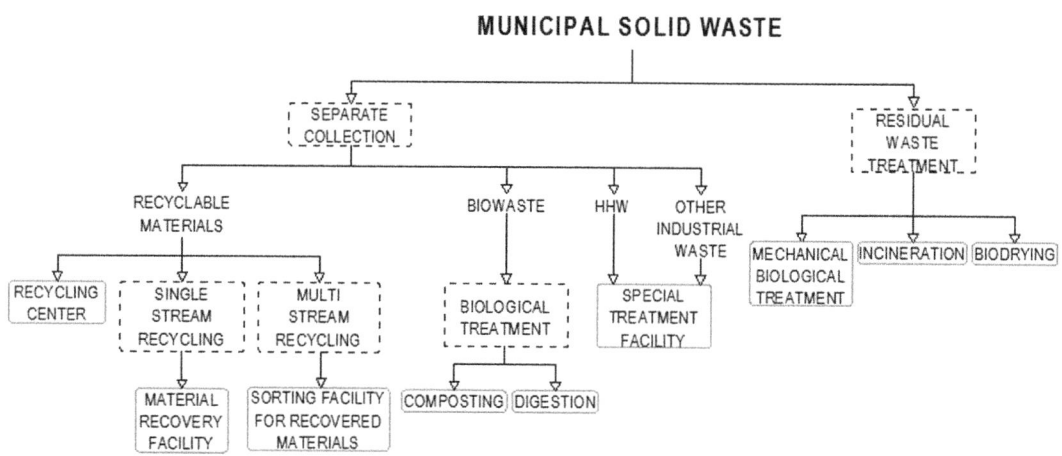

Figure 5.30
Overview of municipal waste management.

In the following paragraphs a short summary of the available waste treatment technologies is presented. It should be mentioned that household hazardous waste, and special type of streams like end-of-life vehicles (ELVs) and waste from electric and electronic equipment (WEEE) require special treatment facilities, and the analysis is beyond the scope of the present chapter. A detailed review on household and hazardous waste management is provided by Inglezakis and Moustakas [80]. Typical waste treatment technologies for MSW are the following:

- **Material recovery facilities**. Source-separated waste is transferred to materials recovery facilities (MRF) for sorting. An MRF is a specialized plant that receives, separates, and prepares recyclable materials for marketing to end-user manufacturers. An MRF accepts recyclable materials that have been separated at source from MSW generated, mainly using mixed recyclable materials bins. Materials are sorted to specifications, then baled, shredded, crushed, or otherwise prepared for shipment to market.

- **Composting facilities**. Composting is an aerobic biological treatment process used for the stabilization of biodegradable waste. Composting facilities can process biowaste collected after the source separation of household organic waste (and potentially green waste) at each household. Biowaste is defined as "biodegradable garden and park waste, food and kitchen waste from households, restaurants, caterers and retail premises, and comparable waste from food processing plants" [81]. Generally, biodegradable waste materials are suitable for composting given that the mixture sent for treatment meets certain conditions such as moisture, pH, C/N ratio, and porosity that will allow air circulation. There are several composting technologies that can be used in a composting facility. The main distinction is between open composting systems (ie, windrows and aerated static piles), reactor type configurations (horizontal and vertical), and channel reactors. Each composting system is unique and constitutes a separate process with its own advantages and disadvantages. Their suitability depends mostly on the planned treatment capacity. Some systems are more flexible for low treatment capacities while others for large-scale composting. The end product of the composting process is the compost which has the potential to be used as a soil amendment in various applications. Compost can substantially improve the fertility, texture, aeration, nutrient content, and water retention capacity of the soil. Due to its beneficial characteristics, compost has a variety of potential applications and can be used by several market segments. The compost should meet certain quality standards in order to be used.

- **Anaerobic digestion (AD)**. In terms of MSW management anaerobic digestion facilities usually process biowaste collected after the implementation of separation at source collection schemes. The end products of an anaerobic digestion facility consist of biogas and digestate. Biogas is a mixture of gases consisting approximately of 60% methane, 40% carbon dioxide and trace components of ammonia (NH_3) and hydrogen sulfide (H_2S). Biogas can be utilized for the generation of electricity and heat that cover primarily the energy needs of the facility. A typical combined heat and power (CHP) unit, produces 35–40% of electric energy and 50–55% of heat, while the remaining energy is lost. Surplus electric energy is supplied to the local grid. Digestate is the solid end product of the process and contains organic matter and nutrients. It can be further treated through a composting step in order to produce compost. Anaerobic digestion offers a significant advantage in comparison to composting through the production of renewable energy, but AD plants are more complex to operate and their capital and operational cost is higher. The decision for or against an AD plant is mainly governed by the subsidies offered for the renewable energy produced. There are many marketable AD technologies, but not all of them are suitable for any type of feedstock. Therefore it is very important when considering a certain AD technology to examine its suitability with respect to the input material, economics, and reliability.

- **Mechanical biological treatment (MBT)**. MBT plants are designed to treat mixed MSW which has not received any prior separation. The plant is divided into a mechanical and a biological section. Sorting of recyclables can be achieved in a similar way to an MRF but with a different and more intense technical configuration since the recyclable fraction has to be first separated from the wet organics and even then several organic and other impurities have to be removed. As a result the final quality of the separated materials is inferior in comparison to source-separated recyclables. The mechanical section includes equipment such as trommel sieves, bag openers, ballistic separators, magnets, eddy-current separators, shredders, etc. in order to separate the organic fraction and recover recyclable materials (ferrous metals, nonferrous metals) and/or refuse derived fuel. The separated organics will have to be treated in the biological treatment section using composting or anaerobic digestion technology. Since an MBT plant is in principle a combination of an MRF and a composting or AD plant, it can receive a wide variety of materials suitable for those processes. This includes apart from mixed MSW, biodegradable materials, packaging waste, industrial waste, sewage sludge, etc.

- **Biological drying**. Biodrying plants are designed to treat mixed MSW. Like MBT plants a biodrying plant is divided into a mechanical and a biological section; however, the treatment sequence is different. The waste is first shredded and then treated biologically (biodrying) in order to remove moisture and increase its calorific value. Afterward in the mechanical section, inert material and metals are removed. The remaining fraction can be utilized as secondary fuel (SRF). The mechanical section includes equipment such as trommel sieves, ballistic separators, magnets, eddy-current separators.

- **Incineration**. The incineration (combustion) of carbon-based materials in an oxygen-rich environment (greater than stoichiometric), typically at temperatures higher than 850°C, produces a gas composed primarily of carbon dioxide (CO_2) and water (H_2O). Other air emissions are nitrogen oxides, sulfur dioxide, etc. The inorganic content of the waste is converted to ash. The object of this thermal treatment method is the reduction of the volume of the waste with simultaneous utilization of the produced energy. The method can be applied for the treatment of mixed solid waste as well as for the treatment of preselected waste. It can reduce the volume of the MSW by 90% and its weight by 75%. The high capital and operational expenditure is a drawback for the construction and operation of such facilities. Only large facilities that achieve economies of scale and serve larger regions can be viable. An alternative to incineration is the implementation of advanced thermal treatment methods such as pyrolysis and gasification. These methods are not yet fully proven for MSW and have higher capital and operational costs.

- **Sanitary landfills**. Landfilling involves the managed disposal of waste on land with little or no pretreatment. Landfilling of biodegradable wastes results in the formation of landfill gas, which is dangerous if accumulated (explosions); as it mainly consists of methane, it represents a potent GHG. A typical landfill consists of a biogas collection system, a piping system for the collection of leachate, and an impermeable material (natural compacted clay and/or synthetic liners) to avoid the penetration of leachate into the aquifer. Landfill gas (LFG) can be flared (burned) on site or utilized in energy recovery systems. Within 1 year of the initial deposit of waste, the LFG produced by the decomposition of waste should be exploitable with a composition of approximately 50% methane and 50% carbon dioxide, with roughly 1% of other organic and inorganic compounds [82]. In 2010, the USEPA reported that there were 519 operating landfills generating 1597 MW [76]. As the least favored option in the waste management hierarchy, landfill should be reserved for stabilized wastes from which no further value may be recovered. All components of MSW are acceptable for landfilling, including residual fractions left over after the separation of materials for recycling and the residues from the treatment processes such as incineration and MBT.

Several combinations of technologies are possible and an example of a scenario is provided in Fig. 5.31.

Costs

Establishing waste management schemes can be very costly, and significant financial resources have to be allocated for waste collection, recycling, treatment, and disposal. The costs for collection and recycling schemes depend mainly on the cost of the local labor. The costs for waste treatment comprise mostly of technology and labor cost, amounting to only 16–30% of the total capital expenditure costs [83]. For instance, open windrow composting capital costs varies from 89 €/ton in Romania to 110 €/ton in

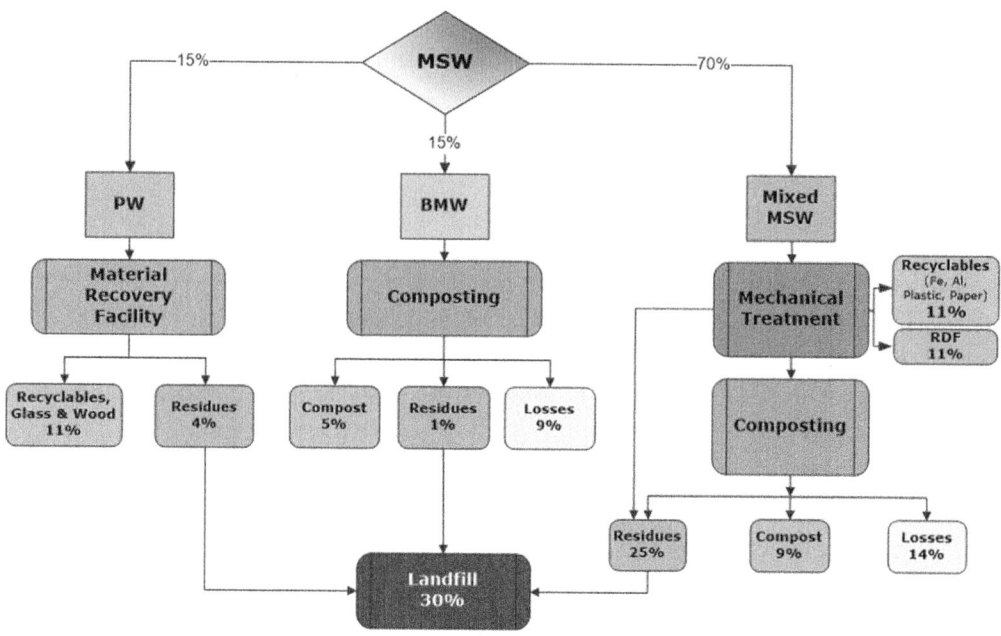

Figure 5.31
Example of an integrated waste management scenario with material balance [79].

Belgium [83]. Moreover, the nominal capacity of a plant determines the final capital expenditure. The unit cost per metric ton of an MBT facility is lower when the capacity increased and economies of scale are achieved. For instance, the investment cost for an anaerobic MBT in the United Kingdom is 340–530 €/ton for a capacity of <50,000 tons/year and 150–417 €/ton for a capacity of >50,000 tons/year. The values reported in Tables 5.5–5.7, are only approximate aiming at giving an overview of the required capital expenditure [58,79,84,85].

The degree and effectiveness of source separation influences both the total amount and the quality of secondary materials. Recyclables recovered from mixed MSW are of lower

Table 5.5: Estimated Waste Management Costs by Income Level in US $/ton [58].

	Low Income	Lower Middle Income	Upper Middle Income	High Income
Collection	20–50	30–75	40–90	85–250
Sanitary landfill	10–30	15–40	25–65	40–100
Composting	5–30	10–40	20–75	35–90
Incineration	—	40–100	60–150	70–200
Anaerobic digestion	—	20–80	50–100	65–150

Table 5.6: Capital and Operational Cost (€/ton) of Various Municipal Solid Waste (MSW) Management Technologies in Europe [79,84,85].

Treatment Technology	Capital Cost (€/ton)	Operational Cost (€/ton)
Anaerobic digestion	366–411	70–78
Composting	98–245	30–72
MRF	87–193	30–46
MBT—AD	150–530	24–104
MBT—composting	42–338	30–104
Biological drying	180–208	55
Incineration	645–691	92–107

MRF, materials recovery facility; *MBT*, mechanical biological treatment; *AD*, anaerobic digestion.

Table 5.7: Capital and Operational Cost (€/ton) of Collection and Transportation of Municipal Solid Waste (MSW) in Romania [83,86].

Parameter	Investment Cost (€/ton)	Operational Cost (€/ton)
Residual waste collection and transport	74–101	42–55
Separate packaging waste collection and transport	140–154	75–81
Separate biowaste collection and transport	68–143	40–73
Bulky waste collection and transport	67–149	46–80
Hazardous (household) waste collection and transport	522–724	1276–1383
Waste transfer station	23–29	6–7
Waste transport from transfer stations to the landfill	10–13	8–10

quality reducing marketing possibilities and value. Furthermore, revenue from recyclables is not always guaranteed as recyclables market can be volatile following the trends of the economy. After the advent of 2008 global economic crisis, recycled materials market values fluctuate widely and thus recycling programs need to have the flexibility to cope with price volatility. However, it should be noted that the indirect revenue coming from avoided disposal costs might be higher than the direct one coming from recyclables market. Indicative recyclables prices are provided in Table 5.8, and they offer only an order of magnitude as general guidance [84–87].

Table 5.8: Recyclables Prices in
Europe [84–87].

Recyclable	€/ton
Plastics	60–535
Metal (mixed)	140–160
Ferrous metals	40–107
Aluminum	100–650
Glass	13–54
Paper	20–308

5.8.4 Legislation

Implementation of waste policies is one of the European Commission's key priorities as confirmed by the 7th Environment Action Programme and the Roadmap to a resource efficient Europe. EU policy on waste sets high targets and is based on the waste hierarchy principle, which prioritizes waste prevention, followed by reuse, recycling, recovery, and final disposal or landfilling as the least desirable option [1,88]. EU waste policy is a key driver to more sustainable consumption and production patterns. The Waste Framework Directive (2008/98/EC) sets the basic concepts and definitions of waste management as well as the certain waste management principles. Waste valorization as material and energy resource is one of the key element toward a resource-efficient and circular economy. Waste prevention is placed at the top of the hierarchy followed by material recycling/recovery and energy recovery. Finally, the least desirable option is the landfilling of waste.

EU waste management legislation is structured in the following three main categories:

- horizontal legislation, establishing the overall framework for the management of wastes
- legislation on treatment and disposal operations, such as landfill or incineration
- legislation on specific waste streams, such as waste oil or batteries

Fig. 5.32 presents the EU legislation related to waste. It is important to mention that legislation regarding issues other than waste, even if indirectly related with waste, are not included. Also, the grouping of legislative acts in this graph is done by the author for convenience and is not the one used by EU.

5.9 Water Resources
5.9.1 EU Water Policy

The implementation of water resources policy in the EU has significantly contributed toward the improvement of the quality of water bodies in Europe, including urbanized

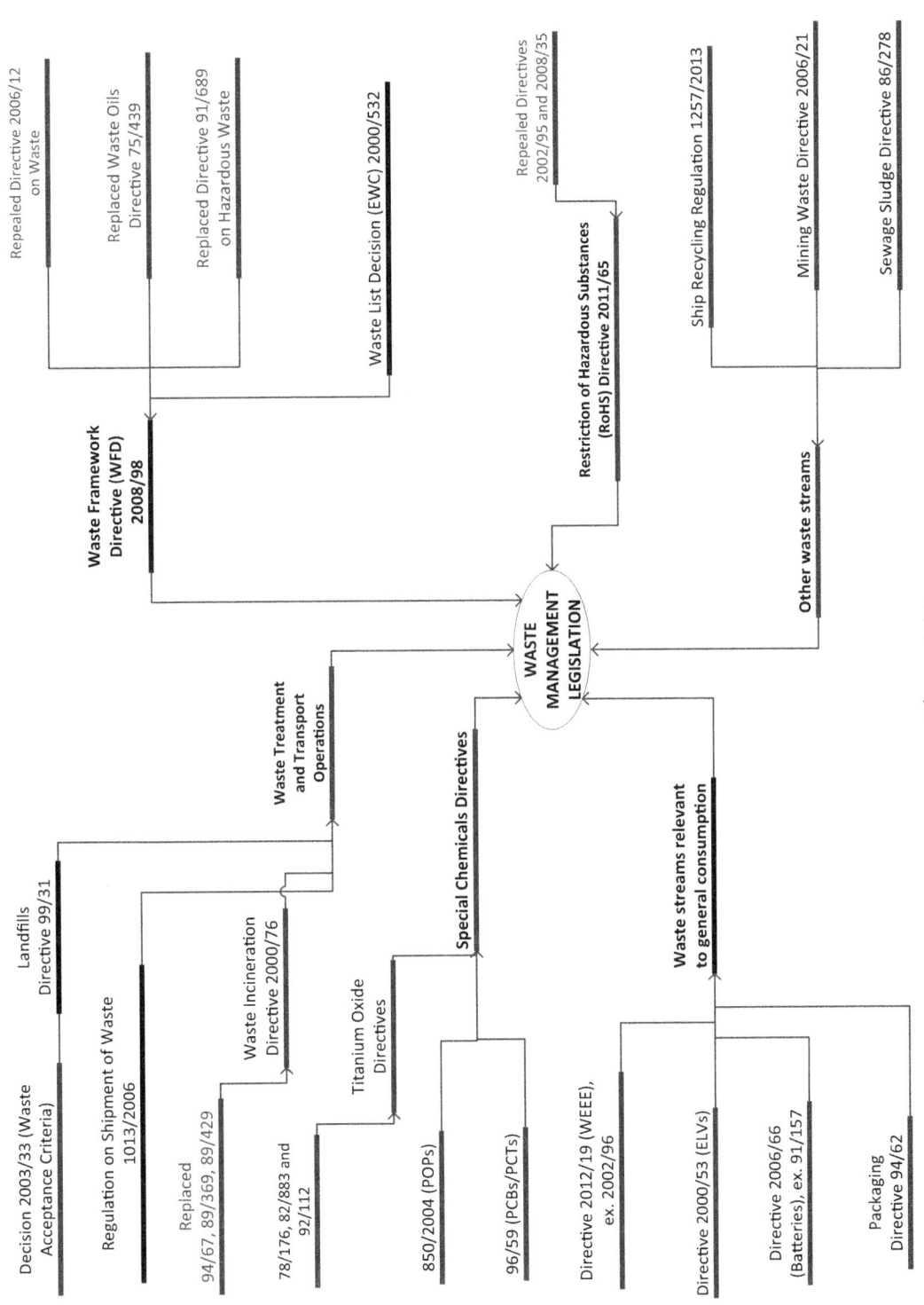

Figure 5.32
Overview of EU waste legislation.

areas. Environmental policies have resulted in significant improvements in drinking water, bathing water quality, and sanitation, while exposure to key hazardous pollutants has been reduced. The EU Water Framework Directive (WFD-Directive 2000/60/EC) [89] takes a pioneering approach to protecting water bodies based on the river basins. This Directive adopts a holistic and innovative approach for the protection and the management of water resources in an effort to integrate previous European Directives related to water resources management. The WFD does not replace or dismiss previous EU water legislation; on the contrary, it forms the water legislative umbrella, integrating earlier EU water legislation that are in place, such as those concerning drinking water, urban wastewater, bathing water, groundwater, and others. The WFD aims to protect all types of water resources, including surface, groundwater, transitional and coastal waters in order to ensure at least good water status up to 2015. The river basin (catchment) is the basic unit for all the actions that need to be taken concerning water. Member States must first identify and analyze European waters, by individual river basins and water districts [90].

The main objectives of the WFD are (1) to prevent further degradation of water resources and improve their quality in order to achieve at least good water status for all water bodies by 2015, (2) to promote the viable management of waters through the long-term protection of water resources, (3) to improve the water quality through the implementation of specific measures concerning the gradual decrease of the discharge of priority pollutants, and (4) to deal with extreme weather conditions, such as floods and droughts.

According to the requirements of the WFD, the surface water quality is assessed based on chemical (good or poor) and ecological (high, good, moderate, poor, or bad) terms, while groundwater quality is assessed based on its chemical (good or poor) and quantitative (good or poor) status. Environmental quality standards (EQS) have been established for 45 chemical pollutants of high concern across the EU in order to define good chemical status. The list of EQS was initially specified by the Directive 2008/105/EC, which included 33 priority pollutants and was amended by the more recent Directive 2013/39/EU, which added 12 more compounds. In terms of groundwater assessment, the EC Member States should use geological data in order to identify distinct water volumes in aquifers and limit their abstraction to only a proportion of their annual recharge. In 2012, the EC published data on the quality of its water bodies. It was reported that 43% of the surface water bodies were in "good status" in 2009, while an increase to 53% was projected by 2015 [91].

The emerging EU environmental legislation paradigm internalizes principles of integrated water management and pollution prevention following a complete concept of sustainability. Brunner and Starkl [92] stated that not only ecological or economic aspects, but also social and institutional ones, such as the participation of stakeholders, are taken into consideration. The European Innovation Partnership on Water [93] aims to enhance water innovation in Europe, to promote the penetration of water innovation by the industry and the public, and to support the implementation of water policy. Water innovation is expected to increase job opportunities in the water sector in the future [94].

In terms of sanitation in cities, the Urban Waste Water Treatment Directive 91/271/EEC [95] has resulted in a significant improvement in the management of sewage. This Directive has set the requirement of collecting and treating sewage in agglomerations having a population equivalent higher than 2000. According to the Directive's requirements all agglomerations with a population equivalent higher than 2000 should apply secondary treatment of sewage, while in the case of agglomerations with a population equivalent higher than 10,000 in designated sensitive areas and their catchments higher level of treatment is required. The Urban Waste Water Treatment Directive sets targets to be achieved concerning the removal of important pollutants, including the biochemical oxygen demand, the chemical oxygen demand, and the total suspended solids. When the population equivalent of the wastewater treatment plant (WWTP) is significant ($>10,000$) and the plant discharges the treated effluent into sensitive recipients, limits for nitrogen and phosphorous also hold. The requirements concerning the pollutants are expressed either as the minimum level of removal efficiency of the pollutant that the plant should accomplish (eg, 80% removal of TSS) or as the limit value of the concentration of the pollutant in the treated effluent, which should not be exceeded. Table 5.9 summarizes the requirements of Directive 91/271/EEC with respect to pollutants removal from sewage discharged after treatment into water recipients [95].

5.9.2 The Impact of Urbanization on Water Resources

A sufficient supply of good quality water for drinking and other uses is vital for cities. There is a wide range of the per capita water consumption in European cities, which ranges from 150 to 400 L/capita/day [96]. The increase of the living standards has resulted

Table 5.9: Requirements of Directive 91/271/EEC Concerning the Discharge of the Treated Effluent to Water Recipients [95].

Parameter	Concentration (mg/L)	Minimum Reduction of Pollutant Load (%)
BOD_5	25	70–90
COD	125	75
TSS (optional)	35	90
N	15 (10,000–100,000 PE) 10 ($>$100,000 PE)	70–80
P	2 (10,000–100,000 PE) 1 ($>$100,000 PE)	80

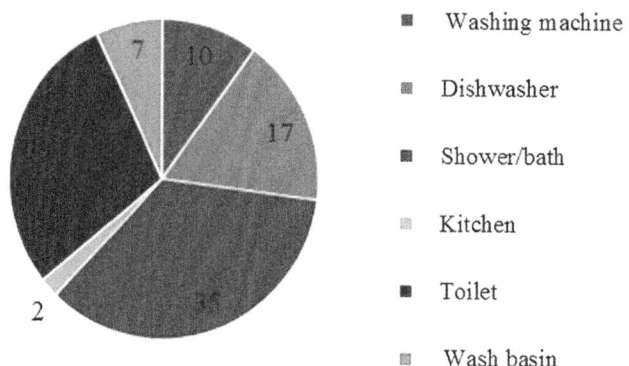

- ■ Washing machine
- ▨ Dishwasher
- ■ Shower/bath
- ▨ Kitchen
- ■ Toilet
- ▨ Wash basin

Figure 5.33
Proportion of different uses for domestic water consumption in a typical household in Europe.

in an increase of the per capita consumption of water. For example, most residents in developed countries have washing machines and dishwashers which increase domestic water consumption [97]. Fig. 5.33 shows the proportion of water consumption in a typical household. It is clearly seen that a significant proportion of household water is consumed for toilet flushing and for the shower/bath use. However, residential use is only a small proportion of the water use. Water is used for several purposes, which include domestic water supply, municipal, commercial, industrial, and agricultural use [98].

The increased urbanization has changed the land cover types in cities from permeable land to anthropogenic impervious surfaces [99]. In cities the high degree of coverage of land by buildings has dramatically increased the volumes of water runoff. Impervious surfaces (eg, constructed surfaces, roofs, roads, parking lots, driveways, sidewalks) alter the natural hydrological conditions by increasing the volume and rate of surface runoff and decreasing groundwater recharge and base flow [100]. This results in more extensive and frequent local flooding and reduced water supplies for urban and suburban areas [101]. Moreover, other direct environmental impacts include degradation of water resources quality when surface runoff transports nonpoint source pollutants from their source areas to receiving lakes and streams [102,103]. Pollutants which are either dissolved or suspended in water or are associated with sediment, including nutrients, heavy metals, oil and grease, and hydrocarbons can accumulate and be washed away from impervious surfaces by runoff.

The changes in landscape due to urbanization result in significant pressure to aquatic ecosystems. The composition of land in a river basin can critically affect the water quality and thus impact on the aquatic ecosystems. The exact way in which the urban environment impacts on ecological conditions is addressed in a few studies [104]. The urban development decreases the land permeability since land that has the ability to drain water is converted to concrete impermeable infrastructure such as roads, pavements, buildings, which are coupled with suitable drainage [105]. Consequently, urbanization removes the natural vegetation that intercepts, decelerates stormwater flow, and returns rainfall to the air through evapotranspiration. During development, the natural surface vegetated soils are removed and the subsoil is compacted. All these processes reduce the amount of water that can infiltrate into the ground and significantly increase the rate of surface runoff. Anthropogenic activities result in the generation of many contaminants which can find their way into the water cycle, detrimentally affecting human health and the environment [106]. Erosion poses additional problems, particularly if the sediments that are washed out originate from contaminated land, construction, or land restoration sites. The loss of topsoil and vegetation deprives an effective, natural mechanism for limiting surface runoff and for filtering and thus treating the produced water volumes through soil infiltration. The conventional drainage systems are simply designed to convey surface runoff away as quickly as possible without any treatment. Consequently, these pollutants are transferred to the receiving water bodies [107].

Urban water cycles are dominated by water runoff. In combined sewer systems, stormwater is collected together with sewage. In such systems the stormwater is a major input to the sewerage network. An increase in the stormwater runoff volume decreases the collection system's capacity with a higher risk of overflow. Moreover, urbanization requires the expansion of wastewater service areas through new development and incorporation of the expanding suburbs [98]. Higher treatment cost required for combined systems has led several countries to turn to the separate collection systems (ie, separate collection of sewage and stormwater) since this is a more cost-effective option [108]. In London, the low capacity of the combined sewer network results in the discharge of millions of cubic meters of untreated, raw sewage into River Thames. In the United States, the enforcement actions for violations of water quality standards due to combined sewer overflows (CSOs) under the Clean Water Act (CWA) have begun to increase in recent years [109]. However, combined systems are still in use in many cities across the world, and the overflow that takes place during floods is a serious environmental burden.

Several studies have determined correlations between builtup land surface, population density, and water quality. Moreover, correlations have been identified between several pollutant loadings and the extent of impervious surface within subdrainage basin with the population density to present apparent correlations with loading rates of various pollutants [110]. The locations of sources of degraded stormwater runoff quality within drainage basins should be mapped in order to identify hazardous areas and protect the water bodies [111]. The use of satellite remote sensing in combination with regression analysis for predicting surface pollutant loadings associated with urban development has been proved to be an effective approach [112,113].

The increase in municipal water demand causes cities to search for new adequate water sources, leading to the creation of sometimes quite complex systems of urban water infrastructure [114,115]. Accounting for urban water infrastructure is essential for estimating the urban population in water stress. McDonald et al. [116] carried out the first global survey of the large cities' water sources. The survey showed that the global hydrologic models which ignored urban water infrastructure significantly overestimated the urban water stress. New developments directly affect the existing drainage infrastructure and the surrounding environment [117]. The increase of the area of paved surfaces reduces infiltration, while causing surface runoff to exhibit higher peak flows, larger volumes, shorter times to peak, fast transport of pollutants, and sediment from urban areas [118−120]. Thus, the latter leads to pollution of the receiving water bodies and increased flood risk within the area of the new development. One option for the reintegration of urban water cycles for sustainable provision of ecosystem services is the strategic implementation of green infrastructure. Examples of green infrastructure include [98,121]:

• Capture and containment of water in earthen, rain barrel, or larger capacity cisterns
• Infiltration of water from the surface into the soil: the water can be stored as moisture in the soil, as groundwater or as plant root uptake.

The urban water supply vulnerability represents the failure of an urban supply basin to meet demands from human, environmental, and agricultural users. In a recent study, a global analysis of urban water supply vulnerability was performed in 70 surface water−supplied cities, with population exceeding 750,000 [122]. According to projections, by 2040, if there are no measures implemented, 44% of the cities are vulnerable due to increased agricultural and urban water demands; out of the 31 vulnerable cities, 13 cities would reduce their vulnerability via reallocating water by reducing environmental flows, and 15 cities would similarly benefit by transferring water from irrigated agriculture [122].

Fig. 5.34 shows the change in the runoff flow as a result of urban development in a region. It is evident that the presence of vegetation and soil results in a slightly higher base flow but in a much lower peak flow which is also less rapid compared to the case where urban development increases land occupation by concrete.

The alteration of natural flow patterns (in terms of both the total quantity of runoff and the peak runoff rates) may lead to flooding and channel erosion downstream of the development. The decrease in percolation into the soil can lead to low base flows in watercourses, reduced aquifer recharge, and damage to in-stream and streamside habitats [105].

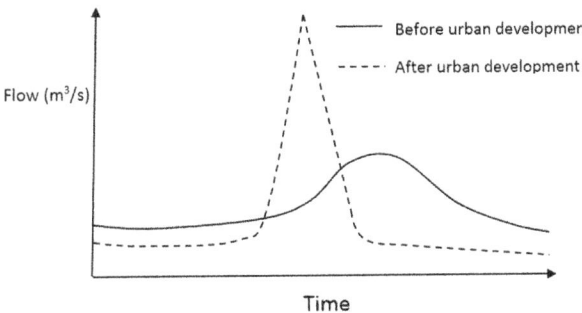

Figure 5.34
Runoff in a region before and after urban development.

The impact of urbanization on water resources can be summarized to the following [106]:

- Stream flow changes: Infiltration into the ground reduces and consequently the drainage speed increases. This causes increased runoff volumes and peak rates, resulting in higher risk of flooding of the downstream areas.
- Stream morphology alternation: The instability of the stream profile increases and as a result the erosion rates increase. The channel becomes wider in order to convey the increased runoff volume; the bed profiles change based on the increased sediment load.
- Disruption of aquatic ecosystems: The new flow conditions (increased flow rates, increased erosion, and sediment transport) can lead to a distortion of the balance of ecosystems, leading to a decrease in biodiversity.
- Undermining of water quality: The increased runoff can wash out several pollutants from impermeable sources such as pavements and roads. The pollutants which are washed away by runoff can include nutrients, heavy metals, hydrocarbons, oil and grease pesticides, litter, pathogens, and others and end up in water recipients causing eutrophication, toxicity, and oxygen-depletion problems.

The above pollutants are collectively termed "urban diffuse pollution" (nonpoint) as they are not produced from a single source or activity, but are the result from all the land use and human activity in the urban area. Rainwater results in runoff volumes which wash out and thus mobilize these pollutants which eventually end up in water recipients. Diffuse pollution produced from urban surface runoff is recognized as one of the major factors that can severely undermine the quality of receiving water bodies [123,124]. The surface water recipients receive the stormwater runoff through the local urban drainage system. The groundwater is affected by the seepage water, which is filtered through the soil. The filtration process is also a natural treatment process, improving the quality of water that ends up in aquifers. The level of purification that is accomplished depends on the retention capacity of the subsurface soil for the specific pollutant. The comparative investigation of the degree of environmental hazards on surface water, seepage water, and groundwater has been investigated through long-term numerical modeling [125]. Table 5.10 summarizes the sources of pollution from impermeable surfaces and the typical pollutant causes.

A significant part of the urban water that is used ends up as wastewater which needs to be properly collected and treated before its discharge. The typical water infrastructure for a city includes (1) water abstraction from a river/lake/reservoir—in many cases this water body is located away from the city, (2) conveyance of the abstracted water to the water treatment plant through the aqueduct including pipes, ditches, canals, tunnels, (3) water treatment plant in order to bring the water up to the standards required for use (ie, drinking water standards), (4) water storage, (5) water distribution system, which distributes the water to households, commercial buildings, industries, and other users, (6) sewerage system for the collection of sewage and its conveyance to the WWTP, and (7) sewage treatment works for the purification of sewage which can then be safely disposed to the water recipients or reused.

An important source of pollution in certain cities and particularly in London are misconnections. Misconnections occur when a drainage or a pipe intended for one type of effluent conveyance has been connected to the wrong sewer system. The first type of misconnection is when sewage is sent to the stormwater collection system instead of the sewerage network.

The second type of misconnection is when the stormwater is connected to the sewerage network, leading to hydraulic overloading as well as placing increased burdens on treatment process downstream. The term misconnection is consequently misleading, as it covers a variety of urban sources including "graywater" (residential sewage without fecal contamination, ie, from sinks, shower, and bath), "blackwater" (ie, wastewater containing feces, urine, and flush water from toilets) discharges, as well as cross-connections between surface water and sewers. Other potential sources of sewage to the stormwater network can be spillages from septic tanks, industrial processes and sewer lines, water from vehicle wash, and contaminated groundwater. A principal focus in the misconnection issue has been on the wrong connection of household sewage discharge into the separate surface water collection system [126]. The polluted stormwater outfall discharges from such misconnections can result in failing to meet the required quality standards and can have an adverse ecological and aesthetic impact. In the United Kingdom and particularly in London urban water pollution by misconnections is one of the most important sources of aquatic pollution [127]. In the United States, similar problems posed to receiving water bodies by such illegal sewage discharge into stormwater collection systems have also been reported [128], and households with misconnections have been targeted as offenders [129].

Table 5.10: Sources of Pollution From Impermeable Surfaces [106].

Source	Typical Pollutants	Main Observation
Atmospheric deposition	Phosphorous, nitrogen, sulfur Heavy metals (lead, cadmium, copper, nickel, zinc, mercury) Hydrocarbons	Industrial activities, traffic air pollution, and agricultural activities all contribute to atmospheric pollution. This is deposited as particulates. Rain also absorbs atmospheric pollutants
Traffic—exhaust	Hydrocarbons MTBE Cadmium, platinum, palladium, rhodium	Vehicle emissions include polycyclic aromatic hydrocarbons (PAHs) and unburned fuel and particles from catalytic converters
Traffic—wear and corrosion	Sediment Heavy metals (lead, chromium, copper, nickel, zinc)	Abrasion of tires and corrosion of vehicles deposits pollutants onto the road or car parking surfaces
Leaks and spillages (eg, from road vehicles)	Hydrocarbons Phosphates Heavy metals Glycol, alcohols	Engines leak oil, hydraulic and deicing fluids and spillages occur when refueling. Lubricating oil can contain phosphates and metals. Accidental spillages can also occur
Roofs—atmospheric depositions, bird droppings, corrosion, and vegetation	Heavy metals (copper, lead, zinc) Bacteria Organic matter	Roof water is often regarded as clean. It can, however, contain significant concentrations of heavy metals resulting from atmospheric deposition or the corrosion of metal roofing or from other coatings such as tar
Litter/animal feces	Bacteria, viruses Phosphorous, nitrogen	Litter typically includes items such as drinks cans, paper, food, cigarettes, animal excreta, plastic, and glass. Some of this will break down and cause pollutants to be washed off urban surfaces. Dead animals in road decompose and release pollutants including bacteria. Pets leave feces that wash into the drainage system

Continued

Table 5.10: Sources of Pollution From Impermeable Surfaces [106].—cont'd

Source	Typical Pollutants	Main Observation
Vegetation/ landscape maintenance	Phosphorous, nitrogen Herbicides, insecticides, fungicides Organic matter	Leaves and grass cuttings are an organic source. Herbicides and pesticides used for weed and pest control in landscaped areas, such as garden, parks, recreation areas, and golf courses, can be a major source of pollution
Soil erosion	Sediment Phosphorous, nitrogen Herbicides, insecticides, fungicides	Runoff from poorly detailed landscaped or other areas can wash onto impervious surfaces and cause pollution of runoff
Deicing activities	Sediment Chloride, sulfate Heavy metals (iron, nickel, lead, zinc), glycol Cyanide Phosphate	Deicing salt is commonly used for deicing roads and car parks. Rock salt used for this purpose comprises sodium chloride and grit. It can also include cyanide and phosphates as anticaking and corrosion inhibitors, heavy metals, urea, and ethylene glycol
Cleaning activities	Sediment Phosphorous, nitrogen Detergents	Washing vehicles, windows, bins, or pressure washing hardstandings lead to silt, organic matter, and detergents entering the surface water drainage
Wrong sewer connections	Bacteria Detergents Organic matter	Wrong connections of foul sewers to surface water sewers where separate sewers exist
Illegal disposal of chemicals and oil	Hydrocarbons Various chemicals	Illegal disposal of used engine oils or other chemicals can occur at small (domestic) or large (industrial) scales

BOD, biochemical oxygen demand; *COD*, chemical oxygen demand; *TSS*, total suspended solids.

Although misconnections are considered to be a major problem in certain cities the severity of the problem is not certain [128,130].

In the United States and the United Kingdom the majority of illegal discharges of sewage into the stormwater network are not detected because of the absence of survey and monitoring data in the urban stormwater pipes. It is very difficult to locate the source of the misconnection and to evaluate the impact of stormwater (polluted with sewage) outflow on the water quality of a receiving water body. In order to eliminate illegal discharges it is important to backtrack the source and apply stringent compliance procedures [126]. These actions must be coupled with public awareness campaigns targeted to households so that the residents become aware of the nature and magnitude of the problem. The cost of the surface water collection system misconnections to the UK water industry is approximately £235 million /per year in terms of asset management and maintenance and equivalent annual value.

Several works have investigated the impact of urban runoff pollution to water recipients. Gnecco et al. [131] investigated stormwater pollution in the urban environment of Genoa by monitoring the water quality produced from road and roof runoff. The authors concluded that the TSS and COD concentration values exceed the water quality standards for road runoff, while for roof runoff the concentration of zinc was significant. Göbel et al. [132] carried out an extensive literature review on the distribution and concentration of surface-dependent runoff water in urban areas. The study concluded that roof materials and traffic densities impact on the pollutant distribution and composition. Out of the examination of 12 different types of roofs, green roofs resulted in the lowest amount of heavy metals in the resulting runoff, with the exception of roofs made of zinc and copper, which resulted in the highest content of heavy metals in the resulting runoff. As far as traffic density is concerned, the results showed that the motorways result in the highest runoff pollution for 13 parameters (BOD_5, COD, Cd, Pb, Ni, Cr, Na, Mg, Ca, K, SO_4^{2-}, Cl$^-$, and mineral oil hydrocarbons) in measurable concentrations. The service roads exhibited the highest PAH concentrations, while pedestrian and cycle ways, as well as yards, exhibited the highest zinc concentrations. Table 5.11 summarizes target substances that have been used as markers for the identification of diffuse urban pollution [133−136].

5.9.3 Water and Wastewater Treatment

Appropriate sanitation is a basic requirement in urbanized communities that protects humans and the environment. Waterborne diseases can be transmitted from water to humans causing serious diseases and even death. These are caused by pathogenic microorganisms which are transmitted to water and can then infect humans during various activities such as bathing, washing, drinking, and the consumption of infected food. Waterborne diseases can be caused by bacteria (eg, *Escherichia coli* infection, cholera,

Table 5.11: Correlation of Substances With the Type of Pollution Based on Data Reported in Literature [131−134].

Type of Pollution	Target Substance	Main Finding	Source
Urban runoff	Cd, Cu, Pb, Ni, Zn, TPH, total PAH	• Degradation of organic pollutants in submerged sediments is slower than in exposed soil • Most pollutants are retained in the top 10 cm of soil	[128]
Urban runoff	pH, electrical conductivity, Dissolved oxygen (DO), TSS, Total petroleum hydrocarbons (TPH)	• Filtering effect of the SUDS is confirmed • High impact registered on water pollutants associated with solid particles	[129]
Urban runoff	COD, TSS, nitrogen, phosphorous	• Pollutant load increased after the treatment during the start-up period. The high loads were attributed to the residues from the construction process and the long antecedent dry period	[130]
Jet fuel−contaminated runoff	Various hydrocarbons	• Subsurface infiltration of jet fuel−contaminated surface runoff is a feasible treatment method • Biodegradation is the main process responsible for the removal of hydrocarbons	[131]

dysentery, leptospirosis, salmonellosis, typhoid fever), parasites (eg, schistosomiasis) protozoa, viruses (Hepatitis A), and even algae. To protect the population, drinking water must be effectively treated, including a final disinfection step in order to destroy pathogens. In the 19th century, several cholera outbreaks were reported in London which were afterward related to the quality of drinking water. In 1993 the cryptosporidiosis outbreak in Milwaukee resulted in 403,000 people becoming ill, with more than 100 deaths. Following this incident, more effective treatment of the water has taken place.

The main stages in a drinking water treatment plant include [137]:

- *Predisinfection*: This takes place only in certain water treatment plants in order to limit the pathogenic microorganisms which enter the plant and eliminate the growth of algae and anaerobic microorganisms in the downstream operations of the water treatment plant. Usually chlorine or ozone is the disinfectant which is added.
- *Coagulation/flocculation*: In this process coagulants are added to water in order to destabilize the colloidal particles that are contained in water and thus enhance

flocculation and the formation of larger colloids which can settle. The coagulants neutralize the negative charge of colloids, allowing them to form larger flocs. Alum $(Al_2(SO_4)_3 \cdot 14H_2O)$, ferric chloride $(FeCl_3)$, ferrous sulfate $(FeSO_4)$, ferric sulfate $(Fe_2(SO_4)_3)$, polyaluminum chloride (PAC), and cationic and anionic polyelectrolytes are the most common coagulants. The mixing of the coagulant with the water takes places in two stages: in the first stage intense agitation takes place to ensure the homogeneous dispersion of the coagulant. This is followed by slow mixing to facilitate the formation of large flocs.

- *Clarification*: After coagulation the clarification process takes place in order to allow the large flocs which have been formed to settle. This is carried out in a settling/clarification tank. In some plants the flocculation and clarification processes steps are carried out in a single process known as clari-flocculation.

- *Filtration*: This step is to remove the remaining suspended solids from water. Filtration is usually carried out at constant flow and head by adjusting the aperture of a valve located downstream of the filters. Filtration is carried out under gravity and the packing material that is used for the filters usually consists of sand and anthracite. The filters are regularly backwashed to clean them from the clogged particles.

- *Disinfection*: This step is to destroy the pathogens that are contained in water. The disinfection process can be chlorination, ozonation, or ultraviolet (UV) radiation. Chlorination is the most widely practiced disinfection method. However, ozonation and UV disinfection are also practiced in several plants. Chlorination has the advantage of residual chlorine which remains in water after chlorine application which acts as a safeguard against additional microbial growth. However, the reaction of chlorine with organic substances can produce trihalomethanes which are considered to be potential carcinogenic.

Advanced treatment of the water may include the use of activated carbon for the removal of specific organic substances such as pesticides and reverse osmosis to removal salts such as sodium chloride from water.

In urban wastewater treatment, as a result of the implementation of Directive 91/271/EEC, many WWTPs were developed in the EC member states. Currently, $\sim 22,000$ municipal WWTPs with treatment capacity higher than 2000 population equivalents are operating in the 28 EU member states, Iceland, Norway, and Switzerland. By 2017, investments in European WWTPs are expected to reach 37.6 billion euro. The Urban Waste Water Treatment Directive is the most important driving Legislation for the construction of new WWTPs and upgrading of existing WWTPs. A typical municipal WWTP consists of the following stages:

- *Pretreatment*: It consists of screening to remove coarse material which could damage downstream equipment, sand removal which is usually accomplished through an aerated chamber to favor settling of sand, and suspension of organics, oil and grease removal.

- *Primary sedimentation*: It removes a significant part of the suspended solids (50–70%) and of the organic matter (30–50%). The primary sedimentation tanks have either a rectangular or a circular plan view. Primary sludge settles at the bottom part of the tank, while oil and grease rise at the top and are skimmed off. The primary sedimentation tank is equipped with mechanically driven scrapers that continually drive the collected sludge toward a hopper located at the base of the tank where it is pumped to sludge treatment facilities.

- *Secondary/biological treatment*: This is the "heart" of the wastewater treatment process. Wastewater comes into contact with microorganisms (mainly bacteria) inside biological reactors. In the presence of oxygen the bacteria oxidize the organic carbon present in wastewater to carbon dioxide and water and thus produce energy (catabolic reactions), which is subsequently used for the bacteria to multiply (anabolic reactions). Oxygen is either provided through blowers and suitable diffusers located at the bottom of the reactor or is induced from the surface through surface aerators. When the microorganisms are suspended in the reactor, the biological process is known as suspended growth, while when the microorganisms are attached to a medium the process is known as attached growth process. The activated sludge process is a suspended growth process and is the most widely applied process for the biological treatment of wastewater. The separation of the microorganisms from the treated effluent is carried out in secondary sedimentation tanks. Attached growth processes that are applied include trickling filters, rotating biological contactors, and moving bed biofilm reactors.

- *Filtration*: The secondary effluent can be further treated by gravity filters or by ultrafiltration/microfiltration membranes in order to completely reject suspended solids.

- *Disinfection*: The final step of a WWTP is the disinfection of the treated effluent before it is discharged. As in the drinking water scheme disinfection can be accomplished by chlorination (most usual), UV irradiation, and ozonation.

- *Sludge treatment works*: The primary and secondary (activated) sludge are further treated. Both types of sludge are thickened to increase their solids concentration and can be treated by anaerobic digestion to produce biogas and recover energy using CHP. The digested sludge is dewatered to further increase their solids concentration. Dewatered sludge can then be further treated by composting, drying, etc. to further stabilize it so that it can then be used as a soil conditioner.

5.9.4 Measures to Cope With the Urban Water Pollution

The main components of the urban water cycle consist of water supply, wastewater disposal, and stormwater drainage (urban water system). Climate change impacts make some of the urbanization consequences more severe, which are difficult to be assessed due to their uncertain nature. In several studies the lack of understanding of the impacts of climate change and urbanization is a strong barrier in establishing an effective integrated urban drainage implementation plan. Enhancing and utilizing the natural water cycle is an important step toward more sustainable urban environments. A limited number of studies have focused on urban catchments [138,139], while urbanization impacts in urban areas have been rarely assessed. Recent studies have revealed that there is a lack of case studies evaluating the drainage system's performance using combined climate change and urbanization-induced drivers [140].

Several EU projects evaluate urban water management policy and infrastructure reforms. SWITCH [141] has examined and developed practices for urban water management focusing on flexibility and sustainability. Transitions to the Urban Water Services of Tomorrow (TRUST) [142] has developed a multinational network towards nine pilot conurbations or urbanized catchments that are grouped into green cities, water scarce regions, and urban/periurban metropolitan areas. The Integrated Urban Watershed Management methodology is most practically achieved in the context of Integrated Catchment Management (ICM) [143,144]. The low impact urban design and development (LIUDD) is coming to incorporate integrated land and water use design on neighborhood-to-catchment scales using an ICM approach [145]. Increased neighborhood water cycle self-sufficiency can be a challenge for planners [146]. Yazdanfar and Sharma [140] summarize selected adaptive measures that are required for the provision of sustainable drainage systems to cope with combined challenges of climate change and urbanization. This provides water planners and decision-makers with state-of-the-art information on technologies and adaptation options to increase drainage systems efficiency under changing climate and urbanization. There is an increasing need to investigate and exploit potential interactions between the main components of urban water [120].

Water management measures and practice for new developments must be based on social, economic, and environmental sustainability aspects [147,148]. The main target of conventional urban water management practices is to meet water demand, while conveying wastewater and stormwater away from urban settings. On the other hand, alternative approaches consider water demands to be manageable and wastewater and stormwater as valuable resources. The development and application of a Decision Support Tool (DST) can facilitate decision making [149]. A DST was developed and applied in the United Kingdom [120], which enables the selection of combinations of water-saving strategies and technologies, supporting the delivery of integrated, sustainable water management for new developments. The introduction of quantitative and qualitative sustainability criteria and indicators facilitated the comparison of alternative composite water management strategies, while preserving the multiobjective nature of the problem. For example, based on the traditional approaches wastewater and rainwater were considered waste streams that should be conveyed away from the urban environment and disposed of. This practice has recently changed; these streams need to be exploited rather than be managed as unavoidable by-products of urbanization. Planning and operating the systems separately creates the potential to treat wastewater and urban stormwater wastewater for beneficial purposes. Another example is related with the water recycling systems and practices that aim to reduce water demand. The reuse of treated wastewater for certain urban applications (eg, watering of parks, cemeteries, etc.) can reduce the demand of a city for fresh water resources. Adoption of strategies that reduce water demand will positively affect the wastewater conveyance systems, while water-recycling techniques will result in the reduction of water demand. Consequently, water supply, wastewater disposal, and stormwater drainage must be considered as interacting components within a single system [150] identifying the future possibilities and the limitations of different systems within the context of sustainable water management for new developments [120]. One major issue in urban water resources management is whether it is feasible to attain sustainable management through improving existing centralized systems or whether it is necessary to develop new decentralized systems or even if there is an optimal combination of decentralized and centralized management [151]. In all cases, in terms of planning and decision making, the holistic approach for sustainable water resources management should consider natural resource, and social, cultural, financial, and technical criteria.

Role of the SUDS: The control of surface runoff is a key element toward urban sustainability. This can be achieved either by collecting rainwater for recycling through harvesting with the advantage of the reduced water demand or by the installation of nonpiped solutions to urban drainage (known as Sustainable Urban Drainage Systems (SUDS)) in the United Kingdom [148,152]. SUDS are technically regarded a sequence of management practices, control structures, and strategies designed to efficiently and sustainably drain surface water, while minimizing pollution and managing the impact on water quality of local water bodies. Centralized SUDS can be applied at large scales to serve a region or a catchment area. On the other hand local SUDS can be used at a local scale to serve a property or a small number of properties. The differentiation between SUDS is mainly based on functionality [120].

A Sustainable Urban Drainage System can address many of the previous issues, thus mitigating many of the adverse effects of stormwater runoff on the environment. This is achieved through:

- reducing runoff rates, thus reducing the risk of downstream flooding [106];
- reducing the additional runoff volumes and runoff frequencies that tend to increase as a result of urbanization, and which can exacerbate flood risk and damage receiving water quality [153,154];
- encouraging natural groundwater recharge (where appropriate) to minimize the impacts on aquifers and river base flows in the receiving catchment;

- reducing pollutant concentrations in stormwater, thus protecting the quality of the receiving water body;
- acting as a buffer for accidental spills by preventing a direct discharge of high concentrations of contaminants to the receiving water body;
- reducing the volume of surface water runoff discharged into combined sewer systems, thus reducing discharges of polluted water to watercourses via CSO spills;
- contributing to the enhanced amenity and aesthetic value of developed areas [155]; and
- providing habitats for wildlife in urban areas and opportunities for biodiversity enhancement [40].

In the United Kingdom there are at least 56 case studies for SUDS, without considering all the systems for individual buildings. An example of a success case study is related with the SUDS built in North Hamilton, Leicester. This system is surrounded by a residential development, located on former agricultural grassland. The aim was to mimic natural drainage patterns in order to remove the need for traditional sewers and therefore reducing costs. The use of swales and retention pond was chosen to help solving existing flooding problems. The results have shown an increase in biodiversity and a benefit for the local flood risk management.

5.10 Conclusion

This chapter has discussed the links between urbanization and transport, land use, water resources, air pollution, climate change, and MSW. Urbanization is the result of two basic phenomena, which are the migration of people from rural and other nonurban areas to urban areas and by the population growth. People tend to cluster in agglomerations and particularly in cities due to more favorable working opportunities and a better way of living. An effective urban transportation network is essential in order to facilitate the commuting of people within the city and especially the everyday transport from their homes to their working places and vice versa. Urban activities require and thus cover part of the available land; land is required for various urban activities, which include housing of people, roads, pavements, offices, etc.; the land use for such activities is not reversible. In many cities urban development results in a decrease of green space and limited parking and open space. The various urban activities also result in emissions that undermine the quality of air. The urban air quality is related to many thousand deaths at a global scale. In the last decade the air quality has improved in many developed cities. However, urban air quality is still a significant environmental challenge which needs to be further addressed. Urban areas consume the majority of the resources worldwide and consequently produce large quantities of MSW which should be properly managed and disposed of. Appropriate waste management planning at all levels is essential. Waste prevention, materials recycling, and energy recovery are placed high in the hierarchy of

MSW management in order to minimize the amount of waste which ends up in landfills. Cities need to meet water demand in both time and space and ensure that the quality of water supplied is of adequate quality for the purposes it is used for. At the same time the appropriate sanitation is required to manage properly municipal sewage in order to protect human health and the environment. Both water and sewage need to be effectively treated before the former are used (eg, drinking water) and the latter are disposed of or reused.

Modern urban cities still face significant challenges that need to be addressed in order to provide a better environment for its citizens and protect water, air, and soil. Even developed cities have still a lot to do in order to become sustainable cities, characterized by minimum requirements for energy, water, and food and by minimum production of waste and air pollution.

Acknowledgments

Steve Kershaw thanks the China University of Mining Technology (Beijing) for facilitating observations of urban development in northern China in October 2015. The Rosendale Allotment Society in south London is warmly thanked for allowing access to its property. Grundon waste management company has a long association with Brunel University and visits to view its operations are gratefully acknowledged. Furthermore the authors would like to thank Professor Jean-Paul Rodrigue for providing copyright permission for two of the Figures included in this chapter.

References

[1] EEA, The European Environment — State and Outlook 2010: Land Use, European Environment Agency, 2010.
[2] B. Cohen, Urbanization, city growth, and the new united nations development agenda, Cornerstone Off. J. World Coal Ind. 3 (2015) 4—7.
[3] H. Giradet (Ed.), Bioregional Solutions for Living on One Planet, Green Books Ltd. for the Schumacher Society, Bristol, UK, 1999.
[4] Royal Commission of Environmental pollution, The Urban Environment, Crown, London, UK, 2007.
[5] T.J. Foxon, Inducing Innovation for a Low Carbon Future, Imperial College Centre for Energy Policy and Technology (ICCEPT). Report for Carbon Trust, London, 2003.
[6] B.R. Gurjar, J. Lelieveld, New directions: megacities and global change, Atmos. Environ. 39 (2005) 391—393.
[7] S. Incecik, U. Im, Air pollution in mega cities: a case study of Istanbul, in: M. Khare (Ed.), Air Pollution — Monitoring, Modelling and Health, 2012.
[8] M.G. Lawrence, et al., Regional pollution potentials of megacities and other major population centers, Atmos. Chem. Phys. 7 (2007) 3969—3987.
[9] B.R. Gurjar, et al., Evaluation of emissions and air quality in megacities, Atmos. Environ. 42 (2008) 1593—1606.
[10] FAO, State of the World's Forests, Food and Agriculture Organization, Rome, Italy, 1997.
[11] FAO/UNEP, Terminology for Integrated Resources Planning and Management, Food and Agriculture Organization/United Nations Environmental Programme, Rome, Italy, Nairobi, Kenya, 1999.
[12] Burgess. http://people.hofstra.edu/geotrans/eng/ch6en/conc6en/burgess.html.

[13] NAAQS. National Ambient Air Quality Standards (NAAQS). http://www3.epa.gov/ttn/naaqs/criteria.html.

[14] I.G. Dyominov, A.M. Zadorozhny, Greenhouse gases and recovery of the Earth's ozone layer, Adv. Space Res. 35 (2005) 1369–1374.

[15] V. Ramanathan, Y. Feng, Air pollution, greenhouse gases and climate change: global and regional perspectives, Atmos. Environ. 43 (2009) 37–50.

[16] D.J. Jacob, D.A. Winner, Effect of climate change on air quality, Atmos. Environ. 43 (2009) 51–63.

[17] P.L. Kinney, Climate change, air quality, and human health, Am. J. Prev. Med. 35 (2008) 459–467.

[18] H. Omidvarborna, et al., Recent studies on soot modeling for diesel combustion, Renew. Sustain. Energy Rev. 48 (2015) 635–647.

[19] M. Hardin, R. Kahn, Aerosols and Climate Change, NASA Earth Observatory, 2010.

[20] R.U. Ayres, E.H. Ayres, Crossing the Energy Divide: Moving From Fossil Fuel Dependence to a Clean-Energy Future, Wharton School Publishing, 2009.

[21] EEA, Performance of Water Utilities Beyond Compliance, Sharing Knowledge Bases to Support Environmental and Resource-Efficiency Policies and Technical Improvements, European Environment Agency, 2014. Technical Report, No 5/2014.

[22] B. Brunekreef, S.T. Holgate, Air pollution and health, Lancet 360 (2002) 1233–1242.

[23] M. Peden, et al. (Eds.), World Report on Road Traffic Injury Prevention, World Health Organization, Geneva, Switzerland, 2004.

[24] WHO, Burden of Disease From Household Air Pollution for 2012, (2012).

[25] M.K. Selgrade, Air pollution and respiratory disease: extrapolating from animal models to human health effects, Immunopharmacology 48 (2000) 319–324.

[26] J. Lee, et al., The adverse effects of fine particle air pollution on respiratory function in the elderly, Sci. Total Environ. 385 (2007) 28–36.

[27] B.B. Jalaludin, et al., Acute effects of urban ambient air pollution on respiratory symptoms, asthma medication use, and doctor visits for asthma in a cohort of Australian children, Environ. Res. 95 (2004) 32–42.

[28] EEA, The European Environment – State and Outlook, Synthesis Report, European Environment Agency, 2015.

[29] C.A. Pope, D.W. Dockery, Health effects of fine particulate air pollution: lines that connect, J. Air Waste Manag. Assoc. 56 (2006) 709–742.

[30] C.A. Pope, et al., Health effects of particulate air pollution: time for reassessment? Environ. Health Perspect. 103 (1995) 472–480.

[31] K. Barrett, et al., Health impacts and air pollution – an exploration of factors influencing estimates of air pollution impact upon the health of European citizens, in: European Topic Centre on Air and Climate Change, Bilthoven, 2008.

[32] C.A. Pope, R.E. Kanner, Acute effects of PM_{10} pollution on pulmonary function of smokers with mild to moderate chronic obstructive pulmonary disease, Am. Rev. Respir. Dis. 147 (1993) 1336–1340.

[33] J. Schwartz, Air pollution and daily mortality: a review and meta-analysis, Environ. Res. 64 (1994) 36–52.

[34] K. Katsouyanni, et al., Confounding and effect modification in the short-term effects of ambient particles on total mortality: results from 29 European cities within the APHEA2 project, Epidemiology 12 (2001) 521–531.

[35] EEA, The European Environment – State and Outlook 2010: Consumption and the Environment, European Environment Agency, Copenhagen, 2010b.

[36] Greater London Authority, Cleaning London's Air: The Mayor's Air Quality Strategy, 2002.

[37] COMEAP, Cardiovascular Disease and Air Pollution, TSO, London, 2006.

[38] WHO, Guidelines for Air Quality, WHO, Geneva, Switzerland, 2000.

[39] M. Krzyzanowski, B. Kuna-Dibbert, J. Schneider (Eds.), Health Effects of Transport-Related Air Pollution, WHO, Geneva, Switzerland, 2005.

[40] DEFRA, Improved Air Quality Can Extend Life Expectancy, Defra News Release, UK, April 5, 2006.

[41] C.K. Chan, X. Yao, Review air pollution in mega cities in China, Atmos. Environ. 42 (2008) 1–42.

[42] J. Hao, et al., Air quality impacts of power plant emissions in Beijing, Environ. Pollut. 147 (2007) 401–408.

[43] EC, Directive 2008/50/EC of the European Parliament and of the Council of 21 May 2008 on Ambient Air Quality and Cleaner Air for Europe, 2008.

[44] EC, Directive 2004/107/EC of the European Parliament and of the Council of 15 December 2004. Relating to Arsenic, Cadmium, Mercury, Nickel and Polycyclic Aromatic Hydrocarbons in Ambient Air, 2004.

[45] G. Hoek, et al., A review of land-use regression models to assess spatial variation of outdoor air pollution, Atmos. Environ. 42 (2008) 7561–7578.

[46] Defra and ONS. The Environment in Your Pocket 2005.

[47] SOER, The European Environment; State and Outlook 2010, 2010.

[48] D.D. Parrish, et al., Air quality progress in North American megacities: a review, Atmos. Environ. 45 (2011) 7015–7025.

[49] D.G. Streets, et al., Air quality during the 2008 Beijing Olympic Games, Atmos. Environ. 41 (2007) 480–492.

[50] M. Fang, et al., Managing air quality in a rapidly developing nation, China, Atmos. Environ. 43 (2009) 79–86.

[51] W. Solecki, et al., Urbanization of climate change: responding to a new global challenge, in: E. Sclar, N. Volavka-Close, P. Brown (Eds.), The Urban Transformation: Health, Shelter and Climate Change, Routledge, 2012, pp. 197–220.

[52] A. While, M. Whitehead, Cities, urbanization and climate change, Urban Cities 50 (2013) 1325–1331.

[53] R.J. Nicholls, et al., Climate change and coastal vulnerability assessment: scenarios for integrated assessment, Sustain. Sci. 3 (2008) 89–102.

[54] F. Biermann, I. Boas, Preparing for a warmer world: towards a global governance system to protect climate refugees, Global Environ. Polit. 10 (2010) 60–88.

[55] CDP, CDP Cities 2013, 2013. Summary Report on 110 Cities.

[56] D. Dodman, Blaming cities for climate change? An analysis of urban greenhouse gas emissions inventories, Environ. Urban. 185 (2009) 185–201. International Institute for Environment and Development (IIED).

[57] D. Satterthwaite, Cities' contribution to global warming: notes on the allocation of greenhouse gas emissions, Environ. Urban. 20 (2008) 539–549. International Institute for Environment and Development (IIED).

[58] World Bank, What a Waste: A Global Review of Solid Waste Management, World Bank, 2012. Urban Development Series Knowledge Papers.

[59] Waste Atlas. http://www.atlas.d-waste.com/, 2015.

[60] T. Karak, et al., Municipal solid waste generation, composition, and management: the world scenario, Crit. Rev. Environ. Sci. Technol. 42 (2012) 1509–1630.

[61] USEPA, Municipal Solid Waste Generation, Recycling, and Disposal in the United States, Tables and Figures for 2012, US Environmental Protection Agency and Office of Resource Conservation and Recovery, 2014.

[62] Eurostat, Recycling – Secondary Material Price Indicator, Eurostat, 2015.

[63] DSE, Waste Generation and Resource Recovery in Australia Reporting Period 2010/11, Department of Sustainability Environment Water Population and Communities, 2014.

[64] D.Q. Zhang, et al., Municipal solid waste management in China: status, problems and challenges, J. Environ. Manag. 91 (2010) 1623–1633.

[65] W. Khan, et al., Use of decision support software tool in assessment of practices in solid waste management in Karachi, Pakistan, in: 5th International Symposium on Energy from Biomass and Waste. Venice, Italy, 2014.

[66] P. Chalmin, C. Gaillochet, From Waste to Resource, an Abstract of World Waste Survey, Cyclope, Veolia Environmental Services, 2009. Edition Economica.

[67] EEA, Managing Municipal Solid Waste—A Review of Achievements in 32 European Countries, European Environmental Agency, 2013. Report No 2.

[68] MDEP, Massachusetts 2010−2020 Solid Waste Master Plan, Pathway to Zero Waste, Massachusetts Department of Environmental Protection. Executive Office of Energy and Environmental Affairs, Massachusetts, 2013.

[69] EC, Council Directive 1999/31/EC of 26 April 1999 on the Landfill of Waste, 1999.

[70] V. Inglezakis, et al., Analysis of current situation in municipal waste management and implementation of decision support software in Astana, Kazakhstan, in: 5th International Symposium on Energy from Biomass and Waste. Venice, Italy, 2014.

[71] J.M. Alhumoud, Municipal solid waste recycling in the Gulf Co-operation Council states, Resour. Conserv. Recycl. 45 (2005) 142−158.

[72] R. Al-Jarallah, E. Aleisa, A baseline study characterizing the municipal solid waste in the State of Kuwait, Waste Manag. 34 (2014) 952−960.

[73] M.O. Saeed, et al., Assessment of municipal solid waste generation and recyclable materials potential in Kuala Lumpur, Malaysia, Waste Manag. 29 (2009) 2209−2213.

[74] N. Gupta, et al., A review on current status of municipal solid waste management in India, J. Environ. Sci. 37 (2015) 206−217.

[75] E. Solano, Integrated Solid Waste Management Alternatives in Consideration of Economic and Environmental Factors: A Mathematical Model Development and Evaluation, in Department of Civil Engineering, North Carolina State University, Raleigh, NC, 1999.

[76] W.A. Worrell, P.A. Vesilind, Solid Waste Engineering, second ed., Cengage Learning, USA, 2012.

[77] ETAGIW, Preparing a Waste Management Plan, a Methodological Guidance Note, Directorate-General Environment. European Commission, 2012.

[78] WWF, Standards of Conservation Project and Programme Management, 2006.

[79] BALKWASTE, Guidelines for Development of Alternative Waste Management Scenarios. LIFE$^+$ Project. LIFE07/ENV/RO/686, 2011.

[80] V.J. Inglezakis, K. Moustakas, Household hazardous waste management: a review, J. Environ. Manag. 150 (2015) 310−321.

[81] EC Environment. Biodegradable Waste. http://ec.europa.eu/environment/waste/compost/index.htm.

[82] V.J. Inglezakis, et al., Comparison between landfill gas and waste incineration for power generation in Astana, Kazakhstan, Waste Manag. Res. 33 (2015) 486−494.

[83] Arcadis and Eunomia, Assessment of the Options to Improve the Management of Bio-waste in the European Union, Annex E: Approach to Estimating Costs, 2009.

[84] Master Plan Neamt, Master Plan for Integrated Waste Management for Neamt County. Rambøll/Ficthner/PM/Interdevelopment. Technical Assistance for the Pipeline of Projects Preparation PHARE. 2005/017−553.04.03/08.01, 2008.

[85] Master Plan Bacau, Master Plan for Integrated Waste Management for Bacau County. ISPA Measure, 2004/RO/16/P/PE/007, Europe Aid/122693/D/SER/RO, 2010.

[86] Eurostat, Recycling − Secondary Material Price Indicator, Eurostat, 2014.

[87] WRAP, Materials Pricing Report: Full Listings, 2015. UK.

[88] EU, Environment Action Programme to 2020, 2015. http://ec.europa.eu/environment/action-programme/index.htm.

[89] EC, Directive 2000/60/EC of the European Parliament and of the Council of 23 October 2000 Establishing a Framework for Community Action in the Field of Water Policy, 2000.

[90] E.P. Chiotelli, Evaluation of the effects of irrigation and drainage practices on the landscape of Lake Pamvotis, Ioannina: implications for landscape management in the context of sustainability, Agric. Agric. Sci. Procedia 4 (2015) 201−210.

[91] WFD. Water Framework Directive. http://ec.europa.eu/environment/pubs/pdf/factsheets/water-framework-directive.pdf, November 2010.

[92] N. Brunner, M. Starkl, Decision aid systems for evaluating sustainability: a critical survey, Environ. Impact Assess. Rev. 24 (2004) 441–469.

[93] European Innovation Partnerships. http://ec.europa.eu/environment/water/innovationpartnership.

[94] EEA and SOER, European Briefings: Waste, European Environmental Agency, 2015.

[95] EC, Council Directive of 21 May 1991 Concerning Urban Waste Water Treatment, 1991.

[96] International Water Association, International Statistics for Water Services, 2008. Vienna.

[97] R.I. McDonald, et al., Implications of fast urban growth for freshwater provision, Ambio 40 (2011) 437–447.

[98] W.D. Shuster, A.S. Garmestani, Adaptive exchange of capitals in urban water resources management: an approach to sustainability? Clean Technol. Environ. Policy 17 (2015) 1393–1400.

[99] US Department of Commerce, Profiles of General Demographic Characteristics. Bureau of the Census Census of Population and Housing, 2000, 2001. Washington, DC.

[100] A.L. Moscrip, D.R. Montgomery, Urbanization, flood, frequency, and salmon abundance in Puget Lowland streams, J. Am. Water Resour. Assoc. 33 (1997) 1289–1312.

[101] J. Harbor, A practical method for estimating the impact of land use change on surface runoff, groundwater recharge and wetland hydrology, J. Am. Plan. Assoc. 60 (1994) 91–104.

[102] N.E. Gove, et al., Effects of scale on land use and water quality relationships: a longitudinal basin-wide perspective, J. Am. Water Resour. Assoc. 37 (2001) 1721–1734.

[103] USEPA, Our Built and Natural Environments: A Technical Review of the Interactions Between Land Use, Transportation, and Environmental Quality, US Environmental Protection Agency, Development, Community, and Environment, Washington, DC, 2001.

[104] M. Alberti, et al., The impact of urban patterns on aquatic ecosystems: an empirical analysis in Puget Lowland sub-basins, Urban Plan. 80 (2007) 345–361.

[105] C.P. Konrad, USGS Fact Sheet FS-076-03, November 2003.

[106] B. Woods-Ballard, et al., The SUDS Manual. CIRIA. C697, London, 2007.

[107] S. Dickie, et al., Planning for SUDS − Making it Happen. CIRIA. C687, 2010.

[108] H. Methods, S.R. Durrans, in: C.T. Waterbury (Ed.), Stormwater Conveyance Modeling and Design, first ed, Haestad Press, USA, 2003.

[109] M. Keeley, et al., Perspectives on the use of green infrastructure for stormwater management in Cleveland and Milwaukee, Environ. Manag. 51 (2003) 1093–1108.

[110] G. Xian, et al., An analysis of urban development and its environmental impact on the Tampa Bay watershed, J. Environ. Manag. 85 (2007) 965–976.

[111] G. Mitchell, Mapping hazard from urban non-point pollution: a screening model to support sustainable urban drainage planning, J. Environ. Manag. 74 (2005) 1–9.

[112] Z.Y. Yin, et al., An analysis of the relationship between spatial patterns of water quality and urban development in Shanghai, China, Comput. Environ. Urban Syst. 29 (2005) 197–221.

[113] H.A. Jeng, et al., Impact of urban stormwater runoff on estuarine environment quality, Estuar. Coast. Shelf Sci. 63 (2005) 513–526.

[114] E. Alcott, et al., in: Natural and Engineered Solutions for Drinking Water Supplies: Lessons from the Northeastern United States and Directions for Global Watershed Management, CRC Press, Boca Raton, FL, 2013.

[115] R.R. Brown, et al., Urban water management in cities: historical, current and future regimes, Water Sci. Technol. 59 (2009) 847–855.

[116] R.I. McDonald, Water on an urban planet: urbanization and the reach of urban water infrastructure, Global Environ. Change 27 (2014) 96–105.

[117] D. Butler, C. Maksimovic, Interactions with the environment, in: C. Maksimovic, J.A. Tejada-Guibert (Eds.), Frontiers in Urban Water Management. Deadlock or Hope, IWA Publishing, London, 2001.

[118] J. Niemczynowicz, Urban hydrology and water management − present and future challenges, Urban Water J. 1 (1999) 1–14.

[119] C. Makropoulos, et al., GIS supported evaluation of source control applicability in urban areas, Water Sci. Technol. 39 (1999) 243–252.

[120] C.K. Makropoulos, et al., Decision support for sustainable option selection in integrated urban water management, Environ. Model. Softw. 23 (2008) 1448–1460.

[121] F. Steiner, et al., The ecological imperative for environmental design and planning, Front. Ecol. Environ. 11 (2013) 355–361.

[122] J.C. Padowski, S.M. Gorelick, Corrigendum: global analysis of urban surface water supply vulnerability, Environ. Res. Lett. 9 (2014) 1–6.

[123] C.A. Arnold Jr., C.J. Gibbons, Impervious surface coverage: the emergence of a key urban environmental indicator, J. Am. Plan. Assoc. 62 (1996) 243–258.

[124] E.T. Slonecker, et al., Remote sensing of impervious surface: a review, Remote Sens. Rev. 20 (2001) 227–235.

[125] J. Zimmermann, et al., Metal concentration in soil and seepage water due to infiltration of roof runoff by long term numerical modelling, Water Sci. Technol. 51 (2005) 11–19.

[126] J.B. Ellis, D. Butler, Surface water sewer misconnections in England and Wales: pollution sources and impacts, Sci. Total Environ. 526 (2015) 98–109.

[127] Defra, Tackling Water Pollution from the Urban Environment, Consultation Report, Department of Environment, Food & Rural Affairs, London, November 2012.

[128] E.D. Brown, et al., Illicit Discharge Detection and Elimination: A Guidance Manual for Program Development and Technical Assessments, Report EPA X-82907801-0, Center for Watershed Protection, Ellicott City, Maryland, US, 2004.

[129] C. Brozozowski, Illicit discharge detection and elimination. Centre for Watershed Protection (CWP), Stormwater J. (2004). March/April, 2004. Ellicott City, Maryland, US.

[130] K. Irvine, et al., Illicit discharge detection and elimination: low cost options for source identification and trackdown in stormwater systems, Urban Water J. 8 (2011) 379–395.

[131] I. Gnecco, et al., Storm water pollution in the urban environment of Genoa, Italy, in: 6th International Workshop on Precipitation in Urban Areas. Genoa, Italy, 2005.

[132] P. Göbel, et al., Storm water runoff concentration matrix for urban areas, J. Contam. Hydrol. 91 (2007) 26–42.

[133] F. Napier, et al., Evidence of traffic-related pollutant control in soil-based sustainable urban drainage systems (SUDS), in: 11th International Conference on Urban Drainage. Edinburgh, Scotland, UK, 2008.

[134] V.C. Andrés-Valeri, et al., Comparative analysis of the outflow water quality of two sustainable linear drainage systems, Water Sci. Technol. 70 (2014) 1341–1347.

[135] S. Perales-Momparler, et al., SuDS efficiency during the start-up period under Mediterranean climatic conditions, Clean Soil Air Water 42 (2014) 178–186.

[136] G.D. Breedveld, et al., Treatment of jet fuel-contaminated runoff water by subsurface infiltration, Bioremediat. J. 1 (1997) 77–88.

[137] A. Andreadakis, Water Treatment: Basic Principles and Processes, Symmetria, Athens, Greece, 2008.

[138] K. Arnbjerg-Nielsen, Quantification of climate change effects on extreme precipitation used for high resolution hydrologic design, Urban Water J. 9 (2012) 57–65.

[139] P. Willems, et al., Climate change impact assessment on urban rainfall extremes and urban drainage: methods and shortcomings, Atmos. Res. 103 (2012) 106–118.

[140] Z. Yazdanfar, A. Sharma, Urban drainage system planning and design — challenges with climate change and urbanization: a review, Water Sci. Technol. 72 (2015) 165–179.

[141] SWITCH. http://www.switchurbanwater.eu/.

[142] TRUST. http://www.trust-i.net/.

[143] P. De Barry, Watersheds: Processes, Assessment and Management, Wiley, Hoboken, NJ, 2004.

[144] I. Heathcote, Integrated Watershed Management Principles and Practice, Wiley, Toronto, 1998.

[145] M.R. van Roon, Emerging approaches to urban ecosystem management: the potential of low impact urban design and development principles, J. Environ. Assess. Policy Manag. 7 (2005) 1–24.

[146] M.R. van Roon, Water localisation and reclamation: steps towards low impact urban design and development, J. Environ. Manag. 83 (2007) 437–447.

[147] R.A. Fenner, et al., Widening engineering horizons: addressing the complexity of sustainable development, Proc. Inst. Civil Eng. Eng. Sustain. 159 (2006) 145–154.

[148] C. Makropoulos, et al., Supporting the choice, siting and evaluation of sustainable drainage systems in new urban developments, in: Urban Drainage Modelling and Water Sensitive Urban Design Conference, Melbourne, Australia, 2006.

[149] A.J. Jakeman, et al., Ten iterative steps in development and evaluation of environmental models, Environ. Model. Softw. 21 (2006) 602–614.

[150] V.G. Mitchell, et al., Modelling the urban water cycle, Environ. Model. Softw. 16 (2001) 615–629.

[151] D. Butler, C. Makropoulos, Sustainable Water Infrastructure: Technological Options and Future Scenarios, UK Environment Agency, 2006.

[152] D. Butler, J.W. Davies, Urban Drainage, second ed., Spon Press, London, 2004.

[153] C. Jefferies, SUDS in Scotland. The Monitoring Programme of the Scottish Universities SUDS Monitoring Group, Scotland & Northern Ireland Forum For Environmental Research, 2004. Report No SR (02) 51.

[154] L. Scholes, et al., A systematic approach for the comparative assessment of stormwater pollutant removal potentials, J. Environ. Manag. 88 (2008) 467–478.

[155] K. Heal, et al., SUDS and sustainability, in: 26th Meeting of Standing Conference on Stormwater Source Control. Dunfermline, 2004.

CHAPTER 6

Energy and the Environment

G. Itskos
Nazarbayev University, Astana, Republic of Kazakhstan

N. Nikolopoulos, D.-S. Kourkoumpas, A. Koutsianos, I. Violidakis, P. Drosatos, P. Grammelis
Chemical Process and Energy Resources Institute, Ptolemais, Greece

Chapter Outline

Environment and Development. http://dx.doi.org/10.1016/B978-0-444-62733-9.00006-X

6.1 Introduction

An important challenge of our society is to reduce global greenhouse gas (GHG) emissions. The largest contributor to these emissions is the energy supply sector. According to the International Energy Agency, 35% of the total anthropogenic GHG emissions were attributed to this sector in 2010 [1]. Despite the coordination spirit of the developed countries realized in the United Nations Framework Convention on Climate Change (UNFCCC) and the Kyoto Protocol, the annual GHG emissions increased from 1.7% per year in the last decade of the 20th century to 3.1% in 2000–10 [1]. This trend is attributed to the rapid economic growth, which is associated to the higher demand for power, heat and transport services, as well as to the increased share of fossil fuels and especially coal to the global fuel mix.

6.2 Fossil Fuels

"Fossil fuels" is a generic term for nonrenewable carbon-based energy sources such as solid fuels, petroleum, and natural gas. They are formed from the remains of plants and animals that lived millions of years ago and that underwent transformation through chemical and physical processes in the Earth's crust [2].

6.2.1 Coal

Coal is formed in swamp ecosystems where plant remains were saved by acidic water and mud from oxidization and biodegradation. It is a sedimentary rock, but the harder forms, such as anthracite coal, can be regarded as metamorphic rocks because of later exposure to elevated temperature and pressure. It is composed primarily of carbon, along with other elements, mainly hydrogen, sulfur, oxygen, and nitrogen. Coals are divided into peat, brown coal (lignite and subbituminous coal), and hard coal (primarily bituminous coal and anthracite) [3]. These ranks correspond to different levels of metamorphic grade and to different fuel calorific values, while anthracite is the highest rank of coal having the highest heat content. Coal was the primary energy source that fueled the Industrial Revolution. Nowadays, it supplies 41% of global electricity generation [4]. Furthermore, coal is an important component of steel production and a significant energy source in cement production.

6.2.2 Petroleum

Petroleum is a complex mixture of gaseous, liquid, semisolid, and solid hydrocarbons found in geological formations in the Earth's crust. Liquid petroleum is called "crude oil" or simply oil [5]. It consists mainly of liquid hydrocarbons as well as organic sulfur compounds, naphthenic acids, and phenols, among other substances. The containing sulfur is an important factor that determines oil quality. Crude oil is rarely used in its natural form, but is refined usually by distillation. The distillation fractions include a wide range of liquid fuels, such as LPG, jet fuel, kerosene, fuel oil, diesel and gasoline, asphalt, and chemical reagents important to the production of plastic materials and pharmaceutical products. Oil has become the primary energy source since the 1950s, and it serves as the primary energy source for all means of transport—in 2004 it accounted for 95% of transport sector energy demand [6].

6.2.3 Natural Gas

Natural gas is a hydrocarbon gas mixture. It contains mainly methane, but commonly includes other higher alkanes, carbon dioxide, nitrogen and hydrogen sulfide as well. Before its usage, natural gas undergoes extensive processing, toward removing almost all other components but for methane. It is found mainly in underground geological formations alone or in association with other fossil fuels, especially petroleum [2]. Biogenic natural gas can be found near the Earth's surface, in marshes, bogs, and landfills. Natural gas is considered the cleanest fossil fuel available, as it produces less carbon emissions per energy unit released than oil and coal and fewer pollutants compared to other fossil fuels [6]. It is used mainly for electricity generation and as a domestic fuel for heating and cooking. Moreover, it can be used as supporting fuel in energy storage systems for integrated intermittent production of energy from renewable sources, i.e., wind and solar plants.

6.3 Role of Fossil Fuels in the Energy Mix

Fossil fuels currently satisfy most of human demands for energy. According to the International Energy Agency, coal has been the fastest-growing global energy source during the first decade of the 21st century. As a result, in 2011 fossil fuels accounted for 82% of the global total primary energy supply (TPES). In particular, coal represented 29%, oil 32%, and natural gas 21% of the world TPES [7]. Especially in the electricity generation sector, the contribution of fossil fuels is considerable, even for the developed countries. For instance, over half (52.3%) of the net EU-27 electricity production in 2012 originates from conventional thermal energy resources—oil, gas, and solid fuels. This can be attributed to the economic advantages that fossil fuels enjoy compared to other technologies, which are difficult to overcome, despite the recent trend of increased fossil fuel prices and decreased renewable energy prices [6]. Globally more than 60% of the electricity output is provided by fossil fuels, 42% of which is based on coal [8].

According to energy policy scenarios involving cautious implementation of commitments and plans undertaken by countries in order to tackle climate change, the global energy demand is expected to increase by 40% in-between 2009 and 2035 [3]. This significant growth could not be covered by nonfossil energy sources, considered as nonemitting. Thus the usage of carbon-based fossil fuels is expected to grow in absolute terms. In particular, the oil demand is expected to increase by 18% between 2009 and 2035, driven mainly by the needs of the transport sector. Coal demand will increase by 25%, as it will continue to be the main fuel for electricity generation in developing countries, whereas the natural gas demand is expected to grow in absolute terms more than 40% in the same time period [3].

The available fossil fuel reserves and resources, namely fuel quantities able to be recovered under existing economic and operating conditions and quantities where economic extraction is potentially feasible, respectively, are expected to cover these needs with ease. Coal reserves are likely to last for around 100 years at current production rates [1]. Natural gas reserves and resources are vast, guarantying an undisturbed fuel supply for decades. On the other hand, oil resources can extend the availability of this fuel considerably, but the peak of oil production will be reached within two decades [1]. Unconventional oil resources could cover the demand for oil, but this would be combined with increased oil prices.

However, regardless of the economic and technical benefits of burning fossil fuels for energy, the resulting environmental impact of carbon dioxide emissions is a critical side effect. Fossil energy use is responsible for about 85% of the annual anthropogenic CO_2 emissions production [6]. In 2010, global energy-related carbon dioxide emissions reached 30.4 Gt, increased by 5.3% compared to 2009, which represents an almost unprecedented annual growth [3]. Furthermore, according to the above-referred energy scenarios, the global energy-related CO_2 emissions will increase by 20% in 2035, which in turn will lead to a long-term rise in the average global temperature in excess of 3.5°C [3]. This is in sharp contrast to the global agreement for decreasing GHG emissions and limiting climate change. Strict energy policy scenarios, aiming to limit the global mean temperature increase up to 2°C, show that CO_2 emissions can decrease by 29% in 2035. In that case the coal and oil demand will decline by 30% and 8%, respectively, by 2035 compared to 2009, whereas natural gas demand will increase by 26%. However, these scenarios involve considerable additional investments, about $15.2 trillion, compared to the indulgent ones [3]. Improving the fossil-fueled power generation, and especially conversion efficiency, could result in substantial CO_2 and fuel savings with relatively low cost, as an efficiency increase of fossil-fueled powered plants by 1% could lead to 2.5% reduction of emitted CO_2 [8].

6.4 State of the Art in Fossil Fuel Power Generation

The classification of technologies used in fossil fuel power generation is based on the following criteria, summarized in Fig. 6.1 [8]:

1. the type of the used fuel;
2. the technology of conversion of fuel chemical energy to thermal energy;
3. the used type of turbine; and
4. the generated steam conditions.

In general, the energy produced from the fuel combustion can be used to heat water in a boiler and produce high-pressure steam. The latter is directed to a steam turbine, causing its rotation and resulting in the generation of electricity, as the turbine is coupled with an electric generator. Alternatively, the flue gases produced by the fuel combustion can pass through a gas turbine to generate electricity. In more advanced power plants, these two types of turbines are combined. In these combined-circle plants, the fuel is combusted in the combustion chamber of a gas turbine. The produced flue gases first generate power by

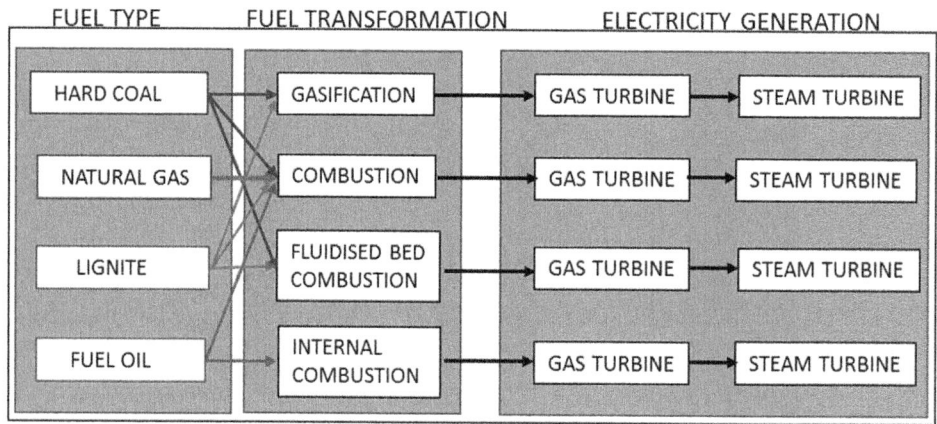

Figure 6.1
State of the art of fossil fuel power generation technology [8].

rotating the aforementioned turbine. In a second stage, they are used to heat water in a boiler and produce steam, which is directed to a steam turbine.

6.4.1 Pulverized Fuel

The main technology used in fossil fuel—fired power plants worldwide as well as in the EU-27 is the pulverized coal (PC) combustion. In these units the fuel is injected in the furnace in the form of very small particles that are easily aeratable by the carrier gas. Thus, the coal combustion can take place within the whole furnace volume. The majority of these plants are now more than 15 years old and operate with subcritical steam. Over the last years, several supercritical (SC) units have been put into operation. The steam conditions in these plants are typically 540°C and 25 MPa [8]. Even higher efficiencies can be achieved in ultrasupercritical (USC) power plants. Electrical efficiency of 47% can be achieved in bituminous coal-fired plants, whereas efficiency of 43% is reported in the lignite-fired ones. In these units the steam conditions are close to 600°C and 30 MPa [8]. Reaching these steam conditions requires complex reheating steam circles. Furthermore steel alloys with better properties are used, especially in relation to corrosion, resulting in higher constructional cost compared to SC plants. Nevertheless, that is counterbalanced by the lower operational cost of the USC plants. However, it should be noted that there are limitations in increasing the plant efficiency by increasing the steam pressure. Additionally, other factors such as the plant geographical location, the availability of cooling water, and the ambient conditions also affect the power plant efficiency. The combination of this type of plants with CO_2 capture technologies can further lower their GHG emissions.

6.4.2 Fluidized Bed

Fluidized bed combustion offers higher fuel flexibility than pulverized coal, as a wide variety of solid fuels can be used. In its basic form the solid fuels are suspended in a hot, bubbling fluidity bed of ash and other materials, such as sand. Jets of air are blown through the bed, resulting in a fast mixing of gas and solids that favors chemical reactions and heat transfer. Another advantage of this technology is the lower emission of pollutants, as up to 95% of them are absorbed before their release to atmosphere by materials included in the bed, such as limestone. On the other hand, the efficiency of this technology is a bit lower compared to PC combustion. Large-scale coal-fired circulating fluidized beds (CFB) are reported to achieve efficiencies as high as 43% [8].

6.4.3 Integrated Gasification Combined Cycle

Another promising fossil fuel transformation technology is the integrated gasification combined cycle (IGCC). This technology involves the gasification of coal and other carbon-based solid fuels to produce syngas, primarily a mixture of carbon monoxide and hydrogen. Syngas can be either used directly as a fuel or hydrogen can be separated and used in other applications. An IGCC power plant with gas and steam turbines, combined with hybrid fuel cells, could reach an efficiency of 60% with almost zero emissions. However, its cost would be higher compared to a PC plant. Until now only demonstration facilities of this type are in operation.

6.4.4 Gas-Fired Power Plant

The share of natural gas in the power generation fuel mixture is steadily increasing during the last 20 years, as it produces the lowest GHG as well as other pollutant emissions compared to the rest fossil fuels. Furthermore, gas-fired power plants are more flexible than coal-fired. In general, air-cooled gas turbines have been reported to achieve efficiencies up to 40% in single circle and over 60% in combined-circle operation [8]. Gas-fired power plants present much lower constructional cost than the coal-fired ones, but their operational cost is higher and is subject to market policies of the gas-supplying countries. On the contrary, newly built oil-fired power plants are quite uncommon. The existing ones are used mainly as peak units and present efficiencies of 32–36%. The general trend is to replace this type of plants with gas-fired ones. Table 6.1 summarizes the key features of the state-of-the-art fossil fuel–fired power generation technologies in Europe.

6.5 Emissions

6.5.1 Emissions With Socioeconomic Indicators

The relations between the CO_2 emissions and the gross domestic product (GDP) or the total population for each individual country constitute general indicators reflecting the constraints and the options followed in its energy policy. Moreover it can reflect the growth rate for an economy, the most active economic sectors, and the degree of implementation of the CO_2 emissions reduction legislation. The CO_2 emissions used in these parameters are defined as the total carbon dioxide emissions from industry electricity production, agriculture, and waste (expressed in kilograms). The GDP is the gross domestic product of a country expressed usually in USD (2005 US dollar) or PPP (purchasing power parity). The latter is an index presenting the adjustments needed to be done on the exchange rate between countries in order for an ideal good to have exactly the same price among all these countries when expressed in the same currency. The CO_2 emissions per population is defined as the ratio of tons of CO_2 per capita.

Based on recent research and worldwide reports, the range of these indexes is quite large, proving economic inequality and wide differences in the way the countries plan and implement their energy mix. Regarding the CO_2 emissions per capita, the divergence between the largest and the smallest emitters is ~ 15.5 tCO_2 per capita, with CO_2

Table 6.1: Main Features of the State-of-the-Art Fossil Fuel—Fired Power Generation Technologies in Europe [8].

Subtechnology	Net Efficiency (%)	Capital Cost (€2008/kW)	Operating Cost Excluding Fuel Cost (€2008/kWh)	Direct CO_2 Emissions (kg/kWh)
Pulverized coal combustion	46—47	1380—1680	0.041—0.050	0.73—0.88
Circulating fluidized bed (CFB) combustion	41—43	2040—2490	0.040—0.048	0.68—0.70
Coal conventional thermal	34—37	2810—3430	0.023—0.028	0.95—1.16
Lignite conventional thermal	32—34	2550—3110	0.037—0.045	0.99—1.21
Integrated coal gasification combined cycle	45—46	2320—2830	0.093—0.113	0.70—0.75
Combined cycle gas turbines	60—61	690—840	0.046—0.056	0.34—0.40
Gas-fired conventional thermal	50—51	430—530	0.049—0.059	0.46—0.56
Gas-fired gas turbine	40—42	560—690	0.131—0.161	0.48—0.58
Oil conventional thermal	32—33	390—470	0.039—0.052	0.74—0.90
Oil-fired gas turbines	35—36	405—475	0.027—0.046	0.65—0.75
Internal combustion engine	41—43	630—820	0.022—0.024	0.71—0.86

emissions per capita being considerably lower in developing countries than in the industrialized ones. More specifically, in 2012 the emissions produced by the United States rose up to 17 tCO_2 per capita, whereas the respective emissions produced by China and India are only 6 and 1 tCO_2 per capita, respectively.

The wide divergence across different counties is also observed for emissions per GDP. These emissions have been reduced during the last decade (between 1990 and 2011) for the five top emitters worldwide. Supplementary socioeconomic indicators are the following [7]:

1. *CO_2 emissions*: These represent the overall amount of carbon dioxide emissions from fuel combustion.
2. *Electricity output*: It is the electricity in terawatt-hour generated from fossil fuels, nuclear, solar, biofuels, etc. by electricity and combined heat and power (CHP) plants.
3. *CO_2/TPES*: It is the ratio of tons of CO_2 emitted using the sectorial approach per tera-joule (TJ) of TPES. The TPES is the sum of all global energy resources, such as coal, oil, gas, nuclear, and hydro.

4. *Per capita CO_2 emissions by sector*: This ratio is expressed in kilograms of CO_2 per capita by sector (transport, road, manufacturing industries, construction, etc.) [9]. In the first ratio the CO_2 emissions from electricity and heat production are calculated separately, while in the second ratio the CO_2 emissions by these sectors have been allocated to final consuming sectors in proportion to the electricity and heat consumed by those sectors [9].

5. *CO_2 emissions per kilowatt-hour*: CO_2 emissions consumed from fossil fuels for generation of electricity per total electricity generated by fossil fuels, biofuels, nuclear, hydro, geothermal, etc. It is clear that this indicator is dependent on the generation mix and varies at different regions worldwide (Table 6.2).

6.5.2 New Developments in Emission Trends and Drivers

According to the International Energy Agency, in 2010 the energy sector contributed globally 49% of all energy-related and 35% of all anthropogenic GHG emissions [1].

Table 6.2: Implied Carbon Emission Factors From Electricity Generation (CO_2/kWh) for Selected Products [7].

Product	gCO_2/kWh
Anthracite*	965
Coking coal*	785
Other bituminous coal	860
Subbituminous coal	925
Lignite	1005
Coke oven coke*	800
Gas works gas*	420
Coke oven gas*	415
Blast furnace gas*	2200
Other recovered gases*	2030
Natural gas	400
Crude oil*	635
Natural gas liquids*	540
Refinery gas*	410
Liquefied petroleum gases*	530
Kerosene*	645
Gas/diesel oil*	715
Fuel oil	670
Petroleum coke*	970
Peat*	745

Average implied carbon emission factors from electricity generation by-products are presented in the following Table, for selected products. Those values are given as a complement of the CO_2 emissions per kilowatt-hourkWh from electricity generation by country presented in the Summary tables. These values represent the average amount of CO_2 per kilowatt-hour of electricity produced in OECD member countries between 2009 and 2011. As they are very sensitive to the quality of underlying data, including net calorific values, and of reported input/output efficiencies, they should be taken as indicative; actual values may vary considerably.
* The electricity output from these products represents less than 1% of electricity output in the average of OECD member countries for the years 2009–11. Values will be less reliable and should be used with caution.

In addition to CO_2, these emissions include methane and N_2O oxides. In spite of the commitment of many governments to take measures in order to control climate change, the GHG emissions of the aforementioned sector increased by 35.7% between 2000 and 2010. The subsector that contributed most to this increment was that of electricity and heat generation, being responsible for the three-quarters of the latter. Fuel production and transmission, and petroleum refining presented smaller shares of 16% and 8%, respectively [1]. The main reasons behind this rise are the global population growth, the increase of GDP and the increase of the sector carbon intensity.

Another key feature of this trend is the nonuniform emission distribution among the countries globally. The developed ones showed a small total primary energy demand increase, which, however, had no negative impact on the GHG emissions, since a small decrease of carbon dependence was achieved during the first decade of the 21st century. On the contrary, developing countries showed an increment of both carbon intensity and primary energy supply. As a result the GHG emissions of their energy sectors showed an annual increase of 5% and in 2010 were more than double than the corresponding ones of the developed countries. Nowadays, Asia is responsible for 41% of the global emissions, with China and India being the main contributors, surpassing OECD countries (countries originally signed the Convention on the Organization for Economic Co-operation and Development) [1].

6.5.3 Distinction of Emissions

Emissions by fuel

The global emissions of CO_2 from the combustion of fossil fuels reached ~ 36 billion tonnes in 2013 [10]. These emissions are considered the most dominant among the total GHG emissions. When a fuel is burned, it emits an amount of CO_2 related to the energy that it produces. Such emissions depend on the fuel carbon content. For instance, the natural gas combustion emits 15 kg carbon per giga-Joule (C/GJ), while for coal and oil this ratio is 26 kg C/GJ and 20 kg C/GJ, respectively. Thus, coal combustion produces higher amounts of CO_2 than oil and natural gas. Moreover, the natural gas–fired combined-cycle power plants achieve higher energy efficiency than coal-fired power plants, almost 15%, because they operate at higher temperatures.

According to the IES, in 2012 the CO_2 emissions breakdown is coal (44%), oil (35%), and gas (18%). The coal share of the world TPES is only 29%, while for oil and gas it is 32% and 21%, respectively. In 2001, the largest emissions share was that of oil (42%), while coal was $\sim 38\%$, and gas was almost the same as in 2012.

Without any further measures, it is expected that CO_2 emissions from coal will reach 15.7 $GtCO_2$ in 2035. However, this scenario can be prevented if coal consumption is reduced,

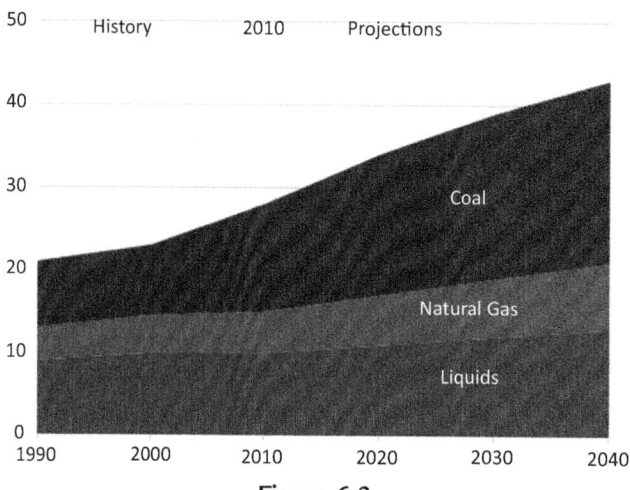

Figure 6.2

Global CO_2 emissions by fuel during the period 1990–2035 (billion metric tons) [11].

thus reducing the CO_2 emissions to 5.7 Gt. This can be achieved by using more efficient power plants, changing the energy mix with increasing renewables, nuclear and the implementation of carbon capture and storage (CCS) technologies. Accordingly, emissions from oil will reach 12.5 $GtCO_2$, mainly due to an increase in the transport sector, while natural gas emissions will grow to 9.1 in 2035 (Fig. 6.2).

Emissions by region

The biggest share of global CO_2 emissions in 2012 and 2013 which is equal to 55% is owed to China (29%), the United States (15%), and the European Union (EU-28) (11%). This share was almost the same as in 2011 (54%). As concerns China, emissions had a historical growth (9.7%) between 2001 and 2011, while in 2012 there was a decrease in this growth with emissions increasing only by 3.1%. This is possibly owed to a slowdown in economic growth that subsequently led to a decline in industry demands in electricity and fuels. In 2013 there was a higher growth than 2012 due to an annual uptake of production increase.

On the contrary, in Africa there was a totally different picture with emissions growth reaching 5.6% in 2012, while in 2011 there was not any increase in CO_2 emissions observed. In Annex II countries, such as North America and Europe, emissions decreased (−2.4 and −4.3% for North America and Europe, respectively) in 2011, while this decline was lower in 2012 (−0.5% and −3.7% for North America and Europe, respectively).

In Asia excluding China, Middle East, and Latin America the growth was kept the same, ie, about 4%, for years 2011 and 2012. As it can be seen the CO_2 emissions in different

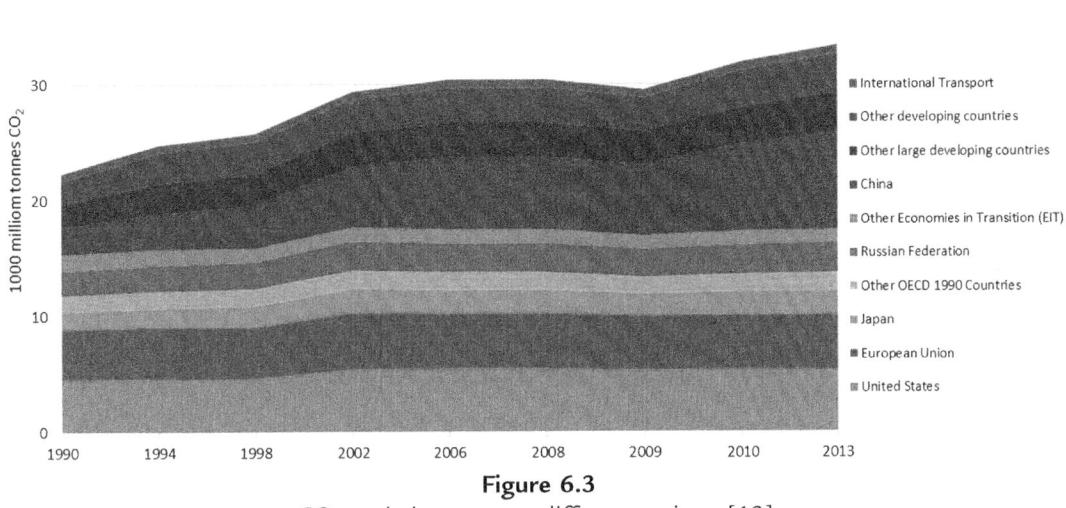

Figure 6.3
CO_2 emissions across different regions [12].

regions change considerably every year. It is quite encouraging that the annual growth in global CO_2 emissions during years 2012 and 2013 are about half of the increases in the previous decade.

It is underlined that all these shares represent the countries' absolute emissions and the picture is totally different when indicators, such as CO_2 emissions per capita or per GDP, are used. This is owed to the existence of the different socioeconomic stratification at different regions worldwide (Fig. 6.3).

Emissions by sector

Another way to allocate global CO_2 emissions to human activities is to break down the emissions to different sectors. In the year 2012 two major sectors were responsible for almost two-thirds of CO_2 emissions worldwide: the (1) electricity and heat generation and (2) transport sector with a share of 42% and 23%, respectively. Other sectors responsible for CO_2 emissions are industry, commercial/public services, agriculture/forestry/land use change, fishing, and waste and energy industries.

The high carbon dioxide emissions of electricity and heat generation are owed to the extensive utilization of coal. This can be mostly observed in Australia, China, Poland, and South Africa. According to IEA, the share of oil in this sector has been reduced since 1990, while that of coal has increased significantly. The share of natural gas increased as well but not considerably. Moreover, it is expected that by 2035 the energy demand will be ~70% higher than currently.

In the transport sector CO_2 emissions are owed to fossil fuels burned for road, rail, air, and marine transportation. The fuel demand of this sector is expected to increase by almost 40% by the year 2035. For this reason some policies should be implemented in order to reduce transport demands and subsequently CO_2 emissions related to this sector. These policies should include: measures to improve the efficiency of vehicles, as already implemented in the European Union and the United States, and measures to encourage citizens to prefer public transportation than cars and also prefer low-carbon fuels, such as electricity and biofuels. Moreover, it is expected that in the future it will be more possible to indirectly capture CO_2 if hydrogen-based technologies will be preferred over traditional fossil-fueled engine technologies.

Types of emissions

Pollutant is any substance that has a harmful effect on the environment. The damages caused by pollutants can be either of short or long term.

The primary air pollutants that result from fossil fuels combustion are the following:

1. **Carbon monoxide (CO)**: It is a colorless, odorless, and tasteless poisonous gas, mainly produced from the incomplete combustion of any fuel. Its effects on human health are varying from just a headache to nausea, unconsciousness, and even death. Moreover, it is the main compound of photochemical smog.

2. **Sulfur oxides (SO_x)**: These oxides are formed by the combustion of sulfur-containing fossil fuels in power plants (73%) and other industrial infrastructures (20%) [13]. The major products are sulfur dioxide (SO_2) and sulfur trioxide (SO_3), with the latter being the most dangerous. More specifically, sulfur dioxide has an irritating smell and forms sulfur trioxide when released to atmosphere. Sulfur trioxide, in turn, is the main compound of acid rain. Moreover, it can form sulfuric acid which is highly corrosive for many materials.

3. **Nitrogen oxides (NO_x)**: This term refers mainly to nitrogen dioxide (NO_2), which is the largest group of nitrogen oxides. It also includes nitrous and nitric acid. Nitrogen dioxide is a product of reaction of nitric oxide (NO) with oxygen when it is released in the air. It is really hazardous and in some cases can cause even death. As sulfuric acid, nitric acid is a component of acid rain that is corrosive for many materials.

4. **Particulate matter (particle pollution or PM)**: It is a mixture of extremely small particles (soot and fly ash), in the order of 10 μm or less, and liquid droplets. They can cause serious damages to human lungs and subsequently lead to breathing problems, like chronic asthma and emphysema. In some cases they can even cause cancer. Among them, soot is a result of the incomplete conversion of carbon to carbon oxides. Fly ash is a residue of PC combustion.

5. **Unburned hydrocarbons (UHCs)**: They are hydrocarbons emitted to the environment during evaporation from fuel tanks or due to leaks or spills from a car engine.

The abovementioned primary pollutants can further generate additional pollutants, i.e., secondary air pollutants, through their interaction with the environment. The secondary air pollutants are the following:

1. **Acid rain**: It is the water rain that has a pH lower than that of natural rainfall. For instance in the United States during summer period the rain falling has a pH of 4 or less in contrast to the natural rain that has a pH of 5.6.
2. **Smog**: It is a substance hybrid of fog and smoke that contains soot particulates from smoke, sulfur dioxide, and other components. For instance, the SO_x aerosols are one of the main contributors to smog. Today smog is most known as photochemical smog.
3. **High ozone levels**: Generally, ozone (O_3) is a colorless odorless gas that is made of oxygen. It can cause severe damage to human health. Moreover, it is corrosive destroying materials, monuments, etc. Some of the factors that affect the ozone levels are the weather, time of the day, and season. For instance, the ozone levels are higher in warm sunny days than in cool or cold days, while the wind can disperse the ozone. Moreover, ozone levels drop during night.
4. **Greenhouse effect**: It is a process by which the atmospheric GHGs absorb a part of infrared radiation of the sun making the atmosphere and thus the surface of the earth warmer. Without the greenhouse effect the earth surface would be almost $33°C$ colder on average.

A GHG is any gaseous compound that traps heat in the atmosphere causing the greenhouse effect. The main GHGs are water vapor (H_2O), carbon dioxide (CO_2), and methane (CH_4).

1. **Methane** (CH_4): It is the main component of natural gas and is emitted by industry, agriculture, and waste management activities.
2. **Water vapor** (H_2O): It is the gaseous phase of water. It is invincible and can be generated with evaporation and removed with condensation.
3. **Carbon dioxide** (CO_2): It is a gas extremely important for plants. It is colorless and odorless.

6.5.4 Emissions Trading Schemes

Emissions trading (or cap and trade) schemes (ETS) are implemented worldwide to reduce pollution by emissions. The term cap or limit is referred to the total amount of GHG emissions that can be released by factories, power plants, and other industrial infrastructures. The main target is to reduce that cap in order to put pollution under control. In this frame, industries receive or buy specific emissions permits that can in turn trade with other industries. Thus, companies that require an increase in their emissions must buy permits from other companies that produce lower emissions than their permits.

At a further step the ones that can reduce their emissions can gain money, whereas firms that need more permits should pay more money.

In some countries such schemes are already operating, such as in the European Union, Australia, New Zealand, Norway, Tokyo, Switzerland, the United States, etc., while in others, such as Korea, China, Kazakhstan, Ukraine, and Chile, they are currently under development. Among them the European Union's Emissions Trading System (EU ETS) is the largest scheme in operation that began in 2005. Moreover, it covers the 28 members of the European Union and three more countries: Norway, Liechtenstein, and Iceland. The carbon emissions of these countries represent almost 45% of Europe's total. In 2008 the European Union was committed to reduce its GHG emissions by 20% by 2020, taking as a reference year 1990.

6.5.5 Potential Emission Reduction From Mitigation Measures

The abrupt climate change is nowadays one of the major problems the humanity needs to address. The major factor contributing to the gradual rise of global temperature is undoubtedly the carbon dioxide and other GHG emissions from the human activities, majorly the electricity production and industry. For decades, several diplomatic and political efforts have been made, aiming to a global sustainable solution by decreasing the CO_2 emissions and promoting the usage of green technologies. Basic and fundamental stages of this procedure was firstly the Kyoto Protocol and afterward other multinational agreements organized by the United Nations. Based on these international environmental treaties, the CO_2 emissions reduction for the first CO_2 emitter, namely China, should rise up to 40% per GDP by 2020 in comparison with the 2005 levels. The next two CO_2 emitters, i.e., the United States and European Union, have agreed to 17% and at least 20% reduction by 2020, respectively, against 2005 levels. Finally, Russia intends to cut CO_2 emissions by 15−25% using as reference the year 1990 and excluding the emissions due to agricultural activities. After 2020, the emissions should be further adjusted, in order for the global temperature to be stabilized, as the Cancun Agreement implies [14].

Recent findings and reports estimate that 42% of the total CO_2 emissions are owed to the electricity production industry, since the main fuel used for combustion is coal. In order to attain the major milestones driven by the international agreements, each country is obligated to choose the most sustainable and evaluated technologies to meet the CO_2 emissions reduction objectives also as far as the electricity production is concerned. Except for the conventional means of electricity production using coal, natural gas, and oil combustion, several other more environmentally friendly technologies have been developed. Characteristic and representative examples of these are nuclear, wind, solar and ocean energy, electricity from biomass combustion, and geothermal heat exploitation. A helpful tool for estimating the potential and the effectiveness of each technological option is the quantification of its life cycle perspective, taking into consideration the pollutant emissions in the fuel chain and the energy conversion technology manufacturing. Since the term GHGs include more species than CO_2, it is necessary to express the contribution of all pollutants to the global heating, implementing the equivalent carbon dioxide measure.

As expected, the largest life cycle GHGs emissions are directly linked to coal combustion. It is estimated that the total emissions originated from the coal-fired power plants vary worldwide from 675 to 1689 gCO_2eq/kWh electricity, whereas the corresponding emissions from oil and natural gas combustion fall to 510−1170 gCO_2eq/kWh. More sophisticated and detailed reports take also into consideration the fugitive methane emissions during mining and transport. Based on them, the modern-to-advanced coal-fired power plants present a range of 710−950 gCO_2eq/kWh electricity, while the natural gas combined-cycle plants show emissions varying from 410 to 650 gCO_2eq/kWh electricity. It is also noteworthy that the systems providing simultaneously heat and power are able to minimize by 25% the pollutant emissions in comparison with the equivalent stand-alone ones [13].

However, the imposition of the temperature stabilization in the second part of the 21st century as primary goal of the Cancun Agreement demands highly sophisticated technologies with even lower emissions, close to zero. In this prospect, a number of power supply technologies offer very low life cycle GHGs emissions. For example, the implementation of CCS technologies are expected to lower even more the pollutants emissions to 70−290 gCO_2eq/kWh and 120−170 gCO_2eq/kWh for the coal and natural gas combustion case, correspondingly. In the latter estimation, it is assumed that a 1% of natural gas leaks during mining, transportation, and storage. According to the literature, the percentage of the natural gas leakage can be between 0.8% and 5.5% [15], resulting in additional equivalent CO_2 emissions ranging from 90 to 370 gCO_2eq/kWh. The upper limit of the equivalent CO_2 emission ranges of these methodologies is directly influenced by the capture efficiency of the system. In the previously mentioned cases, the assumed efficiency is close to 90%. However the manufacturer of the system can adjust the acquired efficiency, lowering the cost but also allowing higher emissions to environment.

Even lower equivalent CO_2 emissions have been ensured using alternative sources, namely solar, wind, nuclear, and ocean energy. The former is subdivided into two main categories: the photovoltaic (PV) and the concentrated solar power systems (CSP). These two technologies present very low life cycle total equivalent CO_2 emissions compared to the previously mentioned cases, ranging from 18 to 180 gCO_2eq/kWh and 9 to 63 gCO_2eq/kWh, respectively [16,17]. For the next two energy sources, it is estimated that the wind energy exploitation results in 7−57 gCO_2eq/kWh emissions, with the upper limit referring to small turbines [18], while the nuclear power causes slightly higher emissions (4−110 gCO_2eq/kWh) [19]. The differences between the upper and the lower limits of these ranges can be justified by the incompleteness or the outdated implemented technology or the poor local weather conditions dominating a region.

Concerning biomass, it is noted that this category is subdivided into three individual parts representing different types of natural input for combustion, ie, (1) crop residues, (2) forest wood, and (3) corn and manure. The emission estimations of all these types of biomass are based on recent literature surveys, assuming an electric conversion efficiency between 30% and 50%. Among the three aforementioned subcategories, the highest pollutant production is noticed in the second, while the lowest in the first. Furthermore, it is interesting that in cold areas affected by seasonal snow cover, the change in surface albedo can be larger than the warming caused by the fugitive CO_2 emissions, and therefore the bioenergy system presents negative impact (cooling).

In conclusion, the recent findings prove that a great number of technologies producing less equivalent CO_2 emissions in comparison with the equivalent ones produced by coal combustion have already been developed. The decrease in pollutants production between these two main categories of electricity sources can rise up to 5%. In the future, it is estimated that even larger decrease can be attained mainly through the improvement of the systems efficiency [20,21].

6.5.6 Cost Assessment of Mitigation Measures

As it has already been mentioned, one of the promising alternative technologies for lower pollutant emissions in the electricity production sector is the CCS power plants. Until now, these systems have not been implemented in large scale due to technological restrictions and high cost. More specifically, these technologies are expected to use approximately 10—40% of the electricity produced by a CCS power plant, thus demolishing all the efficiency gains in coal-fired power plants of the last 50 years and requiring up to 25% more coal input for combustion. Furthermore, even if the CO_2 emissions will be limited by 80—90%, the emissions during mining and transportation are not expected to differ significantly. On the contrary, additional emissions due to leakage from storage tanks can dramatically contribute to the local and global environmental degradation. Furthermore, some recent reports estimate a twofold increase of the total cost of these units compared to the equivalent costs of conventional coal-fired power plants. The vast majority of these additional costs are due to the operation of additional systems and infrastructure for long-term waste products storage and the responsible supervision of all these processes. The related cost for transportation and storage varies from 5USD/tCO_2 to 10USD/tCO_2 [22,23].

Therefore the CCS power plants will be fully operational in large scale and competitive with conventional technologies only if some requirements are fulfilled. For instance, the additional cost of extra systems and equipment, the additional cost for CO_2 transport and storage, and the decreased efficiency of the power plants due to increased operational electricity needs should be compensated by sufficiently high carbon prices and direct financial support. Consequently, the engineers, the managers, and the operators of the CCS units have to deal with multilevel problems in engineering and financing. However, the financial challenges and the high costs do not only refer to the operation and the manufacturing of sophisticated CCS power plants, but also to other aforementioned alternative energy sources, mainly due to their connection to the electrical grid. These costs appear to be much higher in renewable energy sources than they are for conventional systems, ie, coal-fired power plants [24]. These increased costs are caused by three main parameters: (1) flexibility costs in order for the energy systems to instantly fulfill the varying energy needs and maintain a balance between supply and demand, (2) additional capacity adequacy costs to ensure the stabilization and operation of the system even in the peak times of a day, when the base load is not enough, and finally, (3) higher transmission and distribution costs.

The contribution of the renewable energy sources to the meeting of peak demand is less than the installed capacity. This divergence is owed to the lowest system efficiency and the potential nonbeneficial weather conditions dominating a region for some time. The capacity expansion strategy needed to fulfill the peak loads causes also additional costs. It is estimated that these extra costs can vary from 0 to 10 USD/MWh electricity for wind power [3].

One solution to the previously mentioned challenges of the grid balance and the system flexibility is the energy storage and the provision of these supplies whenever it is needed. However, according to surveys carried out for OECD countries this policy directly increases the final cost for wind and solar energy systems by 1−7 USD per MWh electricity. Furthermore, the final cost of an investment in wind or solar energy is highly affected by the transmission infrastructure, especially if these systems are developed as decentralized components far from the demand centers. It is estimated that these additional costs can vary from 0 to 15 USD/MWh, depending on the region, the disposal of the renewable energy (weather conditions), and the survey assumptions. This can be reversed formulating several integrated renewable energy systems closer to demand systems or closer to the developed distribution network. Moreover, in the developing and less developed countries, it is possible the deployment of smaller autonomous sources capable of meeting the demands of isolated systems (Fig. 6.4).

6.5.7 Economic Potential of Mitigation Measures

The quantification of the GHG mitigation measures potential is a challenging and difficult task because of the continuously varying fuel prices, the definition of welfare metrics, and

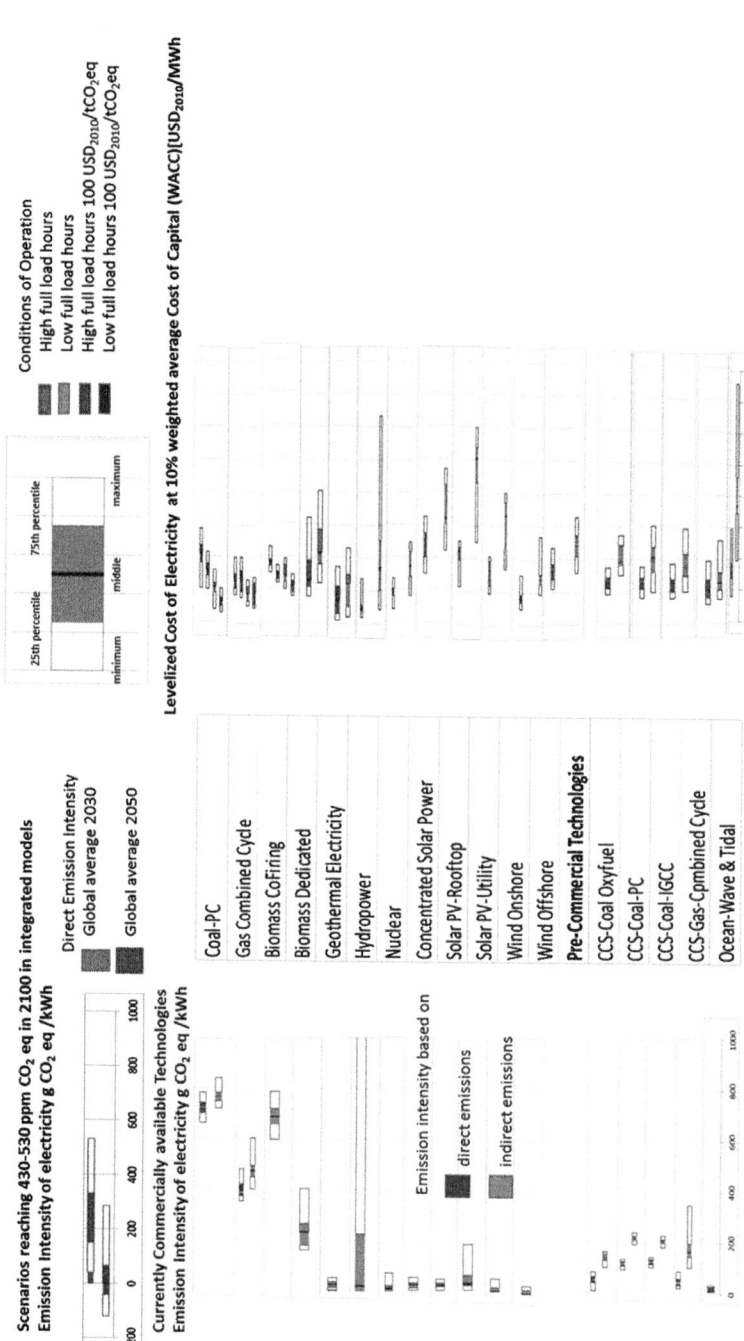

Figure 6.4

Specific direct and life cycle emissions and leveled cost of electricity for various power-generating technologies [25].

the broader interaction between these systems and the political and economic background. In literature, mainly two methodologies are followed to quantify the financial prospect of a mitigation measure. The first approach uses the energy supply cost on an annual or cumulative basis as reference for the comparison among all the potentially implemented measures. The uncertainties associated with this approach therefore include the data unavailability; the divergence between the real, confirmed reserves and the speculative resources; the weather condition's stochastic behavior; and the continuous technological improvements.

A characteristic image summarizing the results of this methodology is shown in Fig. 6.5. This specific graph presents the levelized cost of electricity for commercially available types of renewable energy sources for three cases of discount rates (5%, 7%, 10%). The lower bound of the cost ranges is based on the low ends of the capacity factors, lifetimes and investment, operation, maintenance and feedstock cost ranges, and the high ends of the ranges of conversion efficiencies and by-product revenue. The higher bound of the levelized cost range is accordingly based on the high end of the ranges of investment, O&M and feedstock costs, and the low end of the ranges of capacity factors and lifetimes as well as the low ends of the ranges of conversion efficiencies and by-product revenue.

As it can be seen upon observing Fig. 6.5, the highest cost values among all renewable energy sources are notified in some type of PVs. The high difference between the upper and lower ranges in each examined mitigation measure can be attributed to the uncertainties already discussed.

The second approach is the marginal abatement cost (MAC) curves [27]. This methodology is quite useful as a basis for discussion, but it is not sufficient to base policy decisions on. A MAC curve presents the extra cost and equivalent carbon dioxide reduction for each one of these options relative to a baseline. The base scenario represents the technological behavior under "business-as-usual" circumstances. The Figure used for this methodology comprises several boxes, each for one of the investigated mitigation opportunities. The width of these boxes represents the annual GHGs emissions reduction per year while their height represents the estimated cost to reduce emissions by $1tCO_2eq$. The graph is then ranked left to right from the lowest cost to the highest cost opportunities. A typical example of a MAC curve is displayed in Table 6.3. According to that curve, over 25% of the available technologies under 60 €/tone actually have negative costs, thus representing options much cheaper than the reference scenario. However, as already mentioned, this is only an indication and not a final conclusion. Several other parameters have to be examined to evaluate the implementation potential. One of them is the opportunity on behalf of the investor to capture the additional financial profits.

Nevertheless, even this methodology presents some uncertainties that have to be considered. These uncertainties become even more important for the developing countries, since they

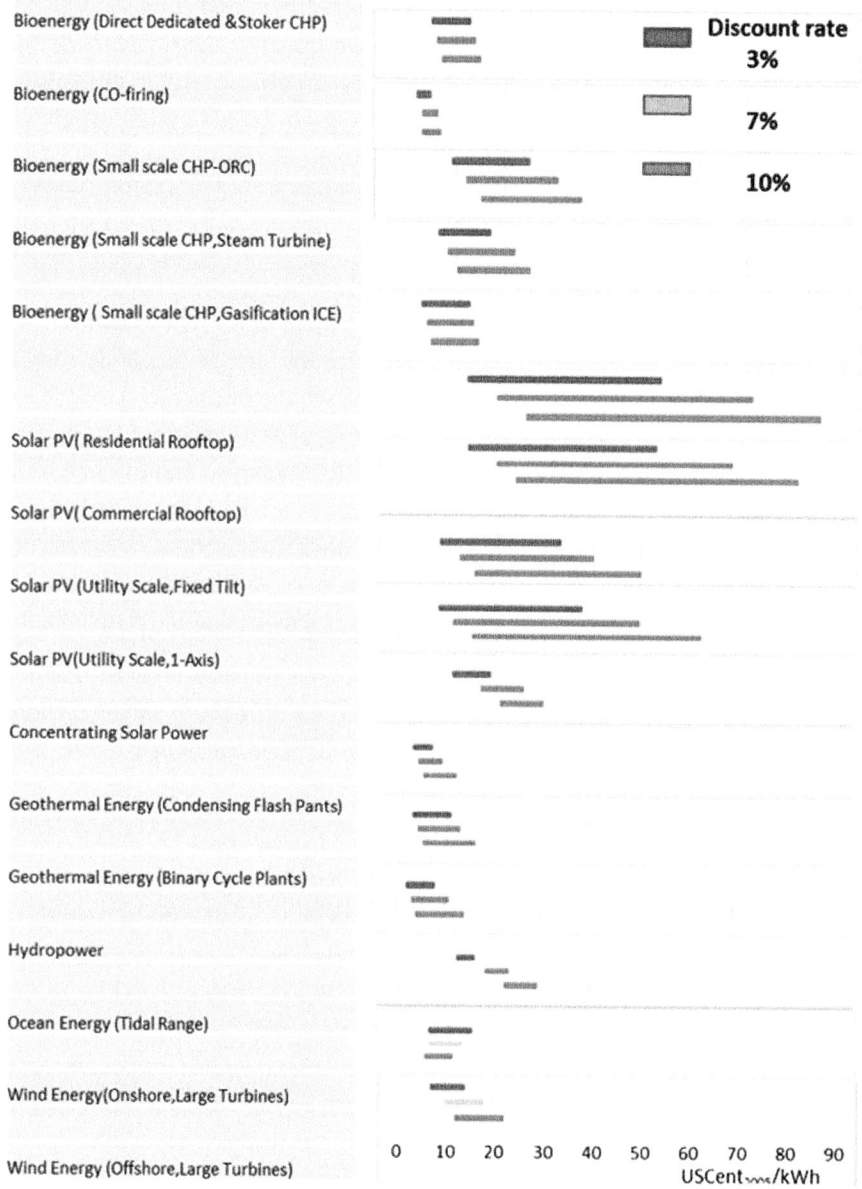

Figure 6.5
Specific direct and life cycle emissions (gCO$_2$eq/kWh) and leveled cost of electricity for various power-generating technologies [26].

Table 6.3: Global Greenhouse Gas (GHG) Abatement Cost Curve [28].

	Abatement Cost (Euro per tCO$_2$ e)	Abatement Potential (Gt CO$_2$ per Year)
Gas plant CCS retrofit	55	0.2
Coal CCS retrofit	41	1
Iron and steel CCS new build	40	0.2
Coal CCS new built	38	0.4
Power plant biomass cofiring	29	0.8
Reduced intensive agriculture conversion	27	1.2
High penetration wind	25	1
Solar PV	15	1.3
Solar CSP	14	1
Low penetration wind	12	2
Cars plug—in hybrid	11	0.2
Degraded forest reforestation	10	1.4
Nuclear	10	2
Pastureland afforestation	10	1
Degraded Land restoration	9	0.4
Second-generation biofuels	8	0.2
Building efficiency new build	7	0.3
Organic soil restoration	6	1
Geothermal	5	0.1
Grass land management	4	1
Reduced pastureland conversion	4	1
Reduced slash and burn agriculture conversion	3	1.6
Small hydro	−4	0.2
First-generation biofuels	−5	0.3
Rice management	−5	0.2
Efficiency improvements other industry	−6	2
Electricity from landfill gas	−13	1.1
Clinker substitution by fly ash	−22	0.5
Cropland nutrient management	−52	0.2
Motor systems efficiency	−65	0.1
Insulation retrofit	−68	0.1
Lightning-switch incandescent to LED (residential)	−99	0.1
Residential electronics	−85	0.2
Residential appliances	−69	0.3
Retrofit residential/hvac	−58	0.3
Tiltage and residue management	−45	0.3
Insulation retrofit (residential)	−32	0.4
Cars full hybrid	−30	0.3
Waste recycling	−15	1

CCS, carbon capture and storage; *PV*, photovoltaic; *CSP*, concentrated solar power system.

show high economic growth rates, high discount rates, and lack of provided data. The most important of these uncertainties, which play a significant role on the results accuracy, is the inventory of a business-as-usual reference baseline. Secondly, similar to the first quantification methodology, an analysis of the future available low- carbon technologies should be conducted. Thirdly, it is necessary to make assumptions regarding the economics of specific alternative CO_2 reduction technologies, such as their lifetime, the risk, the financial profits, and some microeconomic parameters, such as the discount rates.

6.6 R&D Technologies

It is obvious that new technologies in fossil-fueled power generation are required, in order to account for the increased energy demands and keep GHG emissions within acceptable limits at the same time. One of the main objectives of R&D concerning this technological field is the development of commercially viable CO_2 capture and storage (CCS) technologies. In this framework, the Strategic Energy Technology (SET) Plan of the European Union aims to the deployment of cost-effective CCS technologies in coal-fired power plants by 2020–25 [8]. The establishment of a functional emissions trading system would help achieving this goal as well as expanding the usage of CCS systems in all industrial sectors that aggravate the GHG emission balance.

Another field of intensive research is the deployment of the "Advanced 700°C PF Power Plant." It is a project carried out in the framework of European Commission's THERMIE program that focuses on the development of USC power plants with steam conditions of 700°C and 35 MPa and net efficiency of 50–55% [8]. However, the realization of this type of plant is doubtful in the present time, as it includes the overcome of several technological challenges, such as the usage of nickel-base alloys, and the improvement of both turbine blade technology and condenser configuration. Finally, oxy-combustion is a promising technology that would improve the operational features of both retrofitted and newly built coal-fired power plants. A main advantage of this technology is the cost reduction and the efficiency increase of CCS systems, resulting in financially viable zero-emitting fossil-fueled power plants. In the present time only pilot facilities of this type are in operation, as there are many problems to be solved before the commercial implementation of oxy-combustion units [8].

6.7 Process Engineering Tools

6.7.1 Computational Fluid Dynamics

Computational fluid dynamics, commonly known by the acronym "CFD," is regarded as one of the main tools for the in-depth investigation and understanding of scientific or research problems related to fluid flows. CFD approaches are widely used in many engineering aspects of interest in the energy sector. The widespread availability of engineering workstations together with efficient solution algorithms has made CFD approaches applicable in many energy sectors such as the production of energy and/or of high value liquids and gases from fossil fuel energy resources. Such processes among others include combustion, gasification, pyrolysis, fuel cells, and last but not least prediction of any type of GHG emissions. An important advantage of CFD over other approaches is that it is a very compelling, nonintrusive technique with powerful visualization capabilities, and engineers can evaluate the performance of a wide range of applicable system configurations on the computer without the time, expense, and disruption required to make actual changes on-site. CFD costs much lower than the experimental approach, because physical modifications are not necessary, especially taking into consideration that the cost and time for physical changes/modifications increase almost exponentially with the size of the system.

CFD gives an insight into flow patterns that are difficult, expensive, or impossible to study using traditional (experimental) techniques. Compared to experiments, this technique can provide insights into anything related to fluid flow phenomena, (1) for all desired quantities (eg, pressure, temperature, species concentrations), (2) with high resolution both in time and space, (3) values for any governing variable of the actual fluid domain, and (4) conduct simulations and derive results for virtually any problem and realistic operating/boundary conditions. Nevertheless, the results of a CFD simulation can never be considered 100% reliable because (1) the input data may involve too much guessing or imprecision, (2) the mathematical model of the problem at hand may be inadequate, and (3) the accuracy of the results is being limited by the available computing resources.

The main underlying numerical equations, representing any fluid phenomena, are the Navier–Stokes equations, one of which simpler expressions are those of Unsteady-Reynolds-averaged-Navier-Stokes (URANS). The main equations are those of:

Conservation of mass and momentum:

$$\frac{\partial \rho}{\partial t} + \nabla \cdot \rho \vec{u} = 0 \qquad [6.1]$$

$$\frac{\partial \rho \vec{u}}{\partial t} + \nabla \cdot \left(\rho \vec{u} \otimes \vec{u} - \overline{T} \right) = \vec{S_u} \qquad [6.2]$$

where:

$$\overline{T} = -\left(P + \frac{2}{3} \mu \nabla \cdot \vec{u} \right) \overline{I} + \mu \left(\nabla \otimes \vec{u} + (\nabla \otimes \vec{u})^T \right) \qquad [6.3]$$

where $\overline{T}$ is the stress tensor, P the pressure, μ the dynamic viscosity of the fluid, $\overline{I}$ is the unit tensor, and $\vec{S}_u$ is any added source terms.

Any scalar variable φ, such as temperature and species concentration is calculated on the basis of the following equation:

$$\frac{\partial \rho\varphi}{\partial t} + \nabla \cdot (\rho \overrightarrow{u}\, \varphi - \overrightarrow{q}) = S_\varphi \qquad [6.4]$$

where $S\varphi$ represents any added source term and $\overrightarrow{q}$ is the diffusion flux vector usually calculated from a relation of the form:

$$\overline{q} = \Gamma_\varphi \nabla\varphi \qquad [6.5]$$

and Γ_φ is the diffusion factor, usually calculated as $\Gamma_f = \frac{\mu}{\sigma_f}$ (σ_φ is the proper Prandtl number for that scalar variable φ).

This set of equations, along with any appropriate constraints for the case of any numerical problem, is solved iteratively, by assuming an initial field of values, since no analytical solution is available. A typical example of CFD application is in hard coal-fired pulverized fuel (PF) combustors, Fig. 6.6. Such a tool is capable of predicting with high accuracy the combustion characteristics of fuel both in terms of efficiency and emissions (e.g., CO_2, CO, NOx). More information on relevant studies applicable both for PF and CFB boilers can be found in Refs [29–31].

6.8 Process Engineering—Thermodynamic Tools

The term "thermodynamics" in the field of simulation tools refers to the kind of methodology that is based on the calculation of the mass and energy balance without taking into account aspects of the flow like in the case of CFD tools. This section is focused on the thermodynamic tools for fossil fuel—based energy systems simulations. Such energy systems can be sorted in three main categories:

1. **Conventional power plants**: (1) existing or state-of-the- art thermal plants based on Rankine Cycle such as PulverizedF boiler, for solid fossil fuels, or Diesel/Otto Cycles such Internal Combustion Engines (ICE) for gas/liquid fuels or in a combination of Rankine and Bryton Cycle such Gas Turbine Combined Cycle (GTCC) for gas fuels, other cycles (e.g., Stirling) and (2) fuel processing plants such oil refineries, Coal to Liquid (CTL), or Gas to Liquid (GTL) plants
2. **New technologies**: Circulating Fluidized Bed Combustors (CFBC), USC cycles, systems based on Organic Rankine Cycles (ORC)
3. **Advanced technologies**: IGCC, systems with CCS or Utilization (CCU), O_2/CO_2 cycles

6.8.1 Features

Contrary to the CFD tools, the thermodynamic ones have a rather macroscopic point of view. The main objective is to simulate virtually any thermodynamic cycle process in

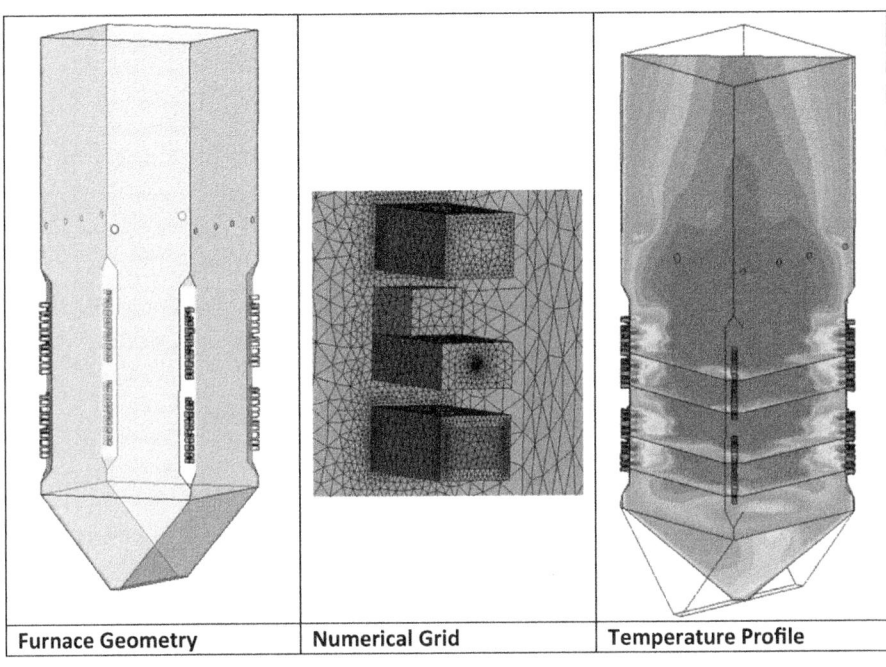

| Furnace Geometry | Numerical Grid | Temperature Profile |

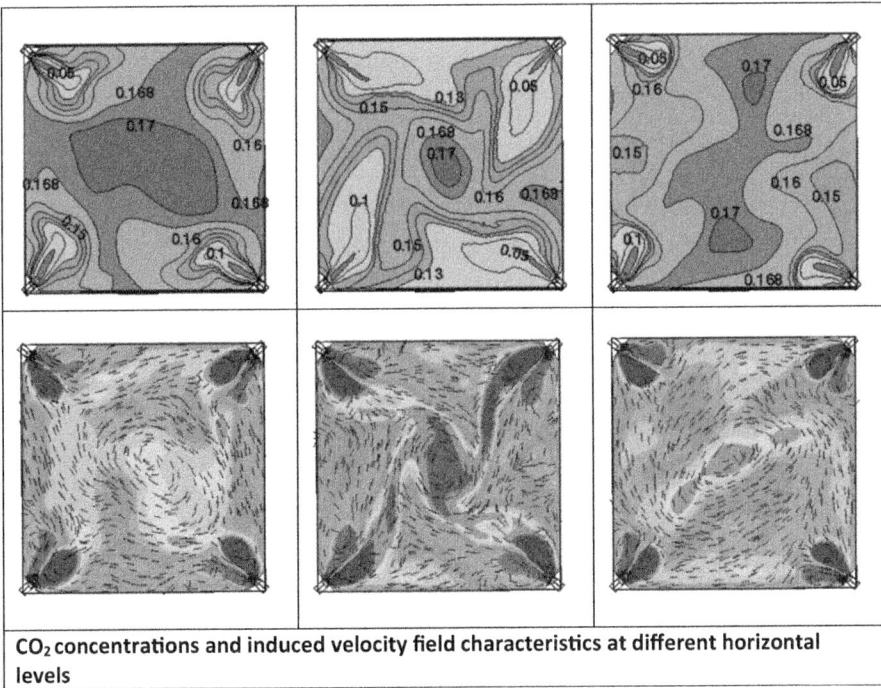

CO_2 concentrations and induced velocity field characteristics at different horizontal levels

Figure 6.6
Example of a computational fluid dynamics analysis for a coal-fired power plant.

order to perform feasibility studies through to detailed dimensioning of the plant. Hence, such approaches can:

1. calculate flow rates in mass, molar, or volume basis;
2. calculate streams composition and consequently the main products properties and emissions rates;
3. calculate useful power generation, power consumptions, and other losses;
4. calculate heat balances and predict design and off-design performance;
5. provide main design parameters such as heat exchanger area, reactors (boiler, gasifier, reformer, etc.), basic dimensions, etc.;
6. monitor and optimize plant performance; and
7. provide the basic input data for the techno-economic evaluation.

Contribution

The basic advantage of the Thermodynamic Tools in contrast to the CFD tools is that, in general terms, the calculated results can be yielded very fast with negligible computational cost, enabling the performance of several runs, test cases, parametric investigations in steady or unsteady conditions (depending on the software used), in off-design loads. These analyses lead to valuable conclusions about (1) the effective operation of the plant in terms of efficiency and the minimization of pollutant emissions, (2) the size and the main dimension of the most significant components, (3) the potential modifications for the plant improvement (if any), and (4) capital and operational cost.

Commercial software

Many organizations, research institutes, and universities have developed their own house-built simulation platforms for thermodynamic calculations. However, there are several commercial simulation tools with Graphical User Interface and well-established algorithms for fast calculations that are employed in both the industrial and the academic area. Some of them are summarized in Table 6.4 (Fig. 6.7).

6.9 Introduction to Renewable Sources

6.9.1 Biomass/Waste

Biomass

Biomass consists of any biological material derived from living, or recently lived, organisms. A definition used by the United Nations Framework Convention on Climate Change [33] is relevant here: nonfossilized and biodegradable organic material originating from plants, animals, and microorganisms. This shall also include products, by-products, residues, and waste from agriculture, forestry, and related industries as well as the nonfossilized and biodegradable organic fractions of industrial and municipal wastes

Table 6.4: Overview of Some Known Commercial Tools for Thermodynamic Simulations of Energy Systems.

	Software	Manufacturer	Focus Area	Custom-Built Models	Unsteady Conditions
1	GateCycle™	General Electric	Thermal power plants	No	No
2	IPSEpro	SimTech	Thermal power plants	Yes	No
3	EBSILON® Professional	STEAG Energy Services	Thermal power plants	Yes	No
4	ASPEN Plus™	Aspentech	Chemical processing	Yes	Through ASPEN dynamics
5	Apros	VTT/Fortum	nuclear and thermal power plants	Yes	Yes
6	ProSimPlus	Pro Sim	Chemical processing	Yes	No

(Fig. 6.8). Biomass also includes gases and liquids recovered from the decomposition of nonfossilized and biodegradable organic materials. The fuel produced by biomass through conversion process is called biofuel. Hence, biofuel is considered any hydrocarbon fuel that is produced from organic matter living or once living material in a short period of time (days, weeks, or even months). Biofuels can be considered upon as a way of energy

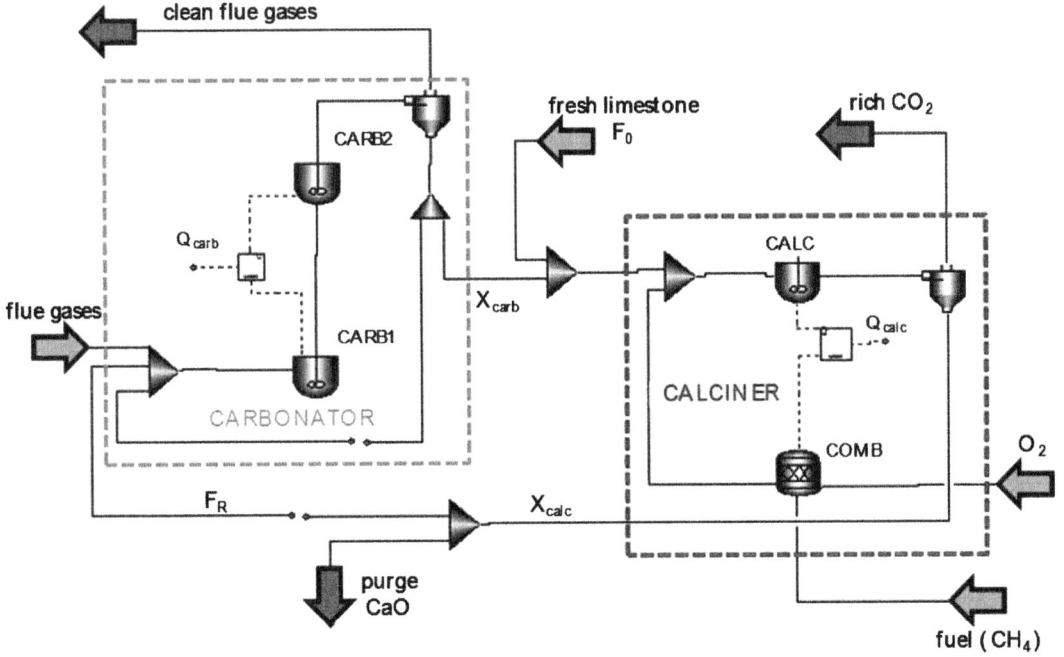

Figure 6.7
Example of an Aspen cycle [32].

| Wood pellets | Tree bark | Straw | Wood residues |
| Wood Chips | Olive kernel | Rice husk | Meat and bone meal (MBM) |

Figure 6.8
Biomass types.

security which stands as a substitute of fossil fuels that are limited in availability, as well as environmental-friendly fuel, since biomass is considered a carbon-neutral or greenhouse gas neutral (GHG neutral) fuel [34].

Biofuels are classified in the following four categories:

1. **First-generation biofuels**. The first-generation biofuels refer to fuels that have been derived from sources like starch, sugar, animal fats, and vegetable oil. The oil is obtained using conventional production techniques. Some of the most popular types of first-generation biofuels are ethanol, biodiesel, other bioalcohols, green diesel, biofuel gasoline, vegetable oil, bioethers, biogas, syngas, and solid biofuels.

2. **Second-generation biofuels**. The second-generation biofuels, also known as advanced biofuels, are fuels that can be derived from various types of biomass, including plant materials as well as animal materials. This means that the feedstock refers to nonfood biomass, in contrast to the first-generation biofuels. Some of the most popular products are hydrotreating oil, bio-oil, FT oil, lignocellulosic ethanol, butanol, and mixed alcohols.

3. **Third-generation biofuels**. The third-generation biofuels refer in the case that the biofuel carbon is derived from aquatic autotrophic organism (eg, algae). Light, carbon dioxide, and nutrients are used to produce the feedstock "extending" the carbon resource available for biofuel production. This means, however, that a heterotrophic organism (using sugar or cellulose to produce biofuels) would not be considered as 3G.

4. **Fourth-generation biofuels**. This is not a very common category, but it is referred to in a number of studies. In fourth-generation production systems, biomass crops are seen as efficient "carbon capturing" components that take CO_2 out of the atmosphere and lock

it up in their branches, trunks, and leaves. The carbon-rich biomass is then converted into fuel and gases by means of second-generation techniques. Crucially, before, during, or after the bioconversion process, the carbon dioxide is captured by utilizing the so-called precombustion, oxyfuel, or postcombustion processes. The fourth-generation biofuels are biohydrogen, biomethane, and synthetic biofuels.

The market for biofuels has been expanded in the time period 2005−2015 especially due to the high demand from the transport sector, as fuel in road vehicles. Biofuels started to be used in increasingly larger scales in aviation to generate electricity for cooking and even in maritime transport. The choice along the aforementioned categories of biofuels depends on a variety of biofuel categories, such as the land use, the competition with food crops, the efficiency of the production process, and the total energy balance.

Waste

According to the 2008/98/EC directive [35], waste is defined as any substance or object which its holder discards or intends or is required to discard. Many items can be considered as waste eg, household waste, green waste, agricultural waste, sewage sludge, wastes from manufacturing activities, packaging items, discarded cars, etc. Thus, all our daily activities can give rise to a large variety of different wastes arising from different sources. The directive 2000/532/EC [36] accompanied with the amendments 2001/118/EC, 2001/119/EC, 2001/573/EC [37−39] defines the categories of waste. The biogenic fraction of waste is directly linked to renewable energy. In specific, according to the directive 2009/28/EC referring to the promotion of the use of energy from renewable sources, "biomass" is the biodegradable fraction of products, waste, and residues from biological origin from agriculture (including vegetal and animal substances), forestry, and related industries including fisheries and aquaculture, as well as the biodegradable fraction of industrial and municipal waste.

6.9.2 Solar/Photovoltaic

Solar energy is harvested from the sun, usually through PVs which are arrays of cells containing an appropriate material, such as silicon, that converts solar radiation into electricity through the PV phenomenon. A wide range of applications use this technology nowadays, ranging from residential rooftop power generation to medium-scale utility-level power generation. Another method for utilizing solar energy is through the technology of concentrated solar power plants (CSPPs). These systems use mirrors or reflective lenses to focus sunlight on a fluid to heat it to a high temperature. The heated fluid flows from the collector to a heat engine for the production of electric energy. An advantage of this technology is that some types of CSP allow electricity generation during the night, as the heat can be stored for many hours [40].

6.9.3 Wind

Wind energy is a form of solar energy stored in the form of motion of air masses, originating from the uneven heating of the atmosphere by the sun, the irregularities of the earth's surface, and rotation of the earth. Wind flow patterns are modified by the earth's terrain, bodies of water, and vegetative cover. This wind flow, or kinetic energy, can be harvested to generate mechanical power or electricity. The mechanical power or electricity which is produced by the wind through this process is called "wind energy" or "wind power." The mechanical power can be used for specific tasks as grinding grain or pumping water, as it has been in the past, or it can be converted through a generator into electricity.

In specific, the kinetic energy is converted into mechanical power and following into electricity through the use of modern wind turbines. Generally wind turbines include a gearbox, although some modern turbines are gearless. The operation of this type of wind turbines is performed by turning the slow-moving turbine rotor through the gearbox into faster-rotating gears, connected with a generator which converts mechanical energy to electricity in a generator. Although less efficient, small turbines can be used in homes or buildings. Wind farms today appear on land and offshore, with individual turbines ranging in size up to 7 MW, with 10 MW planned. High-altitude wind energy capture is also being pursued today by several companies [40,41].

6.9.4 Hydro-Power

Technologies which can capture energy created from flowing water and convert it into electricity is called hydroelectric or hydropower technologies. Water drops gravitationally, driving a turbine and generator, in order to generate electricity [42]. While most of such systems are based on the principle of water falling from dams, some of them generate electricity through water flowing down rivers (run-of-the-river electricity) [40]. Hydroelectric power, using the potential energy of rivers, now supplies 17.5% of the world's electricity (99% in Norway, 57% in Canada, 55% in Switzerland, 40% in Sweden, 7% in the United States). Apart from a few countries with an abundance of it, hydro capacity is normally applied to peak-load demand, because it is so readily stopped and started. It is not a major option for the future in the developed countries because major sites in these countries having potential for harnessing energy from water flow in this way are either being exploited already or are unavailable for other reasons such as environmental considerations. Hydro energy is available in many forms, potential energy from high heads of water retained in dams, kinetic energy from current flow in rivers and tidal barrages, and kinetic energy also from the movement of waves on relatively static water masses. Many ingenious ways have been developed for harnessing this energy but most involve directing the water flow through a turbine to generate electricity [43].

6.9.5 Geothermal

Geothermal energy is simply the extraction of heat from Earth's crust. The vast majority of this internal heat (80%) comes from the radioactive decay of the materials, while the rest (20%) originates from the planet formation [44]. The high pressure values along with the high temperatures exhibited in the outer layers of the eEarth's crust, varying between 93°C and 370°C, heat the water entrapped in porous rocks. This hot water/steam many times finds an appropriate path and manifests itself on the surface through springs. These physical formations and the associated extracted energy can be used to heat a working medium and produce electricity. Steam and hot water from below the Earth's surface have been used historically to provide heat for buildings, industrial processes, and domestic water and to generate electricity in geothermal power plants. The geothermal resources used for these applications are classified into low, intermediate, and high enthalpy resources according to their reservoir temperatures. In power plants, two boreholes are drilled—one for steam alone or liquid water plus steam to filow up, and the second for condensed water to return after it passes through the plant. In some plants, steam drives a turbine; in others, hot water heats another fluid that evaporates and drives the turbine [40,44].

6.10 State-of-the-Art in Renewable Technologies

6.10.1 Biomass/Waste

The main advantage over other major renewable energy sources (RES), such as solar and wind energy, is that biomass can be stored and thus used to produce power and heat. In specific, biomass systems can be used in combination with solar or wind systems, in order to balance the intermittent behavior of an electricity system. Regarding the heating sector, biomass systems can be used in industrial applications or district heating networks, in order to cover the required demands in each case [45,46].

Electricity and/or thermal energy production technologies from biomass can be distinguished into two main categories. The first one corresponds to the direct use of raw biomass for energy production, while the second one corresponds to the production of the advanced biofuels through pretreatment/refining technologies of raw biomass (Fig. 6.9). The produced biofuels can substitute fossil fuels in the energy sector. Two different types of biomass can be grouped into two major categories: thermochemical and biochemical processes. Thermochemical processes are commonly applied to dry biomass fractions, while the biochemical technologies are widespread for the case of wet biomass fractions. Comparing the two categories, thermochemical processes have higher conversion efficiencies than the biochemical technologies, good destructive ability of organic compounds, and require shorter reaction times.

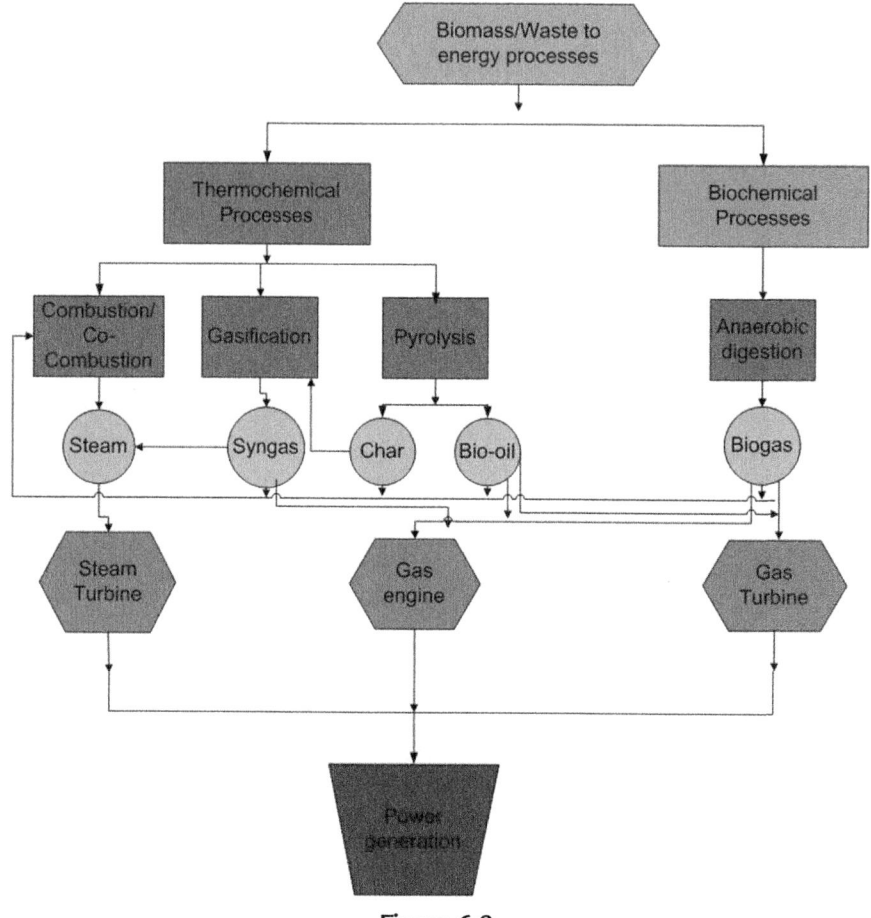

Figure 6.9
Categorization of biomass/waste into energy processes.

Thermochemical processes

The thermochemical processes can be classified in three technologies:

1. direct combustion,
2. gasification, and
3. pyrolysis processes.

Direct combustion is the process that converts the chemical energy stored in biomass into heat through its full oxidation. The hot flue gas produced during the process consists of carbon dioxide, water, nitrogen, excess oxygen, and minor quantities of other compounds, such as nitrogen and sulfur oxides, carbon monoxide, and other partially oxidized chemical species. A flue gas cleaning system exists for the purification of the produced flue gas. The heat generated by combustion can be used to produce electricity [47]. It is

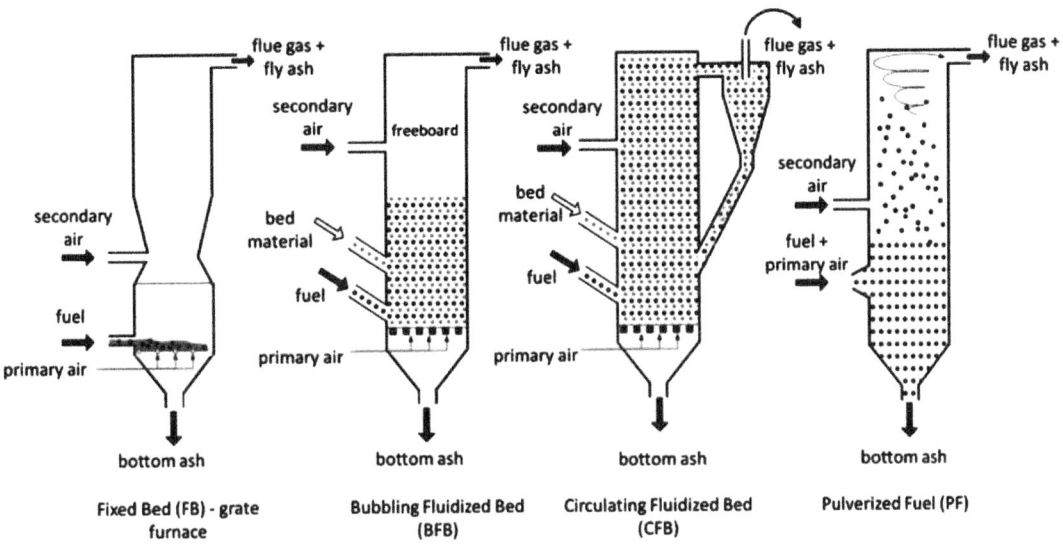

Figure 6.10
Principal combustion technologies for biomass [47].

considered the simplest conversion process with a wide range of applications from one megawatt to a number of hundred megawatts. It is grouped into three types of technologies: fixed beds, fluidized beds, and PF units (Fig. 6.10).

Fixed bed system is the most widespread technology for dedicated biomass combustion in the industrial scale. Combustion in this type of plant usually takes place on a grate located at the bottom. Regarding the movement of biomass fuels on the grate, this can be achieved either by gravity alone (fixed grates) or by applying some sort of movement of the grate, grouping in three categories: vibrating, moving by traveling, and rotating grates [45]. Fixed beds are used in industrial scale. It is an appropriate technology for fuels with high contents of moisture and ash, as well as with large particle size.

Regarding the fluidized bed systems, the bed is filled with an inert bed material. The air is supplied from the bottom of the combustion chamber for the fuel combustion. Provided the air velocity is sufficiently high, the bed and fuel particles are lifted from the bed and are effectively "fluidized." Two categories of fluidized beds are distinguished: bubbling fluidized bed (or BFB) and CFB. The first one corresponds to the case that the air velocity is low and the particles form a well-defined zone in the combustion chamber resembling a boiling fluid, while the second one corresponds to the case that the fluidization velocity is higher and the particles are continuously leaving the combustion zone and need to be "circulated" back after passing through a size-separation cyclone. This type of technology is characterized by high efficiency at low temperatures (650–900°C), low thermal NO_x emissions, minimization of the SO_x emissions, through the use of additive materials, and handling a variety of solid fuels with a wide range of particle size. However, fluidized bed technology requires an efficient dust-cleaning system, while the surfaces exposed to the bed undergo increased erosion. In addition, the phenomenon of agglomeration takes place in this type of reactor by the use of biomass fuels. This phenomenon is summarized as the formation of agglomerates of particles that are too large leading to not sufficient air velocity for their fluidization and they fall to the bottom of the boiler, leading to boiler shutdowns.

PF systems burn a suspension of very fine fuel particles mixed with combustion air. However, as the fine milling of biomass fuels is a quite energy-consuming and difficult process, this technology option for dedicated biomass installations is reserved only for sawdust and other biomass resources that are easily available as fine particles. However, PC units may still use biomass as a co-firing fuel, in a 10–20% thermal share basis, by applying some modifications in the existed boilers.

Biomass/waste gasification is a thermochemical partial oxidation (POX) process in which organic or fossil-based carbonaceous materials are converted into gas consisting mostly of carbon monoxide, hydrogen, and the carbon dioxide in the presence of a gasification agent (air, steam, oxygen, CO_2, or a mixture of these). The gasification process is divided into two main categories, based on the heat source to the reactor: the first one is the autothermal gasification, where heat for the gasification process is provided by the POX of the fuel inside a single reactor, while the second one is the allothermal gasification, where heat is provided by an external source to the reactor.

The reactors used for gasification process can be fixed beds or fluidized beds, such as those described in the previous section (BFBs and CFBs). More specifically, the fixed bed gasifiers can be distinguished as counter-current and co-current gasifiers according to the relative flow configuration between the fuel and the gasification agent. The syngas produced by the cocurrent gasifier has slightly higher lower heating value (LHV) and lower concentration of tars compared to the counter-current configurations. Fluidized bed gasifiers have stricter requirements in terms of fuel particle size but are more flexible in terms of operating load compared to fixed bed gasifiers [48].

Syngas can be combusted directly for power/heat production in internal combustion engines or gas turbines, upgraded for production of a substitute natural gas, or used as feedstock for chemicals production with the Fischer–Tropsch (F-T) synthesis process [34].

In **thermochemical pretreatment processes (pyrolysis processes)**, the organic material of biomass is decomposed by exposure to elevated temperatures in the absence of oxygen. It involves the simultaneous change of chemical composition and physical phase. The main products of the process are gas, liquid, and char. Their relative proportion depends on process parameters such as temperature range and residence time, as well as on feedstock properties.

Torrefaction is a thermochemical upgrade method for the production fuels appropriate for use in power sectors. Torrefaction is performed at low temperatures ($200-315°C$) and low heating rates ($10-70°C$/min) with a typical residence time of $1-3$ h [49]. Torrefaction results in the destruction of the fibrous structure and increases grindability, thus allowing its milling in existing coal mills. Its main product is a solid with higher heating value compared to biomass, with hydrophobic properties, non-biodegradable, and can be easily milled [50].

Biochemical processes

The biochemical conversion processes are distinguished into two categories: the fermentation process and the anaerobic digestion process. The fermentation process is a metabolic process in the absence of oxygen for production of liquid fuels, such as ethanol. This process finds commonly an application in transportation sector. Anaerobic digestion fits directly to the power sector. In specific, the anaerobic digestion of organic material—biomass or organic fraction of waste—is a complex process, involving four degradation steps in which microorganisms break down the biodegradable material in the absence of the oxygen. The main metabolic reactions that take place in the anaerobic digestion process are hydrolysis, acidogenesis, and methanogenesis. The main product of anaerobic digestion process is the biogas which is, depending on the process and feedstock, generally composed of $48-65\%$ methane, $36-41\%$ carbon dioxide, up to 17% nitrogen, $<1\%$ oxygen, and traces of hydrogen sulfide and other gases.

The technology of anaerobic digesters can be classified into the following categories:

1. One-Stage Continuous systems: Low Solid or Wet and High Solid or Dry
2. Two-Stage Continuous Systems: Dry—Wet and Wet—Wet
3. Batch Systems: One Stage and Two Stage

The temperature range is the basic parameter for the selection of reactor. Based on previous research work [45,51], the temperature range can vary in mesophilic or in thermophilic area and in some cases, mainly for experimental purposes, in psychrophilic area [52]. Typical temperature and residence time ranges are given in the Table 6.5.

The biogas yield is affected by many factors including the type and composition of substrate, microbial composition, temperature, moisture, bioreactor design, etc. More specifically, lower temperatures during the process are known to decrease microbial growth, substrate utilization rates, and biogas production. In contrast, high temperatures lower biogas yield due to production of volatile gases such as ammonia which suppresses methanogenic activities. Moreover, a range of pH values suitable for anaerobic digestion

Table 6.5: Temperature and Residence Time Ranges for
Psychrophilic, Mesophilic, and Thermophilic Area.

Thermal Stage	Temperature (°C)	Residence Time (Days)
Psychrophilic	<20	70—80
Mesophilic	30—42	30—40
Thermophilic	43—55	15—20

has been reported by various researchers, but the optimal pH for methanogenesis has been found to be around 7.0. Another factor is moisture. High moisture contents usually facilitate the anaerobic digestion; however, it is difficult to maintain the same availability of water throughout the digestion cycle [53]. Furthermore, the rate of anaerobic digestion affects the biogas yield production, through the type, availability, and complexity of the substrate. Different types of carbon source support different groups of microbes. Nitrogen is essential for protein synthesis and primarily required as a nutrient by the microorganisms in anaerobic digestion. Ammonia in high concentration may lead to the inhibition of the biological process and it inhibits methanogenesis at concentrations exceeding ~ 100 mM. Finally, the C/N ratio in the organic material plays a crucial role in anaerobic digestion. The unbalanced nutrients are regarded as an important factor limiting anaerobic digestion of organic wastes. The C/N ratio of 20—30 may provide sufficient nitrogen for the process.

6.10.2 Solar/Photovoltaic

The solar technologies are generally characterized as either passive or active, depending on the processes and the systems that are used in order to exploit this kind of energy. The passive systems are broadly used for heating, cooling, and lighting the buildings, naturally recirculating the ambient air inside some structures and a wide variety of other commercial and industrial uses [54,55]. This type of exploitation was known since prehistoric times used therefore in a primitive way, since it does not require sophisticated systems and technologies, but only suitable materials and appropriate orientation of the energy receiver to the sun.

The active systems suitably convert the solar energy into other useful outputs. These technologies are majorly distinguished into two subgroups, namely the solar PV and the solar thermal [56]. The PV routes can produce direct current electricity, which in turn is converted into alternating current and transmitted through the electrical grid. At first, the production of electricity using these specific technologies was substantially more expensive compared to other ways of production. However this financial gap is progressively closing, owing to the wide variety of materials capable of producing the PV effect. Consequently, the PV technology is nowadays recognized as very promising for satisfying the global energy needs and contributing to the environmental protection.

The PV technologies are mainly divided into three categories depending on the used materials and the commercial maturity of the technique. The first category uses the wafer-based crystalline silicon (c-Si) technology, either single crystalline (sc-Si) or multicrystalline (mc-Si). The second category includes thin-film systems constructed by one of the following choices: (1) amorphous and micromorph silicon, (2) cadmium telluride (CdTe), or (3) copper indium selenide (CIS) and copper indium-gallium diselenide (CIGS). This specific technology is lately commercialized. The third and final group consists of the concentrating or organic PV cells which are in a vast majority still under research and not in commercial use [57].

The first type of PV cells uses the crystalline silicon as basic manufacturing material. This type of material offers high efficiency around 14–19% in industrial and commercial use. In the market, a number of mature technologies are met that combine the advantages of this specific technology with the advantages of the thin-films constructions (second category) with lower cost and confirmed efficiency at 12–13% [58].

As already mentioned, one type of cells derived from the second group comprises constructions manufactured by amorphous silicon. This technology ensures efficiencies which vary from 5% to 7%. Several adjustments to the formation of these systems, namely the imposition of double- or triple-junction designs raises this efficiency to 8–10%. However, a disadvantage of this technology is the gradual material wear and the consequent efficiency degradation. Also the Staebler–Wronski effect plays a significant role in the system behavior. It gradually changes the material properties and causes problems to the conversion of sunlight into electricity. Further ongoing research efforts propose measures and techniques in order to reduce this phenomenon. Except for the amorphous silicon, another broadly used material for the PV cells of this category is the cadmium telluride (CdTe). The manufacturing of this type of cells follows the close-spaced sublimation methodology, since the latter provides several advantages for large area applications, such as the high deposition rates and the cost reduction [41]. On average, thin-film cells convert 5–13% of the solar energy into electricity. The efficiency is quite lower in comparison with the corresponding efficiency of the crystalline silicon cells.

The technologies of the third demonstrated group are still under development. Few examples already commercialized cannot be yet characterized neither as successful or not, since there is not an adequate number of research reports yet. PV cells are characteristic examples of systems belonging to this group. They rely on quantum dots/wires, quantum wells, or super lattice technologies. Based on the recent literature, these developed constructions and systems provide very high efficiency, since they are able to overpass the physical limitations of the conventional material used in the previously mentioned techniques. However, these technologies are in the fundamental research phase, since they require sophisticated and mature knowledge about nanotechnology. Nevertheless, it is expected that these technologies can give a boom in the development and the integration of the second major category of the solar active systems, namely the solar thermal systems.

These technologies use the solar energy as sole or supplementary source of electricity production in industrial scale. More specifically, a large number of reflectors expanded in a large area can (1) heat a working medium using the solar energy and formulate a Rankine cycle producing electricity or (2) directly convert the solar energy into electricity using appropriate PV modules through the PV effect.

Based on their operation mode, the solar power plants can be characterized as either PV or concentrating. The former are also known as solar farms or solar ranches, especially when the corresponding panels are installed in decentralized or agricultural areas. These systems comprise a vast number of successive ground-mounted PV systems. These can have a specific constant tilt, or they can adjust their inclination using a single or dual axis mechanism for tracking the sun [42]. The extra required systems and modifications provide better efficiency, but raise the overall installation cost simultaneously.

The operating cost of such a unit is low, since no fuel is required. On the other hand, the incomes are mainly affected by the sale price and the final electrical output. The most sophisticated plant of such type is located in Topaz, California, and has a final nominal electrical output of 550 MW [59]. However, the average installed capacity of these systems in operation varies from 1 to 220 MW.

Contrary to conventional PV systems, the concentrator photovoltaic (CPV) technologies use curved mirrors and panels to concentrate the sunlight onto a small but highly efficient solar cell. Except for the electricity production through the PV phenomenon, these systems permit the use of working medium for extra efficiency. Based on research findings, this technological aspect has the potential to be the most competitive solar energy system among all the previously mentioned. Some types of CPVs reach efficiencies up to 42%, while the improvement of the technology can increase this value up to 50% by the mid-2020s [60]. Similar to the previously mentioned category, this type of technology also shows limited costs, since no fuel is used. The main cost is the installation one, lying between 1.40 and 2.20 euros per Watt-peak.

One criterion for distinguishing the several types of CPV technologies using a working medium for extra electrical capacity is the maximum temperature of the working medium. Consequently, the solar thermal power plants of such type can be classified as: (1) low (up to 100°C), (2) medium (up to 400°C), and (3) high (above 400°C) temperature cycles.

The systems classified in the first group use as primarily working mediums organic fluids such as toluene or methyl chloride or refrigerants like R-11, R-11, and R114. The capacity of this technology can reach an upper limit of 50 KW and an efficiency of approximately 7–8% for the Rankine cycle and 25% for the collector. Therefore, the total efficiency of the unit was as low as 2%. This technology was firstly implemented at the early 1970s mainly in Africa. This technology was firstly implemented at the early 1970s mainly in Africa, presenting, though, high operational cost and low efficiency. A way to improve the system behavior and limit the cost was the replacement of the flat-plate collectors with solar ponds [61].

The systems classified in the second group use the line parabolic solar collector to improve the performance and the maximum temperature of the working medium. A parabolic solar collector is comprised of parallel rows of reflectors formulated in parabolic shape. At the focal point of this parabola there is the absorber tube, a black treated pipe covered with a glass tube in order to minimize the thermal losses. A suitable heat transfer fluid is circulated through this pipeline in order to absorb the solar energy. The temperature of this working medium can be greater than 673 K. The biggest application of such a type of CPVs is installed mainly in the United States, in the southern California, with a total installed capacity of 354 MWe. The parabolic trough solar power plant can collect up to 60–70% of the incident solar radiation and has achieved a peak electrical conversion efficiency of 20–25% (net electricity generation to incident solar radiation) [62].

Recent efforts have been focused on the evaluation of mainly two types of concentrating PV power plants in order to improve the attained temperature peak of the working medium. These two types were the paraboloid dish Stirling engine (PDCSSPP) and the central tower receiver (CTRSTPP). The former is capable of producing a total electrical output in kilowattKW, whereas the later in megawattMW. In PDCSSPP systems the concentrator tracks the sun's orbit, using an appropriate configuration, rotating about two axes. At focal point the receiver absorbs solar radiation and transfers the thermal input to the Stirling machine. In CTRSTPP systems the incident radiation is reflected from an array of large mirrors to a receiver, located at the top of a tower. The concentrated radiation heat is absorbed by a working medium flowing through a pipeline, formulating a Rankine or a Brayton cycle. In what concerns the economic behavior of the two systems the unit (kWh_e) electric energy–generating cost for PDCSSPP is estimated between INR 8.76 and INR 11.06, while the same quantity for CTRSTPP varies between INR 10.09 and INR 12.10 assuming the same life span (30 years) of the solar power plants and interest rate (10%) on investment. The unit power generation cost of the PDCSSPP is less than the PTCSTPP and CTRSTPP because its efficiency is high. However, the year-round power output from the PDCSSPP is less compared to PTCSTPP and CTRSTPP; the main reason for this is that the PTCSTPP and CTRSTPP systems have thermal storage facility (Fig. 6.11).

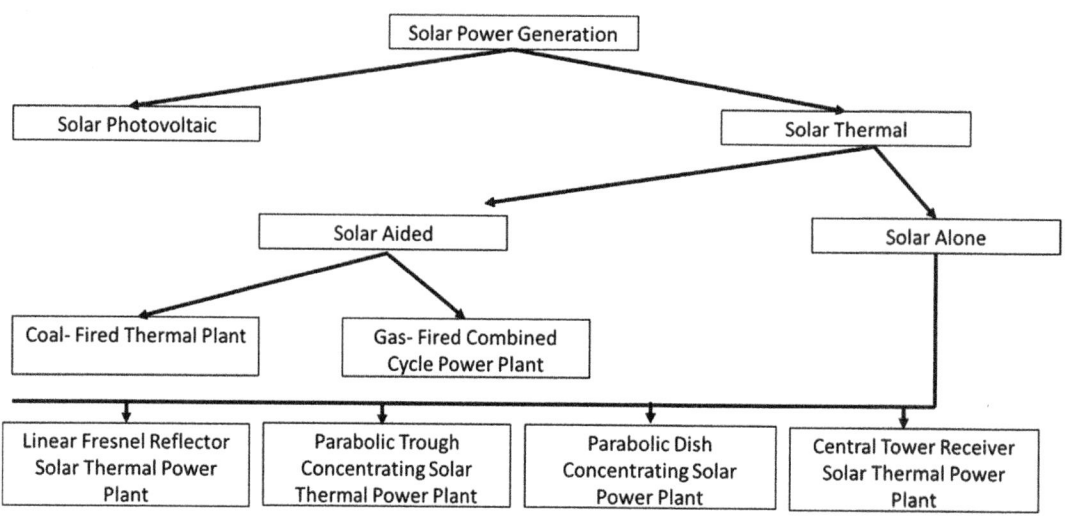

Figure 6.11
Basic types of solar energy systems [62].

6.10.3 Wind

The rapid development of wind energy conversion systems (WECS) took place in early 1990s, by establishing wind energy sources as a significant part in the global renewable energy market [63]. As the wind energy sources' contribution to the energy mix raises, the constructors and engineers declare higher interest in the improvement of the systems efficiency, the reduction of their manufacturing and operating costs, and the guarantee of their reliability. Due to the stochastic behavior of the wind, the efficient energy production from WECS in a wide range of weather conditions and wind velocities is a continuous challenge [64].

The WECS are mainly divided into two main categories based on the flexibility of the turbine blades, rotational speed, and, in particular, if this is varied or constant as the time passes. These two types of wind energy systems are namely the variable speed and fixed speed turbines, respectively. The first developed and commercially used technologies belong to the latter group, due to mainly the technological restrictions [65,66].

One characteristic type of this category, largely applied in Danish wind turbines, is the constant speed WECS with multiple-stage gearbox and a squirrel-cage induction generator (SCIG) directly connected to the grid through a transformer [67]. Their basic advantages are the simplicity, the limited total cost, as well the reliability. However, there are also some basic disadvantages due to mainly the old-fashioned technology that is implied in this type of wind energy systems. Firstly, the rotational speed of the blades is not adjustable and varies in a narrow range around the synchronous speed (1–2%) [67]. As a result, this option is not able to extract maximum energy from the wind in a wide range of velocities far from the synchronous speed. Furthermore, the fluctuations of the rotational speed directly affect the electrical output and impose mechanical fatigue on the structure. Finally, the appliance of multiple-stage gearbox inevitably implies some disadvantages, such as the heat due to friction, high mechanical stresses, and noise.

An evolution of this technology was the replacement of the SCIG with a wound rotor induction generator connected with variable resistor. Nevertheless, as the turbine blade's resistance due to the wind velocity raises, the slip of the applied generator gets higher and extract much more energy from the rotor and consequently from the electrical grid, reducing the system efficiency. The increase of rotational speed implies the necessity for high-rated resistors. The geometrical and the construction limitations of this arrangement allow the blades to rotate in a range of 10% above the synchronous speed [68].

The variable speed generation systems are able to store the varying wind power and maintain optimal energy generators by alternating the rotational speed [69]. One characteristic type of WECS belonging to the variable speed category uses doubly fed induction generator and a partial-scale power converter. The rotor of the generator is connected to the grid, while its stator with a back-to-back converter. This type of generator can deliver the electrical grid with wide varying speed energy, from supersynchronous to subsynchronous speeds, namely ±30% compared to the synchronous speed. The regulation of the energy output to the electrical grid can be obtained basically by adjusting the active power of the rotor-side converter or decoupling the active from the reactive power. However, this technology uses slip rings and brushes, affecting the maintenance cost. Furthermore, the fault ride through capability of this technology is quite low mainly because of the direct connection between the generator stator and the grid [70].

An additional alternative option belonging to this specific category is the wind generator systems comprised of SCIG and a full-scale back-to-back converter. This scheme is quite similar to the first referenced wind turbine type, but it also contains a specific type of converter. By using this additional arrangement the operator can affect the generator speed by controlling the generator-side converter and the energy output by similarly adjusting the grid-side converter. However, a major disadvantage of this particular type is its high cost. A variation of the latter wind turbine system replaces the SCIG with a permanent magnet synchronous generator (PMSG). This system offers high efficiency and robustness. However, it is also the most costly option among all the previously mentioned technologies mainly due to the use of permanent magnetic materials.

All the aforementioned systems use as a primary arrangement and mechanical construction a multiple-stage gearbox. However, as it is known, this mechanical component introduces losses mainly due to friction between the teeth. It is estimated that each stage results in 1% loss from the energy input. The most common type of gearbox used nowadays in the large wind turbines comprise of three stages [71]. As a consequence, the energy output is a priori 97% lower than the wind energy input.

In order to surpass this basic problem and restriction, the designers and the operators develop the direct-drive systems, which do not include any gearbox. Therefore, the connection between the blades and the generation is direct. One representative type belonging to this category consists of an electrically excited synchronous generator (EESG) and a full-scale power converter. Due to the lack of an integrated gearbox system, the generator must be designed with high number of poles. This influences the construction of the component, which gets bigger and heavier. Moreover, the EESGs require slip rings and brushes, thus raising the maintenance cost.

Nowadays, the most prominent technology belongs to the direct-drive systems and includes a PMSG with a full-scale converter. This system does not contain any external excitation and slip rings, so the efficiency and the reliability are high. Furthermore, compared to the previous generator type, the PMSG volume and weight are considerably reduced, mainly because of the permanent magnets usage. Finally, the lack of gearbox improves significantly the noise behavior of the construction. All the previous mentioned types are shown in Fig. 6.12.

6.10.4 Hydropower Systems

The hydropower systems exploit the energy from running or falling water mass flows in order to produce mainly electricity or mechanical work for other human activities and purposes. The human society was familiar with the utilization of this type of energy some hundreds of years ago, however, in a primitive way. Nowadays the hydropower systems are the most important renewable energy sources, providing 17% of the electricity in a global range, since the installed capacity is equal to $\sim 720\,GW$.

The most common hydropower systems exploit the energy from falling water and are categorized according to the electrical output, the size, and the operation. The basic types of this technology are (1) conventional hydroelectric systems, (2) run-of-the-river hydroelectric systems, (3) small hydro systems, (4) microhydro systems, (5) conduit hydroelectric systems, and (6) pumped-storage hydroelectric systems.

The most common type is the conventional hydroelectric. It is used mainly for high electrical output ($>30\,MW$). The most characteristic component of this technology is the dam, which collects huge amounts of water savings at a relative high level compared to the level of the turbine. The height difference between the upper and lower tank is the motive force which rotates the turbine for electricity production. The efficiency of these large installations can reach the levels of 95% [73]. The hydropower system usually operates to fulfill the peak demands, mainly in hot periods of time (summer). Consequently, when the demands do not come close to the critical points (e.g., night hours—off-peak hours), one part of the final electrical production is used to reversibly

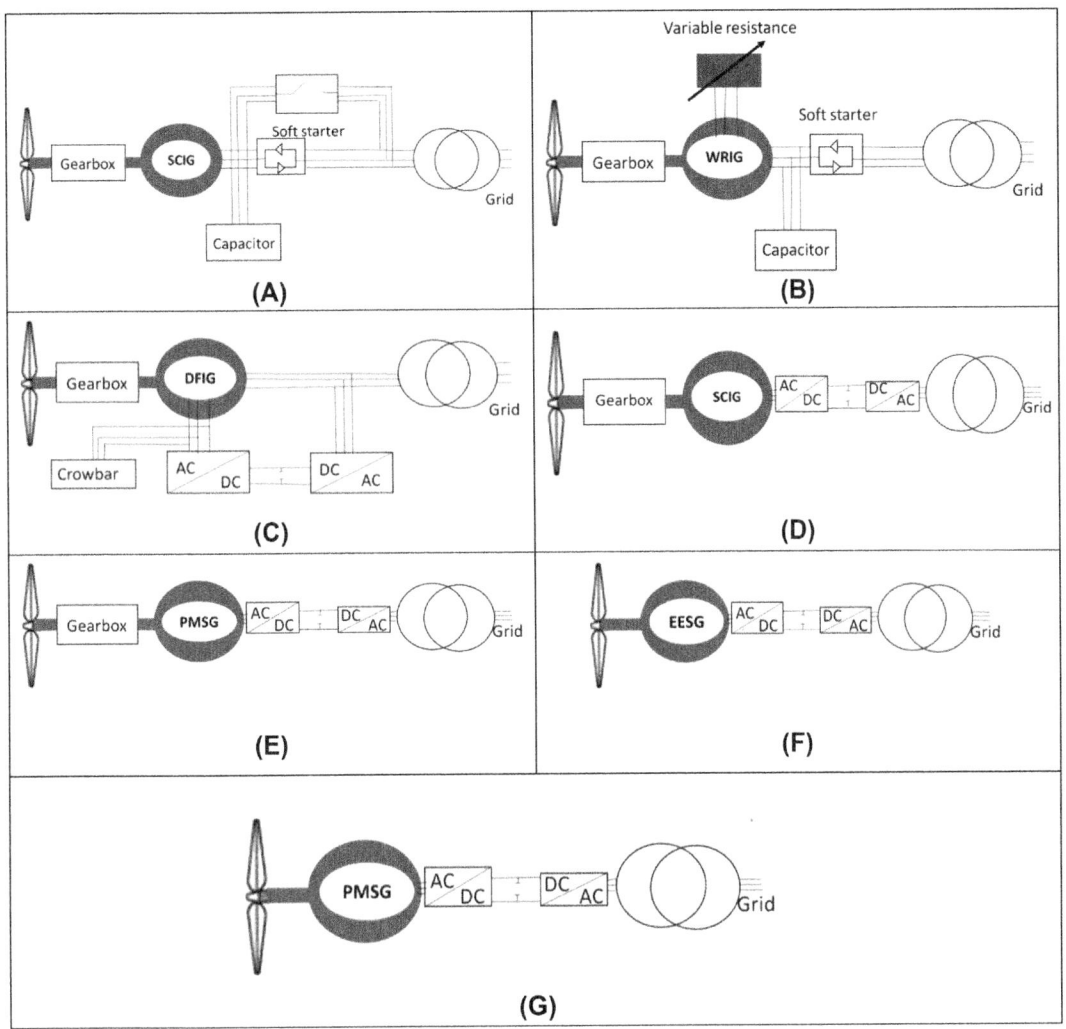

Figure 6.12

Different types of wind energy technologies (A): constant speed WECS with multiple-stage geared SCIG, (B): limited speed WECS with multiple-stage geared WRIG, (C): variable speed WECS with multiple-stage geared DFIG and partial-scale converter, (D): variable speed WECS with multiple-stage geared SCIG and full-scale converter, (E): variable speed WECS with multiple-stage geared PMSG and full-scale converter, (F): direct-drive variable speed WECS with EESG and full-scale converter, (G): direct-drive variable speed WECS with PMSG and full-scale converter) [72]. *WECS*, wind energy conversion system; *PMSG*, permanent magnet synchronous generator; *WRIG*, wound rotor induction generator; *EESG*, electrically excited synchronous generator.

move a quantity of the working medium from the lower tank to the higher. This procedure recycles the water savings and increases the total electricity load. The smaller-in-size constructions compared to previous ones, which produce electrical output between

100 KW and 30 MW or lower than 100 KW, constitute the small hydropower and microhydropower systems, respectively. As the electrical output decreases, also the system efficiency decreases and varies between 80 and 85%, since the extraction of useful power from low mass flow rates is more difficult [74]. Finally, the run-of-the-river hydropower plants are systems which produce electricity without storing water reserves, by exploiting the natural flow of a river. This type of power plants, in contrast with the previous two systems, cannot be used as peaking or base-load plants, but only as intermittent energy sources.

The hydropower technologies are very mature, trustworthy, reliable, and long-term operational [75]. Yet, the first two formats are quite flexible, since the ramp-up and ramp-down rates are quite high (60−90 s from cold phase to full load) [76]. Moreover, they show no pollutant emissions, except for the CO_2 emissions during the construction phase and the methane emissions due to the water reservoirs. However, they show high capital cost, mainly due to the water head component, and remarkable damage on the ecosystem during the formation of higher-level tank.

6.10.5 Geothermal Energy

The developed systems that exploit the geothermal energy are called geothermal power plants. Globally, in 2013, the total installed thermal capacity was equal to ∼11,700 MW. It is estimated that geothermal energy can fulfill around 3% of global electricity demands and 5% of global heating demands by 2050 [77]. The high cost-effectiveness, the sustainability, the environmental-friendly behavior, and the reliability enhance the use of these technologies. In order to improve the system efficiency and the final electrical output, a supplementary source can be used to provide extra heat input for the working medium or extra electricity production. These additional energy sources can be biomass, solar, wind, gas, and oil.

Mainly there are four different types of power plants used to exploit the earth thermal output. The distinction among them is based on the cycle formation and its individual components. The first developed system for this form of energy was dry steam power plants, which begun to operate in 1904. The heated liquid extracted from the inner earth layers is driven through a pipeline system to a turbine, producing with the familiar, conventional way electricity. The cycle closes with the condensation of the used working medium and its return to the earth crust layers (even for further recirculation through the system). The steam power plants produce nowadays a little less than 40% of the US geothermal electricity.

However, the most common power plant type nowadays, producing 45% of the geothermal electricity in the United States, is the flash power plants. This technology is basically an evaluation of the former technology used majorly in cases where the hot mixture extracted from the inner earth surfaces has high pressure and therefore is not vaporized. The derived water gradually loses pressure as it is driven through an integrated pipeline system and as a consequence it abruptly vaporizes. The separation of the two phases is carried out in a separator. The steam is driven to the turbine, while the water can be directly used for buildings' heating. Based on the number of the utilized steam separators, these systems can be characterized as single or double flash power plants.

The technological development in the mid-dle 1980s gave also the opportunity to the formulation of systems which are capable of exploiting the low-temperature geothermal sources, namely the springs that extract low-pressure and low-temperature mixtures (below 74°C). Approximately 15% of the installed geothermal power plants in the United States belong to this category, entitled as binary power plants. In this technology, the extracted hot mixture exchanges energy with an appropriate working medium, used to formulate a Rankine cycle. The two liquids are constantly separated and never mixed. The energy transfer takes place in a properly configured heat exchanger. The geothermal water can constantly recirculate in a closed loop, ensuring the close-to-zero emissions, the reliability of the system, and the extension of the project lifetime. A small difference to the operation of these units results in the formation of new type power plants called two-phase binary systems. Two-phase systems are similar to traditional binary cycles, except the steam flow enters the vaporizer/heat-exchanger, while the geothermal liquid is used to preheat the organic motive fluid. The steam condensate either flows into the preheater or is combined in the geothermal liquid after the preheater.

The fourth and final type of geothermal systems is called flash- binary combined cycle. This technology combines the formation and the organization of the first and third type of geothermal units. More specifically, the derived steam is driven to a high-pressure turbine for electricity production. Afterward, the low-pressure mixture is condensed in a binary system. This hybrid system combines the advantages of the two previously mentioned techniques. It presents higher efficiency as the pressure of the derived mixture is higher.

6.11 Role of the Renewable Energy: Economic Trends, Policies, and Incentives

6.11.1 Green Economy

According to UNEP (2011) [78], green economy is defined as "low-carbon, resource efficient, and socially inclusive." The main objective of the green economy is the "improved human well-being and social equity, while significantly reducing environmental risks and ecological scarcities." It aims at "getting the economy right" by reducing polluting emissions, increasing resource efficiency, preventing the loss of biodiversity, and valuing ecosystem services [79].

The green economy concept was incorporated into the general framework at different levels of EU policy, through EU directives. In this sense, the European Commission has set a long-term goal to develop a competitive, resource efficient, and low-carbon economy by 2050 according to the "Roadmap for moving to a competitive low-carbon economy in 2050" (COM(2011)112 final) [80]. Focusing on industrial policies, the commission works to promote the competitiveness of Europe's primary, manufacturing, and service industries and help them seize the opportunities of globalization and of the green economy.

In order to set a framework for bioeconomy deployment, the European Commission has put forward the European strategy for building a sustainable biobased economy as an opportunity to address several challenges, such as food security, natural resource scarcity, fossil resource dependence, and climate change, with emphasis on the sustainable use of natural resources, competitiveness, socioeconomic and environmental issues. In this sense, policies related to resource efficiency need to be seen as efforts for shifting toward a resource-efficient and low-carbon economy within the global context of green economy transition [81].

Especially power and heat production from biomass is an option that has become relevant in the last decades. The production of liquid and gaseous biofuels for transport and stationary applications is also rising. The bioenergy sector is highly polyparametric due to the wide range of potential feedstocks, as the technical routes for the conversion of biomass to energy. It is mentioned that there are considerable gaps about the assessment of the biomass volumes, which are used for energy production purposes. In 2013, biomass accounted for about 10% of global economy.

6.11.2 Renewable Energy Contribution in the Energy Mix

Consumption of fossil fuels is dramatically increasing at present, following the industrialization of developing countries, the increase of the world population, along with improvements in the quality of life. A major issue arising from this excessive fossil fuel consumption, which leads to an increase in the rate of diminishing fossil fuel reserves, is also the significant adverse environmental impact, resulting in increased health risks and the threat of global climate change [82]. Developed countries are engaging in changes toward environmental improvements as they are becoming more politically acceptable globally. These changes, which are slowly coming in the foreground, involve the search for more sustainable production methods, waste minimization, reduced air pollution from vehicles, distributed energy generation, conservation of native forests, and reduction of GHG emissions [83].

Due to the increasing consumption of fossil fuel, alert for the sufficiency of energy sources has been signaled, which has raised the interest and stimulated the promotion of renewable alternatives to meet the developing world's growing energy needs [84,85]. The

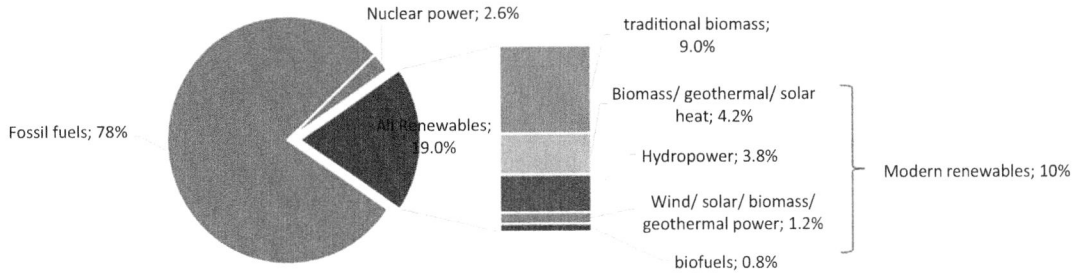

Figure 6.13
Share of global energy consumption for available energy sources in 2012—Breakdown for modern renewable energy [92].

contribution of excessive use of fossil fuels in the global warming through carbon dioxide emissions has made necessary the use of renewable clean energy [86]. To restrict the amount of these greenhouse emissions an international treaty has been signed with the targets of overall pollution prevention, the Kyoto protocol agreement [87,88].

In 2012, an estimated 19% of global final energy consumption was derived from renewable energy provided, a share which continued to grow strongly in 2013 [89]. Of this total share, 9% was traditional biomass (mostly solid biomass eg, wood), which is currently used primarily for cooking and heating in remote and rural areas of developing countries, while modern renewables were estimated to have increased, reaching ∼10% (Fig. 6.13).

The rapid growth of modern renewable energy along with the gradual migration away from traditional biomass and the progressive increase in total global energy demand lead to an almost constant share of renewable energy use compared to 2011 [45,90].

The main renewable energy sources and the options for their usage vary from heat and power generation, fuel production, etc. as given in Table 6.6. In particular, four main

Table 6.6: Main Renewable Energy Sources and Their Usage [91].

Energy Source	Energy Conversion and Usage Options
Hydropower	Power generation
Modern biomass	Heat and power generation, pyrolysis, gasification, digestion
Geothermal	Urban heating, power generation, hydrothermal, hot dry rock
Solar	Solar home system, solar dryers, solar cookers
Direct solar	Photovoltaic, thermal power generation, water heaters
Wind	Power generation, wind generators, windmills, water pumps
Wave	Numerous designs
Tidal	Barrage, tidal stream

sectors are increasingly involved in the use and exploitation of modern renewable energy: power generation, heating and cooling, transport fuels, and rural/off-grid energy services. The breakdown of modern renewables, as a share of total final energy use in 2012, is also presented in Fig. 6.13. In particular, hydropower generated an estimated 3.8%; other renewable power sources comprised 1.2%; heat energy accounted for ~4.2%; and transport biofuels provided about 0.8% [45].

Taking into account the current situation, the share of renewable energy is expected to increase significantly. The global renewable energy scenario, as estimated by 2040, is presented in Table 6.7.

It has to be noted that by the end of 2013, China, the United States, Brazil, Canada, and Germany remained the top countries for total installed renewable electric capacity, with China leading with a share of 24% of the global renewable power capacity, including an estimated 260 GW of hydropower [94,95]. Furthermore, regarding nonhydro renewables, the leading countries were again China, the United States, and Germany, followed by Spain, Italy, and India, with EU-28 concentrating a share of 42% of the global total installed capacity, which is higher than any region [46,95]. Therefore, as expected, the countries with the highest capacity of nonhydro renewables per inhabitant were all in Europe [96]. However, the total share of global power renewables in the European Union is decreasing as renewable electricity markets are expanding worldwide [95].

6.11.3 Renewable Energy Focus on Power Energy

The power sector was the sector with the most significant growth during 2013, with an increase of more than 8% over 2012, as the global capacity exceeded 1560 GW [95].

Table 6.7: Global Renewable Energy Scenario by 2040 [93].

	2001	2010	2020	2030	2040
Total consumption (million tons oil equivalent)	10,038	10,549	11,425	12,352	13,310
Biomass	1080	1313	1791	2483	3271
Large hydro	22.7	266	309	341	358
Geothermal	43.2	86	186	333	493
Small hydro	9.5	19	49	106	189
Wind	4.7	44	266	542	688
Solar thermal	4.1	15	66	244	480
Photovoltaic	0.1	2	24	221	784
Solar thermal electricity	0.1	0.4	3	16	68
Marine (tidal/wave/ocean)	0.05	0.1	0.4	3	20
Total RES	1365.5	1745.5	2964.4	4289	6351
Renewable energy source contribution (%)	13.6	16.6	23.6	34.7	47.7

The increase in hydropower was 4% reaching a capacity of 1000 GW, while in other renewables collectively was nearly 17% to an estimated 560 GW [95]. Globally, hydropower and solar PV each accounted for about one-third of renewable power capacity added in 2013, followed closely by wind power (29%) [95]. Solar PV capacity growth exceeded wind power capacity for the first time, worldwide (Table 6.8) [97].

The global focus of policy support and investment in renewable energy is still primarily on the electricity sector. Consequently, share of electric generation capacity from renewables added globally each year is constantly growing [92]. More than 56% of net additions to global power capacity were from renewables in 2013. Even higher shares of capacity added were achieved in several countries around the world [98,99]. For example, in the European Union, renewables were accounted for the majority of new capacity for the sixth year running [99].

By the end of 2013, 26.4% of the world's power capacity was generated by renewables [100,101]. In this way 22.1% of global electricity was from renewables. The largest share was produced from hydropower, providing about 16.4% as seen in Fig. 6.14 [95,102,103]. Despite the rapid increase rate of renewable capacity, the share of renewables in global electricity generation has a lower rate of increase due to the variable nature of the added renewable capacity and the high levels of increase of overall demand.

The penetration of renewables is high in several countries, yet variable. For example, in Denmark wind power covered 33.2% of electricity demand and 20.9% in Spain, for the year 2013, while in Italy, 7.8% of total annual electricity demand was covered by solar PV [95]. Further, hydropower has been used as a balancing mechanism for systems which incorporate high shares of variable renewables, sometimes with the aid of pumped storage, thus providing the largest share of renewable electricity worldwide. Geothermal power and biopower, which are also nonvariable and can play a similar role of balancing, provide significant shares of total electricity in some countries. For example, in Iceland 29% of

Table 6.8: Power Generation Installed Capacity and Capacity Growth (GW) for the Year 2013 [92].

Power Generation (GW)	Added During 2013	Existing at the End of 2013
Biopower	5	88
Geothermal power	0.5	12
Hydropower	40	1000
Ocean power	~0	0.5
Solar photovoltaic	39	139
Concentrating solar thermal power	0.9	3.4
Wind power	35	318

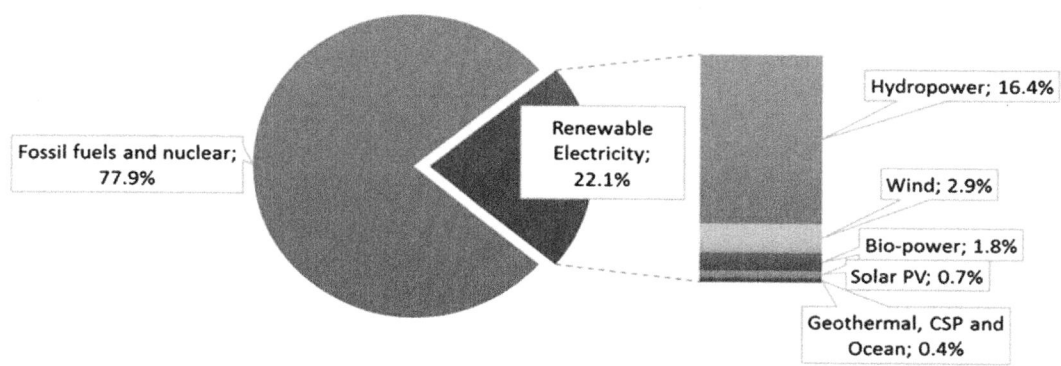

Figure 6.14
Estimated renewable energy share of global electricity production, end of- 2013 [92].

electricity generation and more than 20% in El Salvador and Kenya are from geothermal power [104].

6.11.4 Policy Targets

Policy targets for the increased deployment of renewable energy technologies existed in 144 countries as of early 2014, up from the 138 countries reported in GSR 2013. The types of the policies refer to the following categories: power generation policies, heating and cooling policies, transport policies, green energy purchasing and labeling, and city and local government policies. In the most cases, the targets are set for a future year, which is called target year. However, in some cases a range of years is defined as target time period or no year is reported.

1. **Power generation policies**: A variety of power generation policies are met around the world. More specifically, these policies include feed-in policies, renewable portfolio standards, net metering, tax reduction or exemptions, grants, low-interest loans, and public competitive bidding/tendering mechanisms. Hence, regulatory policies and financing mechanisms have been established in most of the countries in order to promote the renewable power capacity according to the domestic circumstances.
2. **Heating and cooling policies**: Compared to power generation, heating and cooling does not show similar development trend. In general, countries set targets, policies, and incentives for the promotion of renewable heating and cooling technologies. As far as Europe is concerned, the EU-28 Member States have introduced targets for specific shares of renewable heating and cooling. Countries in Africa, Europe, and the Middle East have also set targets for the use of solar water heating.
3. **Transport Policies**: Most policies for increasing renewable energy in transport focus on the support for the production, promotion, or use of biofuels. Common policies include

biofuel production subsidies, biofuel blend mandates, and tax incentives. Many countries are active in promoting "clean fuels," such as the biomethane, renewable hydrogen, and electricity from renewable energy.

4. **Green energy purchasing and labeling**: Green energy labeling provides consumers with the opportunity to purchase "green" electricity as well as "green" gas, heat and transport fuels by evaluating the generation source of available energy supply options. Green power labels are employed in a number of countries and are mostly voluntary. However, in some countries their application is obligatory. Hence, a number of governments require that utilities and/or electricity suppliers offer green power products.

5. **City and local Government policies**: Thousands of cities and towns have active policies, plans, and targets to advance renewable energy. The targets, as set in city policies, are focused on reducing emissions and energy demands through efficiency improvements. Other targets are the creation of a local industry, as well as the independence from the national grid. Local governments made increasing use of their authority to regulate, make expenditure and procurement decisions, facilitate and ease the financing of renewable energy projects, and influence advocacy and information sharing.

6.11.5 Market and Industry Trends

Renewable energy technologies presented a rapid growth rate in installed capacity as well as output energy produced, during the years 2009 through 2013 [95]. Solar PVs presented the fastest capacity growth rates of any energy technology, while wind was the renewable technology with the most power capacity added. On the other hand, the use of modern renewables for heating and cooling showed a steady progress besides the lack of sufficient data for many heating technologies and fuels [56]. A decrease in biofuels production rate for use in the transport sector took place for the years 2010−12, in spite of high oil prices, which increased again in 2013 [105].

Over time, as the situation in the related markets have changed, renewable energy industries have come to a point where they have to address significant challenges—as well as a wide range of opportunities. In Europe, financial support for renewables declined in a growing number of countries, sometimes retroactively, and in a rate faster than the decline in technology costs. These measures are partly owed to the ongoing economic crisis in some member states, the related electricity overcapacity, and rising competition with fossil fuels. Up to 2013, a significant loss of start-up companies (especially solar PV) has emerged in Europe, resulting in widespread financial losses, mostly due to the increased cost of capital and reduced investment which restricts funding of new projects [95]. However, an overview of the share in gross final energy consumption in the European Union (EU28) indicates that a rise of renewables up to 14.1% in 2012 is estimated, compared to 8.3% in 2004 [106].

Table 6.9: New Investment by Renewable Energy Technology in the Period 2004–13 [92].

Renewable Energy Category	2004	2005	2006	2007	2008	2009	2010	2011	2012	2013
Solar power	12.1	16.3	21.7	38.7	59.5	62.9	100.3	157.8	142.9	113.7
Wind power	14.5	25.1	32.1	56.6	69.3	73.0	94.8	85.9	80.9	80.1
Biomass and waste-to-energy	6.2	8.0	10.6	13.2	14.1	13.6	14.2	15.5	11.1	8.0
Hydropower <50 MW	1.7	4.9	5.4	5.5	7.2	5.4	4.8	6.8	6.0	5.1
Biofuels	3.7	9.2	27.6	29.3	19.2	10.4	8.9	9.4	6.6	4.9
Geothermal power	1.3	1.0	1.4	1.9	1.8	2.7	3.5	3.7	1.8	2.5
Ocean energy	0.0	0.1	0.9	0.7	0.2	0.3	0.2	0.3	0.2	0.1

The types of energy sources with sufficient availability in certain areas, such as geothermal, hydropower, and biopower, have offered a financially viable solution for energy production in these areas, while other technologies such as onshore wind and solar PV are also becoming cost competitive in an increasing number of locations [46]. The decrease of levelized costs of generation from onshore wind and, particularly, solar PV over the past 5 years, in contrast to an increase in average global costs from coal and natural gas generation has resulted in an increasing number of wind and solar power projects without public financial support, especially in Latin America, but also in Africa, the Middle East, and elsewhere [46,101]. At the same time, many utilities in Asia, Europe, and North America are investing in wind, solar PV, and other renewables, in addition to hydropower [43,107].

The new investment (annual) in renewable power and fuels has been increased from 39.5 to 214.4 billion USD in the last 10 years (2004–13). The respective increase of the power capacity is summarized in Table 6.9 per category of renewable energy. The total investment growth is shown in Fig. 6.15. In specific, the most significant growth occurred in the power sector with global capacity exceeding 1560 GW in 2013, an increase of more than 8% over 2012.

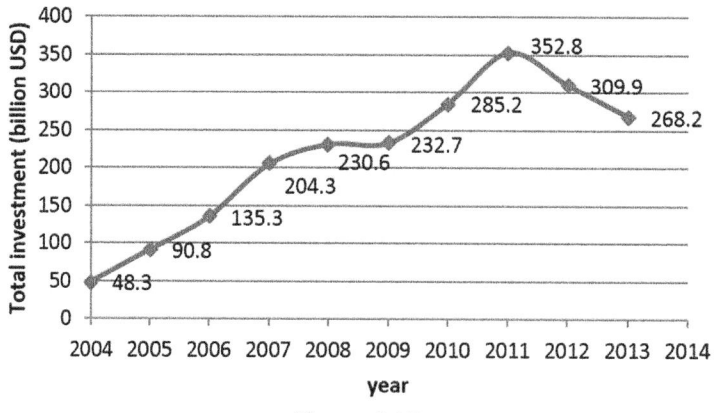

Figure 6.15
Total investment development for renewable energy projects in the period 2004–13 [92].

Around the world, policy support and investment in renewable energy have continued to focus primarily on the electricity sector. Consequently, renewables have accounted for a growing share of electric generation capacity added globally each year. In 2013, renewables made up more than 56% of net additions to global power capacity and represented far higher shares of capacity added in several countries around the world.

Its main advantage over other major RES, such as solar and wind, is that biomass can be stored and thus be used to produce power and heat on demand. As a result, biomass systems can be used to balance the intermittent behavior of an electricity system with a high degree of solar or wind integration as well as to provide heat for industrial applications or district heating networks to meet seasonal demand. IEA forecasts that by 2050, 7.5% of world electricity production could come from biomass [46].

6.12 Life Cycle Methodology

6.12.1 Definition of Life Cycle Methodology

Generally, Life Cycle Management (LCM) is an integrated concept for managing the total life cycle of goods and services toward a more sustainable production and consumption. LCM is applicable for industrial and other organizations demanding a system-oriented platform for the implementation of a preventive and sustainability-driven management approach for a product service system [108]. LCM uses various procedural and analytical tools for different applications and integrates economic, social, and environmental aspects into an institutional context. In this section, a more detailed description regarding the environmental and the economic tool will be given. Hence, the Life Cycle Analysis (LCA) for the environmental aspects and the Life Cycle Costing (LCC) for the economic aspects will be given below.

LCA methodology was introduced to the scientific community in the early 1960s with the aim to evaluate the problems related to raw materials and energy supplies in the industrial sector. It is mentioned that the one of the first studies developed following the LCA methodology was carried out for the Coca-Cola Company to compare the environmental benefits and drawbacks associated with the use of plastic or glass bottles. Within this study, a quantification of the raw materials and fuels required are estimated in order to get the overall environmental footprint for the awareness of the consumers. The study was never been published in its complete version. A summary was given in April 1976 at Science Magazine.

6.12.2 Life Cycle Analysis—Environmental Aspects

LCA can be defined as a method that studies the environmental aspects and potential impacts of a product or system from raw material extraction through production, use, and disposal. The general categories of environmental impacts to be considered include resource use, human health, and ecological consequences [109]. To allow for a consistent comparison between the different scenarios, it is necessary to define a common reference in order to express the results for the same output: this common reference is called the functional unit. The initial typical methodology was proposed by SETAC [110]. In the period 1997–2000, ISO standards introduced the stages of LCA methodology. Now, the ISO standards that are in force are given in the directive ISO 14044:2006 [87].

Stages of Life Cycle Analysis

A typical LCA study consists of the following stages (Fig. 6.16):

1. **Goal and scope definition**. This step includes the objectives of the study, the functional unit, the system boundaries, the data needed, the assumptions, and the limits that must be defined. Particularly, the functional unit is the reference unit used to normalize all the inputs and outputs in order to compare them with each other.
2. **Life Cycle Inventory**. This step refers to the analysis of the material and energy flows and the study of the working system. On the other hand the data collection for the entire life cycle implies the modelization of the analyzed system. Moreover, one of the most critical aspects of this phase is the quality of inputs, which must be verified and validated in order to guarantee the data reliability and correct use. During this stage, a conversion of the available data to appropriate indicators takes place. The indicators are given per functional unit used.

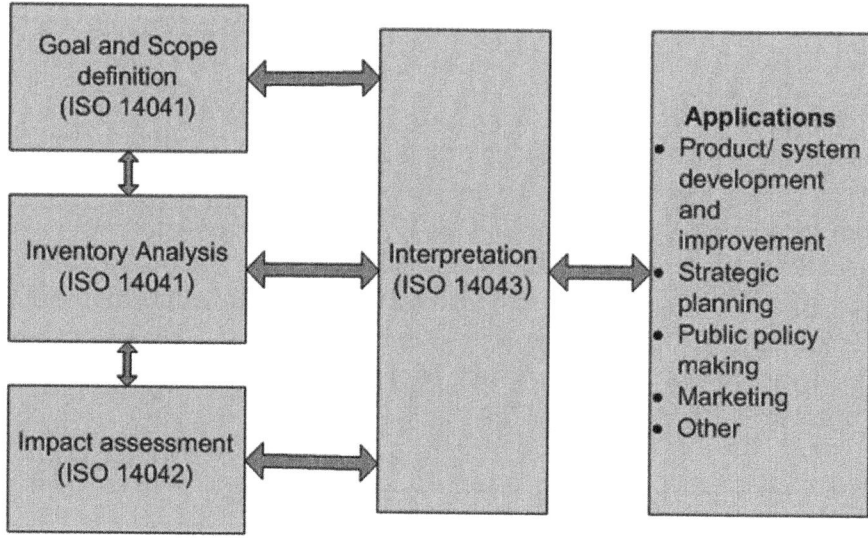

Figure 6.16
Stages of a Life Cycle Analysis study.

3. **Impact assessment**. This step includes the assessment of the potential impacts associated with the identified forms of resource use and environmental emissions. The impact assessment methods, which are used in LCA can be divided into two categories: those that focus on the amount of resources used per unit of product (upstream methods) and those which estimate the emissions of the system (downstream methods).

4. **Interpretation**. In this phase the analyst aims to scrutinize the results and discuss them, giving as much precise information as possible to the decision makers. Moreover, this step may highlight some problems in the LCA development which need a more detailed approach: for instance, it can be decided to improve the quality level of some data collected from the literature, because they describe a process which significantly influences an environmental pressure and therefore a more elevated accuracy of them may guarantee less variability in the results. This mechanism of the LCA assures the improvement of results.

Life cycle impact assessment

The environmental impacts that may be covered by LCA include Climate change, Human respiratory health decrement from particulates, Human health decrement from toxics, Human health decrement from carcinogens, Acidification, Eutrophication, Ecosystem toxicity, Ozone depletion, Smog formation, Habitat alteration, Biodiversity decrease, Resource depletion, Water consumption, Land use, and/or land use change.

The aforementioned impact categories are calculated through the use of some indicators such as the Gross Energy Requirement (GER), Global Warming Potential (GWP$_{100}$), Ozone Depletion Potential (ODP), Acidification Potential (AP), Photochemical Ozone Creation Potential (POCP), Photochemical Oxidation, Eutrophication, Human toxicity.

The different steps during life cycle impact assessment are presented in Fig. 6.17: A description of each process includes the evaluation of the infrastructure needed, such as buildings, asphalt surfaces, machines, infrastructure for pre- and posttreatment, etc. (investment of materials and energy). The materials needed to provide the treating infrastructure were divided by the span of their lifetime in order to obtain the yearly amounts of cement, metals, asphalt, etc. necessary to treat a defined amount of waste. In LCA, all processes, such as raw material extraction, distribution, and manufacturing, could be included up to the moment of building, running, and breaking down the plants. The ecological running costs of the plant included energetic and material parameters such as energy fluxes, parts replaced because of attrition, transports, commodities, etc. as well as the emissions into air and into water caused by the process itself. Generally the emissions can be categorized into three distinct categories: savings, avoided, and direct emissions. A positive number shows emissions to the atmosphere, while a negative number indicates avoidance of emissions. This usually takes place in the evaluation of CO_{2eq} emissions.

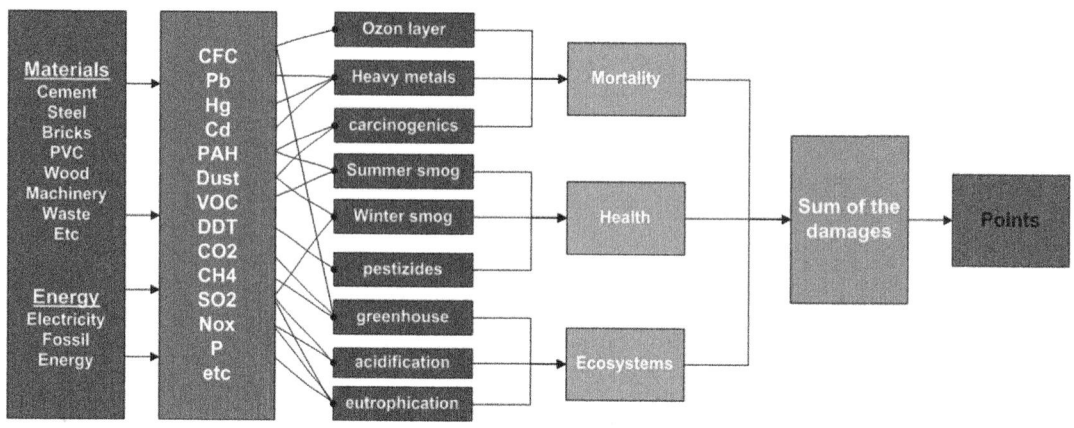

Figure 6.17
Life Cycle Analysis steps for the estimation of total results.

CO_2 savings refer to the GHG emissions avoided by not having to reproduce the recovered materials [111]. CO_2 avoided emissions refer to avoided GHG emissions that otherwise would be emitted, if another method has been realized [112]. CO_2 direct emissions are GHG emissions, emitted directly in the environment from the processes that take place.

Materials and energy consumption cause indirect environmental impacts: The emissions to produce materials and energy for constructing, running, and breaking down the plants were quantified by taking data from the respective database tool used. These impact factors show effects on the impact categories. All impacts caused by the different activities of a waste treating process are first sorted and attributed to the relevant categories. For each damage category, a reference substance has been defined. The impacts are brought to a comparable size by multiplying with a factor corresponding to their relative damage potential. The damages caused by the reference substances of each impact category are weighted for causing mortality, damage to health, and ecosystem impairment. For damage weighting factors, subjective weighting is possible.

Life Cycle Analysis software

In order to apply the aforementioned methodology in a reliable and standardized way, LCA should be performed by means of some commercial software. There are a lot of suppliers of LCA software tools in the market. The software is intended for different types of users and designed for different types of LCA applications [113]. The main differentiations of LCA software is in the database and in the methodology adopted. There are several methods of LCA: Recipe [114], Impact 2002+, Edip 2003, Stepwise 2006 (combination of Impact 2002 and Edip 2003). Impact 2002+ and Edip 2003 methods are second-generation methods, building on previous work (Ecoindicator 1999 [115] and EDIP1997, respectively). A list of LCA tools is available in [116].

The most commonly used LCA packages are Simapro [117], Easewaste [118], Umberto [119], and Gabi [120].

6.12.3 Life Cycle Costing—Economic Aspects

LCC methodology is an economic evaluation technique that determines the total cost of owning, operating, and disposing of a product, service, or technological system throughout its life. This approach implies the recognition of the direct, indirect, recurring, and nonrecurring costs incurred or expected to incur during the phases of design, research, and development, investment, operation, maintenance, shutdown, and other support related to a product during its life cycle. All related costs should be included in this evaluation, regardless of the funding source, the business unit, etc. An environmental LCC methodology takes into account the above four main cost categories plus external environmental costs, as it is shown in Fig. 6.18. The latter may come from LCA analyses on environmental impacts, which measure, for example, the external costs of global warming contribution associated with emissions of different GHGs. Environmental costs can be calculated also in respect of acidification (grams of SO_2, NO_x, and NH_3), eutrophication (grams of NO_x and NH_3), land use (m^2*year), or other measurable impacts.

To be introduced into an "accounting" LCC process, environmental costs must be expressed in monetary terms. In other words, environmental costs should be quantified and monetized so they can be considered as an additional cost input in an LCC analysis.

Monetization is by far the most complicated step of the methodological approach especially when environmental burdens affect nontraded goods such as human life, biodiversity, and so on. These physical impacts are translated to monetary terms by means of different evaluation techniques derived from welfare economics. These are distinguished in direct and indirect approaches [121]. Direct approaches are referred as the Contingent Valuation Method which consists in asking individuals for their willingness to

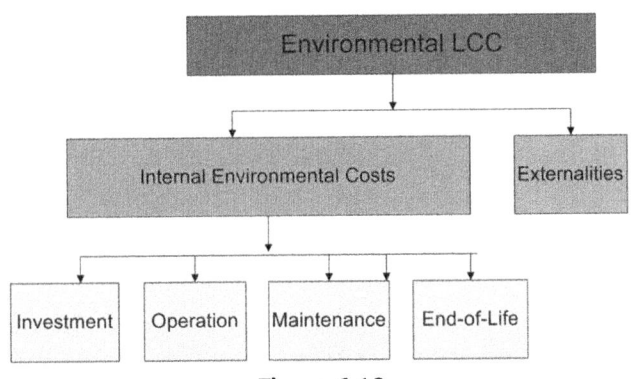

Figure 6.18
Environmental Life Cycle Costing methodology [123].

pay avoid hypothetical scenarios involving reductions in health and environmental risks or effects. Indirect approaches seek to uncover values for the nonmarketed goods by identifying connections to other marketed goods or to specific behaviors that have a well-defined value and act as substitutes or complements of the examined goods [122].

6.13 Introduction to Hydrogen

The production and use of fossil fuel energy is connected with the environmental burdening and the diminution of the energy reserves, endangering the future of the planet and the safety of the energy supply. So, newer and more energy effective production techniques and alternative/renewable fuels should be used. Scientists agree that the solution to these international problems can be found, if fossil fuel is substituted by hydrogen in the current energy production system [124]. Hydrogen is considered an important alternative product and a "bridge" to an energy sustainable future. Hydrogen may not be an energy source, but it indisputably is an important energy "factor." The annual hydrogen production today reaches 0.1 Gt, the 98% of which comes from the fossil fuel reforming process. Internationally, about 48% of hydrogen comes from natural gas, 30% form petroleum, 18% from carbon, and the remaining 4% from water electrolysis [125].

The amount of energy per mass produced during hydrogen combustion is the highest of all fuels. Hydrogen's lower heating value is 2.4, 2.8, and 4 times higher than that of methane, gasoline, and coal, respectively. Hydrogen is the most abundant element on the planet and is considered as a sustainable and clean source of energy, since no GHGs are emitted during its combustion. The hydrogen gas is lighter than air, and so it is rapidly diffused in the atmosphere. Thus hydrogen gas is never found on its own in nature. It has much wider limits of flammability in air than methane and gasoline [126]. Hydrogen is only found in chemical compounds, conjoint with other elements, as carbon, creating various compounds of the general category of hydrocarbons (where fossil fuel belong), as well as with O_2 in the molecule of H_2O.

Hydrogen can be used as fuel directly in an internal combustion engine, in ways that do not differ from car engines that use gasoline [127]. H_2 can also be used in the aforementioned engines, producing positive results such as fast combustion speed, high octane number, zero toxic action, and zero ozone formation probability. However, its use in fuel cells, devices that instantly convert the chemical energy of the fuel into electrical, is more attractive, due to the extremely high performances that can be achieved in these energy conversion systems [128].

All the primary energy sources can be used for the production of hydrogen [124]. The production of hydrogen by fossil fuel is connected to carbon dioxide emissions, and research is currently focused on minimizing the environmental damage caused by GHG emissions. Therefore, the implementation of CCS systems is necessary to reduce GHG emissions. On the contrary, hydrogen that is produced by renewable energy sources, such as wind power, solar energy, and biomass, appears to be the ideal solution for energy production with zero carbon footprint [129].

6.14 Hydrogen Production Methods

Fuel processing technologies transform chemical compounds containing hydrogen into a hydrogen-rich gas mixture. Nowadays, a number of hydrogen production methods are in development, with steam methane reforming being the most common method of producing commercial bulk hydrogen. In this section, well-established hydrogen production methods are summarized.

6.14.1 Hydrocarbon Reforming

Hydrogen gas can be produced from hydrocarbon fuels by three basic technologies: (1) steam reforming (SR), (2) POX, and (3) autothermal reforming [130]. SR process requires an external source of heat. On the contrary, no oxygen is needed. Compared with autothermal reforming and POX, SR takes place at modest temperatures. Syngas (H_2/CO mixture) resulting from SR process is characterized by high H_2/CO ratio, accompanied, however, by high GHG emissions. Table 6.10 summarizes the advantages and challenges of hydrocarbon reforming technologies.

POX is a noncatalytic process, where raw material is gasified in the presence of oxygen and steam. Oxygen and steam amounts are controlled and thus the reaction proceeds without the need for external energy. The specific process takes place at high temperatures resulting in the formation of soot. Syngas H_2 to CO ratio ranges from 1.1 to 2.1, being ideal for the production of synthetic biofuels through the Fischer–Tropsch process.

Autothermal reforming is a combination of SR and POX process, in order to produce a suitable H_2/CO ratio without external energy consumption. It typically takes place at lower

Table 6.10: Comparison of Hydrocarbon Reforming Methods [131].

Technology	Advantages	Disadvantages
Steam reforming	Extensive industry experience Low operation temperature No oxygen demand	High emissions
Autothermal reforming	Lower temperature in comparison to partial oxidation process	Oxygen/air demand Restricted commercialization
Partial oxidation	No catalyst demand	Low H_2/CO Soot formation Extremely high operation temperature

pressures, in comparison to POX process. As mentioned above, there is no need for external energy, as POX is exothermic and SR is thermally neutral. However, a costly oxygen separation unit is required for pure oxygen feeding.

Hydrocarbon reforming reactions

Reforming, water–gas shift, and oxidation reactions are described below [132]:

1. **SR**

$$C_mH_n + {}_mH_2O \leftrightarrow {}_mCO + \left(m + \frac{1}{2}n\right)H_2 \quad \Delta H = \text{depends on hydrocarbon fuel endothermic}$$

[6.6]

$$CH_3OH + H_2O \leftrightarrow CO_2 + 3H_2 \quad \Delta H = +49KJ/mol \qquad [6.7]$$

2. **POX**

$$C_mH_n + \frac{1}{2}mO_2 \leftrightarrow mCO + \frac{1}{2}H_2 \quad \text{depends on hydrocarbon fuel exothermic} \qquad [6.8]$$

$$CH_3OH + \frac{1}{2}O_2 = CO_2 + 2H_2 \quad \Delta H = -1932 \text{ kJ/mol} \qquad [6.9]$$

3. **Autothermal reforming**

$$C_mH_n + \frac{1}{2}mH_2O + \frac{1}{4}mO_2 \leftrightarrow mCO + \left(\frac{1}{2}m + \frac{1}{2}n\right)H_2 \quad \Delta H = \text{depends on hydrocarbon fuel}$$

[6.10]

$$4CH_3OH + 3H_2O + \frac{1}{2}O_2 \leftrightarrow 4CO_2 + 11H_2 \qquad [6.11]$$

4. **Coke formation**

$$C_mH_n \leftrightarrow xC + C_{m-x}H_{n-2x} + xH_2 \quad \Delta H = \text{depends on hydrocarbon fuel} \qquad [6.12]$$

$$2CO \leftrightarrow C + CO_2 \quad \Delta H = +172,4 \text{ kJ/mol} \qquad [6.13]$$

$$CO + H_2 \leftrightarrow C + H_2O \quad \Delta H = +129,7 \text{ kJ/mol} \qquad [6.14]$$

5. **Water–gas shift**

$$CO + H_2O \leftrightarrow CO_2 + H_2 \quad \Delta H = -41,1 \text{ kJ/mol} \qquad [6.15]$$

$$CO_2 + H_2 \leftrightarrow CO + H_2O \quad \text{Reserve Water} - \text{Gas Shift} \qquad [6.16]$$

6. **CO, H$_2$ oxidation**

$$CO + O_2 \leftrightarrow CO_2 \quad \Delta H = -283 \text{ kJ/mol} \qquad [6.17]$$

$$H_2 + \frac{1}{2}O_2 \leftrightarrow H_2O \quad \Delta H = 285.8 \text{ kJ/mol} \qquad [6.18]$$

Enthalpy changes refer to reactions done under standard conditions and assuming that all reactants and products are in the gas phase. Fuel reforming technologies nowadays focus on hydrogen production maximization (Eqs. [6.6]–[6.11]) and reduction of coke formation (Eqs. [6.12]–[6.14]). Under this scope, the most effective catalysts and the optimum operating parameters are examined.

6.14.2 Hydrogen From Water

Many technologies have been explored for the production of hydrogen from water, but it is important to note that only a few have industrial applications [133]. Water splitting can be separated into three categories:

1. Electrolysis
2. Thermolysis
3. Photoelectrolysis

Electrolysis

Water electrolyzers apply for a wide range of power, using three main technologies: alkaline electrolyzers, solid oxide electrolyzers, and proton exchange membrane electrolyzers. The most mature technology is alkaline electrolysis, offering the advantage of simplicity [134,135].

The core of an electrolysis unit is an electrochemical cell, which is filled with pure water and has two electrodes connected with an external power supply. At a critical voltage, the electrodes start to produce hydrogen gas at the negatively biased electrode (anode) and oxygen gas at the positively biased electrode (cathode). A scheme of an electrochemical cell is presented in Fig. 6.19.

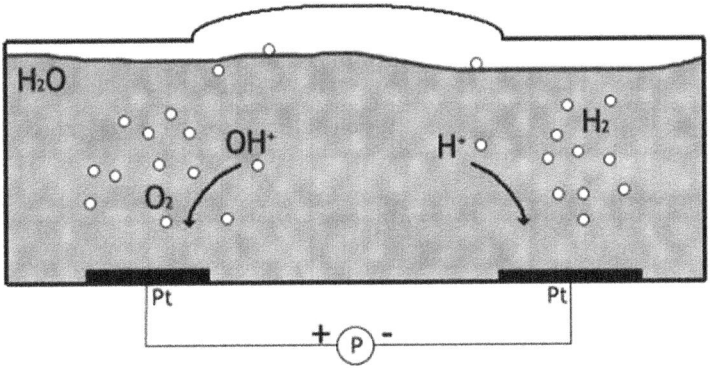

Figure 6.19
Sketch of an electrochemical cell [136].

Alkaline electrolysis

Alkaline electrolyzers are typically composed of electrodes, a microporous separator, and an aqueous alkaline electrolyte of ∼30 wt% KOH or NaOH [132]. Water decomposes at the cathode to H_2 and OH^-. The latter migrates through the electrolyte and a separating diaphragm, and discharges at the anode, liberating the O_2. The following reactions describe alkaline electrolysis process:

$$\text{Electrolyte} \quad 4H_2O \rightarrow 4H^+ + 4OH \qquad [6.19]$$

$$\text{Cathode} \quad 4H^+ + 4e^- \rightarrow 2H_2 \qquad [6.20]$$

$$\text{Anode} \quad 4OH^- \rightarrow O_2 + 2H_2O + 4e^- \qquad [6.21]$$

$$\text{Total} \quad 2H_2O \rightarrow O_2 + 2H_2 \qquad [6.22]$$

The most common cathode material is nickel coated with platinum. The anode is usually made of nickel or copper coated with metal oxides, such as manganese, tungsten, or ruthenium.

Proton exchange membrane electrolysis

Proton exchange membrane (PEM) electrolysis is the reverse process of a PEM fuel cell process that will be analyzed in the following text below. However, the materials are typically different from PEMFC. PEM-based electrolyzers typically use Pt black, iridium, ruthenium, and rhodium for electrode catalysts and a Nafion membrane. The heart of a PEM electrolyzer is a PEM (or solid polymer) electrolyte.

In PEM electrolyzers, water is introduced at the anode, where it is split into protons and oxygen. The protons travel through the membrane to the cathode, where they are recombined into hydrogen. The O_2 gas remains behind with the unreacted water [137]. The following reactions take place during the PEM electrolysis:

$$\text{Cathode} \quad 4H^+ + 4e^- \rightarrow 2H_2 \qquad [6.23]$$

$$\text{Anode} \quad 2H_2O \rightarrow O_2 + 4e^- + 4H^+ \qquad [6.24]$$

$$\text{Total} \quad 2H_2O \rightarrow O_2 + 2H_2 \qquad [6.25]$$

Compared to traditional alkaline electrolysis, in which corrosive potassium hydroxide (KOH) solution is used as electrolyte, PEM electrolysis has more advantages, such as ecological cleanness, high degree of gases purity, and easy maintenance [138].

Solid oxide electrolysis

A single solid oxide electrolysis cell (SOEC) is comprised of an electrolyte layer and two electrodes. The key components of an SOEC are a dense ionic conducting electrolyte and two porous electrodes [139]. SOECs operate at much higher temperatures than the other electrolysis cells and consume much less electricity, due to superior energy conversion efficiency. Steam is fed to the porous cathode. When required electrical potential is fed to the SOEC, water molecules diffuse to the reaction sites and are dissociated to form hydrogen gas and oxygen ions at the cathode–electrolyte interface.

The materials used in SOEC are typically similar to those used for solid oxide fuel cells. For cathode, the most common used material is a porous cermet composed of YSZ and metallic nickel [140], while for anode the most common material used to date is the lanthanum strontium manganite (LSM)/YSZ composite [140]. Regarding the electrolyte, the most frequently used material is a dense ionic conductor, consisting of ZrO_2 doped with 8 mol% of Y_2O_3 (YSZ).

This technology, besides offering the highest faradic efficiency, also offers a possibility of direct electrolyzing of CO_2. Apart from that, the technology goes further and coelectrolyzes H_2O and CO_2 simultaneously.

Thermochemical water splitting

Generally speaking, thermal decomposition is defined as a chemical reaction, where a chemical substance breaks up into at least two chemical substances, when heated. During thermal water splitting specifically, heat is used to decompose water to hydrogen and oxygen [141]. Even partial decomposition of water demands high temperature, at least 2500°C. However, it is clear that stable materials at this temperature and sustainable heat sources are not easily available. Furthermore, the lower total pressure favors the higher partial pressure of hydrogen, making it difficult to build a reactor that works below the atmospheric pressure, at a very high temperature. The separation of product gases is also difficult, at this temperature. Finally, finding a heat source to provide such a high temperature is a problem too. A solar furnace system that concentrates sunlight using large mirrors is the only feasible source of heat. The specific technology nevertheless is at an early stage.

Since direct thermal decomposition of water is difficult, several ways of splitting water using heat have been proposed [137]:

1. Thermochemical cycles
2. Hybrid systems combining thermolysis and electrolysis
3. Direct catalytic splitting of water
4. Chemical splitting of water using plasma technology

Thermochemical water splitting deals with water decomposition to hydrogen and oxygen, using multiple step chemical reactions. The primary energy source for thermochemical hydrogen production is medium temperature heat from solar and nuclear energy. The term "cycles" is used because apart from water, hydrogen, and oxygen, the chemical compounds used in these processes are continuously recycled.

Lower temperatures are required to split water in comparison to direct thermal decomposition. Specifically, Iodine/Sulfur and Bromine/Calcium cycles operate at temperatures below 1000°C [142]. The specific feature makes thermochemical cycles technology particularly tempting. Since the conceptual idea of thermochemical water splitting was proposed by Funk [143], numerous thermochemical cycles have been researched.

Photoelectrolysis

Photoelectrolysis, as reflected by the name, describes electrolysis by the direct use of light. Actually, photoelectrolysis integrates solar energy collection and water electrolysis into a single photoelectrode. This device eliminates the need for a separate power generator and electrolyzer, reducing overall cost and increasing efficiency. Semiconductor materials similar to those used in PVs are used [144]. Semiconducting electrodes are used in a photochemical cell to convert light energy into chemical energy. The semiconductor surface serves two functions, to absorb solar energy and to act as an electrode. Sunlight produces electron–hole pairs in the semiconductor in contact with an aqueous electrolyte. The electrons combine with protons and produce hydrogen at the counter electrode, while the holes oxidize water to oxygen at the semiconductor electrode. Although many semiconductor materials have shown photocatalytic activity, most of them suffer from limitations, including photocorrosion, poor solar spectrum absorption, and the need of external bias [145].

Generally, the efficiency of a photoelectrolysis system is determined by photoelectrode materials and semiconductors substrate [146]. Current photoelectrodes used in photoelectrochemical cells (PEC) that are stable in aqueous solutions have a low efficiency for using photons, in order to split water. The target efficiency is >16% solar energy to hydrogen [144]. The critical challenge for the development of this technology is finding a robust semiconductor to satisfy the competing requirements of nature. Solar photons are primarily visible light, a wavelength that requires semiconductors that require small bandgaps <1.7 eV—for efficient absorption.

6.14.3 Hydrogen From Biomass

Hydrogen production methods using biomass can be divided into two main categories, regarding the mechanisms behind the transformations: thermochemical and biological.

Thermochemical processes

Regarding hydrogen production, biomass can be thermally treated through pyrolysis [147], combustion, gasification [148], and liquefaction. Due to limitations in efficiency and operation, both combustion and liquefaction are not well fitting with the production of hydrogen and have not been extensively developed. Thus the specific processes will not be further analyzed in this section.

So, the main thermochemical processes that are being used and investigated regarding hydrogen production are pyrolysis and gasification processes.

Pyrolysis

Pyrolysis is a thermal process, in which biomass is heated at a temperature ranging from 625 to 800K, in the absence of oxygen and air [126,149]. During this process, the biomass breaks into different phase state products, coexisting solids, liquids, and gasses at the end of the process. The ratio of products is heavily depending on the velocity of the process. The slower the speed of the process is, the higher the solid products ratio becomes. Conventional pyrolysis (slow) is associated with high charcoal product, so the interest is on fast processes, when focusing to hydrogen production. This is called fast (or flash) pyrolysis, where biomass is treated in time ranges of the order of a few seconds (or even under 1 s in flash pyrolysis), and where the main products are in liquid and/or in gas phase.

In general, pyrolysis of biomass samples produces gaseous and liquid products and a solid residue richer in carbon content, char. The expected products depending on the biomass processing are presented in Table 6.11.

Hydrogen production can be obtained directly from the fast or flash pyrolysis processes by the following reaction:

$$\text{Biomass} + \text{Heat} \rightarrow H_2 + CH_4 + CO + \text{other products} \qquad [6.26]$$

Beyond the generation of hydrogen directly from the pyrolysis process, more amount of hydrogen can be obtained indirectly by further processing of gaseous and liquid products

Table 6.11: Product Distribution Yielded From Different Processes [150].

Thermochemical Process	Residence Time (s)	Upper Temperature (K)	Product Yield (wt%)		
			Char	Liquid	Gas
Slow pyrolysis	200	600	32−38	28−32	25−29
	120	700	29−33	30−35	32−36
	90	750	26−32	27−34	33−37
	60	750	24−30	26−32	35−43
	30	950	22−28	23−29	40−48
Flash pyrolysis	5	700	22−27	53−59	12−16
	4	750	17−23	58−64	13−18
	3	800	14−19	65−72	14−20
	2	850	11−17	68−76	15−21
	1	950	9−13	64−71	17−24
Gasification	1500	1250	8−12	4−7	81−88

obtained in the pyrolysis. The gaseous hydrocarbon products (CH_4 and others) can be postprocessed by the SR process:

$$CH_4 + H_2O \rightarrow 3H_2 + CO \qquad [6.27]$$

Hydrogen production can be increased by means of the water—gas shift reaction:

$$CO + H_2O \rightarrow H_2 + CO_2 \qquad [6.28]$$

Apart from obtaining a great percentage of gas in the final products, the yields of hydrogen from the pyrolysis and the air—steam gasification increase with increasing temperature [149].

The use of different catalytic materials within the pyrolysis process is another way to increase hydrogen production and eliminate organic impurities (tars) in the fuel gas [151]. Various biomass species have pyrolyzed in the presence of different materials (eg, zeolites. $ZnCl_2$, K_2CO_3, Na_2CO_3, $CaCO_3$, Al_2O_3, TiO_2 [152]).

Gasification

Biomass gasification is a thermochemical process, which results in a high production of gaseous products and small quantities of char and ash. The process involves partial combustion of biomass with an oxidant (air, oxygen, steam, or CO_2) in substoichiometric concentrations, in order to produce synthesis gas (syngas), which is composed primarily of CO and H_2. Syngas also contains gaseous products such as CO_2, CH_4, sulfur compounds, solid impurities (dust), and other light hydrocarbons. The composition of the syngas depends on the feedstock, the type and amount of co-reactant used (air, H_2O, or CO_2), and the reaction conditions (temperature, heating rate, etc.). The gaseous products of the biomass gasification need to be cleaned from various types of impurities such as (1) solid impurities (dust); (2) inorganic impurities such as nitrogen compounds (NH_3 and HCN), sulfur compounds (H_2S), ash, and metal compounds; and (3) organic impurities (tars) [153]. Several groups of catalysts have been evaluated for the elimination of these impurities [154]. The most important groups of catalysts are minerals, chars, zeolites, alkali metal—based catalysts, and transition metal—based catalysts [151].

The main types of biomass gasifiers are updraft, downdraft, fluidized bed, and entrained flow gasifiers. The updraft gasifiers present the highest tar production, while the downdraft gasifiers show the lowest. Fluidized bed gasifiers show intermediate tar production. For large-scale applications, the preferred type is the entrained flow gasifier, while for small-scale applications the downdraft gasifier is the most appropriate. The bubbling and CFB gasifiers can be competitive in medium-scale applications.

Biological processes

The detected biological processes for producing hydrogen are the following: biophotolysis, photofermentation, dark fermentation, and two-stage process (integration of dark fermentation and photofermentation). Within these processes, different types of microorganisms are responsible for hydrogen production. The first two types directly decompose water into molecular hydrogen and oxygen by means of light. Anaerobic bacteria use organic substances as a unique source of energy (and electrons), leading to fermentation or photodecomposition of feedstock.

All these biological processes are fundamentally governed by hydrogen-producing enzymes [155]. Hydrogenase and nitrogenase are the most widely used enzymes to catalyze the biological production of hydrogen [156].

Biophotolysis

Biophotolysis process is observed in cyanobacteria and microalgae. The same processes are found in algal and plants photosynthesis, adjusted for hydrogen gas production, and not in carbon-containing biomass.

Biophotolysis can be further subclassified into direct and indirect processes. The direct biophotolysis process is a process that takes place under light irradiation. Microorganisms absorb sunlight through the photosynthetic systems, PSI and PSII. The two photosynthetic systems responsible for photosynthesis process are (1) photo system I (PSI), which produces reductant for CO_2, and (2) photo system II (PSII), which splits water to evolve O_2. The two photons obtained from the splitting of water can either reduce CO_2 by PSI or form H_2 in the presence of hydrogenase [157]. Apart from hydrogenase, nitrogenase can also be used, utilizing the carbohydrates stored in the heterocyst or transferred from neighbor cells in a two-step process [158].

The process of indirect biophotolysis is similar to the direct one (two-step process), employing fermentation instead of light, in the second stage of the transformation.

Photofermentation

Photosynthetic bacteria are able to generate hydrogen through the nitrogenase. The energy required is obtained again in this process by light, and the feedstocks are organic acids or biomass. The presence of molecular nitrogen or nitrogen compounds directs the reaction route toward ammonia formation.

Organic substrate is oxidized to CO_2 in the cycle of tricarboxylic acids (TCA). Electrons generated in this process are transferred to nitrogenase, via many carriers (eg, NAD and ferredoxin). Nitrogenase reduces protons to molecular hydrogen. The photosynthetic apparatus acts simultaneously with TCA cycle, transforming light into chemical energy.

Dark fermentation

As mentioned above, dark fermentation process is light-independent. Through this process, hydrogen is produced by anaerobic bacteria, grown on carbohydrate-rich substrates.

Dark fermentation process has been examined in different temperature ranges, mesophilic (25−40°C), thermophilic (40−65°C), extreme thermophilic (65−80°C), and hyperthermophilic (>80°C).

Except for hydrogen, CH_4, CO, and H_2S are produced, depending on different system and feedstock [159]. Environmental conditions are the major parameters to be controlled in hydrogen production. Medium pH affects hydrogen production yield, biogas content, type of the organic acids produced, and the specific hydrogen production rate [130].

Dark fermentation can produce H_2 all day, being a continuous process, since there is no need for light and oxygen.

Two-stage process (integration of dark fermentation and photofermentation)

This process combines photofermentation and dark fermentation to achieve higher H_2 yields. The combined fermentation allows the organic acids produced during dark fermentation of waste materials, to be used as substrate in the photofermentation process [160], a critical step taking into account that substrate otherwise fails to achieve a complete conversion, due to thermodynamic limitations [161]. The overall reactions of the process are as follows:

$$\text{Dark fermentation} \quad C_6H_{12}O_6 + 2H_2O \rightarrow 2CH_3COOH + 2CO_2 + 4H_2 \qquad [6.29]$$
$$\text{Photofermentation} \quad 2CH_3COOH + 4H_2O \rightarrow 8H_2 + 4CO_2 \qquad [6.30]$$

In accordance to the stoichiometry of the reactions, 12 mol of H_2 could be produced per mole of glycose [162].

6.15 Fuel Cells

Fuel cells convert chemical energy in fuels into electrical energy directly [163]. The overall process is the reverse of water electrolysis. The main unit of the fuel cell is the unit cell, where the conversion of chemical energy into electrical energy takes place.

6.15.1 Principle of Operation

An electrolyte with a negative electrode (anode) and a positive electrode (cathode) constitutes a unit cell. Hydrogen and oxygen (air) are supplied to cathode and anode, respectively. When hydrogen is led to the anode, the hydrogen molecules are split into their basic elements, a proton and an electron. The protons migrate through the electrolyte to the cathode, where they react with oxygen to form water. At the same time, the electrons are forced to travel around the electrolyte to the cathode side, because they cannot pass the electrolyte. This movement of electrons thus creates an electrical current.

A schematic representation of a unit cell is depicted in Fig. 6.20.

A fuel cell power system comprises of the following components [164]:

1. Unit cell
2. Stacks, in which individual cells are modularly combined by electrically connecting the cells to form units with the desired output capacity
3. Balance of plant which consists of components that provide feedstream conditioning (including a fuel processor if needed), thermal management, and electric power conditioning, among other ancillary and interface functions

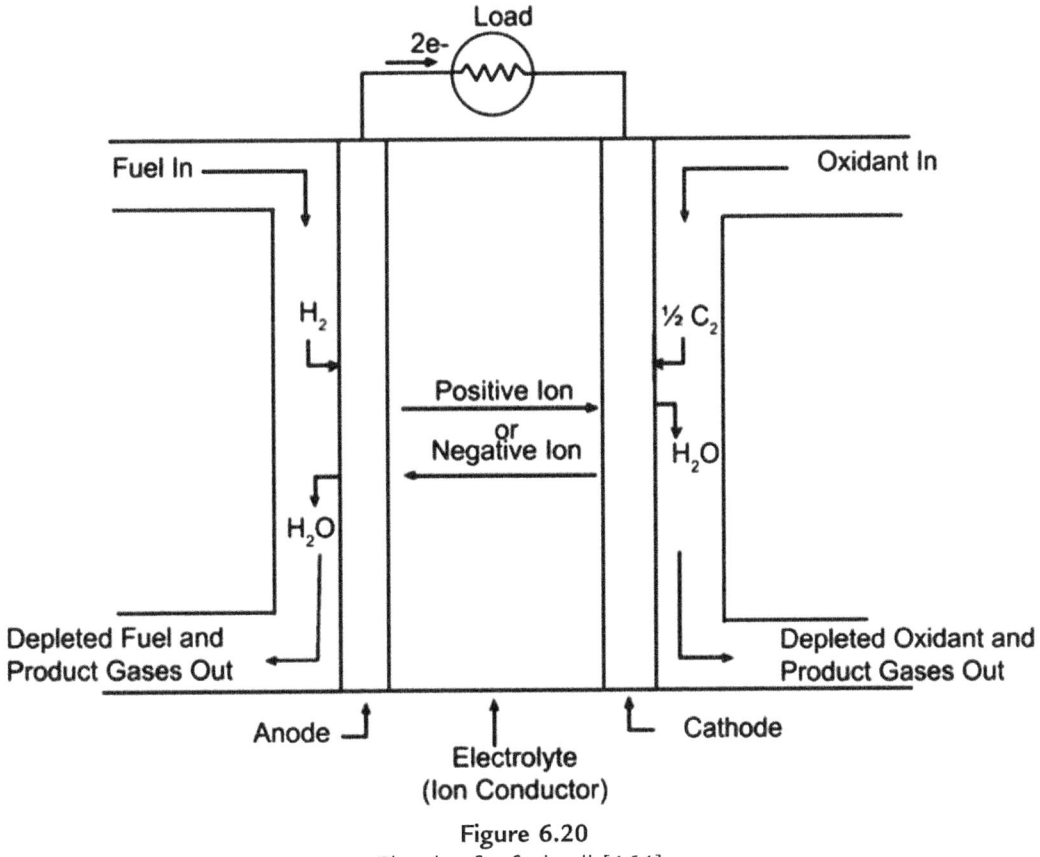

Figure 6.20
Sketch of a fuel cell [164].

6.15.2 Types of Fuel Cells

There are eight main types of fuel cells, based mainly on the type of electrolyte [165]:

1. PEMFCs, proton exchange membrane or polymer electrolyte membrane fuel cells
2. AFCs, alkaline fuel cells
3. PAFCs, phosphoric acid fuel cells
4. MCFCs, molten carbonate fuel cells
5. SOFCs, solid oxide fuel cells
6. DMFCs, direct methanol fuel cells
7. DAFCs, direct ammonia fuel cells
8. DCFCs, direct carbon fuel cells

Apart from DAFCs, DMFCs, and DCFCs other types of fuel cells are fed with hydrogen.

The main features of each type of fuel cell are summarized in Table 6.12 [165−169].

6.15.3 Fuel Cell's Applications

Fuel cells have three main applications: transportation, portable uses, and stationary installations.

In the future, fuel cells could power our cars, with hydrogen replacing the petroleum fuel that is used in most vehicles today. Many vehicle manufacturers are actively researching and developing transportation fuel cell technologies.

Stationary fuel cells are the largest, most powerful fuel cells. They are designed to provide a clean, reliable source of on-site power to hospitals, banks, airports, military bases, schools, and homes.

Fuel cells can power almost any portable device or machine that uses batteries. Unlike a typical battery, which eventually goes dead, a fuel cell continues to produce energy as long as fuel and oxidant are supplied. Laptop computers, cellular phones, video recorders, and hearing aids could be powered by portable fuel cells, where PEM and DMFC play a major role due to their reduced operating temperature.

Fuel cells have strong benefits over conventional combustion-based technologies currently used in many power plants and cars. They produce much smaller quantities of GHGs and none of the air pollutants that create smog and cause health problems. If pure hydrogen is used as a fuel, fuel cells emit only heat and water as a by-product. Hydrogen-powered fuel cells are also far more energy efficient than traditional combustion technologies.

The biggest hurdle for fuel cells today is cost. Fuel cells cannot yet compete economically with more traditional energy technologies, though rapid technical advances are being

Table 6.12: Major Types of Fuel Cells.

Type	PEM	AFC	PAFC	MCFC	SOFC	DMFC	DAFC	DCFC
Operating temperature (°C)	60–140	100–200	150–200	600–700	500–1200	30–80	300–700	500-1000
Fuel	H_2	H_2	H_2	CH_X, CO, H_2	CH_X, CO, H_2	Methanol	NH_3	Carbon
Catalyst	Platinum	Platinum	Platinum	Electrode material	Electrode material	Platinum/ruthenium	Nickel/silver	
Electrical efficiency (%)	40–50	60	40	45–50	$\geq$50	30–40	$\geq$60	$\geq$70
Anode reaction	$H_2 \rightarrow 2H^+ + 2e^-$	$H_2 + 2OH^- \rightarrow 2H_2O + 2e^-$	$H_2 \rightarrow 2H^+ + 2e^-$	$H_2 + CO_3^{2-} \rightarrow H_2O + CO_2 + 2e^-$	$H_2 + O^{2-} \rightarrow H_2O + 2e^-$	$CH_3OH + H_2O \rightarrow CO_2 + 6H^+ + 6e^-$	$NH_3 \rightarrow N_2 + 3H^+ + 3e^-$	$2O^2 + C \rightarrow CO_2 + 4e^-$
Cathode reaction	$\frac{1}{2}O_2 + 2H^+ + 2e^- \rightarrow H_2O$	$\frac{1}{2}O_2 + H_2O + 2e^- \rightarrow 2OH^-$	$\frac{1}{2}O_2 + 2H^+ + 2e^- \rightarrow H_2O$	$\frac{1}{2}O_2 + CO_2 + 2e^- \rightarrow CO_3^{2-}$	$\frac{1}{2}O_2 + 2e^- \rightarrow O^{2-}$	$\frac{1}{2}O_2 + 6H^+ + 6e^- \rightarrow 3H_2O$	$\frac{1}{2}O_2 + 3H^+ + 3e^- \rightarrow \frac{3}{2}H_2O$	$O_2 + 4e^- \rightarrow 2O^{2-}$

PEM, proton exchange membrane; AFC, alkaline fuel cell; PAFC, phosphoric acid fuel cell; MCFC, molten carbonate fuel cell; SOFC, solid oxide fuel cell; DMFC, direct methanol fuel cell; DAFC, direct ammonia fuel cell; DCFC, direct carbon fuel cell.

made. Although hydrogen is the most abundant element in the universe, it is difficult to store and distribute. Canisters of pure hydrogen are readily available from hydrogen producers, but as of now, you cannot just fill up with hydrogen at a local gas station.

6.16 Emissions

Over the last years, the environmental concerns mainly referred to gas pollutants that contribute to the greenhouse effect are related to energy production in mobile and immobile applications. Recently, an increased interest in biofuel use has been noted. This interest is expressed by international organizations, due to the increase in cost of conventional energy forms and to the inevitable environmental problems that are related to them. Currently, the need for the energy policy operators to redesign the future's energy map is dominant, responding to the relevant energy and environmental challenges [131].

Biodiesel, biochemical products, transport biofuels, and biogas for power production or for further process for transport fuels, such as hydrogen, are included in the potential bioproducts [170]. There are various reasons why biofuels should be reckoned as an ideal perspective in the developing, as well as in the developed countries. These reasons include the energy supply's safety, the reduction of environmental impact, the trade balance, and the socioeconomic issues that are related to the agricultural sector [131]. The share of biofuels in the fuel market will rapidly increase in the next decade, due to their environmental performance [171].

Hydrogen can be produced by biorenewable energy sources, with zero carbon footprints or with zero carbon, such as solar energy or wind power. This way, the use of hydrogen could eventually lead to the elimination of harmful gas pollutants that are related to the use of fossil fuels. The vehicles, using hydrogen as fuel, have zero carbon emissions and thus subsequent benefits for the atmosphere. Hydrogen-fed fuel cells can contribute to the reduction or to the elimination of CO_2 and other GHG emissions. There is an increasing interest in the future role of hydrogen in the energy sector, based on energy systems, especially in the transports sector. Given that fossil fuels are evidently connected to harmful gas emissions, hydrogen engines appear to be an attractive alternative future solution [131]. Almost all these pollutants can be eliminated with the use of hydrogen as fuel. The environmental problems that can arise during hydrogen's use appear during its production but not during its use. The use of renewable energy sources for the production of hydrogen eliminates environmental problems [172]. If hydrogen is produced with no CO_2 or other GHG emissions, it could be the basis of an actually sustainable and viable energy system. Therefore, the products of biomass pyrolysis and gasification reactors can be checked in advance for atmospheric pollutants. There are many strategies for the control of gas emissions and particulate pollutants of the thermochemical processes, for the energy exploitation of biomass, and for those depend on the needs of the occasional process.

The use of fossil fuel is connected with harmful environmental effects. During production, refining, transport, and storage of crude oil and similar petrochemical products, there are emissions and leaks of gas pollutants that cause the pollution of water resources and atmosphere. Most environmental effects by fossil fuels are related with combustion, since the huge amounts of various gases, fumes, and ash are released into the atmosphere [173]. The hydrogen vehicles do not directly or indirectly produce large amounts of CO_2, HCs, particles, SO_x, sulfuric acids, ozone and other oxidants, benzene and other carcinogenic aromatic compounds, formaldehyde and other aldehydes, Pb and other toxic metals, smoke or CO_2, and other GHGs. The only pollutant that is concerning is NO_x. If hydrogen is produced by H_2O using renewable energy sources, then hydrogen's production and distribution produces no pollution [149]. The hydrogen fuel cell vehicles may make the use of biomass in transports possible, with zero or minimum atmospheric pollution and very low CO_2 emission levels, in their entire life circle, if the biomass raw materials are cultivated in a renewable manner [150].

Scientists estimate that the probable climatic effects of the devices that use hydrogen as fuel for energy production will be more moderate than of those of the devices based on fossil fuel. However, these effects greatly depend on the production, storage, and final use of hydrogen [149]. If an international hydrogen economy replaces the current economy that is based on fossil fuel and has a leak rate of 1%, then there will be a 0.6% effect on the climate internationally, compared to the current energy system that is based on fossil fuel. If the leak rate is increased to 10%, then the effect to the climate internationally will be of 6% of the fossil fuel system [174]. An important advantage of hydrogen is that its combustion does not create gas pollutants, compared to a number of harmful substances that are emitted by petrol and diesel engine vehicles. Hydrogen combustion is clear and is comprised by water and small amounts of NO_x. The only by-product of hydrogen combustion is water.

Certain measures are thought to reduce the quantity of NO_x, reaching even 1/200 of the respective emissions by diesel engines [175]. The NO_x emissions are created due to the high temperatures in the combustion chamber. The high temperature favors the interaction of N_2 and the air's O_2, leading to the creation of NO_x. The quantity of NO_x evolved depends on the ratio of air/fuel, the engine's compression ratio, the engine's speed, the ignition time, and the thermal dilution [127].

Biohydrogen offers some technological and environmental benefits, compared to conventional fossil fuel, that make it preferable as an alternative solution in transports sector. These benefits include the reduction of GHGs emissions, contributing this way to the realization of the environmental targets set by international organizations, concerning the amelioration of the fuel sector, biodiversity, sustainability, and the creation of a new market for agricultural products.

Hydrogen has not been used widely, due to certain technological and financial weaknesses. In the case of developing countries, the relevant research and development programs that are realized, mostly on an international cooperation level, could contribute to the import of new hydrogen technologies that are becoming more and more competitive. The developing countries face the following dilemma: in which ways should they invest in hydrogen research and development technologies that lead to an international hydrogen economy. Most developing countries may not comprise important development factors of those technologies. It is possible though for them to benefit from an approach to the hydrogen economy, at least as much as the respective developed countries will, since they generally encounter more severe urban pollution problems. International organizations are expected to play an important role in helping these countries develop and adopt a policy based on the hydrogen market and on other relevant clear energy systems. The national organizations should also support with funds and other motives the developing countries, so that they can participate in the hydrogen economy.

As the targets for GHG emissions that have been set at the world climate conferences are determined, relevant studies have focused on the increasing use of hydrogen in the energy balance, as the fuel that will realize the vision for clear energy. The research and development investments have already led to remarkable technological advancements concerning hydrogen, in the European Union, the United States, Canada, and Japan. Several companies participate in the development and commercialization of hydrogen technologies, while British Petroleum (BP) and Shell are both involved in hydrogen technologies research programs. Shell in particular has invested 1 billion dollars in hydrogen research and development and in commercialization activities. BP provides the hydrogen distribution infrastructure (hydrogen refueling stations for transports and other uses) for the realization of H_2 technologies adoption in transports demonstrative programs, in 10 cities of the world, trough CUTE program (Clear Urban Transport for Europe). The goal of the demonstrative programs is to promote hydrogen technologies from the research and development stage, up to the market and trade stage, and also increase the identifiability and public acceptance of the H_2 technologies by the consumers. BP also constructs two large hydrogen production facilities, through the reformation of fossil fuel with parallel geological storing of the emitted CO_2 in Peterhead, Scotland, and in Carson, California. The program in California is of industrial scale and uses coke as raw material for hydrogen power production, while at the same time it dramatically reduces the GHG emissions, constraining the CO_2 and safely and permanently storing it in appropriate underground geological spaces. Besides the abovementioned energy behemoths, other companies, such as Stuart Energy Systems Corp, Linde AG, Air Products and Chemicals Inc., that provided the necessary infrastructure for H_2 also participate in these demonstrative programs.

6.17 Nuclear Energy Production

The present status of nuclear energy technology is the result of over 50 years of development and operational experience. The latest designs for nuclear power plants take into account technological developments to offer enhanced safety and performance. Nuclear power is a mature low-carbon technology that is already available today for a wider deployment [176].

Nuclear energy is produced when an atom's nucleus is split into smaller nuclei by a process called fission. The three most used atoms are Uranium 235, Plutonium 239, and Thorium 232, with Uranium being the most dominant. The fission of such large atoms produces a great amount of energy. Actually, the fission of 1 g of Uranium 235 produces the same amount of energy as the combustion of 3 tons of coal. The main uses of the energy, or heat, produced by the fission are electricity production, space craft propulsion, and empowering of weapons such as the atomic bomb [177]. The advantage of using nuclear energy is that fission does not produce soot and potentially harmful gases such as CO_2. However, during mining, transportation, and refinement of uranium such gases are emitted and their amounts are comparable to renewable forms of electricity generation, such as wind and hydropower. The difference lies in the fact that these wastes are radioactive, and thus more complicated to handle [178].

6.17.1 Energy Production and Waste Disposal

Nuclear fuel can be used in a reactor for several years. The remaining used fuel must be stored and either disposed of or recycled to make new fuel. However, it is much more practical to do this with used nuclear fuel than with the wastes and emissions from fossil fuels, as the amount of fuel used to produce electricity is substantially smaller than in coal-fired plants. Radioactive waste is divided into different categories depending on its origin and characteristics [179–181]:

1. **Spent nuclear fuel**: The uranium fuel used in a reactor and that is no longer so efficient in fission and therefore heat generation. In addition to high radiation levels, a second hazard is the extremely remote possibility of an accidental self-sustained fissioning and splitting of the atoms of uranium and plutonium.
2. **High-level waste**: Highly radioactive waste created by spent fuel reprocessing (mainly for defense purposes).
3. **Transuranic (TRU) waste**: A relatively low-activity radiowaste that contains more than a certain level of long-lived elements heavier than uranium (primarily plutonium). Elements that have an atomic number greater than uranium are called TRU ("beyond uranium"). In the United States, TRU waste is generated almost entirely by nuclear weapon production lines.

4. **Low-level waste**: Includes items contaminated with radioactive material or items that have become radioactive through exposure to neutron radiation. This type of waste typically includes contaminated protective clothing, equipment, and tools, while discarded parts from nuclear reactors and small gauges containing radioactive material are the most dangerous ones. In general, low-level waste contains relatively low amounts of radioactivity that decays rather quickly.

5. **Uranium mill tailings**: Sandy uranium ore residues. Such tailings have very low radioactivity but extremely large volumes that can pose a hazard, particularly from radon emissions or groundwater contamination.

Therefore, Nuclear Regulatory Commission (NRC) requires strict control in the process of waste handling to ensure minimization of potential hazards. The disposal of nuclear waste differs depending on the particular category that the waste type belongs to. A portion of the spent fuel is reused. The methods that the remaining quantity along with other high-level waste are disposed of are the wet and dry storage. TRU as well as low-level waste and uranium mill tailings are most commonly geologically disposed of [182].

Some of the isotopes found on spent nuclear fuel are extracted and reused in certain industrial applications such as food irradiation and radioisotope thermoelectric generators. Moreover, there have been proposals about reactors that consume nuclear waste and transmute it to other, less-harmful nuclear waste. This process is termed nuclear waste transmutation. In particular, there was proposed such a reactor that could consume TRU waste (the Integral Fast Reactor—IFR), but after rigorous testing it was proven unreliable and the project was canceled. Reuse does not eliminate the need to manage radioisotopes, but it reduces the quantity of waste produced.

Another disposal method is the wet storage where the remaining used fuel along with other high-level waste are stored in rods in especially designed pools at individual reactor sites. These rods lie at least 6 m beneath water, which provides with adequate shielding from radiation for anyone approaching the pool. However, this storage option is limited by the size of the pool that need to be constructed and the need to keep a safety distance between rods in order to avoid potential interaction and nuclear reaction. Pool storage requires a greater and more consistent operational surveillance to ensure the satisfactory performance of the container and to avoid possible accidents.

When pool capacity has been reached, the use of above-the-ground dry storage casks is preferred. In this method, spent fuel is surrounded by inert gas inside a container called the "cask." The casks are mostly made of lead or concrete and some can be used for both storage and transportation. Dry storage is simpler; it uses fewer support systems and offers fewer possibilities for unpredictable human or mechanical error, but it is not appropriate until fuel has been removed out of the reactor for a few years and the amount of heat produced by radioactive decay has been reduced. Monitored retrievable storage (MRS) facilities have been constructed under the enforcement of restrictions applied by the Nuclear Waste Policy Act (NWPA).

Low-level waste may be stored to allow short-lived radionuclides to decay to innocuous levels and to provide safekeeping when access to disposal sites is not available. Although storage can be safe for a short period of time, on the long-term disposal is preferable. Low-level radioactive waste is packaged in containers (casks) appropriate to its level of hazard. Nuclear power plants may store waste in special buildings that provide an extra degree of shielding. Hospitals typically keep their waste stored in special containers or separate rooms. Radioactive waste storage areas are posted to identify the radioactive waste so that workers and the public will keep the required safe distance.

There are disposal facilities constructed in tunnels, vaults, or silos that can accept a wide range of low-level wastes. They primarily accept low-level waste with small concentrations of radioactive material that are generated after the facility has shut down permanently, and the removal of a large quantity of contaminated material is needed—relative examples can be contaminated soil or debris from demolished buildings—in preparation for license termination. The low-level wastes are buried a few hundred meters below ground level, usually in the containers in which they were shipped [183]. However, a geological disposal facility could receive all types of radioactive waste with appropriate design.

In addition to the methods previously presented, space disposal is highly researched. Space disposal is appealing because it removes radioactive nuclear waste from the planet. However, it has significant drawbacks. Due to the amount of waste that needs to be disposed of, multiple launches may be necessary. Therefore, the cost associated with the operation could severely increase whereas a possible launch failure could result in a catastrophic dispersion of radioactive material into the atmosphere and around the globe [184].

As for now, the main policy of Belgium, China, France, Germany, India, Japan, Russia, Switzerland, and the United Kingdom is nuclear waste reprocessing in contrast with the direct disposal policy that is followed by Canada, Finland, South Korea, Spain, Sweden, and the United States. However, most of these countries are either considering alternating their policy or have already initiated the construction of the required facilities [185].

Despite the waste management issue, 443 nuclear plants are currently operating worldwide, with the capacity of 394 of them being over 5000 MW (Fig. 6.21). 66 plants have been scheduled and are under construction at the moment [186]. It is noted that nuclear power plants can operate for multiple months uninterrupted and without maintenance, providing reliable and accurately pre-estimated supplies of electricity.

Total net electrical capacity [MW] -plants over 5000 MW

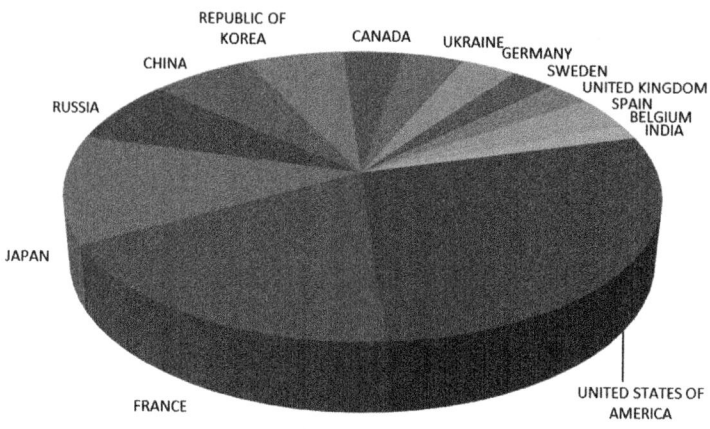

Figure 6.21
Total net electrical capacity (MW)—plants over 5000 MW [186].

6.17.2 Environmental Impacts

The impact of nuclear energy production on the environment can be divided into three categories: impact of mining, impact of energy production, and impact of nuclear waste processing.

Due to the substantial global expansion of nuclear power, the demand for fuel resources is growing in fast pace. However, studies have shown that uranium and other metal reserves are adequate for the present and future projected global power needs [185,187].

Although there is no concern about the stock amount of fuel used, the main hazards are the radiation released during uranium mining and reaction, as well as storage and further process of the waste produced [188]. The environmental effects of nuclear production are found in worker and public health and safety, in water quality, in bioaccumulation, and in issuing the waste rock and the tailings from uranium extraction [189].

Uranium is relatively harmless as long as it remains out of the body of a living organism. However, during mechanical extraction of uranium, miners are potentially exposed to fine particles of uranium in addition to possibly inhaling by-product radioactive gases, such as radon. Mining operations are conducted under the country's safety arrangements. For instance, in Australia there is the country's Code of Practice and Safety Guide for Radiation Protection and Radioactive Waste Management in Mining and Mineral Processing and in Canada the Canadian Nuclear Safety Commission regulations. These arrangements set the health standards for gamma radiation and radon gas exposure, as well as for ingestion and inhalation of radioactive materials. Standards apply to both workers' and public health. The product of uranium mining is normally uranium oxide concentrate—U_3O_8—which is barely radioactive, but has chemical toxicity similar to lead (Pb), so occupational hygiene precautions are taken similar to those in a lead smelter. Most of the radioactivity from the ore ends up in the tailings [190].

Furthermore, health risks for human population in proximity to uranium mines can be possibly posed by the heaps, tailings, and evaporation ponds, should they have not been properly taken care of. Groundwater contamination and radioactive dust transfer through the wind are also considerable threats. Radioactive particles can also enter the body by drinking or cooking with contaminated water [191].

Abandoned mines, commonly flooded with water, may also pose health risk. Water quality issues are critical to managing the environmental impact of a mining facility, given the need for the proper treatment and disposition of surface and groundwater affected by operational use. This is often more important for open-cut mines due to the size of the catchment area affected and the potential need for water to control dust [189]. A recentstudy [192] suggests that uranium, as a heavy metal, may also damage the DNA of living organisms, independent of its radioactive properties. When cells are exposed to uranium, this may bind to DNA and the cells could undergo mutations.

Some radiation effects are not immediately visible and thus cannot be observed or quantified as indicators of damage. Radiation is carcinogenic and is related to virtually every type of human cancer. As a result, the International Atomic Energy Agency (IAEA) requires a rigorous assessment of radiation levels through nuclear power generation for protecting the people and the natural environment [193].

6.17.3 Case Studies: Three Mile Island, Chernobyl, and Fukushima Accidents

Overall, the main hazards of energy production are related to the process of disposing nuclear waste with the minimum effect on humans and the environment. Another principal argument against nuclear production is the danger of accidents. Strict regulations on nuclear plant operation are in place and they get constantly updated and reassessed. 99 operational accidents in nuclear power stations have been globally reported as of today. Fifty-seven accidents have occurred since the Chernobyl disaster, and 56 of all nuclear-related accidents have occurred in the United States [194]. Extremely serious cases of nuclear power plant accidents include the Fukushima Daiichi nuclear disaster (2011), Chernobyl disaster (1986), and the Three Mile Island accident (1979).

The Three Mile Island Unit 2 (TMI-2) reactor, in Pennsylvania of the United States, partially melted down on 1979. This was the most serious nuclear power plant accident in the United States, although its radioactive releases were relatively restricted and had no detectable health effects on plant workers or the public. However, it resulted in sweeping changes about emergency response planning, reactor operator training, radiation protection, and many more areas affected during nuclear power plant operations. The reactor meltdown was attributed to equipment malfunctions and human errors [195–197].

The Chernobyl disaster (1986) was probably the worst possible accident in a nuclear power plant. It was the biggest catastrophe ever happened since the beginning of operating nuclear power stations. It started by a total meltdown of the reactor core. The explosion and the consequent reactor fire, burning for 10 days, resulted in vast emissions of radioactive material, early deaths of 31 people and adverse consequences for the public and the environment [198]. The Chernobyl disaster provided multiple invaluable lessons. One is the recognition that even near zero radiation dosage can lead to cancer, death, and hereditary disorders. Another lesson learned was that medical practitioners were lacking the necessary knowledge and preparation for nuclear emergencies. That was one of the factors that increased public anxiety and stress and often led to social panic and overreactions [199,200].

Twenty years after Chernobyl, the Fukushima accident occurred. A great earthquake (the Great East Japan Earthquake) on March 11, 2011 triggered a tsunami which affected the TEPCO's Fukushima Daiichi nuclear power plant. According to IAEA [201], the defense in depth provisions at the power plant were insufficient to provide the appropriate levels of protection for critical safety systems. Consequently, "there was a failure of the power supplies needed to provide ongoing support to key safety functions, including cooling of the reactor and spent fuel. This led to severe core damage and the release of significant quantities of radioactive material to the atmosphere and to the ocean. The released radioactive material exposed the local population to radiation in a number of ways, including external radiation exposure from radioactive material in the air and from radioactive material deposited on the ground, and internal radiation exposure from inhalation of radioactive material in the air and from ingestion of radioactive material in food or water" [201].

Although accidents had occurred before, the Chernobyl accident set the base for the majority of countries depending on nuclear power production to question the benefits compared to the hazards of a total catastrophe. Italy started phasing out the nuclear power plants and for many countries this was the call to create a federal ministry particularly dedicated to the environment. It also led scientists to further research the health impacts, not only on those directly affected, but also on those affected through the contamination of the environment. In Fukushima, the scale of disaster was comparable to Chernobyl, albeit it influenced public opinion and the decision makers more substantially [202]. Germany responded by accelerating its plans to shut all its nuclear reactors down [203]. France, Taiwan, Japan, and Belgium decided to reduce their energy dependence on domestic nuclear stations. Multiple countries, including Italy, Spain, and Switzerland, abandoned their plans to construct new nuclear power plants. In an Italian referendum, 94% of the electorate voted against the construction of new nuclear reactors, which is indicative of the accident's impact on the public acceptance of nuclear energy production. Consequently, IAEA revised to half its estimates of additional nuclear generating capacity by 2035 [204].

Both Chernobyl and Fukushima accidents led to deeply researching the relationship between humanity, energy production/consumption, and the environment, at a universal level. Nuclear production is highly reliable from the technological point of view and does not pollute the environment through direct gas emissions as much as coal-, gas-, or oil-fired power production may do, but in case of poor reactor maintenance or a natural disaster, the result may be an unprecedented catastrophe.

References

[1] T. Bruckner, et al., 2014: energy systems, in: Climate Change 2014: Mitigation of Climate Change. Contribution of Working Group III to the Fifth Assessment Report of the Intergovernmental Panel on Climate Change Cambridge University Press, Cambridge, United Kingdom and New York, NY, USA, 2014.
[2] Fossil Fuels, 2015. http://www.dulabab.com/energy-transportation/fossil-fuels/.
[3] World Energy Outlook 2011, International Energy Agency, Paris, France, 2011.
[4] World Coal Association, Coal Facts 2014, London, UK, 2014, http://www.worldcoal.org/bin/pdf/original_pdf_file/coal_facts_2014(12_09_2014).pdf.
[5] M.R. Riazi, Characterization and Properties of Petroleum Fractions, American Society for Testing and Materials (ASTM), Philadelphia, PA, USA, 2005.
[6] R.E.H. Sims, et al., Energy supply, in: B. Metz, O.R. Davidson, P.R. Bosch, R. Dave, L.A. Meyer (Eds.), Climate Change 2007, Mitigation of Climate Change − Contribution of Working Group III to the Fourth Assessment Report of the Intergovernmental Panel on Climate Change, Cambridge University Press, Cambridge, United Kingdom and New York, NY, USA, 2007.
[7] International Energy Agency, CO_2 Emissions from Fuel Combustion, 2013. Paris, France.
[8] 2011 Technology Map of the European Strategic Energy Technology Plan (SET-Plan), Technology Descriptions. European Commission Joint Research Centre, Institute for Energy and Transport Luxembourg, 2011.
[9] International Energy Agency, Per Capita CO_2 Emissions by Sector, Paris, France, 2012, http://www.oecd-ilibrary.org/energy/data/iea-co2-emissions-from-fuel-combustion-statistics/per-capita-co2-emissions-by-sector_data-00434-en?isPartOf=/content/datacollection/co2-data-en.
[10] Carbon Dioxide Information Analysis, 2013 Global Carbon Project, Tennessee, U.S.A, 2013, http://cdiac.ornl.gov/GCP/carbonbudget/2013/.
[11] U.S. Energy Information Administration, International Energy Outlook, Washington D.C., U.S.A, 2013, http://www.eia.gov/forecasts/ieo/pdf/0484%282013%29.pdf.
[12] J. Olivier, Trends in Global CO_2 Emissions: 2014 Report, PBL Netherlands Environmental Assessment Agency, Hague, The Netherlands, 2014.
[13] M. Pehnt, Environmental impacts of distributed energy systems—the case of micro cogeneration, Environ. Sci. Policy 11 (2008) 25−37.
[14] United Nations, Framework Convention on Climate Change, 2011. http://unfccc.int/resource/docs/2011/awglca14/eng/inf01.pdf.
[15] A. Burnham, et al., Correction to life-cycle greenhouse gas emissions of shale gas, natural gas, coal, and petroleum, Environ. Sci. Technol. 46 (2012) 7430.

[16] D.D. Hsu, et al., Life cycle greenhouse gas emissions of crystalline silicon photovoltaic electricity generation, J. Ind. Ecol. 16 (2012) S122–S135.

[17] J.J. Burkhardt, G. Heath, E. Cohen, Life cycle greenhouse gas emissions of trough and tower concentrating solar power electricity generation, J. Ind. Ecol. 16 (2012) S93–S109.

[18] A. Arvesen, E.G. Hertwich, Assessing the life cycle environmental impacts of wind power: a review of present knowledge and research needs, Renew. Sustain. Energy Rev. 16 (2012) 5994–6006.

[19] E.S. Warner, G.A. Heath, Life cycle greenhouse gas emissions of nuclear electricity generation, J. Ind. Ecol. 16 (2012) S73–S92.

[20] M. Caduff, et al., Wind power electricity: the bigger the turbine, the greener the electricity? Environ. Sci. Technol. 46 (2012) 4725–4733.

[21] A. Arvesen, E.G. Hertwich, Environmental implications of large-scale adoption of wind power: a scenario-based life cycle assessment, Environ. Res. Lett. 6 (2011).

[22] H.J. Herzog, Scaling up carbon dioxide capture and storage: from megatons to gigatons, Energy Econ. 33 (2011) 597–604.

[23] S.T. McCoy, E.S. Rubin, An engineering-economic model of pipeline transport of CO_2 with application to carbon capture and storage, Int. J. Greenhouse Gas Control 2 (2008) 219–229.

[24] L. Hirth, The market value of variable renewables: the effect of solar wind power variability on their relative price, Energy Econ. 38 (2013) 218–236.

[25] T. Bruckner, I.A. Bashmakov, Y. Mulugetta, Energy Systems. Climate Change 2014, in: Mitigation of Climate Change, 2014, pp. 511–597.

[26] M. Fischedick, R. Schaeffer, Mitigation Potential and Costs, IPCC Special Report on Renewable Energy Sources and Climate Change Mitigation, 2011, pp. 791–864.

[27] C. Clapp, et al., National and Sectoral GHG Mitigation Potential: A Comparison Across Models, International Energy Agency, 2009.

[28] McKinsey & Company, Pathways to a Low-Carbon Economy — Version 2 of the Global Greenhouse Gas Abatement Cost Curve, 2009.

[29] E. Karampinis, et al., Numerical investigation Greek lignite/cardoon co-firing in a tangentially fired furnace, Appl. Energy 97 (2012) 514–524.

[30] N. Nikolopoulos, et al., Parametric investigation of a renewable alternative for utilities adopting the co-firing lignite/biomass concept, Fuel 113 (2013) 873–897.

[31] A. Nikolopoulos, et al., High-resolution 3-D full-loop simulation of a CFB carbonator cold model, Chem. Eng. Sci. 90 (2013) 137–150.

[32] K. Atsonios, et al., Calcium looping process simulation based on an advanced thermodynamic model combined with CFD analysis, Fuel 153 (2015) 370–381.

[33] UNFCCC, Clarifications of Definition of Biomass and Consideration of Changes in Carbon Pools Due to a CDM Project Activity, EB-20, Appendix 8, July 2005.

[34] P. Basu, Combustion and Gasification in Fluidized Beds, CRC Press, Taylor and Francis Group, FL, USA, 2006.

[35] Directive 2008/98/EC of the European Parliament and of the Council of 19 November 2008 on Waste and Repealing Certain Directives. European Commission, 2008.

[36] Commission Decision of 3 May 2000 Replacing Decision 94/3/EC Establishing a List of Wastes Pursuant to Article 1(a) of Council Directive 75/442/EEC on Waste and Council Decision 94/904/EC Establishing a List of Hazardous Waste Pursuant to Article 1(4) of Council Directive 91/689/EEC on Hazardous Waste, European Commission, 2000.

[37] Commission Decision of 16 January 2001 Amending Decision 2000/532/EC as Regards the List of Wastes, European Commission, 2001.

[38] Commission Decision of 22 January 2001 Amending Decision 2000/532/EC Replacing Decision 94/3/EC Establishing a List of Wastes Pursuant to Article 1(a) of Council Directive 75/442/EEC on Waste and Council Decision 94/904/EC Establishing a List of Hazardous Waste Pursuant to Article 1(4) of Council Directive 91/689/EEC on Hazardous Waste, European Commission, 2001.

[39] Council Decision of 23 July 2001 Amending Commission Decision 2000/532/EC as Regards the List of Wastes, European Commission, 2001.

[40] M.Z. Jacobson, M.A. Delucchi, Providing all global energy with wind, water, and solar power, Part I: technologies, energy resources, quantities and areas of infrastructure, and materials, Energy Policy 39 (2011) 1154−1169.

[41] C.S. Ferekides, et al., High efficiency CSS CdTe solar cells, Thin Solid Films 361−362 (2000) 520−526.

[42] C.R. Landau, Optimum Tilt of Solar Panels, 2001. http://www.solarpaneltilt.com/.

[43] R. Pernick, C. Wilder, J. Belcher, Clean Energy Trends 2014, March 2014, p. 10.

[44] Classification of geothermal resources an engineering approach: Proceedings of 21st Workshop on Geothermal Reservoir Engineering, Stanford University Conference, 1996.

[45] E. Karampinis, D.-S. Kourkoumpas, G. Panagiotis, E. Kakaras, New power production options for biomass and cogeneration, Wiley Interdiscip. Rev. Energy Environ. 4 (2015) 471−485, http://dx.doi.org/ 10.1002/wene.163.

[46] Technology Roadmap: Bioenergy for Heat and Power, International Energy Agency, Paris, France, 2012.

[47] S.v. Loo, J. Koppejan, The Handbook of Biomass Combustion and Co-firing, Earthscan, London, Sterling VA, UK and USA, 2008.

[48] M. Asadullah, Biomass gasification gas cleaning for downstream applications: a comparative critical review, Renew. Sustain. Energy Rev. 40 (2014) 118−132.

[49] T.G. Bridgeman, et al., Torrefaction of reed canary grass, wheat straw and willow to enhance solid fuel qualities and combustion properties, Fuel 87 (2008) 844−856.

[50] N. Nikolopoulos, et al., Modeling of wheat straw torrefaction as a preliminary tool for process design, Waste Biomass Valorization 4 (2013) 409−420.

[51] A. Khalid, et al., The anaerobic digestion of solid organic waste, Waste Manage. 31 (2011) 1737−1744.

[52] N.M.C. Saady, D.I. Massé, Psychrophilic anaerobic digestion of lignocellulosic biomass: a characterization study, Bioresour. Technol. 142 (2013) 663−671.

[53] D. Kourkoumpas, et al., An environmental assessment for anaerobic digestion of biowaste based on life cycle analysis principles, in: 3rd International Exergy, Life Cycle Assessment, and Sustainability Workshop & Symposium (ELCAS3) Nisyros, Greece, 2013.

[54] D.G. Karalis, D.I. Pantelis, V.J. Papazoglou, On the investigation of 7075 aluminum alloy welding using concentrated solar energy, Sol. Energy Mater. Sol. Cells 86 (2005) 145−163.

[55] C. Philibert, The Present and Future Use of Solar Thermal Energy as a Primary Source of Energy, 2005.

[56] U. Gangopadhyay, S. Jana, S. Das, State of art of solar photovoltaic technology, in: Conference Papers in Energy, 2013, p. 9.

[57] M.A. Green, et al., Crystalline silicon on glass (CSG) thin-film solar cell modules, Sol. Energy 77 (2004) 857−863.

[58] C.P. Lund, et al., Field and laboratory studies of the stability of amorphous silicon solar cells and modules, Renew. Energy 22 (2001) 287−294.

[59] E. Wesoff, Topaz, the Largest Solar Plant in the World, Is Now Fully Operational, 2014. http://www. greentechmedia.com/articles/read/550-megawatts-AC-to-be-exact.

[60] Technology Roadmap: Solar Photovoltaic Energy, International Energy Agency, Paris, France, 2014.

[61] S.P. Sukhatme, J.K. Nayak, Solar Energy: Principles of Thermal Collection and Storage, 2011. New Delhi.

[62] S.V. Reddy, et al., State-of-the-art of solar thermal power plants—a review, Renew. Sustain. Energy Rev. 27 (2013) 258−273.

[63] C. Zhe, J.M. Guerrero, F. Blaabjerg, A review of the state of the art of power electronics for wind turbines, Power Electron. IEEE Trans. 24 (2009) 1859−1875.

[64] N. Masseran, Markov chain model for the stochastic behaviors of wind-direction data, Energy Convers. Manage. 92 (2015) 266−274.

[65] L.L. Freris, Wind Energy Conversion System, Prentice Hall, London, 1990.

[66] S. Heier, Grid Integration of Wind Energy Conversion Systems, Wiley, New York, 1998.

[67] H. Li, Z. Chen, Overview of different wind generator systems and their comparisons, Renew. Power Gener. IET 2 (2008) 123–138.

[68] P. Thoegersen, F. Blaabjerg, Adjustable speed drives in the next decade: future steps in industry and academia, Electr. Power Compon. Syst. 32 (2004) 13–31.

[69] A.O. Ibrahim, et al., A fault ride-through technique of DFIG wind turbine systems using dynamic voltage restorers, Energy Convers. IEEE Trans. 26 (2011) 871–882.

[70] R.S. Semken, et al., Direct-drive permanent magnet generators for high-power wind turbines: benefits and limiting factors, Renew. Power Gener. IET 6 (2012) 1–8.

[71] S.M. Xu, X.D. Huang, R. Du, An investigation of the solar powered absorption refrigeration system with advanced energy storage technology, Sol. Energy 85 (2011) 1794–1804.

[72] M. Cheng, Z. Ying, The state of the art of wind energy conversion systems and technologies: a review, Energy Convers. Manage. 88 (2014) 332–347.

[73] U.S. Department of Energy, Types of Hydropower Plants, Washington D.C., U.S.A, 2015, http://energy.gov/eere/water/types-hydropower-plants.

[74] BISYPLAN (Web-Based Handbook), 2012. http://bisyplan.bioenarea.eu.

[75] I. Yüksel, Hydropower in Turkey for a clean and sustainable energy future, Renew. Sustain. Energy Rev. 12 (2008) 1622–1640.

[76] H. Susskind, et al., Combined Hydroelectric Pumped Storage and Nuclear Power Generation, Brookhaven National Laboratory, Upton, N.Y, 1970.

[77] Cambridge University Press, Renewable energy sources and climate change mitigation: special report of the intergovernmental panel on climate change, in: Intergovernmental Panel on Climate Change Conference. New York, 2012.

[78] Towards a Green Economy: Pathways to Sustainable Development and Poverty Eradication, United Nations Environment Programme, 2011.

[79] Green Economy Initiative, United Nations Environment Programme, 2014.

[80] COM (2011) 112: A Roadmap for Moving to a Competitive Low Carbon Economy in 2050 (08 Mar 2011) European Commission, 2011.

[81] N. Scarlat, et al., The role of biomass and bioenergy in a future bioeconomy: policies and facts, Environ. Develop. 15 (July 2015) 3–34.

[82] S. Farhad, M. Saffar-Avval, M. Younessi-Sinaki, Efficient design of feedwater heaters network in steam power plants using pinch technology and exergy analysis, Int. J. Energy Res. 32 (2008) 1–11.

[83] R.E. Sims, Bioenergy to mitigate for climate change and meet the needs of society, the economy and the environment, Mitigation Adapt. Strategies Global Change 8 (2003) 349–370.

[84] I. Youm, et al., Renewable energy activities in Senegal: a review, Renew. Sustain. Energy Rev. 4 (2000) 75–89.

[85] G. Hiemstra-van der Horst, A.J. Hovorka, Fuelwood: the "other" renewable energy source for Africa? Biomass Bioenergy 33 (2009) 1605–1616.

[86] D. Hall, Cooling the greenhouse with bioenergy, Nature 353 (1991) 11–12.

[87] J.B. Holm-Nielsen, T. Al Seadi, P. Oleskowicz-Popiel, The future of anaerobic digestion and biogas utilization, Bioresour. Technol. 100 (2009) 5478–5484.

[88] N. Panwar, S. Kaushik, S. Kothari, Role of renewable energy sources in environmental protection: a review, Renew. Sustain. Energy Rev. 15 (2011) 1513–1524.

[89] European Commision Decision of 18 July 2007 Establishing Guidelines for the Monitoring and Reporting of Greenhouse Gas Emissions Pursuant to Directive 2003/87/EC of the European Parliament and of the Council (Notified under Document Number C(2007) 3416), 2007.

[90] World Energy Outlook 2013, op. cit. note 1, International Energy Agency, 2013, p. 200.

[91] A. Demirbaş, Global renewable energy resources, Energy Sources 28 (2006) 779–792.

[92] A. Zervos, Renewables 2013 Global Status Report, Paris, France, 2013.

[93] I. Kralova, J. Sjöblom, Biofuels-renewable energy sources: a review, J. Dispers. Sci. Technol. 31 (2010) 409–425.

[94] G. Wang, et al., Pretreatment of biomass by torrefaction, Chin. Sci. Bull. 56 (2011) 1442–1448.

[95] J.L. Sawin, F. Sverrisson, Renewables 2014 Global Status Report, REN21 Secretariat, Paris, France, 2013.

[96] Commission, E, COM (2011) 112: A Roadmap for Moving to a Competitive Low Carbon Economy in 2050 (08 Mar 2011), 2011.

[97] G. Masson, S. Orlandi, M. Rekinger, Global Market Outlook for Photovoltaics 2014–2018. European Photovoltaic Industry Association (EPIA).

[98] E. Karampinis, et al., Co-firing of biomass with coal in thermal power plants: technology schemes, impacts and future perspectives, WIREs Energy Environ. 3 (4) (2013) 384–399.

[99] G. Corbetta, T. Miloradovic, Wind in Power: 2014 European Statistics, February 2015.

[100] World Energy Outlook 2013, op. cit. note 1, International Energy Agency, 2013, p. 574.

[101] A. McCrone, Global Trends in Renewable Energy Investment 2013, Frankfurt School-UNEP Centre/BNEF, 2013.

[102] M. Salomón, et al., Small-scale biomass CHP plants in Sweden and Finland, Renew. Sustain. Energy Rev. 15 (2011) 4451–4465.

[103] Monthly Energy Review, April 2014, U.S. Energy Information Administration, 2014.

[104] Orkutölur 2013, Orkustofnun (Energy Statistics in Iceland 2013), Reykjavik, April 2014.

[105] op. cit. note 1, World Energy Outlook 2013, 2013, p. 199.

[106] Eurostat, Renewable Energy in the EU28-Share of Renewables in Energy Consumption up to 14% in 2012, Press Release Brussels, March 10, 2014.

[107] M. Osborne, Hareon Solar Teaming With Shanghai Electric Power on 800 MW of PV Projects, PV Tech, March 13, 2014.

[108] United Nations Environment Programme. http://lcinitiative.unep.fr/default.asp?site=lcinit&page_id=11A26B55-8A61-4FDA-AE7F-47C13119E384.

[109] H.J. Bjarnadótti, et al., Guidelines for the Use of LCA in the Waste Management Sector, Nordest Report, 2002.

[110] SETAC: http://www.setac.org/.

[111] Resource Savings and CO₂ Reduction Potential in Waste Management in Europe and the Possible Contribution to the CO₂ Reduction Target in 2020. Prognos AG in Co-Operation with Ifeu – Institut für Energie- und Umweltforschung Heidelberg GmbH and Institut für Umweltforschung (INFU) (Institute for Environmental Research), University of Dortmund, 2008.

[112] Directorate General Environment Refuse Derived Fuel, Current Practice and Perspectives European Commission.

[113] A.K. Jönbrink, et al., LCA Software Survey, Swedish Industrial Research Institutes' Initiative, 2000.

[114] Recipe. http://www.lcia-recipe.net/.

[115] Pre. Eco-Indicator 99. http://www.pre-sustainability.com/content/eco-indicator-99.

[116] European Commission Tools. http://lca.jrc.ec.europa.eu/lcainfohub/toolList.vm.

[117] SimaPro. http://www.simapro.co.uk/.

[118] EASEWASTE. http://www.easewaste.dk/.

[119] Umberto. http://www.umberto.de/en/.

[120] Gabi. http://www.gabi-software.com/international/index/.

[121] D.W. Pearce, R.K. Turner, Economics of Natural Resources and the Environment, Harvester Wheatsheaf, Hemel Hempstead, Herts, 1990.

[122] J. Vail Farr, Systems Life Cycle Costing: Economic Analysis, Estimation and Management.

[123] Environment, Life – Cycle Costing. http://ec.europa.eu/environment/gpp/lcc.htm.

[124] G. Pekridis, et al., Hydrogen production in solid electrolyte membrane reactors, Int. J. Hydrogen Energy 32 (2007) 38–54.

[125] A. Konieczny, et al., Catalyst development for thermocatalytic decomposition of methane to hydrogen, Int. J. Hydrogen Energy 33 (2008) 264–272.

[126] M. Balat, Potential importance of hydrogen as a future solution to environmental and transportation problems, Int. J. Hydrogen Energy 33 (2008) 4013–4029.

[127] M. Balat, Hydrogen in fueled systems and the significance of hydrogen in vehicular transportation, Energy Sources Part B 2 (2007) 49–61.

[128] G.E. Marnellos, et al., Integration of hydrogen energy technologies in autonomous power systems, in: Hydrogen Based Autonomous Power Systems. Technoeconomic Analysis of the Integration of Hydrogen in Autonomous Power Systems, Springer, 2008, pp. 23–82.

[129] S. Yolcular, Hydrogen production for energy use in European Union countries and Turkey, Energy Source A 31 (2009) 1329–1337.

[130] I.K. Kapdan, F. Kargi, Bio-hydrogen production from waste materials, Enzyme Microb. Technol. 38 (2006) 569–582.

[131] A. Demirbas, Biofuels sources, biofuels policy, biofuel economy and global biofuel projections, Energy Convers. Manage. 49 (2008) 2106–2116.

[132] B. Sorensen, Hydrogen and Fuel Cells Emerging Technologies and Applications, Elsevier Academic Press, New York, 2005.

[133] P. Haussinger, R. Lohmuller, Encyclopedia of Industrial Chemistry, 2011.

[134] K. Zeng, D. Zhang, Recent progress in alkaline water electrolysis for hydrogen production, Prog. Energy Combust. Sci. 36 (2010) 307–326.

[135] J. Ogden, Developing an infrastructure for hydrogen vehicles: a Southern California case study, Int. J. Hydrogen Energy 24 (1999) 709–730.

[136] C. Neagu, et al., The electrolysis of water: an actuation principle, Mechatronics (2000) 571–581.

[137] National Academy of Science, the Hydrogen Economy: Opportunities, Costs, Barriers, and R&D Needs, National Academies Press, Washington D.C., U.S.A, 2004.

[138] S.A. Grigoriev, V.I. Porembsky, V.N. Fateev, Pure hydrogen production by PEM electrolysis for hydrogen production, Int. J. Hydrogen Energy 31 (2003) 171–175.

[139] M. Ni, M. Leung, D. Leung, Technological development of hydrogen production by solid oxide electrolyzer cell (SOEC), Int. J. Hydrogen Energy 33 (2008) 2337–2354.

[140] A. Hauch, et al., Highly efficient high temperature electrolysis, J. Mater. Chem. 18 (2008) 2331–2340.

[141] A. Steinfeld, Solar thermochemical production of hydrogen-a review, Sol. Energy 78 (2005) 603–615.

[142] M.A. Rosen, Advances in hydrogen production by thermochemical water decomposition: a review, Energy 35 (2010) 1068–1076.

[143] J.E. Funk, R.M. Reinstrom, Energy requirements in the production of hydrogen from water, Ind. Eng. Chem. Process Des. Dev. 5 (1966) 336–342.

[144] J.D. Holladay, et al., An overview of hydrogen production technologies, Catal. Today 139 (2009) 244–260.

[145] Hydrogen energy and photoelectrolysis of water, in: Proceedings of the Technical Sessions Conference. Sri Lanka, 2004.

[146] V.M. Aroutiounian, V.M. Arakelyan, G.E. Shahnazaryan, Metal oxide photoelectrodes for hydrogen generation using solar radiation-driven water splitting, Sol. Energy 78 (2005) 581–592.

[147] R.K. Jalan, V.K. Srivastana, Studies on pyrolysis of a single biomass cylindrical pellet-kinetic and heat transfer effects, Energy Convers. Manage. 40 (1999) 467–494.

[148] Q. Yan, L. Guo, Y. Lu, Thermodynamic analysis of hydrogen production from biomass gasification in supercritical water, Energy Convers. Manage. 47 (2006) 1515–1528.

[149] A. Demirbas, Hydrogen-rich gases from biomass via pyrolysis, Energy Sources Part A 31 (2009) 1728–1736.

[150] M. Balat, et al., Main routes for the thermo-conversion of biomass into fuels and chemicals. Part 1: pyrolysis systems, Energy Convers. Manage. 50 (2009) 3147–3157.

[151] Z. Abu El Rub, E.A. Bramer, G. Brem, Review of catalysts for tar elimination in biomass gasification, Ind. Eng. Chem. Res. 43 (2004) 6911–6919.

[152] M. Ni, et al., An overview of hydrogen production from biomass, Fuel Process. Technol. 87 (2006) 461–472.

[153] P. Simell, J.S. Bredenberg, Catalytic purification of tarry fuel gas, Fuel 69 (1990) 1219–1225.

[154] D. Sutton, B. Kelleher, J.R.H. Ross, Review of literature on catalysts for biomass gasification, Fuel Process. Technol. 73 (2001) 155−173.

[155] P.C. Hallenbeck, J.R. Benemann, Biological hydrogen production; fundamentals and limiting processes, Int. J. Hydrogen Energy 27 (2002) 1185−1193.

[156] State of the art of biological hydrogen production processes, in: World Hydrogen Energy Conference. Lyon, 2006.

[157] D. Das, N. Veziroglu, Hydrogen production by biological processes: a survey of literature, Int. J. Hydrogen Energy 26 (2001) 13−28.

[158] J. Yu, P. Takahashi, Biophotolysis-based Hydrogen Production by Cyanobacteria and Green Microalgae, Hawaii Natural Energy Institute, University of Hawaii, 2007.

[159] R.P. Datar, et al., Fermentation of biomass-generated producer gas to ethanol, Biotechnol. Bioeng. 86 (2004) 587−594.

[160] Y. Tao, et al., High hydrogen yield from a two-step process of dark- and photo-fermentation of sucrose, Int. J. Hydrogen Energy 32 (2007) 200−206.

[161] K. Nath, D. Das, Improvement of fermentative hydrogen production: various approaches, Appl. Microbiol. Biotechnol. 65 (2004) 520−529.

[162] D. Das, N. Khanna, N. Veziroglu, Recent developments in biological hydrogen production processes, Chem. Ind. Chem. Eng. Q. 14 (2008) 57−67.

[163] J. Vielstich, H. Gasteiger, A. Lamm, Handbook of Fuel Cells—Fundamentals, Technology, Applications, Wiley, New York, 2003.

[164] E.G. Technical, Fuel Cell Handbook, Morgantown, 2004.

[165] U. Lucia, Overview on fuel cells, Renew. Sustain. Energy Rev. 30 (2014) 164−169.

[166] S. Giddey, et al., A comprehensive review of direct carbon fuel cell technology, Prog. Energy Combust. Sci. 38 (2012) 360−399.

[167] Q. Ma, et al., A high-performance ammonia-fueled solid oxide fuel cell, J. Power Sources 161 (2006) 95−98.

[168] M.Z. Kamaruddin, et al., An overview of fuel management in direct methanol fuel cells, Renew. Sustain. Energy Rev. 24 (2013) 557−565.

[169] J.C. Ganley, An intermediate-temperature direct ammonia fuel cell with a molten alkaline hydroxide electrolyte, J. Power Sources 178 (2008) 44−47.

[170] P. Tseng, J. Lee, P. Friley, Hydrogen economy: opportunities and challenges, Energy 30 (2005) 2703−2720.

[171] S. Kim, B.E. Dale, Life cycle assessment of various cropping systems utilized for producing biofuels: bioethanol and biodiesel, Biomass Bioenergy 29 (2005) 426−439.

[172] W. El-Osta, J. Zeghlam, Hydrogen as fuel for transportation sector: possibilities and views for future use in Libya, Appl. Energy 65 (2000) 165−171.

[173] H.J. Plass, et al., Economics of hydrogen as a fuel for surface transportation, Int. J. Hydrogen Energy 15 (1990) 663−668.

[174] R. Derwent, et al., Global environmental impacts of the hydrogen economy, Int. J. Nucl. Hydrogen Prod. Appl. 1 (2006) 57−67.

[175] J. Ma, et al., Simulation and prediction on the performance of a vehicle's hydrogen engine, Int. J. Hydrogen Energy 28 (2003) 77−83.

[176] International Energy Agency. Key Findings. https://www.iea.org/Textbase/npsum/nuclear_roadmapsum.pdf.

[177] World Nuclear Association, Electricity Generation − What Are the Options?, 2015. http://www.world-nuclear.org/Nuclear-Basics/Electricity-generation-what-are-the-options-/.

[178] J. Baek, A panel cointegration analysis of CO_2 emissions, nuclear energy and income in major nuclear generating countries, Appl. Energy 145 (2015) 133−138.

[179] World Nuclear Association, Radioactive Waste Management, 2015. http://www.world-nuclear.org/info/Nuclear-Fuel-Cycle/Nuclear-Wastes/Radioactive-Waste-Management/.

[180] M. Holt, CRS Report for Congress: Civilian Nuclear Waste Disposal, Congressional Research Service, Washington D.C., U.S.A, 2013.

[181] V.V. Rondinella, Handbook of Advance Radioactive Waste Conditioning Technologies, Woodhead Publishing, Cambridge UK, 2011.
[182] R.L. Murray, K.E. Holbert, Nuclear Energy: An Introduction to the Concepts, Systems and Applications of Nuclear Processes, seventh ed., Elsevier and Book Aid International, Oxford, GB, UK, 2015.
[183] Radioactive Waste: Production, Storage, Disposal, U.S. Nuclear Regulatory Commission, Washington D.C., U.S.A, 2002.
[184] Committee on Disposition of High-Level Radioactive Waste Through Geological Isolation, Board on Radioactive Waste Management, Disposition of High-Level Waste and Spent Nuclear Fuel: The Continuing Societal and Technical Challenges, National Research Council, National Academies Press, Washington D.C., U.S.A, 2001.
[185] E.M. Harper, et al., The criticality of four nuclear energy metals, Resour. Conserv. Recycl. 95 (2015) 193–201.
[186] International Atomic Energy Agency, Operational & Long-Term Shutdown Reactors, April 04, 2015. http://www.iaea.org/PRIS/WorldStatistics/OperationalReactorsByCountry.aspx.
[187] World Nuclear Association, Supply of Uranium, 2014. http://www.world-nuclear.org/info/Nuclear-Fuel-Cycle/Uranium-Resources/Supply-of-Uranium/.
[188] A.S. Paschoa, Environmental Effects of Nuclear Power Generation, in Interactions: Energy/Environment, Encyclopedia of Life Support Systems (EOLSS), . Eolss Publishers, Oxford ,UK, 2004.
[189] OECD, Managing Environmental and Health Impacts of Uranium Mining, OECD Publishing, Paris, France, 2014.
[190] World Nuclear Assosiation, Occupational Safety in Uranium Mining, 2014. http://www.world-nuclear.org/info/Safety-and-Security/Radiation-and-Health/Occupational-Safety-in-Uranium-Mining/.
[191] R.L. Blanchard, et al., Potential health effects of radioactive emissions from active surface and underground uranium mines, Nucl. Saf. 23 (1982).
[192] Canadian Nuclear Safety Commission, Radiation Health Effects, Canada, 2015, http://nuclearsafety.gc.ca/eng/resources/radiation/introduction-to-radiation/radiation-health-effects.cfm.
[193] Safety of Nuclear Power Plants - Design: Specific Safety Requirements No. SSR-2/1, International Atomic Energy Agency, 2012.
[194] K. Shrader-Frechette, Fukushima, flawed epistemology, and black-swan events, Ethics Policy Environ. 14 (2011) 267–272.
[195] S. Wing, et al., A reevaluation of cancer incidence near the Three Mile Island nuclear plant: the collision of evidence and assumptions, Environ. Health Perspect. 105 (1997) 52–57.
[196] M.C. Hatch, et al., Cancer near the Three Mile Island nuclear plant: radiation emissions, Am. J. Epidemiol. 132 (1990) 397–412.
[197] A.C. Cilliers, Benchmarking an expert fault detection and diagnostic system on the Three Mile Island accident event sequence, Ann. Nucl. Energy 62 (2013) 326–332.
[198] Environmental Consequences of the Chernobyl Accident and Their Remediation: Twenty Years of Experience. Chernobyl Forum Expert Group "Environment", International Atomic Energy Agency, Vienna, Austria, 2006.
[199] I. Kovalchuk, et al., Molecular aspects of plant adaptation to life in the Chernobyl zone, Plant Physiol. 135 (2004) 357–363.
[200] Health Effects of the Chernobyl Accident and Special Health Care Programmes. UN Chernobyl Forum Expert Group "Health", World Health Organization, Geneva, Switzerland, 2006.
[201] IAEA Report on Radiation Protection after the Fukushima Daiichi Accident: Promoting Confidence and Understanding, International Atomic Energy Agency, Vienna, Austria, 2014.
[202] T. Ohnishi, The disaster at Japan's Fukushima-Daiichi nuclear power plant after the march 11, 2011 earthquake and tsunami, and the resulting spread of radioisotope contamination, Radiat. Res. 177 (2012) 1–14.
[203] A. Breidthardt, German Government Wants Nuclear Exit by 2022 at Latest, Reuters, 2011.
[204] Reference Data Series, vol. 1, International Atomic Energy Agency, 2012.

CHAPTER 7

Extraterrestrial Environment

V.J. Inglezakis

Nazarbayev University, Astana, Republic of Kazakhstan

Chapter Outline

7.1 Introduction

The Earth is by all means an astonishing planet, an extremely rare marvel inhabited by a uniquely intelligent animal. Left to its own devices, Earth's life expectancy is about

Environment and Development. http://dx.doi.org/10.1016/B978-0-444-62733-9.00007-1

4—7 billion years. As solar luminosity steadily increases, in less than a billion years plant and animal life will die off and oceans will evaporate. Gradually, Earth's surface temperature will be immense and eventually the planet will be swallowed by the swollen Sun.

The most intelligent inhabitant of the blue planet relentlessly tries to take his future in his own hands and avoid being killed by whom he takes life from. Man develops sophisticated theories and technologies, but his amazing works come with consequences able in some ways and to some extent to alter the very future of his home planet and seal his own fate as well.

Humans always turn eyes to heavens when dreaming or philosophizing, seeking for answers to right or even wrong questions. Looking at the sky on a clear moonless night is stunning. No one has counted all the stars in the night sky, but considering all directions around Earth the upper end on the estimates is about 10,000 visible stars. It seems as if between us and the stars is an almost empty pristine space untouched by the hazards of modern civilization. But something invisible to our eyes and in vast numbers is out there; the space between us and the stars is no more empty as a potentially hazardous cloud of "space junk" is orbiting the Earth. Space is getting dirty and things might turn even worse if the dumb idea of disposing high-level radioactive waste in space someday is materialized. Another, weird idea is already a reality: space burials available for humans and pets.

And what about climate change? Is it only Earth's privilege? Is it a curse or a blessing? To answer these questions we need to travel not far away and ask the planets and moons, named after Greek gods and goddesses; Venus, the beautiful Aphrodite of Greek mythology, which suffers in an inferno under a thick 92-bar atmosphere and furnace-like blazing temperatures; Red planet, Mars, who as a true warrior breaths elegantly under harsh but bearable conditions and prays to avoid Moon's and Mercury's fate, which suffer from wild variations of surface temperature due to the absence of the greenhouse effect; Titan, the mystery moon of Saturn, the only natural satellite known to have a dense nitrogen atmosphere, full of methane—ethane clouds and the only known celestial body aside from Earth that has liquid bodies, seas, and lakes filled with liquid methane and ethane; Pluto, the dwarf planet that travels deep in the oblivion of a space odyssey, covered by iced-water oceans and carrying a thin and fragile nitrogen-rich atmosphere, which dies and resurrects periodically; and Triton, Neptune's largest moon only 35 K from absolute zero.

Life on other planets and humans as well, if forced to seek for a new home in the event of destroying our splendid blue planet, needs water. When, rightfully, almost everybody is in the search of liquid water in our solar system, few are looking for humble dirty magic rocks, hydrated porous minerals such as aluminosilicates. These amazing water-related minerals have been observed at different solar system objects such as meteorites, asteroids, terrestrial planets like Mars, and some icy planets' satellites as Europa, Ganymede, and Enceladus. The implications of the existence of such minerals in planets are multiple; they can offer the right environment for essential-to-life chemical and biological reactions, act as reservoirs of liquid water, influence the atmospheric composition of planets, and can be even utilized in situ by humans in several life-supporting applications.

This chapter is not an ordinary one; it deals with some aspects of the extraterrestrial environment, which is relevant—directly or indirectly—to human development.

It is a trip to a known unknown full of unknown unknowns.

7.2 Space Waste

7.2.1 Space Exploration and Space Junk

Waste is everywhere; air, water, and soil on Earth gradually became polluted as a result of human activities, but until late '80s ago no one could have imagined that space waste would also become a problem. At the first decades of space exploration, space was empty, so no one was considering the impact of our actions there. On the contrary, and surprisingly, some thought that it would be not such a bad idea to deposit radioactive waste into space (see Section 7.2.2). These ideas, fortunately, were abandoned long ago. But there is another type of waste that threatens the space environment in our neighborhood, namely orbital debris. Pollution from orbital debris is imminent, and the analysis focuses on this type of space waste, but as human activities expand further than Earth's orbit, the problem of debris will certainly spread. For example, the Moon is already littered with over 100 tons of man-made waste, while the Mars Reconnaissance Orbiter that entered orbit around Mars in 2006, had to maneuver in order to avoid man-made objects orbiting Mars.

The facts and the numbers

Vanguard 1, a 1.47-kg aluminum sphere 16.5 cm in diameter, is the oldest piece of space junk still orbiting the Earth. It was launched by the United States in 1958 and was operational until 1964, but will continue orbiting Earth for 240 years. In January 1997 the fuel tank of a Delta II rocket, used for a US Air Force satellite launch in 1996, was crashed into the Earth's atmosphere. Half an hour later, a metallic piece of this tank hit a resident of Tulsa in Oklahoma, the United States, making her the only person in the world known to have been hit by man-made space debris.

It might sound as black humor, but space burials became a popular idea since the 1990s. The first memorial service was held in 1997, when a private company carried the ashes of 24 people, including Gene Roddenberry's, the creator or Star Trek, into orbit. Since then, hundreds of human remains samples, stored in small capsules of a size of an AAA battery, were sent into space. This service is widening and a Houston-based company in 2014 announced a program for the remains of dogs and cats. Space burial objects are certainly tiny but quite wired kind of man-made space articles that do not serve any practical function.

Orbital debris generally refers to material that is in orbit as the result of space missions, but does no longer serve any function [1]. More generally, this term is used to describe a collection of defunct objects in orbit around Earth. Space waste sizes from few millimeters, as paint fragments and miscellaneous items left behind by astronauts and cosmonauts, to meters and more, as defunct satellites and spent rocket stages. In 1965, the first American spacewalker, Ed White, lost a spare glove, in 2005 spacewalker Piers Sellers lost a spatula, in 2006 the Atlantis astronauts made their own contributions when a couple of bolts escaped the international space station, and in 2008 Heide Stefanyshyn-Piper lost a tool bag while doing a spacewalk. Dried-up urine, toothpaste, and shaving cream are in orbit as well. These are some of the hundreds of thousands unusual trash that circle the Earth. In some occasions, though it may seem funny, the reality is that many of these items can kill astronauts and damage or even destroy satellites and other expensive instruments. Additionally, orbital debris might impact the Earth upon reentry, endangering human lives and the environment by dispersing toxic pollutants and other materials over long areas. All in all, orbital debris entails risks both for the future of human in space and on Earth, and unlike most pollution, space junk is self-perpetuating and the question is not "if" but "when" the exponential growth of the space debris population is going to start [2].

Since the launch of Sputnik on October 4, 1957, some 50 nations now own and/or operate spacecrafts, and 6000 satellites are put into orbit, with more than 1000 of which being still in use for several applications, corresponding to over 30,000 tons of weight [3−6]. According to an approximate count made in April 2010 by the Union of Concerned Scientists (UCS), 41% of the world's active satellites are commercial, 17% are military, 18% are government, and the remaining 24% are either multiuse or used for research and scientific purposes [7]. All these activities generate a considerable amount of waste of different kinds and sources: discarded hardware, abandoned end-of-life satellites, material released during deployments and separations such as separation bolts, lens caps, clamp bands and adapter shrouds, material resulting from degradation due to solar radiation and heating such as paint flakes, and bits of insulation and object breakup caused by explosions, including satellites deliberately detonated, and collisions with other objects in space, mostly debris [1]. It is estimated that more than 15,000 large objects can be tracked by ground-based radars and telescopes, with around 6% of which being active satellites [8]. These objects are space debris, but only of a trackable size and thus, cataloged; smaller debris is not visible to ground-based radars and telescopes and a rough estimation gives more than 6000−7000 tons of space debris, 3000 tons of which are found in low Earth orbit, at an altitude lower than 2000 km (Table 7.1) [6,9,10].

Table 7.1: Types of Orbits.

Orbit	Altitude (km)	Use
Low Earth orbit (LEO)	160–2000	Earth monitoring, military surveillance, and some communications satellites
Medium Earth orbit (MEO)	2000–36,000	Mainly navigation satellites such as the Global Positioning System (GPS)
Geostationary orbit (GEO)	36,000	Satellites at this altitude orbit the Earth in exactly 1 day, so they are always above the same spot on the Earth's surface. This makes them useful for communications, as receivers on Earth can always point in the same direction
High Earth orbit (HEO)	>36,000	Little use by satellites

It is estimated that 56–65% of the cataloged objects originate from breakups in orbit—more than 240 explosions—as well as fewer than 10 known collisions [1,4]. Another 38% of the cataloged orbital population can be attributed to decommissioned satellites, spent upper stages and mission-related objects, and only 6% of them represent operational satellites [4]. While prior to 2007 the primary source of orbital debris was explosions of spent rocket engines, collision events gradually became one of the largest contributors [7]. The first in-orbit satellite fragmentation occurred on June 29, 1961, when a US Ablestar upper stage exploded into nearly 300 large pieces, overwhelming the official total Earth orbital population of only 54 objects [11,12]. The first deliberate destruction of a spacecraft by its operator is that of Cosmos 50, which was fragmented into at least 94 cataloged pieces on November 5, 1964 [13]. In 1996, a French satellite was hit and damaged by debris from a French rocket that had exploded a decade earlier. The first in-orbit collision between two satellites occurred on February 10, 2009, at 776 km altitude above Siberia, where an American satellite, Iridium 33, and a Russian satellite, Kosmos-2251, collided at a relative speed of 11.7 km/s resulting in complete destruction of both satellites and the generation of more than 2,200 trackable fragments [1]. One of the Chinese antisatellite missile test was conducted on January 11, 2007, when a weather satellite at an altitude of 865 km with a mass of 750 kg was destroyed resulting in the largest recorded creation of space debris in history with over 2,000 pieces of trackable size (golf ball size and larger) and an estimated 150,000 smaller debris particles [5]. If this event had not occurred, the increase in the historical cataloged debris count would have been only 6% (vs 27%) and the increase in on-orbit debris would have been only 8% (vs 69%) [11].

As a rule of thumb, low altitude, cataloged debris are assessed to be larger than 10 cm in diameter while at higher altitudes objects less than 1 m in diameter may be undetectable [11]. Generally, small-in-size and difficult-to-be-tracked and quantified space waste was overlooked, and no measures were taken to deal with that rising problem. It is important to mention that small debris is much more numerous and probably the principal threat to space operations [7]. Another reason is that waste at *Low Earth orbit* (LEO) at 2,000 km does not threaten military, communications, and reconnaissance satellites, as the latter are in *Geosynchronous orbit* (GEO), at about 35,785 km above the Earth's surface. It should be noted that studies have shown that the instability of the debris population in the upcoming centuries is only limited to LEO while at GEO region the buildup of debris is progressing at a slower rate [12]. Overall, threats from orbital debris are greater in the LEO region due to a combination of high debris concentration, large number of crossings, and high relative velocities [4].

Numbers are astonishing: from 1989 to date the number of tracked large objects with diameters ranging from several meters to about 10 cm increased from 7,000 to 29,000 [1,14−16]. Other publications report that the number of objects larger than 10 cm is around 23,000 and information concerning 17,000 of them is publicly available. The remaining 6,000 objects are of unknown origin or the data remain confidential [10]. These objects are trackable and cataloged (Table 7.2; Figs. 7.1 and 7.2). Although the majority of detectable fragmentation debris has already fallen out of orbit and the effects of 40% of all fragmentations have completely disappeared, it is estimated that the total number of space debris objects in orbit larger than 1 cm is around 670,000−740,000 and those larger than 1 mm are more than 170 million (Table 7.3) [1,10,11].

Figs. 7.3−7.5 are computer-generated images of objects in Earth's orbit that are currently being tracked and are regularly published by NASA and ESA. These images provide a good visualization of where the greatest orbital debris populations exist.

The problems and the risks

Concerning the ground risks, space junk has been known to fall onto the Earth, especially into the oceans, but in most cases they are small enough to burn up in the atmosphere upon entrance. Sophisticated monitoring systems, special shields, and mission controllers are designed so as to avoid or minimize potential impacts. North American Air Defense Command and United States Space Command do monitor man-made objects in space using radars. They also track when space debris falls into the Earth's atmosphere or onto the Earth. However, the real and growing problem is mostly related to human activities in space and in particular, satellites.

Table 7.2: Satellites and Debris by Orbit [5,17].

Orbit	Satellites	Trackable Debris
Low Earth orbit	600	22,000
Medium Earth orbit	115	1000
Geosynchronous region	450	1000

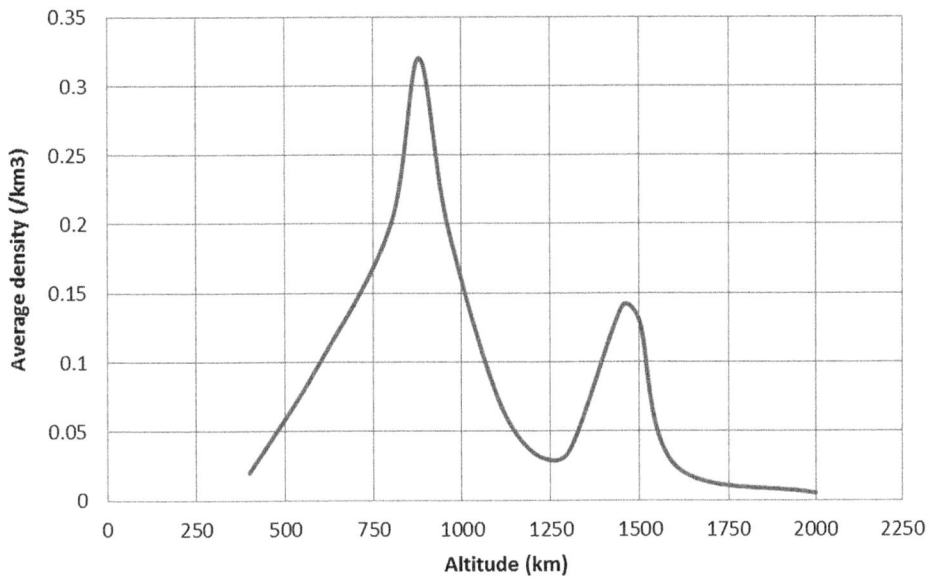

Figure 7.1

Cataloged debris average density (multiplied by 10^7) versus altitude. *Approximate data adapted from Ref. [2].*

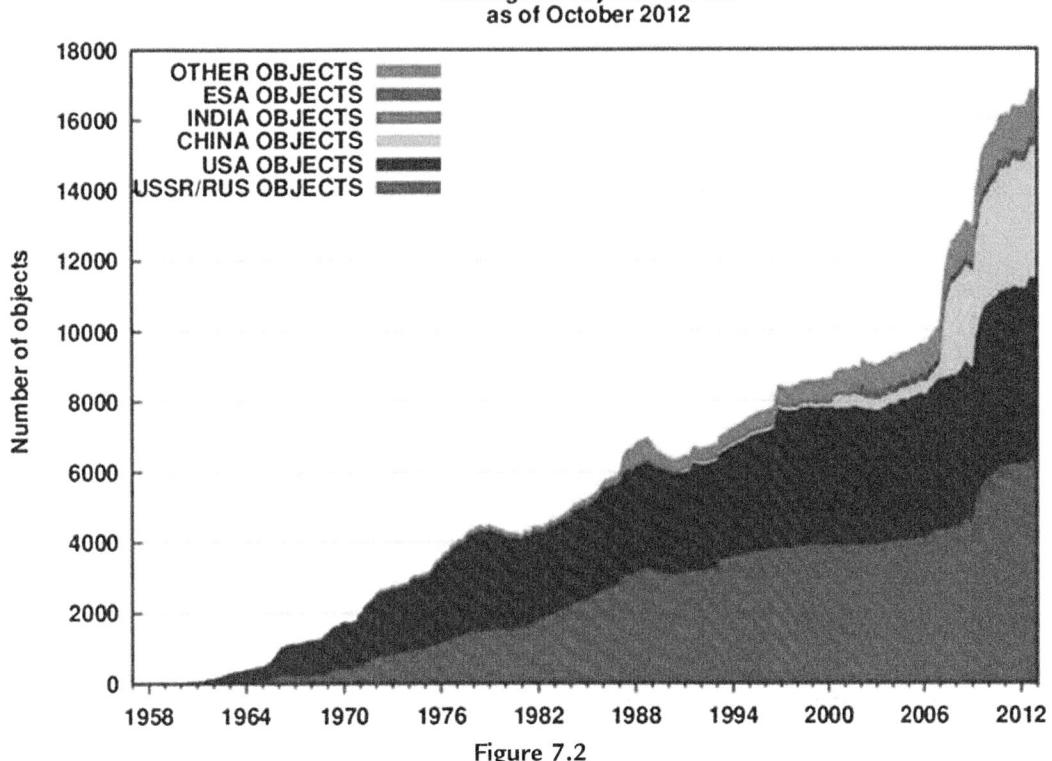

Figure 7.2

Cataloged objects in orbit as of October 2012. *Copyright: ©ESA. European Space Agency (ESA). http://www.esa.int/ESA (accessed March 2015).*

Table 7.3: Number and Size of Debris and Potential Damage to Man-Made Objects [1,4].

Size Class	Quantity	Notes
>10 cm	Thousands	Tracked and cataloged Catastrophic damage
5–10 cm	Tens of thousands	Lower limit of tracking Catastrophic damage
1–5 cm	Hundreds of thousands	Most cannot be tracked Major damage
3 mm–1 cm	Millions	Cannot be tracked Localized damage—upper limit of shielding
1–3 mm	Tens of millions	Cannot be tracked Localized damage

Modern society and economy is increasingly dependent upon satellites for the provision of services such as communications, meteorological data and remote sensing, and this dependence requires detailed risk assessment [10]. Orbital debris generally moves at very high speeds relative to operational satellites, and thousands of objects are bigger than a baseball ball, travelling at speeds that reach or exceed 25,000 km/h, namely 10–20 times faster than a bullet. In LEO, the average relative velocity at impact is 10 km/s (36,000 km/ h) while in GEO, is much lower, about 200 m/s (720 km/h). The energy these objects carry

Figure 7.3
Debris in the low Earth orbit—LEO (2000 km altitude). *Courtesy of Orbital Debris Program Office, NASA/Johnson Space Center, Houston, Texas, USA.*

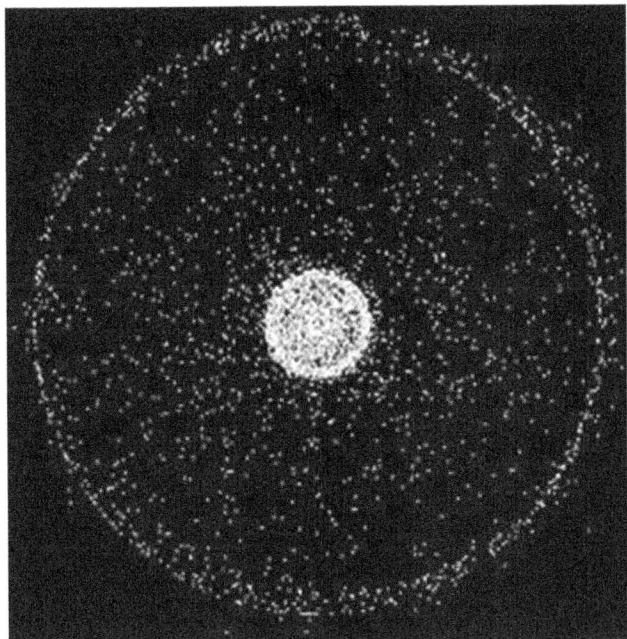

Figure 7.4

GEO Polar image is generated from a vantage point above the north pole showing the concentrations of objects in low Earth orbit—LEO (2000 km altitude) and in the geosynchronous region—GEO (around 35,785 km altitude). *Courtesy of Orbital Debris Program Office, NASA/Johnson Space Center, Houston, Texas, USA.*

Figure 7.5

Distribution of debris. *Copyright: ©ESA. European Space Agency (ESA). http://www.esa.int/ESA (accessed March 2015).*

Table 7.4: At 10 km/s Typical Low Earth Orbit (LEO) Impact Speed [1].

Debris size	Mass (g) Aluminum Sphere	Kinetic Energy (J)	Equivalent TNT (kg)	Energy Similar to:
1 mm	0.0014	71	0.0003	Pinched baseball
3 mm	0.038	1910	0.008	Bullets
1 cm	1.41	70,700	0.3	Falling anvil
5 cm	177	8,840,000	37	Hit by bus
10 cm	1413	70,700,000	300	Large bomb

is enough to seriously damage satellites and spacecrafts. A fragment 10 cm long is roughly comparable to 25 sticks of dynamite (Table 7.4; Fig. 7.6) [1]. Depending on the velocity, debris smaller than 1 mm in size can erode sensitive surfaces, but in general it does not pose a serious hazard to spacecraft functionality; from 1 mm to 1 cm in size may or may not penetrate a spacecraft depending on the shielding, while larger fragments between 1 and 10 cm in size will penetrate and damage most spacecrafts [1]. In terms of velocity, at low velocity plastic deformation normally prevails, with increasing velocities a crater on the target is formed, while in hypervelocity impacts velocity exceeds the speed of sound within the target material and the resulting shock wave can lead to local stress levels that exceed the material's strength, causing thus cracks and/or the separation of parts at significant velocities (Fig. 7.7) [1].

Each time a shuttle goes on a mission, it runs a one in a million chance of being hit by a trackable piece of space debris. Space debris shields for both manned and unmanned

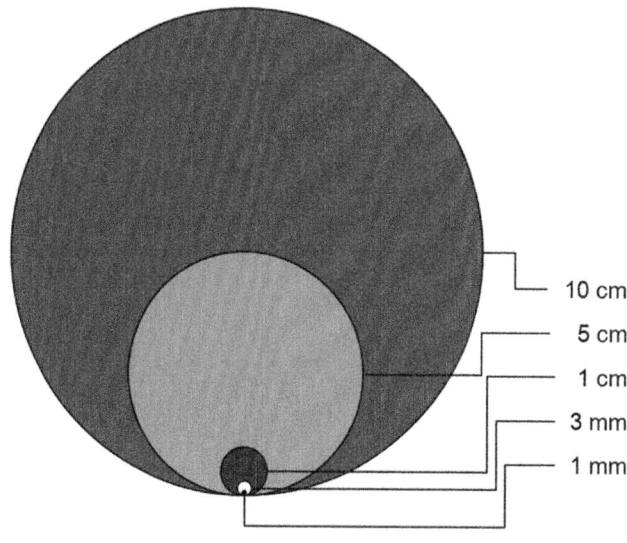

Figure 7.6
Debris size class comparison. *Credit: Art used by permission of The Aerospace Corporation.*

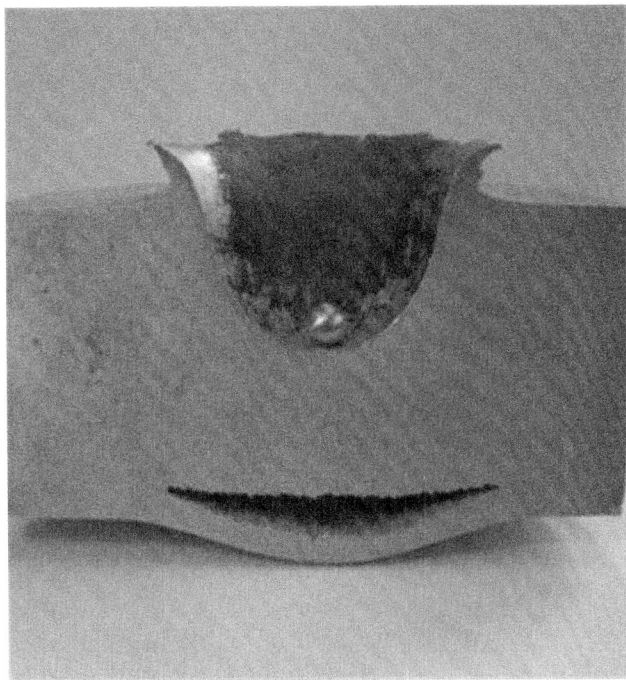

Figure 7.7

Hypervelocity impact. *Copyright: ©ESA. European Space Agency (ESA). http://www.esa.int/ESA (accessed March 2015).*

spacecraft can be quite effective against small particles of 0.1—1 cm in size, while protection against particles 1—10 cm in size can be achieved through special features in the design of space systems [18]. Physical protection against particles larger than 10 cm is technically difficult if not impossible. Concerning the risk for astronauts, NASA estimated a 1-in-89 chance of a fatal injury during the operations on Hubble Space Telescope repair mission in May 2009 [19]. Based on the above and as research has shown, the debris size range of greatest potential danger is 1—10 cm. It is believed that debris in this size range constitutes 50 tons of mass [10]. For debris smaller than 1 cm, shielding could be a solution and for those larger than 10 cm it is generally possible to maneuver a spacecraft to avoid colliding with them [20]. It is important to note that particularly in LEO at around 800 km in altitude, the majority of remote sensing satellites operates, and it is this altitude where 1—10 cm debris average density reaches its peak (Fig. 7.8).

Aerospace debris models clearly show that the future population will be dominated by collision debris (Fig. 7.9). Long-term forecasting predicts approximately from 20 to more than 83 catastrophic collisions during the next 200 years [17,21]. Older models have shown that this number could be much higher reaching 50 major collisions until the year

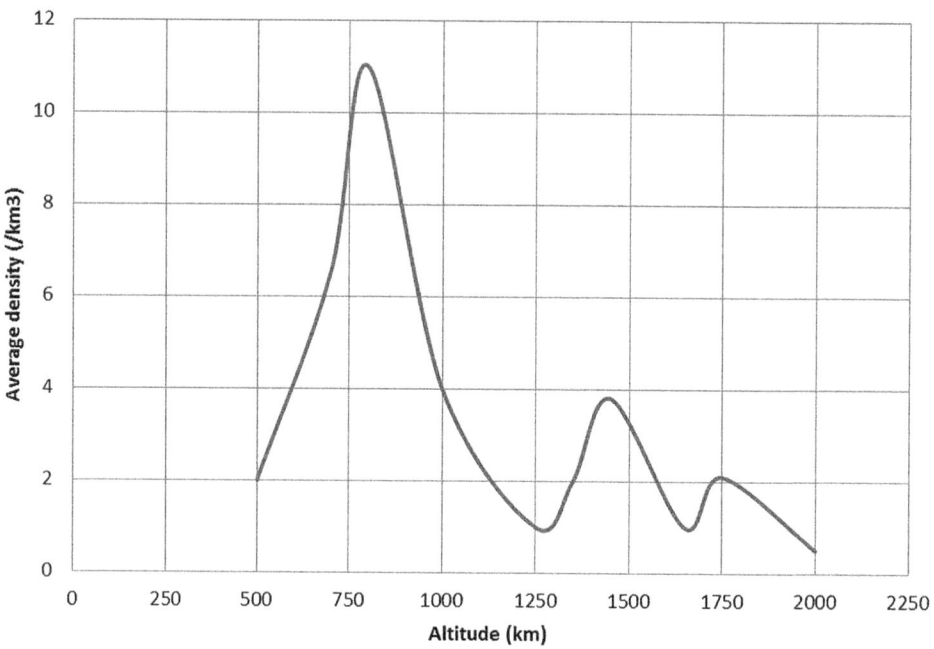

Figure 7.8

Debris average density (multiplied by 10^7) versus altitude. *Approximate data adapted from Ref. [6].*

2095 [18]. With today's annual launch rates of 60–70 new satellites per year, and with future breakups continuing to occur at the average historic rates of four to five per year, the number of debris objects in space will steadily increase and so the probability of collisions will [1]. The net growth in the debris population in LEO is estimated to be at an average rate of ∼5% per year [1]. This cascade effect is known in the related literature as Kessler syndrome, which predicts that after a certain debris density has been reached, its population will continue growing even without the launch of new objects [4,21]. It should be noted that although the number of debris objects is many times higher for the small-sized debris, nearly 95% of the mass of the LEO debris is concentrated in the large objects (>10 cm). This means that the potential of formation of large clouds of new, smaller, high-speed debris is very high, thus adding substantially to the problem [4]. Moreover, the increase in the number of individual objects launched along with the trend in space technology, which is moving toward smaller satellites, is the factor that is expected to multiply the relevant problems. It is also obvious that as satellites become smaller, they will become more vulnerable to the smaller debris impacts.

If serious attempts to clean up low-orbit space are not made in a systematic and timely manner, an artificial asteroid ring could form around Earth with unforeseen consequences.

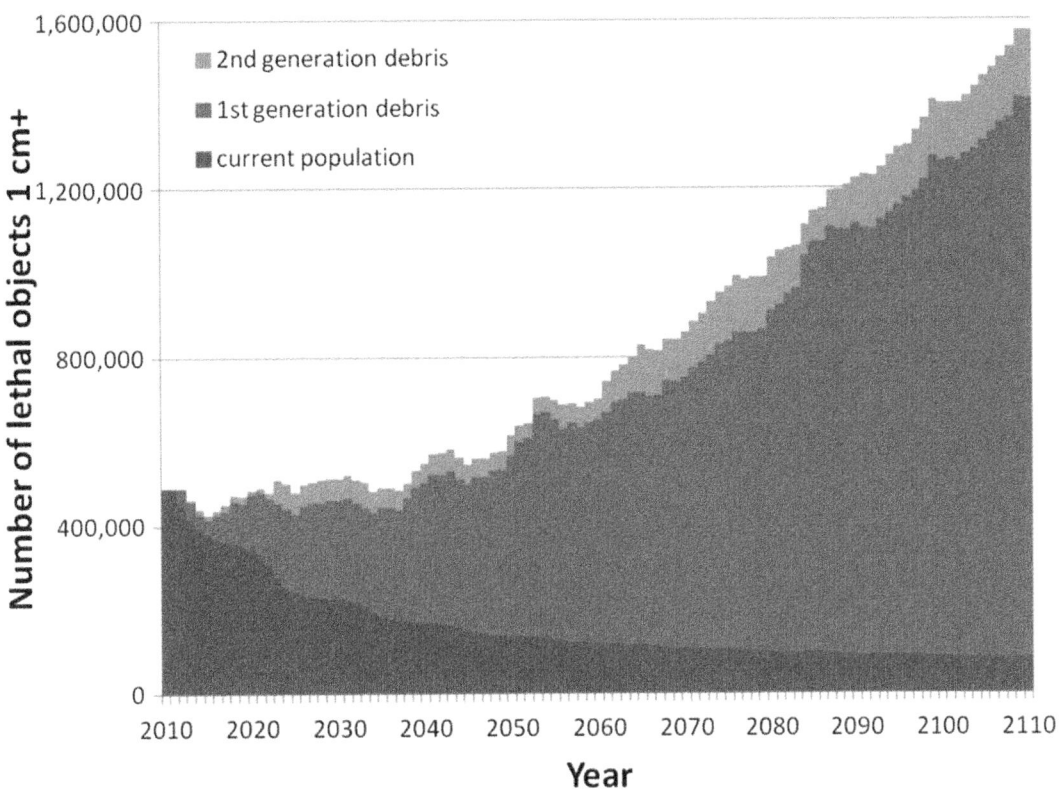

Figure 7.9

Future trends of space debris. *Credit: Art used by permission of The Aerospace Corporation.*

It is estimated that if mitigation and remediation measures are not applied as soon as possible, then some orbit regions, particularly at the valuable 800–1400 km altitude, may experience a collisional cascading process that could render these regions too dangerous for space activities within a few decades [1].

The responses: mitigation and remediation

Once trash is put into space, it is nearly impossible to remove it, and the only way to keep the LEO clean is to prevent waste from getting there in the first place, a policy that the two major space waste contributors, the United States and the Soviet Union, were reluctant to follow during the Cold War, because it would have prohibited both sides from testing antisatellite weapons [14–16]. Waste prevention seems to be the optimum solution; however, debris mitigation usually increases mission cost unless the relevant procedures are specified early in the development phase, eg, proper design of deployment procedures and passivation of explosion-prone components at the end of useful life. The most effective short-term means of reducing the space debris growth rate is through the prevention of in-orbit explosions [1]. Mitigating space debris falls into two main categories: protecting satellites from debris by avoiding collisions and shielding, and reducing the amount of new debris created [22].

Recent studies have suggested that the current LEO environment has already reached a point where the debris population is unstable and growth will continue in spite of implementing the commonly adopted mitigation measures [23]. According to these studies, the removal of 5–10 large objects per year from LEO region can prevent the debris collisions from cascading [8]. Recent modeling has shown that even if no future launches occurred, collisions between existing satellites would increase the 10-cm and larger debris population faster than atmospheric drag would remove them. This scenario highlights the need for the remediation of the existing debris population, also known as Active Debris Removal or ADR [24]. Concerning future ADR methods, taking into account several technical reasons, studies indicate that [12]:

1. for controlling future debris population growth or reducing collision activities, removal operations should focus on large objects (at least of several meters), and
2. for reducing mission-ending threats to operational spacecraft, the focus should be on the 5-mm-to-1-cm debris.

Thus, the only effective long-term measure is through the removal of large objects from regions with high object densities and long orbital lifetimes, ie, remediation measures. In the more distant future, it may be feasible to completely remove all large objects from orbit, but new technologies should be first developed. Several techniques have been proposed, all targeting at pushing debris to a new orbit or to cause it to reenter the atmosphere [22]: attaching propulsion devices, using robotic grappling devices on another spacecraft, using a momentum exchange tether that acts like a swing, using an electrodynamic tether which causes a drag on the satellite due to the magnetic field of the Earth, and slowing objects using high-powered lasers fired from Earth [20].

The idea of orbital debris removal was first suggested almost 30 years ago [25]. Since that time, several methods have been proposed but due to technical and budget challenges orbital debris removal has never been viewed as feasible [12]. The odds of developing an economic incentive mechanism for removing space debris in LEO are small because there is little direct economic value in LEO, as nearly all the economic activity in space takes place in GEO, where the vast majority of insured satellites, worth a total value of about $18.3 billion, are operating [10,26]. As the debris situation in this region is not yet as serious as in LEO, many insurers perceive space debris to be a manageable risk and in any case lower than the ones posed by space weather and technical malfunctions [10]. As Sarah Laskow, a journalist of *Boston Globe*, puts it "*Space, in other words, is a classic 'tragedy of the commons' problem, in which many individuals benefit from a collective resource, but no one has an incentive to bear the cost of maintaining it.*" [27].

Currently, the only mechanism for removal in LEO is orbital decay through atmospheric drag, which ultimately leads to reentry [1]. In contrast to GEO altitude where atmospheric drag is unimportant, LEO is continually cleansed by atmospheric drag. This mechanism is only effective in a restricted range of low Earth orbits, since in high orbits it takes hundreds to thousands of years for objects to reentry. Also, the relevant risk posed to the ground must be assessed and mitigated; although most debris burns up in the atmosphere, larger objects can reach the ground intact and an average of one cataloged piece of debris has been falling back to Earth each day for the past 50 years [28]. Objects at higher altitudes can either perform postmission maneuvers to drop the perigee (orbital point closest to Earth) further down into the atmosphere or if this is not feasible then to place them in disposal orbits, also called quasi-non-decaying graveyard orbits [1,4,29]. Already by the end of 2011, more than 325 spacecrafts in GEO were boosted into a higher disposal orbit at the end of mission life [1,9]. European Space Agency has also reorbited all GEO spacecrafts controlled by the Agency.

Since 2012, the European Space Agency has designed a mission scheduled for launch by 2021 to remove large space debris, heavier than 4000 kg, from LEO [1]. Furthermore, Swiss Space Center has initiated the Clean-mE project, which consists in developing technologies for nanosatellites, which should remove debris in orbit around the Earth. The plan is to match the target satellite's orbital plane of 630−750 km and grasp it, and stabilize it while moving at 28,000 km/h. Once nanosatellites have captured their target, both of them will head out of orbit and move toward the Earth, where they will both burn up in the atmosphere [30].

Legislative and other regulatory issues

Space waste management should be much like waste management on Earth where responsibilities are shared between the public and the private sector. However, it is the nature of space waste that requires international public−private partnership capable of addressing technical, financial and regulatory challenges [4].

It is true that most countries have agreed not to pollute the space environment, but this is merely a suggestion and there are no direct legal consequences or financial penalties for polluting the outer space [7]. Moreover, there is no definition of space debris in the international legislation, which makes it impossible to recognize which objects can be removed, but even if such definition were available, removal would be complicated due to international regulations that apply to space objects [4]. The Outer Space Treaty (1966) provides the basic framework on international space law [31]. However, the treaty was written before wide awareness of the debris problem and so it does not contain specific provisions about it; in fact, space debris is not even mentioned in the treaty [4,32]. Nevertheless, some of space law provisions can be applied, as for example, the responsibility of parties to conduct activities *"with due regard to the corresponding interests of all other States Parties to the Treaty*" and the responsibility to ensure that any exploration "of outer space, including the Moon and other celestial bodies" is undertaken in a manner "*so as to avoid their harmful contamination*" [10]. Liability for space objects is governed by Article VII of the Outer Space Treaty and Article VII's extension through the Liability Convention (1972). Art III of the Liability Convention establishes that "*In the event of damage being caused elsewhere than on the surface of the earth to a space object of one launching State or to persons or property on board such a space object by a space object of another launching State, the latter shall be liable only if the damage is due to its fault or the fault of persons for whom it is responsible.*" [33]. Liability Convention at its present form does not impose any clear obligation upon the States to prevent the space debris creation or to undertake mitigation measures [4]. Article III of Liability Convection together with article VII of Outer Space Treaty declaring that "Each State Party to the Treaty that launches or procures the launching of an object into outer space, including the Moon and other celestial bodies, and each State Party from whose territory or facility an object is launched, is internationally liable for damage to another State Party to the Treaty or to its natural or juridical persons by such object or its component parts on the Earth, in air space or in outer space, including the Moon and other celestial bodies" establishes a regulatory framework that does not facilitate debris removal, since each debris should be identified and its removal should be negotiated first with the launching State that is the only one that has jurisdiction and control over that object [4,34]. Additionally, defunct space objects still belong to the launching party and as a result they cannot be removed from orbit without permission [22]. Thus, as it is frequently impossible to determine who is ultimately responsible for a debris collision, it is technically illegal for one country to remove another country's debris without permission. Concluding, the present international space law conventions and instruments fail in creating a legal regime for ADR [4].

Apart from the Outer Space Treaty and the Liability Convention, there are other nonbinding initiatives in the form of guidelines on space debris. In 1995 NASA published a comprehensive list of orbital debris mitigation guidelines that future US space missions should observe, and US corporations have adopted these measures. If the spacecraft does not conform to these rules, it is not allowed to launch [7]. Other countries and organizations have then followed issuing their own orbital debris mitigation guidelines. In 2002, the Inter-Agency Space Debris Coordination Committee (IADC) published the IADC Space Debris Mitigation Guidelines, a reference document for technical guidelines designed to mitigate the growth of the orbital debris population. These guidelines served as a baseline for the United Nations Space Debris Mitigation Guidelines, endorsed by the UN General Assembly in 2008, which are voluntary technical guidelines targeting space debris in LEO. Europe's Network of Centers on Space Debris, a grouping of the Italian, British, French, and German space agencies plus ESA, has prepared the European Code of Conduct for Space Debris Mitigation in 2005, which is voluntary and should be applied by the European Space Agency, by national space agencies within Europe and their contractors [35]. It has been noted by stakeholders that international mitigation measures must be harmonized to avoid unfair competition [10].

The above guidelines provide a framework for what needs to be done, but the implementation should be specified in international standards or binding national requirements. International standards enable global stakeholders to communicate with a common language and expectations when setting requirements. ISO 24113:2011 defines the primary space debris mitigation requirements applicable to all elements of unmanned systems launched into, or passing through, near-Earth space, including launch vehicle orbital stages, operating spacecrafts and any objects released as part of normal operations or disposal actions [36]. Other relevant standards are shown in Table 7.5. It should be kept in mind that as is true for any ISO standard, ISO standards on space debris mitigation will remain nonbinding unless they are imposed via national and/or international legislation.

7.2.2 Nuclear Waste Disposal in Space: A Terrible Idea

The problem of nuclear waste disposal is growing and threatens the environment, people, and future generations as well. In the United States several projects have been proposed for the safe disposal of nuclear waste since '80s, the most famous being the Yucca Mountain project in northern Nevada. However, scientists cannot guarantee absolute safety as this waste needs thousands of years to decay harmlessly (the half-life of certain radioactive wastes can be in the range of 500,000 years or more) and no one can exclude leakages, let alone the fact that the waste containers are only designed to contain waste for 300—1000 years. Not surprisingly, the US Department of Energy in 2011 had withdrawn the licensing application for the Yucca Mountain Repository, leaving the nuclear industry responsible for the safety of its waste [37]. So, the question seems to be obvious: why not

Table 7.5: ISO Standards Relevant to Space Debris.

ISO 26872:2010	Disposal of Satellites Operating at Geosynchronous Altitude
ISO 24113:2011	Space debris mitigation requirements
ISO 14200:2012	Guide to process-based implementation of meteoroid and debris environmental models (orbital altitudes below GEO + 2000 km)
ISO 11227:2012	Test procedure to evaluate spacecraft material ejecta upon hypervelocity impact
ISO/TR 16158:2013	Avoiding collisions with orbiting objects
ISO 16126:2014	Assessment of survivability of unmanned spacecraft against space debris and meteoroid impacts to ensure successful postmission disposal
ISO 16164:2015	Disposal of satellites operating in or crossing low Earth orbit
ISO/PRF TR 18146	Space debris mitigation design and operation guidelines for spacecraft

to put high-level radioactive waste far out in space where it will not endanger anyone on Earth?

The idea of disposing nuclear waste in space is not new. Russian scientist Pyotr Kapitsa, a Nobel Prize—winning physicist, suggested in 1959 that outer space may be an option for dealing with nuclear waste [38]. NASA's Lewis Research Center concluded that this idea was technically feasible in 1974, and the Space Systems Technical Committee of the American Institute of Aeronautics and Astronautics endorsed it in 1981 [39]. A 1982 report to NASA estimated that 750 missions and a phenomenal cost would be required to dispose off the waste from the used fuel rods that will have been accumulated by the year 2003.

Probably the most comprehensive and detailed report on the subject was published in 1978 by NASA [40]. The report, among others, investigates High Earth Orbit, Lunar Soft Landing, Solar Orbit, and Solar System Escape as potential disposal areas. These and other studies, published in academic journals, offered solid evidence supporting the idea of disposal of nuclear waste in space. Some papers even found this idea as promising, practical and economically plausible. One such article, published in 1980, states that *"the promise brightens if we consider use of more effective second generation systems such as fully reusable single-stage-to-orbit transports"* [41]. According to Coopersmith [39], the waste is first solidified, then placed in an explosion-proof vehicle, which is launched first into Earth's orbit, and then further away in a range of possible destinations, including orbits inside Venus, Earth—moon libration points, lunar landings, and destinations outside the solar system [39]. A article published in 2007 concludes that space disposal *"could be a very attractive long term solution from a safety point of view and with an acceptable cost"* [42].

Coopersmith revisits the same idea in 2005 [43]. The author, first analyses the cost of space flights ($50—500 million per launch) and after some math he concludes that a chicken-and-egg situation exists; as space flight remains very expensive, payloads will be small (currently ~200 tons/year), and as long as payloads remain small, space flights will be expensive. How to bring the space flights cost down? To answer this question he brings back the terrible idea of packing the rockets with high-level nuclear waste. The author finds three good reasons to send nuclear waste into space: *"First, it is safe. Second, space disposal is better than the alternative, underground burial. Third, it may finally open the door to widespread utilization of space."* On the safety, he writes, *"Can nuclear waste be safely launched into earth orbit? The answer is yes."* The author sets three criteria for a safe operation, namely the use of alternative launch systems, new unbreakable containers, and multiple layers of safety precautions. In few words, the launch system stays on the ground avoiding putting it on the vehicle, so the spacecraft is almost all payloads eliminating the in-flight danger of an explosion. These ground-based launch systems are based on laser propulsion devices that are able to propel capsules far away into space. The capsules can be designed to maintain their integrity despite all potential threats. From the financial point of view, the expected cost of space disposal is billions of dollars, not very far from the estimations made for burying the waste underground, which is $10–60 billion.

According to Coopersmith's analysis, the capsules would be placed at least 1100 km above the Earth, claiming it as being "a nuclear-safe orbit" and next, a space tug would drive the capsules to their final destination orbiting around the Sun or, alternatively, to a solar orbit inside Venus, which would retain the option of retrieving the capsules [43]. However, it is not difficult to imagine the grave risks of storing the waste around the planet. Also, putting radioactive waste into a heliocentric orbit would only reduce, not eliminate, the risks of putting it into a geocentric orbit. Even though the waste containers would be farther away, collisions with meteors and asteroids are probable, spreading the dangerous contaminants all over a huge area without any guarantee that some would not ultimately reach the Earth.

Fortunately, these ideas were never materialized as they have not received strong support from governments or private sectors. World Nuclear Association investigates the possible disposal and storage options of nuclear waste, and concerning space nuclear waste it mentions that *"investigations are now abandoned due to cost and potential risks of launch failure"* [44]. Apart from formidable technical challenges, there are legal ones as analyzed in an excellent article published by Dusek in 1997 [38]. According to the author, *"The international treaties, cases, and principles do not seem to indicate a clear answer to the question of whether disposal of nuclear waste in outer space is permissible. They do indicate, however, that the country that disposes the waste would be liable for any harm that is caused by such disposal"* [38].

As long as nuclear energy is used by humanity, nuclear waste will continue to pose serious risks and challenges, and the underground disposal on Earth cannot be a permanent solution. Thus, if an option exists through which the waste is permanently sent into outer space or destroyed by incineration on the Sun, this option must be considered and evaluated. However, if the gamble is wrong and the waste contaminates our space neighborhood posing a potential threat of radionuclides rain on Earth, this "promising" solution may turn to be our last and grave mistake. Concluding, the only ultimate solution to the problem of nuclear waste might be to abandon the use of nuclear energy.

7.3 Greenhouse Effect in the Solar System

Furnace-like temperatures on Venus under 92 bars of carbon dioxide, a very thin atmosphere on Mars with rich water past, methane rains in Titan's frozen atmosphere, atmosphere temperature inversion on Pluto, and global warming on Triton; research on other terrestrial planets and moons can provide insights on the greenhouse effect and atmospheric interactions here on Earth. Indeed, the strong greenhouse effect on Venus provides insights into Earth's future, the warm Mars' past provides a comparative bridge to Earth's past, the role of the condensable methane on Titan's atmospheric temperature structure provides an analog to water vapor on Earth, and the wild variations of surface temperature on Moon and Mercury, illustrate how beneficial the greenhouse effect could be.

7.3.1 Planetary Evolution and Climate

A planet's climate is basically decided by its mass, its distance from the Sun, and the composition of its atmosphere. Planets are heated by radiation from the Sun and will heat up or cool off until their temperature is such that they emit exactly as much thermal energy as they receive from the Sun, reaching an equilibrium temperature, which is the theoretical temperature that the planet would be at when considered simply as if it were a blackbody being heated only by its parent star. In this model, the presence or absence of an atmosphere is not considered. However, some planets and moons do have atmospheres that influence the temperature and so the climate. The difference between the actual and the equilibrium temperature gives the greenhouse warming effect on the planet. In summary, the phenomenon is as follows: Clouds reflect part of the solar flux while the rest is reaching the surface, which in turn emits infrared radiation of its own. Certain gases and water vapor are transparent in visual wavelengths and opaque in the thermal infrared. The greenhouse effect describes the condition where infrared radiation is trapped and absorbed by gases and vapor in the atmosphere, which is then heated more than it would otherwise be without the greenhouse effect [45,46]. Another important and relevant phenomenon is that rock and soil layers absorb and reflect solar radiation at the surface, but ices, while opaque in the infrared, are partially transparent in the visible spectral range. The result is the so-called solid-state greenhouse effect, a phenomenon similar to the atmospheric greenhouse effect, resulting in subsurface energy deposition. This phenomenon may play an important role in the energy balance of icy surfaces in the solar system [47]. Apart from this, it allows the possibility of liquid water pockets under icy surfaces that are too cold and/or atmospheric pressure is too low for liquid water to survive.

When it comes to the greenhouse effect, the important question that arises is whether the planet has the ability to retain an atmosphere. This is an important question with a complicated answer. How fast the atmosphere of the planet is lost depends on how the average gas speed, which depends on the molecule mass and temperature (or the distance from the Sun and planetary albedo), compares to the escape velocity of the planet, which in turn depends on the mass of the planet (Table 7.6). As a rule of thumb, if the escape velocity is about 10 times larger than the average gas speed, the gas is retained over immense time periods. If it is roughly five to nine times the gas velocity, the gas will be partially retained, and if the escape velocity is lower than five times the gas velocity, the gas will escape into space quickly [48].

Table 7.6: Summary of Key Data of Terrestrial Planets and Moons With Substantial Permanent Atmosphere Able to Develop Weather and Clouds.

	Earth	Venus	Mars	Pluto	Titan	Triton
Perihelion (million km)	147.1	107.5	249.2	7311	1353 (Saturn)	4459 (Neptune)
Mean radius (km)	6371	6051	3396	1186	2576	1353
Mean density (g/cm^3)	5.514	5.243	3.934	1.88	1.88	2.061
Escape velocity (km/s)	11.19	10.36	5.03	1.212	2.64	1.455
Main atmospheric components	78% N_2 21% O_2	96.5% CO_2 3.5% N_2	96% CO_2 3.8% N_2 and Ar	97% N_2 2.5% CH_4 0.5% CO	98.4% N_2 1.4% CH_4	100% N_2 CH_4, CO_2 (traces)
Average surface atmospheric pressure (Pa)	101,000	9,200,000	636	0.3–2.4	147,000	1.4–1.9
Average surface temperature (°C)	15	460	−48	−229	−180	−238
Equilibrium surface temperature (°C)	−17	−33	−53	−233	−192	−
Greenhouse warming	32	493	5	4	12	−

Jupiter's moon Io and Saturn's moon Calypso are not included as they have extremely thin atmosphere.

Gas retention depends on the selection of temperature and results might be considerably different, as for example, in the case of Venus (Fig. 7.10). Nevertheless, it is clear that H_2 cannot be retained by terrestrial planets while CO_2 frequently does. Although these figures give an idea on the relevant phenomena, it should be noted that apart from body's escape velocity, gas molecular weight and atmospheric temperature, several other factors also affect the retention of a gas in a body's atmosphere. In particular, the analysis above ignores the effect of solar winds (sputtering effect), especially in the case of the absence of a magnetic field; for instance, this is one of the reasons why Venus lost its water. Furthermore, atmospheric chemical reactions are capable of breaking heavier compounds into lighter ones; for instance, water vapor is photodissociated by ultraviolet radiation in the upper atmosphere to hydrogen, which easily escapes the terrestrial planets. Also, if the upper layers of the atmosphere are colder than lower layers, then condensation might take place as these layers act as a barrier preventing the gas from escaping; for example, this is happening on Earth, contributing to the retention of vapor water which otherwise could have been photodissociated in the upper atmosphere of the planet.

Each planet has its own atmospheric temperature profile where variations might be wild and unexpected (Fig. 7.11). Furthermore, while on Earth, Venus, and Titan the main atmospheric constituent is only in the gas phase, on Mars, Pluto, and Triton the main atmospheric constituent exists in both vapor and condensate (ice) states. The phase

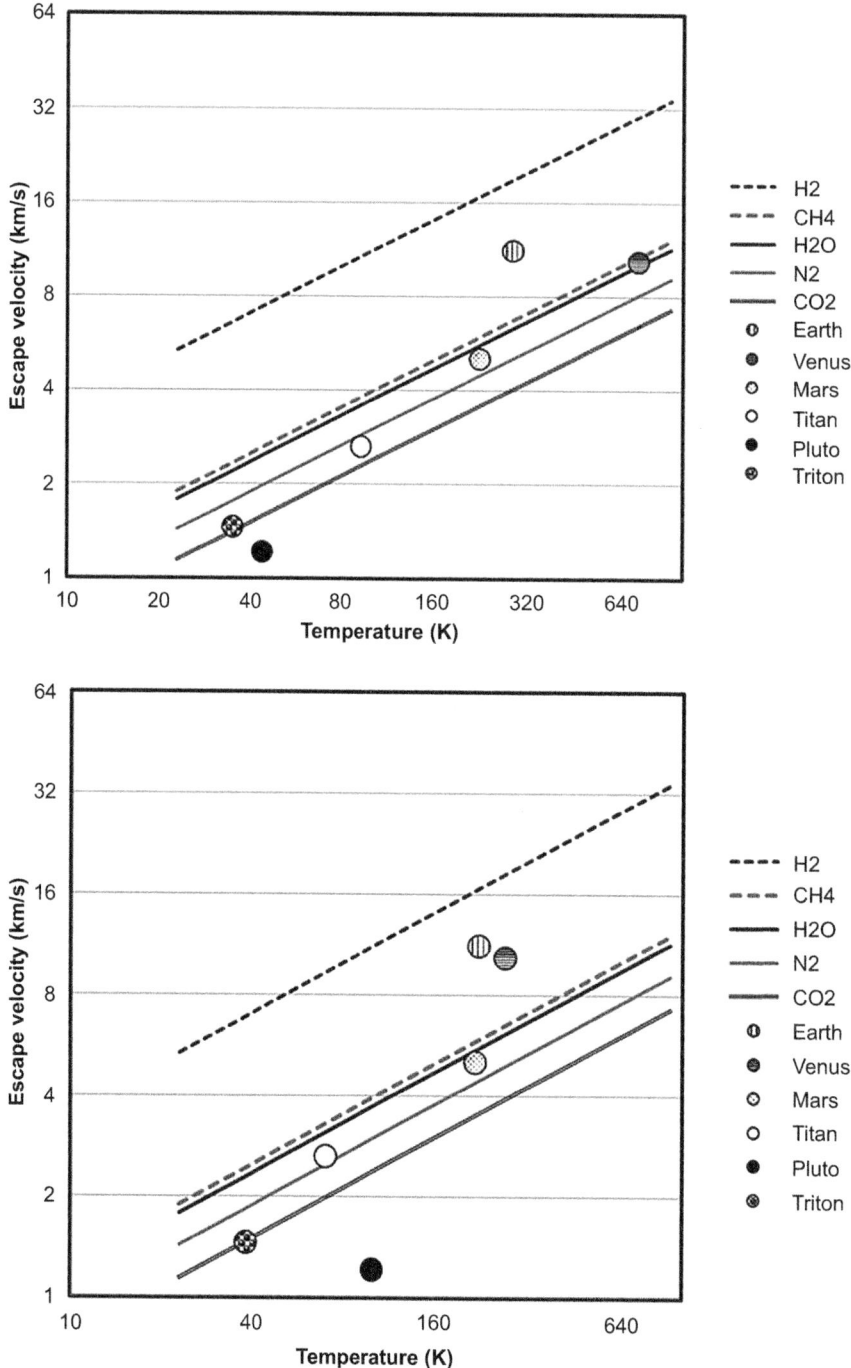

Figure 7.10

A line on the graph represents 10 times the average gas velocity and its variation with temperature. Circles represent the escape velocity and temperature coordinates of the bodies. Average surface temperature is used in the upper graph and average atmospheric temperature in the lower graph at the altitude of 10 km for Earth, Mars, and Pluto, 30 km for Titan, and 60 km for Venus. These are the altitudes where clouds or weather is formed.

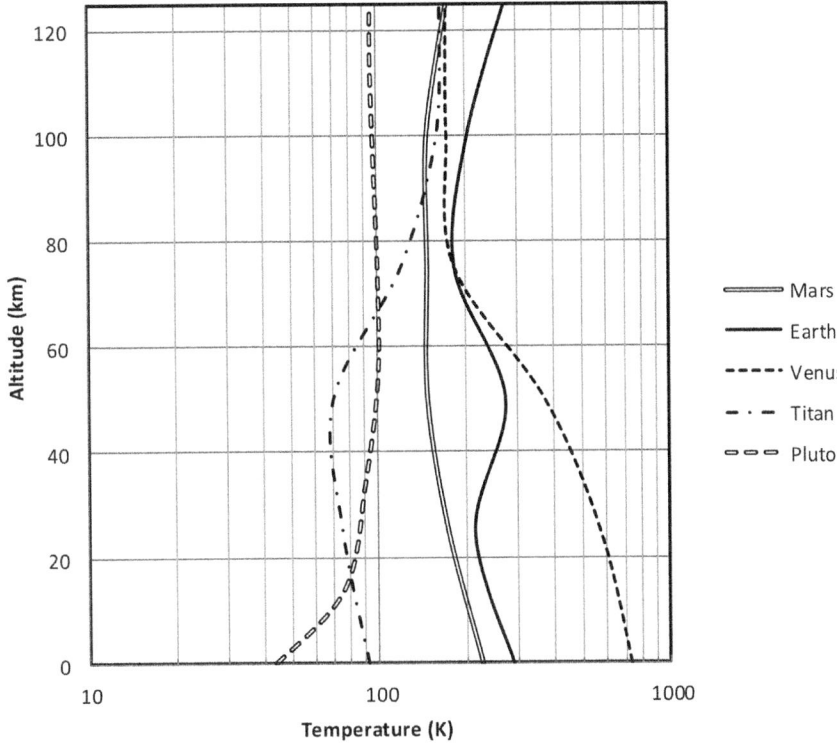

Figure 7.11

Vertical atmospheric temperature profiles of Earth, Venus, Mars, Pluto, and Titan.

equilibrium between vapor and solid phases has profound effects on the atmospheric structure. For instance, an antigreenhouse effect occurs when the surface of Pluto is cooled because of sublimation of its icy surface components resulting in an intense temperature inversion. For bodies on which the main atmospheric constituent is condensable, the atmospheric pressure is governed by the saturated vapor pressure, which is an exponential function of the surface temperature (Fig. 7.12). This means that the atmospheric pressure is very sensitive to surface temperature variations. Moreover, the sublimation/deposition cycle alters the surface albedo and density of clouds, both having a strong effect on sunlight adsorbed/reflected and thus on the temperature [49]. Even in atmospheres where condensable components are in low concentrations, evaporation/condensation may have important consequences, as for example, on Earth and Titan, where climate is heavily influenced by water and methane vapors, respectively. However, when the main atmospheric component is condensable, these effects are severe.

Water vapor is the most powerful greenhouse molecule and its phase forms and existence on other planets are of great interest. The subject is complex and only some aspects are discussed, in particular, the thermodynamics of water, its role in the evolution of

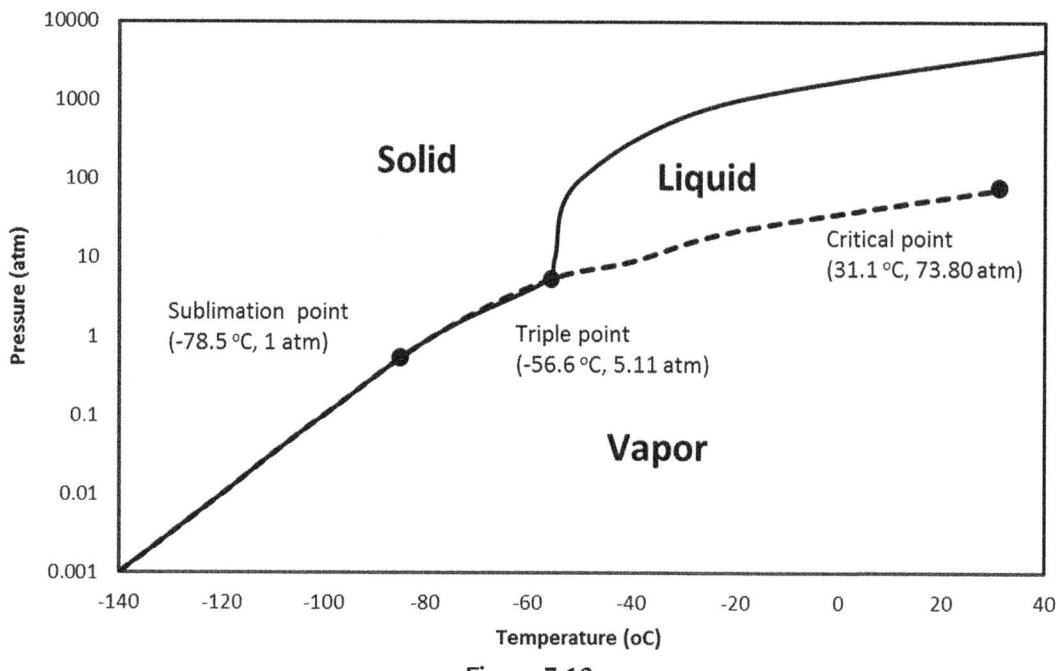

Figure 7.12
Temperature—Pressure phase diagram of pure CO_2. Note that for $T = -78.5°C$ the saturated vapor pressure of solid CO_2 is 1 atm, and sublimation (rapid evaporation analogous to boiling) takes place when the external pressure equals this value. In bodies where atmospheric pressure is close to the vapor pressure of the main component, sublimation leads to rapid evaporation.

greenhouse effect, and eventually the possibility of existence of water on other planets. The phase diagram of water is presented in Fig. 7.13. Possible phase transitions are from liquid to vapor (evaporation/condensation), solid to liquid (melting/freezing), and solid to vapor (sublimation/deposition). Triple point occurs where the three phase lines meet and all phases coexist in equilibrium. Furthermore saturation vapor pressure of a pure substance is only a function of temperature.

According to simplified models, before the development of an atmosphere the initial surface temperature of a planet would be the same as the equivalent blackbody temperature of the planet. Assuming that the planetary atmosphere evolves through volcanic outgassing, water vapor is produced building up pressure [50]. The atmospheric vapor traps the outgoing solar radiation contributing to a greenhouse effect and warming of the planet. The evolution of Earth, Mars, and Venus, according to this simplified model is shown in Fig. 7.13. The hypothetical climate trajectories emanate from the initial effective temperature of the planets, and as atmosphere evolves they intersect the phase transition curves containing the planetary climate [50]. According to this model, Earth's

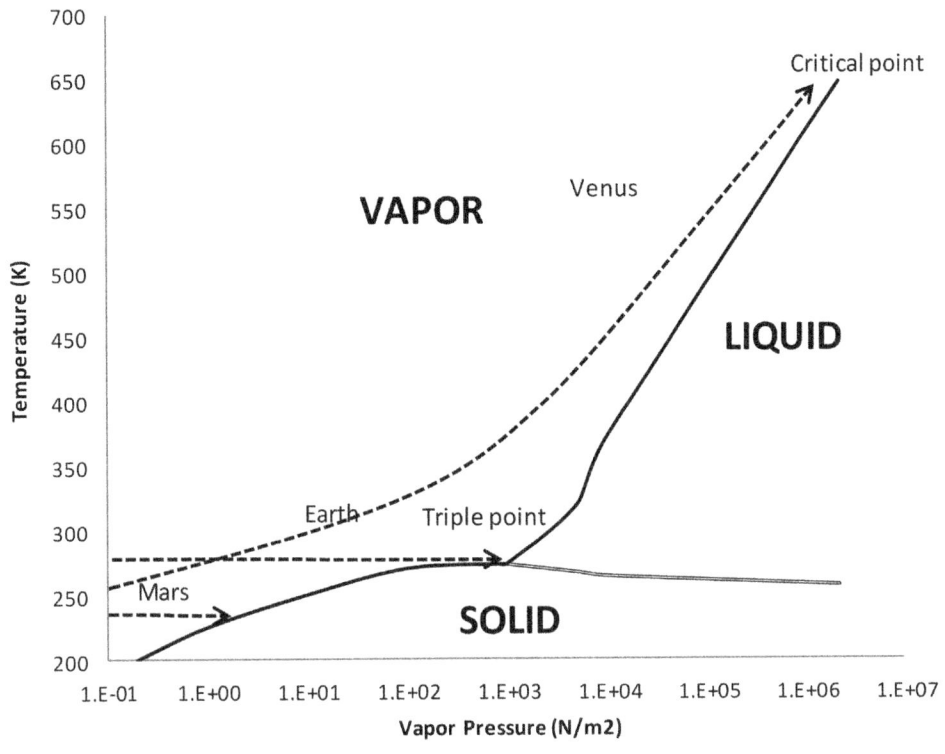

Figure 7.13

The phases of water as function of the partial pressure and temperature. *Solid curves* are the phase transition curves between water, vapor, and solid (ice) phases. *Dotted lines* are the climate trajectories of planets. *Adapted from Ref. [50].*

climate trajectory encounters the triple point allowing all three water phases to coexist. Venus missed all phase transition curves probably due to a water-driven runaway greenhouse effect, driving water to superheated and even supercritical region, while Mars' climate trajectory collided with the vapor–solid equilibrium line allowing only ice and vapor to coexist in equilibrium. Thus, available water becomes vapor, solid, and water and the relevant distribution of phases depends on thermodynamics as well as on the ability of the planet to retain its vapors and gases.

7.3.2 Earth—An Uncertain Future

Greenhouse effect on Earth is caused by water vapor ($<5\%$), carbon dioxide ($0.03-0.04\%$), methane, nitrous oxide, ozone, and CFCs (chlorofluorocarbons). The contribution of each gas to the greenhouse effect is affected by the properties of the gas, its concentration, and other indirect effects it may cause. For example, methane is about

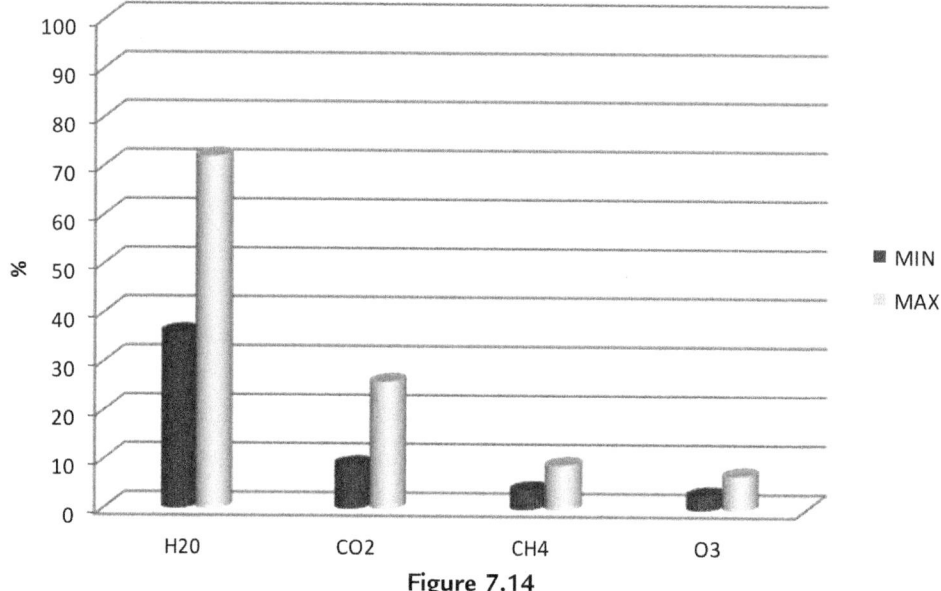

Figure 7.14

Greenhouse gases ranked by their direct contribution to the greenhouse effect on Earth. *Adapted from Ref. [51].*

72 times stronger than the same mass of carbon dioxide, but its atmospheric concentration is much smaller, and thus its total direct effect is smaller (Fig. 7.14).

Earth's greenhouse is relatively mild, currently providing 32°C of greenhouse warming and keeping the surface temperature to a life-supporting average of 15°C, critical for maintaining the Earth as a habitable planet [45,46,52]. This natural greenhouse effect is beneficial for the Earth and without it the planet would have been probably frozen and unable to support liquid water and thus life. The extremely thin atmosphere and thus the absence of greenhouse effect on Moon and Mercury contribute to the wild average surface temperature variations from −153 to 123°C and −173 to 427°C, respectively. Thus, the existence of greenhouse effect can stabilize the temperature and provide conditions for liquid water and consequently, habitable environments. But, although water vapor cannot escape Earth's gravitational field (Fig. 7.10), it could have been lost due to the effect of photodissociation. Fortunately, the Earth has a fairly complicated climate which enables it to retain water vapor and a major contributor is the tropopause cold trap, which does not allow water vapor to reach the upper atmosphere [50]. Furthermore, Earth is oxygen rich (owned to life and biological activity), which allows the formation of an ozone layer and offers protection from UV radiation. Finally, the planet possesses a magnetic field, which protects the atmospheric gases from the solar wind.

Water vapor forms clouds, which reflect light from the sun and have a temperature-lowering effect. On the other hand, water vapor absorbs infrared radiation even more strongly than carbon dioxide does, influencing thus greatly the greenhouse effect. Also, a large fraction of solar energy is converted to latent heat by evaporation of water into the atmosphere, and as water condenses from atmospheric air, large quantities of heat are released, increasing the temperature [45,52,53]. Although water vapor is such a powerful greenhouse molecule, it is not substantially influenced by human activities and it has a short atmospheric life of some days in contrast to CO_2, which stays in the atmosphere for tens of years and is heavily influenced by human activities. Furthermore, water vapor creates the so-called "positive feedback loop" in the atmosphere amplifying the greenhouse effect caused by other gases. This positive feedback loop is caused by increased temperatures, which result in increased evaporation of water and thus further increase of temperature due to its strong greenhouse effect. However, positive feedback effects do not necessarily lead to runaway effects, at least until the Sun becomes too hot in about 1 billion years, when the average surface temperature of the Earth is estimated to reach 47°C, causing such a runaway effect (see the case of Venus). But a runaway process could be also triggered by another hidden source, frozen methane in hydrate form (methane clathrate) [54,55]. This hypothesis states that an increase in sea temperatures and/or falls in sea level can trigger the sudden release of methane from methane clathrate compounds buried in seabed. The impact of such a release would be immense, since even if only 10% of the trapped methane was released within a few years, it would have an impact equivalent to a factor of 10 times an increase in atmospheric CO_2. The potential climate impact could be significant on geologic timescales as it is estimated that the release could reach the levels of the carbon released by fossil fuel combustion [54].

The presence of sedimentary rocks and apparent absence of glacial deposits on Earth about 4 billion years ago suggest that early Earth was possibly warmer than today [53]. This is problematic because the solar flux is believed to have been about 25% lower in the past [56]. Several explanations have been offered in the related literature, one of which is a much larger greenhouse effect than today due to the higher concentrations of CH_4, CO_2, and H_2O and NH_3, the presence of an organic haze layer, an increased cloud cover, and increased pressure broadening from a significantly larger N_2 partial pressure and total atmospheric pressure [53,57,58].

7.3.3 Venus—Burned by Love

Venus has a very thick atmosphere composed of 96.5% carbon dioxide and 3.5% nitrogen and is shrouded in thick clouds made of water and sulfur compounds (dioxide, trioxide and sulfuric acid). Greenhouse effect and clouds are sensitive to concentration changes of these compounds [59,60]. In 1960, Carl Sagan studied Venus and suggested that it could be a greenhouse effect furnace [61]. Indeed, the combination of a dense atmosphere (92 bars), high CO_2 concentrations (about 300 times as much carbon dioxide in its atmosphere as on Earth and Mars), and thick clouds results in an extremely strong greenhouse effect that keeps the planet's surface temperature at about 460°C, rather than its equilibrium temperature of −33°C [53,62,63].

Some billions of years ago it is likely that Venus did have oceans and a very different atmosphere compared to the present, but as solar activity gradually increased, during millions of years large amounts of carbon were released from the ground with simultaneous increase of water vapor from the oceans. This caused a runaway greenhouse effect, where the oceans evaporated and disappeared into space, leaving behind an atmosphere with water vapor content of only 30 ppmv and a huge amount of carbon dioxide [56,64,65]. However, the runaway reaction on Venus cannot be triggered by increased CO_2 alone and a possible explanation is typically thought to be an increased solar flux [66]. This is only a simplified picture of the phenomenon, but it provides insight into Earth's future as the solar flux continues to increase. More elaborated models study the interactions between the mantle and atmosphere, ie, coupling of climate and plate tectonics. According to these models, an increase in surface temperature (e.g., due to outgassing) leads to a reduction of mantle heat flux, an increase in partial melting, and hence an increase in atmospheric density and surface temperature [60,67,68]. The mechanism is illustrated in Fig. 7.15.

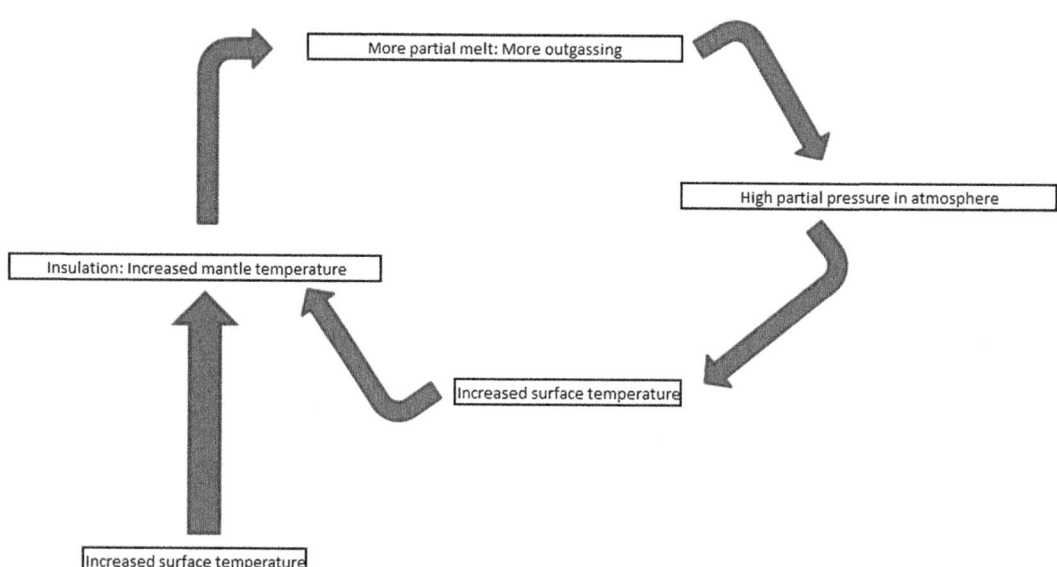

Figure 7.15
The runaway greenhouse effect on a planet like Venus according to Ref. [67].

According to Fig. 7.10, water should have been retained in Venusian atmosphere; however, the planet is almost bone dry. The history of water on Venus is largely unknown and explanations range from Venus losing its water gradually over time to water being lost rapidly during an early magma ocean stage, probably due to a weak magnetic field [68]. What probably happened is that the runaway greenhouse effect drove all available liquid water up to the atmosphere, where it was gradually photodissociated and lost as hydrogen to space. Another contributor to the escape of hydrogen is the lack of an ozone layer and of a magnetic field, which means that it has no protection from the solar wind that can strip atoms from the upper layers of the atmosphere. There is one more observation that further supports this theory and has to do with a hidden property of water. Water is made of two hydrogen and one oxygen atoms, but heavy water, which contains deuterium or tritium, heavier isotopes of ordinary hydrogen, is also possible. On average, the ratio of ordinary hydrogen to deuterium is 1/10,000 and thus, the ratio of heavy to ordinary water is 1/5,000. This is true all around the solar system and beyond but not on Venus, where this ratio is 100 times larger. This can happen because deuterium is heavier than ordinary hydrogen and thus its escape velocity is 1.43 times slower, making it more difficult to escape the planet.

7.3.4 Mars—The Tragedy of the Absence of a Greenhouse Effect

Mars' surface pressure is ~ 6 mbar and its atmosphere is composed of 95% carbon dioxide. However, Mars' greenhouse effect is only 5°C, though evidence suggests that it was substantially higher in the past [69]. The absence of a stronger greenhouse effect, despite the atmospheric composition, is explained by the distance of Mars from the Sun (it receives about 57% less solar energy in comparison with Earth) and the extremely low density of its atmosphere (100 times less dense than the atmosphere on Earth), which is simply too thin to absorb much of the outgoing heat. The loss of Martian atmosphere can be, at least partly, explained by its low escape velocity, which is less than 50% of that on Venus and Earth. Moreover, Martian atmosphere is relatively unprotected from solar winds, as the planet does not have a strong magnetic field. The net result is that Mars is very cold, with a surface temperature of about $-48°C$ and huge variations from -143 to $35°C$.

At this point, it is interesting to mention that the solid-state greenhouse effect may play an important role in the energy balance on Mars. When dry ice (CO_2) is underneath surface ice, then upon heating it turns into gas (sublimation) and as the pressure builds up, eventually escapes to the surface explosively, creating geysers [47]. According to other models, the same effect on translucent ice, which resides in underneath layers, leads to water ice melting (instead of sublimation in case of CO_2) [70]. This water can remain liquid for long periods. This is a process that takes places on Earth also, for instance in Antarctica, where the surface ice remains frozen, but a layer of liquid water exists below the surface

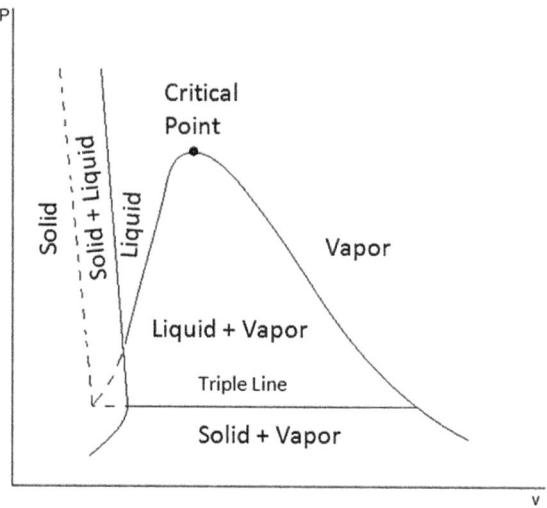

Figure 7.16
Generic pressure-specific volume phase diagram of a pure substance. The almost vertical constant temperature lines become parallel to specific volume axis when in the phase equilibrium envelope. At very low temperatures only solid—vapor equilibrium is possible.

(blue ice) [71]. Although the underneath ice exists, the chances of considerable amounts of liquid water on the surface of Mars are slim. This is because under very low temperatures only solid—vapor equilibria exist and if the total external (atmospheric) pressure is very low, either rapid evaporation (sublimation) or rapid freezing of water can take place, limiting thus the possibility of liquid water presence on the surface of Mars (Fig. 7.16). However, on Monday September 28, 2015, NASA scientists announced a breakthrough in science, reporting the existence of liquid water flows during summer months, probably only when the surface of Mars rises above $-23°C$. This liquid water can exist in these frigid conditions because of salts, which lower its freezing point. On the origin of this water, one of the suggested possibilities is porous rocks under the Martian surface, which might hold frozen water that melts in the summer months and seeps up to the surface. This water might be the ghost of the ancient liquid water oceans of Mars. In fact, the existence of clays and sulfates on Mars supports this theory and provide clues on the water history on the planet; clay formation takes a lot of time and requires a lot of liquid water also, where sulfates form quickly during evaporation of water (for hydrated minerals on Mars see Section 7.4.2). A tremendous change is believed to have taken place around 3.5 billion years ago when liquid water disappeared leaving behind clays and sulfates as trails of its path.

Thus, there is strong evidence that early Mars was warm and wet, with liquid water available (although in unknown quantities), presumably due to the greenhouse effect, providing thus ideal for life conditions of 1 bar atmospheric pressure and 27°C surface temperature [53,62,69,72]. However, this water was gradually lost. According to Fig. 7.10, Mars' escape velocity is low enough to allow some water vapor molecules to escape, but water loss should have been further accelerated due to the photodissociation effect, which is intensified by the lack of ozone layer and the very thin atmosphere as well as by the absence of a strong magnetic field. Outgassing through extensive volcanism might have produced a relatively dense CO_2 atmosphere in the past of the order of 0.1—1 bar [73]. Volcanic outgassing is also one of the main sources of other volatiles, and degassing of interior potentially influenced the early Martian climate [72]. However, CO_2 alone or in combination with H_2O cannot warm early Mars to such high temperatures (even with a CO_2 atmosphere of several bars) and the possible warming role of N_2, CH_4, NH_3, H_2S, and SO_2 as well as dust has been discussed [72—75]. Furthermore, it seems that the release of CH_4 is probably not sufficient to account for appreciable greenhouse warming, and sulfur volatiles could add significantly to an early Martian greenhouse effect [72]. Other studies propose a combination of 10—20% H_2 and 3-bar 95% CO_2, able to produce surface temperatures above 290 K. These CO_2—H_2 atmospheres are considered by recent reports as the mechanism that appears capable of maintaining such conditions for extended periods [76]. As in the case of Mars, the Earth was warmer beyond the limits of CO_2 greenhouse warming and this similarity provides some insights to our planet's past.

Some scientists suggested that, at least theoretically, Martian atmosphere can be altered by adding strong greenhouse gases, such as octafluoropropane (C_3F_8). Models show that if the concentration of this gas in the Martian atmosphere was 300 ppm, a runaway greenhouse effect could be triggered, evaporating polar ice, which is composed of water and carbon dioxide. That extra CO_2 would lead to even more melting and warming, thickening the Martian atmosphere and increasing atmospheric pressure. However, this amount of octafluoropropane represents 25,700 times Earth's annual production of fluorine- and carbon-based gases, rendering this idea unrealistic [77].

7.3.5 Pluto—Vitalstatistix's Fear Come True

Vitalstatistix, the chief of Gauls in Asterix, has only one fear; that the sky may fall on his head tomorrow. He is safe as long as he is not walking on Pluto.

Pluto's extremely thin atmosphere is composed mainly of nitrogen, some methane, lower amounts of carbon monoxide, and traces of water [78,79]. Furthermore, the latest data gathered by the New Horizons spacecraft show that Pluto's surface consists of a spatially heterogeneous mix of solid N_2, CH_4, CO, C_2H_6, and an additional component that may not be ice [80]. The nitrogen-dominated atmosphere is supported by the formation of vapor above the surface ice (sublimation/condensation under solid—vapor phase equilibrium), which in turn depends on the surface temperature [81].

It is well known that Pluto's surface temperature varies because of its extremely elliptic orbit, which can send the planet as close as 30 astronomical units (AU), or as far away as 50 AU during its 248-Earth-year journey around the Sun. Thus, there is the possibility that the atmosphere freezes out and disappears slowly or following a spectacular sudden collapse as temperature drops. The low temperature of Pluto surface, which is about 10 K colder than expected, is also strongly influenced by the sublimation of surface nitrogen ice to vapor with a cooling effect due to the received sunlight heat. This cyclical sublimation and freezing is complemented by the escape of Pluto's atmosphere into space due to the low escape velocity (Fig. 7.10). This cooling antigreenhouse effect is in contrast to the natural greenhouse effect on planets like Venus and Earth, where sunlight energy striking the surface is absorbed and heats up the surface. Another phenomenon contributing to the atmospheric temperature profile is linked to methane, present in a mixing ratio $0.5 \pm 0.1\%$, which plays a crucial role in the heating processes in the atmosphere and can explain the elevated atmospheric temperature. The net result is a sharp temperature inversion, with average temperatures 36 K warmer 10 km above the surface reaching temperatures of even 100 K (Fig. 7.11). Even before its detection, methane had been recognized to be the key heating agent in Pluto's atmosphere, able to produce a sharp thermal inversion [79,82].

7.3.6 Titan—The Battle Between Methane and Haze

Titan is the largest moon orbiting Saturn and receives only about 1% as much sunlight as Earth. Also, it is the only moon known to have a substantial atmosphere, composed of 95% nitrogen, 3% methane, 0.3% hydrogen, and traces of other gases. Earth and Titan share some common characteristics, including a similar surface pressure and the presence of a condensable greenhouse gas, which forms clouds and results in organic rains [83]. Like water on Earth, Titan's methane is an abundant molecule close to its triple point and condensable, and therefore capable of modifying the moon's temperature profile (Fig. 7.11).

Atmospheric methane creates a greenhouse effect on Titan's surface without which Titan would be far colder. However, greenhouse effect on Titan follows very different mechanisms in comparison with Earth and Mars [84—86]. Titan's atmospheric methane absorbs some of the infrared heat energy given off by the surface, but the moon also has a high altitude layer of organic haze (aerosol), which blocks sunlight while allowing transmission of thermal infrared, counteracting some of the greenhouse effect. Titan's hydrocarbon haze is generated from ultraviolet photolysis and electron-initiated dissociation of methane, nitrogen, and other trace species and keeps the surface roughly

10% cooler than it would be without its presence. The net result of these opposing phenomena is about 12°C net greenhouse effect [84,87–89].

This antigreenhouse effect has also been proposed for the early Earth, resulted by haze, which might be formed if the early Earth atmosphere had higher than present day levels of methane and carbon dioxide, even with CH_4/CO_2 as low as 0.1 [87,89]. Relevant studies have shown that early Earth organic aerosol might have had a profoundly harsher antigreenhouse effect than their Titan counterparts. Moreover, a similar antigreenhouse effect is observed on Earth after giant volcanic eruptions, which are able to produce and inject huge amounts of ash to the stratosphere.

7.3.7 Triton—35 K From Absolute Zero

Triton, Neptune's largest moon, has a sparse nitrogen-rich atmosphere with trace amounts of carbon monoxide and small amounts of methane near the surface [90]. Triton offers a large-scale laboratory for climate scientists as relevant phenomena can be easily studied because of its simple, thin atmosphere and the absence of oceans. Also, Earth and Triton share some contributing factors to global warming, such as the amount of methane and carbon monoxide in the atmosphere.

Although extremely thin, Triton's atmosphere is active, with clouds of condensed nitrogen that lie between 1 and 3 km from the surface and winds in the troposphere that rises up to an altitude of 8 km. Triton is a unique example of a vapor pressure atmosphere, ie, its main atmospheric component (N_2) is condensable and a vapor–solid equilibrium governs the atmospheric pressure. This leads to a uniform global surface temperature and near surface atmospheric pressure [49]. The temperature of Triton's upper atmosphere, at 95 ± 5 K, is higher than that at the surface, due to the cooling effect of nitrogen sublimation on the surface, the heat absorbed from solar radiation, and Neptune's magnetosphere. A haze permeates most of Triton's troposphere, which is thought to be composed largely of hydrocarbons and nitriles created by the action of sunlight on methane.

A paper published in Nature in 1998 entitled "Global warming on Triton" [91] states that Triton has undergone a period of global warming since 1989 and according to the most conservative estimates, the atmosphere was doubled in bulk every 10 years. The explanation given is that permanent polar caps on Triton play a dominant role in regulating seasonal atmospheric changes, as it happens on Pluto [91]. The increase in temperature between 1989 and 1998 is estimated to 5% and several explanations are offered, including a change of surface ice patterns, which leads to change in ice albedo and thus to an increase of the absorbed heat. These changes seem to be driven by seasonal changes in its polar ice caps, which happen every few hundred years.

7.4 Hydrated Minerals in Extraterrestrial Environments and Potential Implications

7.4.1 Introduction

Water-related minerals have been observed at different solar system objects such as meteorites, asteroids, terrestrial planets like Mars, and some icy planets' satellites as Europa, Ganymede, and Enceladus [92,93]. The implications of the existence of such minerals in planets and moons are multiple; they can offer the right environment for essential-to-life chemical and biological reactions, act as reservoirs of liquid water, influence the atmospheric composition of planets, and can be even utilized in situ by humans in several life-supporting applications. This section is focusing on the cases of Moon, Titan, and Mars where the right components and environmental conditions seem to occur and presents the implications of such discoveries. Asteroids are also addressed, as they are directly or indirectly involved in the formation or even "seeding" of hydrated silicates on planets and moons in our solar system.

7.4.2 Occurrence of Hydrated Minerals in Space

Asteroids

Water-related minerals have been observed in asteroids [92,93]. Asteroids are involved in the formation and/or "seeding" of hydrated silicates on terrestrial planets as they are the primary source of meteorites, which collide with planets and moons since billions of years. It is believed that asteroids began as mixtures of water ice and anhydrous silicates, and through heating events early in solar system history magma was created, which through several processes led to the formation of hydrated minerals. These minerals, particularly clay minerals, occur rapidly and easily in environments, where anhydrous rock and water are found together. Serpentine-group, smectite-group, and chlorite-group minerals are the most common phyllosilicates in meteorites [93]. Studies on asteroids show that the presence of phyllosilicates in interplanetary dust particles suggests that they can survive atmospheric friction and impact, as well as millions of years in vacuum, potentially leading to "seeding" of hydrated minerals on moons and planets [93]. However, the conditions generated during impact of meteorites on solid surfaces are more important. Impact craters host important characteristics, which include the formation of clays and zeolites as secondary hydrothermal minerals [94]. These minerals are formed by aqueous fluids that fill the heated rocks within the crater and react with the shock-derived aluminosilicates and impact glasses under high pressures [94]. Almost 62 terrestrial impact structures have been studied for such hydrothermal alterations, and it is found that altered minerals can contribute up to 25% of the mineralogy of the breccia formed in a crater [94].

Moon and Titan

Up to the 1970s, it was believed that either the Moon is totally dry or uncertainties were too large to draw clear conclusions, and it was only until September 24, 2009, when *Science* magazine reported that water had been detected on the Moon [95]. This water is mainly found in dark regions in the form of underground ice, but also it might have been trapped in porous silicates. Several visible-near-infrared reflectance spectra have demonstrated that the lunar surface is hydrated and that the H_2O content of the lunar mantle may be as much as several hundred parts per million [96]. Furthermore, direct measurements of the lunar surface using multispectral thermal emission mapping showed spectral signatures consistent with basalt compositions and highly evolved silica-rich soils [97]. There is evidence of highly silicic materials such as quartz, silica glass of similar composition, sodium- and potassium-rich varieties of feldspar. Taking into account the numerous impacts of meteorites on Moon, it would not be surprising that such minerals, even in small amounts, exist on the Moon. Furthermore, zeolites and other minerals can be relatively easily synthesized by using raw materials found on Moon. Ming and Lofgren claimed in 1990 that if zeolites and other ion-exchange minerals existed on the lunar surface, they would have a variety of potential applications in life support systems [98]. Ming and Lofgren synthesized in the laboratory analcime, sodalite hydrate, and zeolite A, under mild hydrothermal conditions from basaltic glass, which is abundant in lunar soil.

Titan is the second largest moon in our solar system and the largest among Saturn's known satellites. Titan is primarily composed of surface frozen and liquid hydrocarbons followed by thick layers of water ice, and finally a rocky silicate core [99,100]. Titan's bulk density along with solar system formation models indicates the possibility of existence of considerable water as well as some OH-bearing minerals like serpentine [101]. Some models of the early history of Titan suggest that a liquid ammonia-rich water layer is present at the subsurface of Titan that could have reacted with silicates to put substantial quantities of sodium and potassium into solution, subsequently leading to a cryogenic equivalent of volcanism [102,103]. The development of biotic systems on Titan is considered as possible, since the chemical ingredients are in place and geothermal activity has been projected to exist [104]. Crucial elements in these processes are fluid—solid interfaces, zones of fluid accumulation or entrapment, and areas enriched with material that could act as a catalyst [104,105]. Obviously, natural porous materials like hydrated aluminosilicates could play this role by trapping molecules together so that they can react. Discoveries with the Huygens probe suggest there may be mud on Titan and the soil is found to have a texture resembling that of wet sand or clay with a thin solid crust. The composition is primarily a mix of dirty water ice and hydrocarbon ice with the possibility to contain some kind of aluminosilicates, such as clays and zeolites. However, scientists believe that unlike Earth's sand, which is made of silicates, Titan's is likely composed of solid hydrocarbons that have precipitated out of the moon's thick atmosphere.

Mars

The most interesting case of extraterrestrial hydrated minerals' existence and activity is Mars. In fact, clays and sulfates exist in considerable quantities on Mars and they are strongly connected to the history of the water on the planet. Moreover, Mars has all components and conditions needed for the formation of another magic rock: zeolite. First of all, volcanic materials with a significant glass component are abundant on the planet's surface as evidenced by several observations. Liquid water and saline brines, necessary components of zeolite formation, clearly have been a part of Martian geologic history [106,107]. On the basis of examples from Earth and on thermodynamic grounds, as well as spectral data, it has been argued that zeolites exist as components of the Martian regolith [106,108−111]. Regolith is a layer of loose, heterogeneous material covering solid rock and consists of dust, soil, broken rock, and other related materials.

Studies in the period of 2002−03 demonstrated that surface dust is dominated by silicate material with the possibility of existence of feldspar component, hydrous components, and small amounts of carbonates with the possibility of zeolites being also present [106,112,113]. A Thermal Emission Spectrometer (TES) graph, acquired by the Mars Exploration Rover Spirit's mini-TES instrument, shows the spectral signature of silicates (Fig. 7.17). The bound water feature at 1630 cm^{-1} and a transparency feature at 830 cm^{-1} can be attributed to feldspars and zeolites [114]. Later in the decade more spatially widespread and mineralogically diverse alteration minerals were identified including Fe/Mg-smectites, Al/Mg-phyllosilicates, serpentine, and analcime (Fig. 7.18) [111,115].

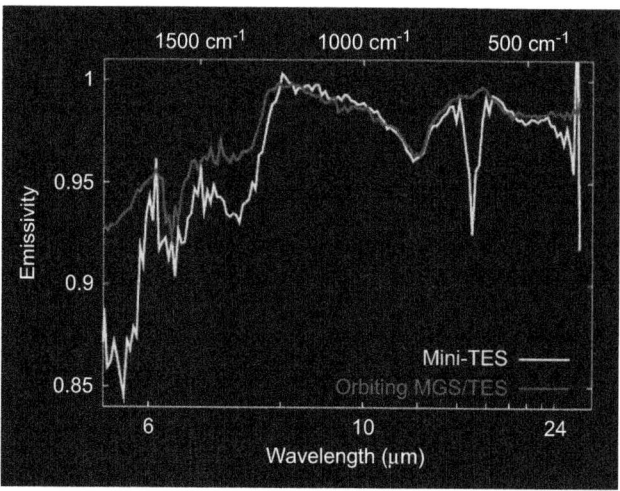

Figure 7.17
TES graph. *Image Credit: NASA/JPL/Arizona State University.*

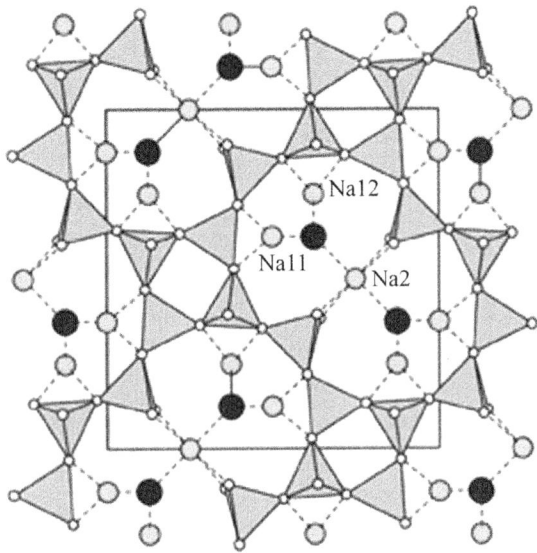

Figure 7.18

Analcime crystal structure: Na ions in *yellow*, H_2O molecules in *blue*. *International Zeolites Association, IZA Commission of Natural Zeolites. http://www.iza-online.org/natural/Datasheets/Analcime/analcime.htm.*

In the beginning of the last decade, Odyssey spacecraft observed heterogeneously distributed hydrogen at Martian low-to-middle latitudes, suggesting that large areas of the near equatorial highlands contain near surface deposits with up to about 10% water-equivalent hydrogen (WEH) [113,116,117]. Earlier infrared spectroscopic data also provided evidence of chemically and/or physically bound H_2O and/or OH on Mars [113]. There are three dominant reservoirs of water on Mars: the polar regions, the atmosphere, and the subsurface [110].

As the existence of water on Mars in ice and aqueous salts had been firmly established, the attention was drawn to another important source of constituent water, namely zeolites, and other porous aluminosilicates, which bound and retain water in their structure [112]. The potential effects of zeolites on the planet's atmosphere are supported by observations made in 2004 by the High Resolution Stereo Camera (HRSC) on board the Mars Express spacecraft. In particular, water fog was observed and was attributed to the presence of hydrated minerals [118]. David Bish supported the idea of water storage in Mars in zeolites and other hydrous materials. In an interview published in *Astrobiology Magazine* in 1998, he claimed that mineral storage is a viable mechanism for binding water indefinitely in a form that will not evaporate or sublimate under current Martian surface conditions [119]. The absence of water at the surface of the planet could be attributed to this mechanism, and apart from zeolites a number of potentially hydrous silicate phases, notably clay minerals, could also contribute. Zeolites and clays are common terrestrial alteration products of hydrovolcanic basaltic ashes and palagonitic material comparable to those that may be widespread on Mars [113,116,117,120,121]. Beyond the search for life, extraction of water from ice or zeolite minerals would be vital for the support of human activity on Mars. In fact, hydrous minerals in the regolith are likely to be more widely distributed than water ice, but the challenge would be to find the energy required to extract water from hydrous minerals, as it is highly dependent on the minerals present and their hydration states [116,117]. Further investigations, namely quantitative thermodynamic data of the hydration of clinoptilolite and smectite, illustrate that some varieties of these common terrestrial minerals would be at least partially hydrated under the environmental conditions found on the Martian surface [116,117]. The shallow (<1 m) occurrences of H_2O ice near the Martian equator are particularly enigmatic because H_2O ice (or bulk liquid) is not stable at these latitudes and under the present Martian atmospheric temperature (130–308 K) and pressure (0.4–0.87 kPa) and thus this water was referred to as adsorption water and in the absence of water ice, the most likely H_2O reservoir is hydrated minerals [113,118,121].

The possibility of existence of large zeolite deposits may have a profound effect on evolution of the Martian atmosphere as zeolites may store and exchange reversibly CO_2, SO_2 and H_2O [106,108]. Furthermore, experimental studies have suggested that zeolites, if present on the Martian surface, could undergo a strong diurnal cycle of hydration and dehydration with possible impact on the atmosphere. However, another study (2005) shows that the diurnal atmospheric water cycle is unlikely to be affected by zeolites on the Martian surface [122]. The atmosphere—surface interaction has implications on the possibility of water adsorption by minerals. The main component of the rather thin Martian atmosphere is CO_2 with a surface pressure of about 6 mbar, while water vapor has a partial pressure of only 0.001 mbar. Thus, water has to be adsorbed in a more than 1000-fold surplus of carbon dioxide and the CO_2 adsorption cannot be neglected [118]. Experimental results published in 2007 show that smectites and zeolites can remain hydrated under the present Martian atmospheric conditions and adsorb from about 2.5–25% of water at a partial water vapor pressure of 0.001 mbar and in a temperature range of 257–333 K [118]. Similar laboratory studies showed that these minerals can remain hydrated at water vapor pressures lower than 0.01 mbar and temperatures between 257 and 333 K, while nuclear magnetic resonance (NMR) measurements of adsorbed water in porous minerals suggest that adsorbed water remains mobile in some minerals at temperatures down to 200 K. Another experimental study showed that the much higher vapor pressure of carbon dioxide in the Martian atmosphere enables chabazite or clinoptilolite to accommodate about 8–10% carbon dioxide at 273 K, but in the absence of water. For thermodynamic reasons, the adsorption of carbon dioxide in the presence of water is placed at a disadvantage. The NMR measurements reveal partial mobility of the adsorbed water, which may be important for adsorption water—related chemistry on Mars [109,118].

On Monday September 28, 2015, NASA scientists announced the existence of surface liquid water, which according to some interpretations could have come from porous rocks under the Martian surface; could these rocks contain hydrated minerals? Probably yes, but further research is required before drawing such a conclusion.

7.4.3 Hydrated Minerals From the Origins of Life to Space Stations

Hydrated minerals and the origins of life

Life, as we know it, needs liquid water, and hydrated minerals can store it and through adsorption—desorption cycles make it available. However, the role of aluminosilicate minerals extends far beyond water supply. Mineral surfaces and crystalline surfaces of common rock-forming minerals are likely to have played important role in the origins of life [123]. Adsorption and catalysis at mineral surfaces might generate replicating biopolymers, the essential building blocks of life, from simple chemicals supplied by meteorites, volcanic gases, and photochemical gas reactions [124,125].

A major problem of existing life evolution theories is that too much water is involved, which poses a challenge for the organic chemistry to take place. Several essential reactions cannot simply happen by mixing chemical compounds dispersed in aqueous solutions, and the assembly of the necessary complex biomolecules by random collisions in dilute aqueous solutions is implausible. Furthermore, several biochemically significant polymers are prone to photochemical destruction by solar radiation. Adsorption of these molecules in mineral surfaces would concentrate organic solutes and also offer a protective environment from dispersion into dilute solutions, hydrolysis, and destructive UV radiation [125,126].

Silica-rich volcanic glasses should have been abundant on the early Earth and other terrestrial planets, ready for crystallization into zeolites and feldspars. Although most silicate minerals are hydrophilic, organophobic and unsuitable for catalytic reactions, some silica-rich surfaces of partly dealuminated feldspars and zeolites are organophilic and potentially catalytic [125]. Dealumination of high-silica zeolites like mordenite, tschernichite, and mutinaite is possible during hot/cold/wet/dry cycles and it is true that catalytic activity increases with Al—OH substitution for Si [124]. A small amount of remaining aluminum would provide the catalytic centers for assembling organic molecules into polymers.

Life support systems and waste management in space stations

The potential applications of hydrated minerals, especially zeolites in extraterrestrial environments, and in particular in planetary stations or human settlements, require the existence of such minerals in the soil of these planets or in the absence of such deposits, the possibility of synthesis of these minerals in situ by use of existing resources. It is not difficult to imagine that in case this is possible, the applications of such minerals open an extraordinary potential for research and technological advances.

The discussion on possible applications of zeolites in outer space is old. The research interest is focused on the development of life support systems in extraterrestrial stations. An effective life support system should be able to supply oxygen, water and food, while at the same time it should be able to remove carbon dioxide, water vapor, and trace contaminants, and recycle other waste. Such integrated systems were studied since mid-1980s by NASA and a list of technologies that could be used for space applications is presented in the related literature [127−130]. D.W. Ming in his publication in 1988 identified zeoponics, wastewater recycling and gas separation and purification, as the major applications with a great potential [131]. Adsorption processes are able to support such recycling systems, and some examples of successful research are the removal of CO_2 from breathing air and recycling of wastewater [132−134]. A recent article describes a zeolite system for carbon dioxide removal integrated into a closed air regeneration cycle aboard a spacecraft for life support of the Mars exploration crew [134]. Furthermore, the extraction of atmospheric water vapor via zeolite molecular sieve adsorption at Mars conditions is considered as feasible [110].

Zeoponics is of the most promising applications and they were first developed by NASA to provide a plant growth medium for long-term uses in outer space [135]. This is a plant growth system based on zeolite-containing artificial soils. The zeoponic substrate slowly releases plant growth nutrients such as nitrogen and other cations into "soil" solution where they become available for plant uptake [127]. The benefits of such systems are obvious as they are used for growing plants, which are source of food, and can also be used for converting CO_2 back into O_2, for eliminating excess humidity, and through evapotranspiration for converting wastewater into potable water [109]. The use of plants for bioregenerative life support for space missions was first studied by the US Air Force in the decade 1950−60 and Russian researchers in the period 1960−80 at the Institute of Biophysics in Krasnoyarsk, Siberia, and the Institute for Biomedical Problems in Moscow [138]. In 1984, a Russian-Bulgarian group created the first Space Greenhouse named SVET, and in 1990 the first experiment began after the equipment was launched to the MIR Orbital Station inside the Krystal module, and the first fresh vegetables, radishes and Chinese cabbage, were produced in space during this experiment [136]. The nutritive medium used for plant cultivation was the, so-called, Balkanine substrate made of natural zeolite clinoptilolite loaded with mineral salts [137]. In the 1990s, in the framework of the Johnson Space Center's Advanced Life Support Program, the construction of the Human-Rated Test Facility (HRTF) began. The HRTF was a larger facility that combined people, plants, microorganisms and physicochemical processes to recycle air and water, produce food, and process wastes [127]. Zeoponic substrates have flown on two Shuttle missions (STS-60, STS-63) as part of a series of experiments, called ASTROCULTURE plant-growth experiments, being conducted in collaboration with the Wisconsin Center for Space Automation and Robotics (WCSAR) [127,130].

Concluding, the potential applications of zeolites and similar minerals in space applications are numerous and promising, and these humble rocks may play a crucial role in the space adventures of humans.

References

[1] European Space Agency (ESA). http://www.esa.int/ESA (accessed March 2015).
[2] J.C. Dolado-Perez, C. Pardini, L. Anselmo, Review of uncertainty sources affecting the long-term predictions of space debris evolutionary models, Acta Astronauti. 113 (2015) 51−65.
[3] T. Hitchens, The United Nations and its efforts to develop treaties, conventions, or guidelines to address key space issues including the de-weaponization of space and orbital debris, in: J.N. Pelton, R.S. Jakhu (Eds.), Space Safety Regulations and Standards, Elsevier, Oxford, GB, UK and Burlington, MA, USA, 2010.
[4] M. Emanuelli, et al., Conceptualizing an Economically, Legally and Politically Viable Active Debris Removal Option, 64th International Astronautical Congress, Beijing, China, 2013.
[5] Secure World Foundation (SWF), Space Sustainability: A Practical Guide, 2014.
[6] T. Ebisuzaki, et al., Demonstration designs for the remediation of space debris from the International Space Station, Acta Astronaut. 112 (2015) 102−113.
[7] D. Baiocchi, W. Welser, Confronting Space Debris, Strategies and Warnings From Comparable Examples Including Deepwater Horizon, National Defense Research Institute, 2010.
[8] M.M. Castronuovo, Active space debris removal—A preliminary mission analysis and design, Acta Astronaut. 69 (2011) 848−859.
[9] Inter-Agency Space Debris Coordination Committee (IADC), Space Debris. Assessment Report for 2011, 2013.
[10] P.A. Slann, Space debris and the need for space traffic control, Space Policy 30 (2014) 40−42.
[11] NASA Orbital Debris Program Office, History of On-Orbit Satellite Fragmentations, fourteenth ed., 2008.
[12] NASA Orbital Debris Program Office, Orbital Debris Q. News 15 (July 2011).
[13] NASA Orbital Debris Program Office, Orbital Debris Q. News 19 (January 2015).
[14] J. Eberhart, Tallying orbital trash, Sci. News 138 (1990).
[15] S.J.J. Goldstein, R.M. Goldstein, Some properties of millimetric space debris, Astron. J. 107 (1994) 367−371.
[16] S.J.J. Goldstein, R.M. Goldstein, Flux of millimetric space debris, Astron. J. 110 (1995) 1392−1396.
[17] B. Weeden, Current Issues in Space Sustainability and Space Traffic Management, AIA Space Council, Arlington, VA, June 26, 2014.
[18] United Nations, Technical Report on Space Debris, New York, 1999.
[19] W. Matthews, Trackers of Orbiting Junk Sound Warning, Defense News, June 10, 2009.
[20] S. Shuangyan, J. Xing, C. Hao, Cleaning space debris with a space-based laser system, Chin. J. Aeronaut. 27 (2014) 805−811.

[21] D.J. Kessler, et al., The Kessler syndrome: implications to future space operations, in: AAS Paper Number 10-016: 33rd Annual AAS Guidance and Control Conference. Breckenridge, Colorado, February 6–10, 2010.

[22] Parliamentary Office of Science of Science and Technology (POST), Space Debris, Postnote no 355, 2010.

[23] J.C. Liou, N.L. Johnson, Risks in space from orbiting debris, Sci. News 311 (2006) 340–341.

[24] NASA Orbital Debris Program Office. http://www.orbitaldebris.jsc.nasa.gov/ (accessed March 2015).

[25] NASA Conference Publication 2360, Orbital Debris, 1985.

[26] B. Weeden, Overview of active debris removal, in: Active Debris Removal Symposium. Leiden, Netherlands, June 21, 2012.

[27] S. Laskow, The Economics of Space Junk – It's a Classic Tragedy of the Commons—This Time, in Orbit, The Boston Globe, October 13, 2013.

[28] M. Brown, Orbital Debris Frequently Asked Questions, 2012. http://orbitaldebris.jsc.nasa.gov/faqs.html.

[29] M.K. Macauley, The economics of space debris: estimating the costs and benefits of debris mitigation, Acta Astronaut. 115 (2015) 160–164.

[30] M. Richard, et al., Uncooperative rendezvous and docking for microsats – the case for cleanspace one, in: 6th International Conference on Recent Advances in Space Technologies, RAST 2013. Istanbul, Turkey, June 2013, pp. 12–14.

[31] United Nations, Treaty on Principles Governing the Activities of States in the Exploration and Use of Outer Space, Including the Moon and Other Celestial Bodies, 1966.

[32] J. Wheeler, The current legal framework associated with space debris mitigation, in: Proceedings of the Institution of Mechanical Engineers, Part G: Journal of Aerospace Engineering, vol. 221, 2007, pp. 911–914.

[33] United Nations, Article III, the Convention on International Liability for Damage Caused by Space Objects, Resolution 2777, 1972.

[34] United Nations, Article VII, Treaty on Principles Governing the Activities of States in the Exploration and Use of Outer Space, Including the Moon and Other Celestial Bodies, 1967.

[35] L. Anselmo, C. Pardini, Compliance of the Italian satellites in low Earth orbit with the end-of-life disposal guidelines for space debris mitigation and ranking of their long-term criticality for the environment, Acta Astronaut. 114 (2015) 93–100.

[36] International Organization for Standardization, ISO 24113:2011, Space Systems – Space Debris Mitigation Requirements, 2011.

[37] M.B. Schaffer, Toward a viable nuclear waste disposal program, Energy Policy 39 (2011) 1382–1388.

[38] R. Dusek, Lost in space? The legal feasibility of nuclear waste disposal in outer space, William Mary Environ. Law Policy Rev. 22 (1997) 181–218.

[39] J. Coopersmith, Disposal of High-Level Nuclear Waste in Space, 1999.

[40] R.E. Burns, et al., Nuclear waste disposal in space, NASA Tech. Pap. 1225 (1978).

[41] R. Salkeld, R. Beichel, Nuclear waste disposal in space: implications of advanced space transportation, Acta Astronaut. 7 (1980) 1373–1387.

[42] D. Iranzo-Greus, et al., Nuclear waste disposal in space: a long term solution, in: 7th International Symposium on Launcher Technology Session: Missions Conference, 2007.

[43] J. Coopersmith, Nuclear waste in space? Space Rev. (August 22, 2005). http://www.thespacereview.com/article/437/1.

[44] World Nuclear Association (WNA). http://www.world-nuclear.org/ (accessed September 2015).

[45] R.W. Boubel, et al., Fundamentals of Air Pollution, third ed., Academic Press, 1994.

[46] R.M. Harisson, Understanding Our Environment, an Introduction to Environmental Chemistry and Pollution, Royal Society of Chemistry, UK, 1999.

[47] E. Kaufmann, A. Hagermann, Penetration of solar radiation into pure and Mars-dust contaminated snow, Icarus 252 (2015) 144–149.

[48] University of Nebraska-Lincoln (UNL). http://astro.unl.edu/naap/help/general_overview.html (accessed October 2015).

[49] D.K. Cruikshank, M.S. Matthews, A.M. Schumann, Neptune and Triton, The University of Arizona Press, 1995.

[50] J.P. Webster, The role of hydrological processes in ocean-atmosphere interactions, Rev. Geophys. 32 (1994) 427−476.

[51] J.T. Kiehl, K.E. Trenberth, Earth's annual global mean energy budget, Bull. Am. Meteorol. Soc. 78 (1997) 197−208.

[52] S.E. Manahan, Fundamentals of Environmental Chemistry, CRC Press LLC, Boca Raton, FL, 2001.

[53] I. de Pater, J. Lissauer, Planetary Sciences, Cambridge University Press, 2007.

[54] D. Archer, Methane hydrate stability and anthropogenic climate change, Biogeosciences 4 (2007) 521−544.

[55] J.P. Kennett, et al., Methane Hydrates in Quaternary Climate Change: The Clathrate Gun Hypothesis, American Geophysical Union, Washington, DC, 2003.

[56] C. Goldblatt, K.J. Zahnle, Faint young sun paradox remains, Nature 474 (2011) E3−E4.

[57] A.A. Pavlov, et al., Greenhouse warming by CH4 in the atmosphere of early Earth, J. Geophys. Res. 105 (2000).

[58] C. Goldblatt, et al., Nitrogen enhanced greenhouse warming on early Earth, Nat. Geosci. 2 (2009) 891−896.

[59] M.A. Bullock, D.H. Grinspoon, The recent evolution of climate on Venus, Icarus 150 (2001) 19−37.

[60] L. Noack, D. Breuer, T. Spohn, Coupling the atmosphere with interior dynamics: implications for the resurfacing of Venus, Icarus 217 (2012) 484−498.

[61] C. Sagan, The planet Venus, Science 133 (1961) 849−858.

[62] J.B. Pollack, Climatic change on the terrestrial planets, Icarus 37 (1979) 479−553.

[63] V.A. Krasnopolsky, Atmospheric chemistry on Venus, Earth, and Mars: main features and comparison, Planet. Space Sci. 59 (2011) 952−964.

[64] J.F. Kasting, Runaway and moist greenhouse atmospheres and the evolution of Earth and Venus, Icarus 74 (1988) 472−494.

[65] V.S. Meadows, D. Crisp, Ground-based near-infrared observations of the Venus nightside: the thermal structure and water abundance near the surface, J. Geophys. Res. 101 (1996) 4595−4622.

[66] R. Kippenhahn, A. Weigert, A. Weiss, Stellar Structure and Evolution, Springer-Verlag, Berlin, Heidelberg, New York, 1994.

[67] R.J. Phillips, M.A. Bullock, S.A. H II, Climate and interior coupled evolution on Venus, Geophys. Res. Lett. 28 (2001) 1779−1782.

[68] P. Driscoll, D. Bercovici, Divergent evolution of Earth and Venus: influence of degassing, tectonics, and magnetic fields, Icarus 226 (2013) 1447−1464.

[69] M.H. Carr, J.W. Head III, Geologic history of Mars, Earth Planet. Sci. Lett. 296 (2010) 185−203.

[70] D.T.F. Möhlmann, Temporary liquid water in upper snow/ice sub-surfaces on Mars? Icarus 207 (2010) 140−148.

[71] J.-G. Winther, et al., Melting, runoff and the formation of frozen lakes in a mixed snow and blue-ice field in Dronning Maud Land, Antarctica, J. Glaciol. 42 (1996) 271−278.

[72] M. Grott, et al., Volcanic outgassing of CO_2 and H_2O on Mars, Earth Planet. Sci. Lett. 308 (2011) 391−400.

[73] P. von Paris, et al., N2-associated surface warming on early Mars, Planet. Space Sci. 82−83 (2013) 149−154.

[74] C. Sagan, G. Mullen, Earth and Mars: evolution of atmospheres and surface temperatures, Science 177 (1972) 52−56.

[75] J.F. Kasting, CO_2 condensation and the climate of early Mars, Icarus 94 (1991) 1−13.

[76] N. Batalha, et al., Testing the early Mars H_2-CO_2 greenhouse hypothesis with a 1-D photochemical model, Icarus 258 (2015) 337−349.

[77] M.M. Marinova, C.P. McKay, H. Hashimoto, Radiative-convective model of warming Mars with artificial greenhouse gases, J. Geophys. Res. Planets 110 (2005).

[78] P. Blondel, J.W. Mason, Solar System Update, Springer, 2006.

[79] E. Lellouch, et al., Pluto's lower atmosphere structure and methane abundance from high-resolution spectroscopy and stellar occultations, Astron. Astrophys. 495 (2009) L17–L21.

[80] D.P. Cruikshank, et al., The surface compositions of Pluto and Charon, Icarus 246 (2015) 82–92.

[81] A.S. Bosh, et al., The state of Pluto's atmosphere in 2012–2013, Icarus 246 (2015) 237–246.

[82] B. Sicardy, et al., Drastic expansion of Pluto's atmosphere as revealed by stellar occultations, Nature 10 (2003).

[83] A. Coustenis, Formation and evolution of Titan's atmosphere, Space Sci. Rev. 116 (2005) 171–184.

[84] C.P. McKay, J.B. Pollack, R. Courtin, The greenhouse and antigreenhouse effects on Titan, Sci. News 253 (1991) 1118–1121.

[85] J.L. Hunt, et al., Collision-induced absorption in the far infrared spectrum of Titan, Icarus 55 (1983) 63.

[86] R. Wordsworth, F. Forget, V. Eymet, Infrared collision-induced and far-line absorption in dense CO_2 atmospheres, Icarus 210 (2010) 992.

[87] C.P. McKay, R.D. Lorenz, J.I. Lunine, Analytic solutions for the antigreenhouse effect: Titan and the early Earth, Icarus 137 (1999) 56–61.

[88] T. Tokano, Meteorological assessment of the surface temperatures on Titan: constraints on the surface type, Icarus 173 (2005) 222–242.

[89] C.A. Hasenkopf, et al., Optical properties of Titan and early Earth haze laboratory analogs in the mid-visible, Icarus 207 (2010) 903–913.

[90] A. Masters, et al., Neptune and Triton: essential pieces of the solar system puzzle, Planet. Space Sci. 104 (2014) 108–121.

[91] J.L. Elliot, et al., Global warming on Triton, Nature 393 (1998) 765–767.

[92] O. Prieto-Ballesteros, V. Muñoz-Iglesias, L.J. Bonales, Comparative mineralogy in the solar system: water-related minerals and habitability, EPSC Abstr. 7 (September 23–28, 2012). Number EPSC2012-926 European Planetary Science Congress. Madrid, Spain.

[93] A.S. Rivkin, et al., Hydrated minerals on asteroids: the astronomical record, in: W.F. Bottke (Ed.), Asteroids III, University of Arizona Press, 2002, p. 235.

[94] C.S. Cockell, The origin and emergence of life under impact bombardment, Philos. Trans. R. Soc. B 361 (2006) 1845–1856.

[95] C.M. Pieters, et al., Character and spatial distribution of OH/H_2O on the surface of the Moon seen by M3 on Chandrayaan-1, Science 326 (2009) 568–572.

[96] S. Li, R.E. Milliken, Quantitative mapping of lunar surface hydration with moon mineralogy mapper (M3) data, paper number 1337, in: 44th Lunar and Planetary Science Conference, 2013.

[97] B.T. Greenhagen, et al., Global silicate mineralogy of the Moon from the diviner lunar radiometer, Science 329 (2010) 1507.

[98] D.W. Ming, D.E. Lofgren, Crystal morphologies of minerals formed by hydrothermal alteration of synthetic lunar basaltic glass, Dev. Soil Sci. 19 (1990) 463–470.

[99] A. Coustenis, F. Taylor, Titan: The Earth-Like Moon, Series on Atmospheric, Oceanic and Planetary Physics, vol. 1, World Scientific Publishing Company, 1999.

[100] I. Müller-Wodarg, et al., Titan: Interior, Surface, Atmosphere, and Space Environment, Cambridge University Press, 2014.

[101] T.B. McCord, et al., Composition of Titan's surface from Cassini VIMS, Planet. Space Sci. 54 (2006) 1524–1539.

[102] S. Engel, J.I. Lunine, D.L. Norton, Silicate interactions with ammonia-water fluids on early Titan, J. Geophys. Res. Planets 99 (1994) 3745–3752.

[103] G. Tobie, et al., Titan's internal structure inferred from a coupled thermal-orbital model, Icarus 175 (2005) 496–502.

[104] S.H. Abbas, D. Schulze-Makuch, Synthesis of biologically important precursors on Titan, J. Sci. Explor. 21 (2007) 673–687.

[105] D. Schulze-Makuch, D.H. Grinspoon, Biologically enhanced energy and carbon cycling on Titan? Astrobiology 5 (2005) 560–567.

[106] S.W. Ruff, Spectral evidence for zeolite in the dust on Mars, Icarus 168 (2004) 131−143.

[107] M.R.M. Izawa, et al., Basaltic glass as a habitat for microbial life: implications for astrobiology and planetary exploration, Planet. Space Sci. 58 (2010) 583−591.

[108] D.G. Towell, A. Basu, Zeolite cement in volcaniclastics, paper no 6149, in: The Fifth International Conference on Mars. Pasadena, California, July 1999, pp. 18−23.

[109] V.J. Inglezakis, A.A. Zorpas, Natural Zeolites Handbook, Bentham Science Publishers, 2012.

[110] O.J. Igbinosun, S.E. Wood, A.P. Bruckner, Interfacial water as a Mars ISRU objective: detection of microscopic water using fiber optic sensors, paper number AIAA 2013-0437, in: 51st Aerospace Sciences Meeting Including the New Horizons Forum and Aerospace Exposition. Grapevine, Texas, January 7−10, 2013.

[111] Z. Eldar, et al., Aqueous mineralogy and stratigraphy at and around the proposed Mawrth Vallis MSL landing site: new insights into the aqueous history of the region, MARS, Int. J. Mars Sci. Explor. 6 (2011) 32−46.

[112] G.G. Kochemasov, Topics in comparative planetology, in: 36th Vernadsky-Brown Microsymposium. Moscow, Russia, October 14−16, 2002.

[113] C.I. Fialips, et al., Hydration state of zeolites, clays, and hydrated salts under present-day martian surface conditions: can hydrous minerals account for Mars Odyssey observations of near-equatorial water-equivalent hydrogen? Icarus 178 (2005) 74−83.

[114] NASA, Rover Senses Silicates, 2004. Press Release on January 9, 2004, Mars exploration rover mission, http://www.jpl.nasa.gov/missions/mer/images.cfm?id=340.

[115] B.L. Ehlmann, et al., Identification of hydrated silicate minerals on Mars using MRO-CRISM: geologic context near Nili Fossae and implications for aqueous alteration, J. Geophys. Res. Planets (1991−2012) 114 (2009).

[116] D.L. Bish, et al., Stability of hydrous minerals on the martian surface, Icarus 164 (2003).

[117] D.L. Bish, et al., Can hydrous minerals account for the observed mid-latitude water on Mars? Paper no 3066, in: Sixth International Conference on Mars, 2003.

[118] J. Jänchen, D.T.F. Möhlmann, H. Stach, Water and carbon dioxide sorption properties of natural zeolites and clay minerals at martian surface temperature and pressure conditions, Stud. Surf. Sci. Catal. 170 (2007).

[119] D. Bish, Water from a stone (interview), Astrobiol. Mag. (November 15, 2004). http://www.astrobio.net/interview/water-from-a-stone/.

[120] P.J. Jakes, D. Rajmon, Zeolites − fate of martian water, Abstract #1627, in: Lunar and Planetary Science XXIX Conference. Houston, Lunar and Planetary Institute, 1998.

[121] J. Jänchen, et al., Investigation of the water sorption properties of Mars-relevant micro- and mesoporous minerals, Icarus 180 (2006) 353−358.

[122] T. Tokano, D.L. Bish, Hydration state and abundance of zeolites on Mars and the water cycle, J. Geophys. Res. 110 (2005) 1−18.

[123] R.M. Hazen, D.A. Sverjensky, Mineral surfaces, geochemical complexities, and the origins of life, Cold Spring Harbor Perspect. Biol. 2 (2010) a002162.

[124] J.V. Smith, Biochemical evolution. I. Polymerization on internal, organophilic silica surfaces of dealuminated zeolites and feldspars, Proc. Natl. Acad. Sci. U.S.A. 95 (1998) 3370−3375.

[125] I. Parsons, M.R. Lee, J.V. Smith, Biochemical evolution II: origin of life in tubular microstructures on weathered feldspar surfaces, Proc. Natl. Acad. Sci. U.S.A. 95 (1998) 15173−15176.

[126] D. Deamer, et al., Self-assembly processes in the prebiotic environment, Philos. Trans. R. Soc. B 361 (2006) 1809−1818.

[127] J. Gruener, Planting the Planets, Beyond LEO (Low Earth Orbit), JSC Exploration Office, NASA, 1995. http://ares.jsc.nasa.gov.

[128] D.W. Ming, D.L. Henninger, Lunar Base Agriculture: Soils for Plant Growth, American Society of Agronomy, Madison, Wisconsin, 1989.

[129] D.W. Ming, et al., Zeoponic Plant Growth Substrates, Paper no AIAA 94-4571, AIAA Space Programs and Technologies, Huntsville, Alabama, 1994.

[130] R.C. Morrow, et al., The ASTROCULTURE flight experiment series: validating technologies for growing plants in space, paper no F4.1-M.1.04, in: 29th COSPAR Scientific Assembly. Washington, DC, 1992.

[131] D.W. Ming, Applications for special-purpose minerals at the lunar base, in: The Second Conference on Lunar Bases and Space Activities of the 21st Century. Huston, Texas, April 5—7, 1988.

[132] H. Ghobarkar, H. Schiif, U. Guth, Zeolites — from kitchen to space, Prog. Solid State Chem. 27 (1999) 29—73.

[133] L.A. Dallbauman, J.E. Finn, Adsorption processes in spacecraft environmental control and life support systems, Stud. Surf. Sci. Catal. 120 (1999) 455—471.

[134] I.F. Chekov, Life support of the Mars exploration crew. Control of a zeolite system for carbon dioxide removal from space cabin air within a closed air regeneration cycle (in Russian), Aviakosmicheskaia i Ekologicheskaia Meditsina 43 (2009) 37—45.

[135] R.D. Andrews, et al., Zeoponic materials allow rapid greens grow-in, Golf Course Manag. 67 (1999) 68—72.

[136] T.N. Ivanova, Y.A. Bercovich, A.L. Mashinskiy, G.I. Meleshko, The first vegetables have been grown up in the "SVET" greenhouse by means of controlled environmental conditions, Microgravity Q. 2 (1992) 109—114.

[137] T. Ivanova, et al., Zeolite gardens in space, in: G. Kirov, et al. (Eds.), Natural Zeolites Sofia '95. Proceedings of the Sofia Zeolite Meeting '95, June 18—25, 1995, 1997. PENSOFT, 3 pp.

[138] R.M. Wheeler, J.C. Sager, Crop Production for Advanced Life Support Systems, Purdue University, Technical Reports, Paper 1 (2006). http://docs.lib.purdue.edu/nasatr/1.

CHAPTER 8

Environment and Development

N.M. Katsoulakos, L.-M.N. Misthos, I.G. Doulos, V.S. Kotsios

Metsovion Interdisciplinary Research Center, National Technical University of Athens, Athens, Greece

Chapter Outline

Environment and Development. http://dx.doi.org/10.1016/B978-0-444-62733-9.00008-3

499

8.1 Introduction

Development is, clearly, not a neutral procedure. Despite the different concepts attributed to development over time [1], in the field of economics, development is related to increase in production of products and services within an economy. The basic indicator used for development is the gross domestic product, defined as "an aggregate measure of production equal to the sum of the gross values added of all resident, institutional units engaged in production (plus any taxes, and minus any subsidies, on products not included in the value of their outputs)" [2]. The undeniable relation between development and economic growth produces significant interactions between development and the environment, both physical and man-made. As Braudel [3] states: "The actual situation of a civilization depends, to a significant extent, on the advantages or disadvantages of its geographical space (surrounding environment)."

In a world functioning through free-market economies and in an era dominated by finance capital, the fact that economic growth is the main focus of development policies is inevitable. However, the rise of neoliberal ideas and policies coincided with the invigoration of environmental/ecological movements [4,5]. These movements contributed to awakening about increasing pollution problems, irrational use of natural resources, etc. Moreover, the intensifying inequalities between the developed and the developing world (reflected in major humanitarian disasters such as those in Africa between 1981 and 1984, when over a million people starved to death because of continuous draught) also aroused global concern. Hence, the leaders of the world were driven to adopt a new development paradigm, focused not only on economy. Sustainable development was introduced as a global priority in the so-called "Earth Summit," in Rio de Janeiro, in 1992. Then, in 2002 and 2012 most of the world's countries refreshed their commitments to achieve sustainability. Sustainable development consists of three pillars: economic development, social development, and environmental protection [6].

Indeed, sustainable development—even as a global ideological framework—comprised a major progressive step for the world. Various actions and measures in a wide variety of sectors have led to great improvements in the social and the environmental field. Nevertheless, a lot of great challenges still remain. Our world is a field of inequalities, while global environment is under constant threat. This is a major reason leading to criticism on sustainable development from various points of view. A usual accusation is that sustainable development has become a catch phrase rather than a factual motivation for action [7].

The objective of this chapter is to sum up the concept and implications of development and the related policies, at global level. By realizing a critical study of the extensive literature and data sets related to development, a concise "guide" to the basic trails of development is composed. It is aimed to provide the reader with views and information, which will make comprehensible the basic issues of development and its environmental and social implications. In this way, awareness about these important matters could be raised. Moreover, by presenting views doubting on current development strategies, we aim to strengthen critical approaches to these issues; this is a fundamental step toward the direction of creative change.

As far as the structure of this chapter is concerned, Section 8.2 deals with the notion of development and its evolution over time. Moreover, the dominant development strategy, that of sustainable development, is analyzed and concerns about it are put forward. In Section 8.3 implications between environment and development are analyzed. The analysis is supported by several case studies and some basic indicators reflecting the interactions between development/economic growth and the environment. Section 8.4 refers to the major issues of poverty, whose alleviation is the main target of all, major development policies. Some basic tools supporting development planning are presented in Section 8.5. In the same section, a synopsis of major, global policies regarding environment and development is made. In Section 8.6, alternative approaches to development are searched. General concerns and ethical issues are noticed and a theoretical framework of an alternative pathway to current development policies that of integrated, worth-living development is presented.

8.2 Conceptual Approaches to Development and Principles of Sustainable Development

"Development" is a contested concept, implying that it has meant different things from one historical situation to another and from one actor to another [8]. However, during the past century the word "development" has been used in numerous contexts because of its ability to guide thought and behavior [9]. As Rist [10] argues, the strength of development discourse comes of its power to seduce, in every sense of the term, to charm, to please, to fascinate, to set dreaming, but also to abuse, to turn away from the truth, to deceive.

Table 8.1: Meanings of Development Over Time [1].

Period	Perspectives	Meanings of Development
1800s	Classical political economy	Remedy for progress, catching up
1850>	Colonial economics	Resource management, trusteeship
1870>	Latecomers	Industrialization, catching up
1940>	Development economics	Economic growth—industrialization
1950>	Modernization theory	Growth, political and social modernization
1960>	Dependency theory	Accumulation—national, autocentric
1970>	Alternative development	Human flourishing
1980>	Human development	Capacitation, enlargement of people's choices
1980>	Neoliberalism	Economic growth, structural reform, deregulation, liberalization, privatization
1990>	Postdevelopment	Authoritarian engineering, disaster
2000	Millennium development goals	Structural reforms

What is actually the interpretation of the word "development"? It is a fact that there is no consensus of the scientific community on how the term can be understood. The various theories of development, the different sociopolitical and philosophical viewpoints, and perceptions highlight a series of concepts with an objective hypostasis, such as growth, development, movement, alteration, radical change, progress, management, reform, modernization, amendment/modification, transformation, action—reaction, which characterize the quantitative and qualitative figures of development [11]. In Table 8.1 the meanings of development over time can be seen.

Furthermore, Potter et al. and Gasper [12,13] citing Thomas [14] recognize a number of different usages of the word development in the development studies literature. These are worth noting here as they effectively expand upon a simple dictionary:

- development as a fundamental or structural change—for example, an increase in income,
- development as intervention and action, aimed at improvement, regardless of whether betterment is, in fact, actually achieved,
- development as improvement, with good as the outcome, and
- development as the platform for improvement—encompassing changes that will facilitate development in the future.

Moreover, Rokos [15] defines development as a new, improved dynamic balance between human relations and systems of land use, production, employment, consumption, and distribution, which aims at the optimal use of physical and socioeconomic resources, according to the average social consciousness of the citizens, the specific social dynamics, and the political will of authorities. Talmage [16] points out that development is an effective change process aimed toward positive impact that is facilitated through the efficient use of resources.

However, during the previous century Western perceptions about the world and history have created a broad trend, according to which development is associated with something positive, something desirable, regardless if development refers to societies, regions, or specific population groups. As highlighted by Schumacher [17], in every branch of modern thought, the concept of development plays a central role.

For many people, ideas of development are linked to concepts of modernity [18]. Some of them interpreted this diffusion of modernity as development and progress, while others connected it with the alienation of cultural practices, the destruction of habitat, and loss of quality of life. This is what Horkheimer and Adorno proposed in their work: *Dialectic of Enlightenment* [19], in which they argue that the logic behind the rationalism of the Enlightenment is logic of domination and oppression. The desire for dominance over nature meant domination over men and could only ultimately lead to "nightmarish situation of sovereignty over ourselves."

However, despite the reservations, development was the dominant ideology of the previous century. Describing this situation Wolfgang Sachs [20] notes: "Like a towering lighthouse guiding sailors towards the coast, development stood as the idea which oriented emerging nations in their journey through post-war history."

This lighthouse of development was created just after the World War II. Development became significant when the Western world confronted the new challenge of rebuilding countries, especially in Europe, a continent that had been shattered by war [21]. As Potter et al. [12] reported, many development theorists support that the modern era of development started with a speech made by President Truman in 1949, in which he employed the term "underdeveloped areas" to describe what was soon to be known as the Third World concluding with the duty of the West to bring development to such relatively underdeveloped countries and urging other countries to follow the Western development policy [20]. In this general context, the core meaning of development was catching up with the advanced industrialized countries [1].

The foregoing case has been a very useful mental tool for the United States. As a result, development prevailed as the one-dimensional economic development and in fact, as Rokos [22] pointed out, in its most vulgar version, that of growth. The classical equation that regurgitated ever since was:

$$Development = Natural\ Resources + Capital + Labor$$

This equation is marked, in the course of humanity, in boundary moments, by the "extreme" historical precedents of:

- colonialism, the plundering of our planet's natural and human reserves of dependence, underdevelopment, external debt, predatory lending in the form of assistance, and deficits of the "Third World" countries and poor countries in general,

- financial and economic "incentives" of tax exemptions and the rampant speculation of the investors and the usurious financial system, and
- exploitation of man by man, of immigrants, of women, and child labor, etc.

Under those circumstances, according to Willis [18], an idea of an impasse became increasingly common. In the 1960s and 1970s the contrasting approaches of modernization theories and dependency theories represented differing perspectives on development. However, the global economic problems of the 1980s and the awareness that in many senses existing development theories had not been translated into practical success led theorists to stop and think about what development was and how it could be achieved. While neoliberal thinking now dominates development policy-making, the post-1980s period has been associated with the recognition of much greater diversity within conceptions of development. This has included greater awareness of environmental concerns, gender equity, and grassroots approaches.

8.2.1 Sustainable Development

The concept and the term of "sustainability" expressed the concerns about the issues of development and environment. Although sustainability was not an unknown idea, it came up in the 1970s and 1980s. At international institutional levels, the main milestones in the growth of the concept of sustainable development are the UN Conference on the Human Environment (1972), the World Commission on Environment and Development (1987), the Summit on Environment and Development (1992), the Special Session of the UN for Environment and Development (1997), and the World Summit on Environment and Development (2002). The main feature of these international meetings was the attempt to link the aspirations of humanity. Sustainable Development, emphasizing on the most recent dual concerns on environment and development, is typical of such attempts.

With the Brundtland's Commission [23], the concept of Sustainable Development acquired international interest and became a keyword for politicians, decision makers, development actors, academics, and environmental groups. The goal of the World Commission was to identify practical ways to address environmental and development problems around the world. In particular:

- to reconsider the critical environmental and development issues and to formulate realistic proposals,
- to propose new forms of international cooperation on the issues that will influence policies and actions toward the necessary changes, and
- to improve the comprehension and commitment levels so that individuals, voluntary organizations, businesses, foundations, and governments act more intensively.

Thus, the commission supported an approach to the development that would take into consideration the relationship between the ecological, economic, social, and technological issues. This approach is called "sustainable development," defined as the kind of development that meets the needs of the present without compromising the ability of future generations to meet their needs [23].

According to this definition, sustainable development promises economic growth in its traditional sense of the increase of per capita income, in conjunction with reduction of poverty and social inequalities and on condition that natural resources will not be exhausted [24]. In other words, increasing the prosperity of the people should not be at the expense of the welfare of future generations. There are two key concepts in this definition:

- the concept of needs, especially the needs of the poorest of this world, in which priority should be given and
- the concept of limits, as set by the state/level of technology and social structures to the ability of the environment to satisfy both current and future needs.

The report recommends urgent action on key issues to ensure that development is sustainable. These are population and human resources, industry, food security, species and ecosystems, urban challenge, and energy. The meaning of sustainable development was then analyzed further, as evidenced by the issue of two more reports: "Caring for the Earth. A Strategy for Sustainable Living" [25] and "Agenda 21," the action plan adopted at the UN Summit in Rio in 1992.

Agenda 21 is a comprehensive program of actions that would have to initiate in 1992 and to continue during the 21st century by governments, UN agencies, development agencies, nongovernmental organizations, and the private sector. These actions relate to every field of human activity that has an effect on the environment and should be adopted at global and national level. In other words Agenda 21 defines the projects that are necessary to promote a prosperous, fair, and sustainable Earth.

Ten years after the Earth Summit in Rio de Janeiro, the heads of state and governments around the world committed themselves again in fighting against poverty and environmental protection at the Johannesburg Summit in 2002. The Johannesburg Summit produced results in three levels: (1) the Political Declaration, which is now referred to as the "Johannesburg Declaration on Sustainable Development," (2) the Johannesburg Plan of Implementation, and (3) "Type II nonnegotiated partnerships" between governments, enterprises, and NGO's. However, environmental policies remain very different among countries and largely dependent on the level of economic development and environmental awareness. The Johannesburg Summit results were rather scanty. Even though the adoption of the Declaration on Sustainable Development was finally accomplished, the General

Assembly, during the 57th Session, was limited to vague references on the results of the conference [26].

Meanwhile, the phrase sustainable development has been continually redefined. Aspects of government policy, business strategy, and even lifestyle decisions have been shaped around the concept. As Mawhinney [27] notes, sustainable development as a concept promises many things to many people. The implicit vagueness of the Brundtland Commission's definition along with transcendence, which is connected with the concept, has stimulated interest from different academic fields, which have attempted to interpret the meaning of sustainable development according to their field of knowledge. Blewitt [28] states that many people are coming to sustainable development with little understanding of the key issues and debates. Basiago [29] points out that sustainability is susceptible to varying interpretations by different disciplines. In biology, sustainability is connected with the interaction between human and natural systems.

The biological definition of sustainability concerns itself with the need to save natural capital on behalf of future generations, particularly the genetic diversity contained in plant and animal species, or "biodiversity." In economics, "sustainability" encompasses instruments to internalize the environmental costs of industrial activity in the economy by way of public intervention in private markets. In sociology, "sustainability" refers to the way: certain human interest groups make decisions over the use of natural resources, other groups are affected, and the equity issues that are raised. The urban definition of "sustainability" seeks to reduce notions of "sustainability" to the practical planning of regions, communities, and neighborhoods. The ethical definition of "sustainability" probes the domain where humans ponder whether they are a part of, or apart from, nature, and how this should guide moral choice.

It is a fact that the interpretation of the concept varies also in different cultural contexts, even if it expresses the same or different connotations. In English the term sustainable refers to a process the rate of which should be maintained. It is a dynamic, nonstatic concept, which introduces a long-term vision [30] cited in Ref. [31]. The term sustainable development takes different forms in different societies and environments and is the process by which societies are driven in a dynamic equilibrium condition called sustainability [32]. However, in practice, there are discrete, different interpretations of sustainability. Hopwood et al. [33] mapped different trends of thought on sustainable development, based on combining environmental and socioeconomic issues. They present three broad views on the nature of the changes, necessary in society's political and economic structures and human–environment relationships to achieve sustainable development: that it can be achieved within the present structures; that fundamental reform is necessary but without a full rupture with the existing arrangements; and that as the roots of the problems are the very economic and power structures of society, a radical transformation is needed.

Supporters of the status quo recognize the need for change but see neither the environment nor society as facing insuperable problems. Adjustments can be made without any fundamental changes to society, means of decision making, or power relations. This is the dominant view of governments and business, and supporters of the status quo are most likely to work within the corridors of power talking with decision makers in government and business. Development is identified with growth, and economic growth is seen as part of the solution. Those who take a reform approach accept that there are mounting problems, being critical of current policies of most businesses and governments and trends within society, but do not consider that a collapse in ecological or social systems is likely or that fundamental change is necessary. They generally do not locate the root of the problem in the nature of present society, but in imbalances and a lack of knowledge and information, and they remain confident that things can and will change to address these challenges. They generally accept that large shifts in policy and lifestyle, many very profound, will be needed at some point. The key is to persuade governments and international organizations, mainly by reasoned argument, to introduce the needed major reforms. They focus on technology, good science and information, modifications to the market, and reform of government.

Transformationists see mounting problems in the environment and/or society as rooted in fundamental features of society today and how humans interrelate and relate with the environment. They argue that a transformation of society and/or human relations with the environment is necessary to avoid a mounting crisis and even a possible future collapse. Reform is not enough, since many of the problems are viewed as being located within the very economic and power structures of society because they are not primarily concerned with human well-being or environmental sustainability. While some may use the established political structures and scientific arguments, they generally see a need for social and political action that involves those outside the centers of power such as indigenous groups, the poor and working class, and women. The transformationists include those who focus primarily on either the environment or the socioeconomic field and those who synthesize both.

8.2.2 The Political Dimension of Sustainability

Ambiguity is observed in the relationship between development and sustainability, as these concepts separately neither specify nor restrict the kind of relationship between them. For some scholars and, generally, all the national and international institutions, sustainability and sustainable development are used interchangeably, while for others, the two terms are not absolutely synonymous [34]. The flexibility of the concept has given rise to questions about what it is supposed to mean: sustainability of what, for whom, and why? [11,35].

Since the conclusion of the Brundtland Commission [23]—in itself something of a political compromise—the two competing notions of strong and weak sustainability have dominated the theoretical debate on sustainable development. Loosely speaking, strong sustainability argues that we must live within the environmental and ecological limits that the planet clearly has. Weak sustainability argues that humanity will replace the natural capital we have used and that we depend on, with human-made capital [36].

Parker et al. [37: 278] argue that, although the notion of sustainable development has done much to raise public debate and attention on environmental issues, many feel that it does not go far enough in challenging the practices that have led to environmental degradation. A "strong" view of sustainability would hold that economic growth is incompatible with the earth's finite resources. From this perspective, the notion of sustainable development is a contradiction in terms, a smoke screen used by government and business to pay lip service to environmental issues while maintaining their commitment to economic development. Devkota [38] citing Pearce and Barbier, Pearce and Atkinson, and Serafy [39,40,41] reports weak sustainability as a correct measurement of income, and hence, sustainability need not bother distinguishing between natural and other forms of capital.

Strong sustainability considers natural capital as a provider of some functions that are not substitutable from the human-induced capital. These functions are highlighted by defining sustainability as a mortgage to the future generations of a stock of natural capital not less than that enjoyed by the present generation. Sustainability is expressed in terms of nondeclining natural capital. Contrary to this notion is the concept of mild sustainability. After the definition proposed by Pearce and Atkinson [40], an economy is considered sustainable if the saving rates are larger than the combined devaluation percentage of natural and human capital. In this sense, sustainability is equivalent to a nondecreasing total capital reserve. This is called weak sustainability, and it does not take any limitation of substitutability between natural and human capital. From the weak sustainability perspective, "an economy is considered to be sustainable if its savings rate is greater than the combined depreciation on natural and man-made capital" [42].

Verstegen and Hanekamp [43] by examining the different definitions of sustainable development indicated that all this controversy is defined by two different worldviews of Western society. According to the first theory, idealism expresses clear opposition to the prevailing politics. Under this worldview, economic growth cannot be continued because in less than 100 years the natural resources will be exhausted and the system will collapse. Even if the resources are not exhausted, the collapse will occur either from excessive pollution or overpopulation. Growth cannot be continued. Therefore, either it will be limited voluntarily or we are led to a system crash. This approach is characterized as pessimistic. Nevertheless, pessimism must be treated as a political tool in the search for "good society." Unlike idealism, the conformist perspective is the most optimistic. It does not introduce a separation in relation to the past. It is the perspective of political and economic cohesion. The optimism in this approach is based on technological progress which can broaden the natural limits to the point that there are not any. Therefore, the population will continue to grow over the next 200 years with decreasing rate of increase but with increasing per capita income. The poorest people will greatly benefit by continuous economic growth through the development of new technologies. Both sides claim that the other strategy is impossible.

Eventually, sustainability "is a word that hides more than it reveals" [44]. Anyone can redefine the term and interpret it according to his purposes. Potter et al. [12] citing O'Riordan [45] argue that the concept of sustainable development can mean anything or everything you want. This may be unintentional or, worse, it can be utilized to "disguise" or to "green" socially or environmentally destructive activities. Kates et al. [46], despite this criticism, argue that each effort to determine sustainability is an important parameter in an ongoing dialogue on the concept of sustainability. "In fact, sustainable development owes a large part of its appeal, strength and creativity to the lack of clarity. These challenges of sustainable development are at least heterogeneous and complex as the diversity of human societies and natural ecosystems are, around the world" [46]. Hartman [47] reminds us that the ideas for sustainable development are inevitably controversial, since their supporters are of different values and interests and they wish to support different sets of ecological, environmental, and social relations. Cooperation among them is a challenge with insurmountable difficulties.

8.3 Economic Growth and Environmental Implications

8.3.1 Introductory Remarks

The air, water, and soil pollution; the desertification of large areas; and the greenhouse effect are some of the most serious environmental problems, nowadays. The need of investigating the causes but also finding solutions to these problems has been widely acknowledged.

The natural environment is a dynamic system, in which the following four functions interact: Firstly, it is a source of raw materials and resources. Secondly, it provides space for waste accumulation and storage. Thirdly, it constitutes an effective machine for the assimilative—regenerative processes with regard to chemically and biologically active wastes. Finally, it determines the health level and general quality of life for all organisms that live within it [48]. In this aspect, two key questions are raised nowadays, regarding the economic growth and environmental implications: Will continued economic growth bring ever greater harm to the Earth's environment? Or do increases in income and wealth sow the seeds for the amelioration of ecological problems?

The neoclassical economists interpreting the environmental problems either as cases of market imperfection and/or failure or as a consequence of market lack for natural resources. Therefore, they suggest a state interference in order policies, which will lead to the market refinement as well as to an effective and sustainable utilization of natural resources, to be established [49,50]. On the other hand, the majority of Marxists agree with the fact that pollution and natural resources depletion in modern capitalist societies are inextricably linked with the capitalist class process, ie, with the production and distribution of the surplus value [51,52].

8.3.2 Theories Based on Orthodox Economic Perspective

The neoclassical economic theory of environment and natural resources is based on (1) the notions of individual preferences and subjective assessment of the value of consumer goods and factors of production (ie, marginal utility, subjective cost), (2) the technology, and finally (3) the market mechanism, which allocates the limited resources to alternative uses in order for the individual choices to be satisfied. The concepts of external economies, as well as those of public and free goods, are central to the interpretation of the market mechanism failure to protect the environment and natural resources. Therefore, state regulation measures are proposed, in order to address market failures so as effective and sustainable use of natural resources and the environment to be achieved [53].

Specifically, according to the "orthodox economists" the ecological deterioration is presumed as a market failure. The market is unable to orient enterprises toward the proper use of environmental capital if the latter is not fully integrated into the market system through a rational price structure. The neoclassical environmental economists are inherently based on a three-step process. First, they analyze the environment in specific goods and services, which are separated from the biosphere in such a manner so that they are capable of transforming those into commodities. Subsequently, through the generation of supply and demand curves, an imputed value is attributed to goods and services, a fact that allows economists to determine the optimal level of environmental protection. Finally, market mechanisms and policies are designed either in order to change the values in existing markets or to create new markets [54–56]. In Box 8.1, there are some basic notes regarding the issue of determining the optimum level of environmental protection.

It is a fact that the neoclassical approach is based on the transformation of the environment into a set of goods. The explicit goal is to overcome market failures in the environmental field by creating alternate markets for environmental products. This particular approach considers that the environmental pollution occurs because the environment does not function according to the laws of economic supply and demand. The nature of this approach arises from an attempt to interpret the whole society and ecosystems in their entirety on the basis of the concept of market goods [56].

Box 8.1 Estimating the Optimal Level of Environmental Protection

Great attention is given to the demand curves generation, which are built based on the estimation of the willingness of consumers to pay. Given that there are no real markets for environmental goods willingness to pay is presumed in two ways.

Creative Pricing
According to this approach, consumer preferences are disclosed by the demand for goods and services which are directly related to an environmental good. For instance, consumer willingness to pay for a quiet neighborhood is estimated by comparing the prices among a house next to an airport and a similar residence which is located in a most tranquil area.

Contingent Valuation Method
In this case, hypothetical markets are built and the consumers are invited to indicate their preferences through a market survey. In these surveys, consumers are asked how much they would pay for a certain protection level of an environmental good.

Having determined—with the aforementioned approaches—the optimal level of environmental protection, the economists continue by resolving the issue of creating new markets. Typically, two approaches are used: (1) either taxes enforcement, which will increase the cost of causing environmental damage, or subsidization which will increase the benefit of environmental upgrades and (2) creation of new autonomous markets through state interventions (eg, tradable pollution permits).

A different approach which has been released since 1990, among the circles of environmental economists, is the notion that the natural tendency of the capitalist economy toward "dematerialization" is a key response to all the environmental problems. In particular, the increased energy efficiency and the development of the "new economy" in the developed capitalist economies decoupling economic growth from energy and materials use as well as from waste disposal, minimizing the environmental impacts of any further GDP growth. According to this view, in fact, no measures are necessary to reduce the environmental impact of growth. The continuous innovation as well as the market laws resolves the problem. Ideally political decisions should simply accelerate the trend toward "dematerialization" and ensure that the environment is integrated into an innovative economy, which is more knowledge oriented [57,58].

The aforementioned hypothesis is presented with terms of the environmental Kuznets curve (Fig. 8.1). The Environmental Kuznets Curve (EKC) is often used to describe the relationship between economic growth and environmental quality. It refers to the hypothesis of an inverted U-shaped relationship between economic output per capita and some measures of environmental quality. The shape of the curve can be explained as follows: As GDP per capita rises, so does environmental degradation. However, beyond a

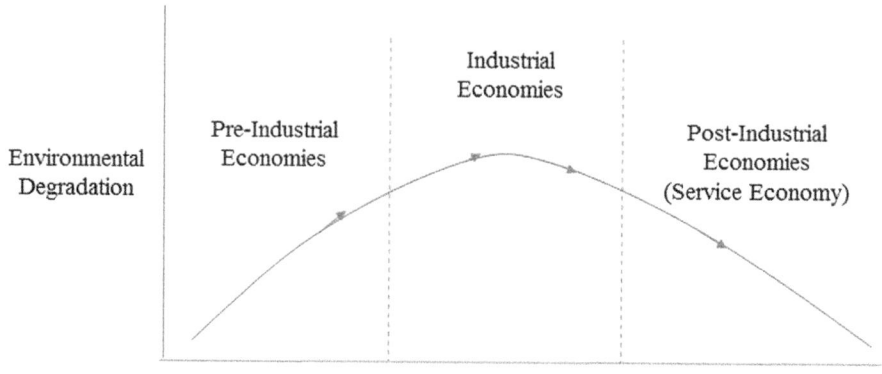

Figure 8.1
The environmental Kuznets curve [59].

certain point, increases in GDP per capita lead to reductions in environmental damage [59–61]. Specifically:

- at low incomes, pollution abatement is undesirable as individuals are better-off using their limited income to meet their basic consumption needs,
- once a certain level of income is achieved, individuals begin considering the trade-off between environmental quality and consumption, and environmental damage increases at a lower rate, and
- after a certain point, spending on abatement dominates as individuals prefer improvements in environmental quality over further consumption, and environmental quality begins to improve alongside economic growth.

8.3.3 Marxist Political Economy for the Dipole Economic Growth—Environmental Problems

The Eco Marxism—Eco Socialism constitutes an alternative critical view regarding nature's ownership under capitalism. This particular approach interprets the ecological degradation and environmental problems as issues inextricably linked to the economic, political, and ideological organization of capitalist societies.

The main feature of capitalism is that it is a self-expanding value system, in which the accumulation of economic surplus value should be realized in an ever-growing scale. Simultaneously, capitalism represents an expansion process which catalyzes all qualitative interactions converting them into quantitative, in terms of exchange value [62]. The general formula for capital, as explained by Marx, is the one where money is exchanged for a commodity (or the means of production of a commodity) which is then sold for money once again, this time at a profit. This procedure reflects the overarching objective of capitalism which is the expansion of monetary value. It is obvious from the foregoing, the interminable expansion which characterizes this system. As noted by the conservative economist Schumpeter "capitalism is a process, the stagnant capitalism would be *contradictio in adjecto*" [63].

Given that under the capitalist production procedure the natural resources (1) constitute the fixed capital assets and (2) sustain human life, ie, ensure the existence and reproduction of the direct producers, whose labor consists the source of value and surplus value under capitalism, become obvious the negative environmental impacts of the aforementioned interminable procedure of economic growth [53]. Nevertheless, many orthodox economists adopt the aspect according to which the growth of the man-made capital value compensates for natural capital losses. However, according to the concept of "strict version of sustainability" the man-made capital cannot always replace the natural capital due to the fact that there is the notion of "the critical natural capital," ie, the natural capital, which is crucial for the biodiversity reservation [63]. Furthermore, despite the obvious link between the capitalist economic development and environmental degradation, there are economists who argue that increasing scarcity of natural resources will result into the market reorientation to the rationale of conservation. However, according to the radical ecologist Rudolf Bahro: "The rising cost of land was unable to suspend the cementation of space" [64].

Finally, it is worth mentioning the customary solution—for environmental problems—promoted by the capitalist economies. According to capitalist approach, switching technology in environmental friendly direction (ie, energy efficiency, replacing fossil fuels, recycling) constitutes a solution to the environmental problems. Nowadays the prevailing view is that everything has to be done in order on the one hand energy efficiency technologies to be promoted and on the other hand ecological practices to be introduced into the production process. Thereby, the economic growth without further environmental degradation is presumed. However, according to William Jevons, a known British economist, pioneer of neoclassical economic analysis, increased efficiency in the use of a natural resource leads to increased demand for this resource rather than decrease [65].

From the abovementioned definitions, it is clear that the capitalist economy aims exclusively to the expansion of profits and to an interminable economic growth through the extreme exploitation of all production factors. It is granted that this process is accompanied by the absorption of energy and materials and the disposal of an increasing number of wastes into the environment, leading to further environmental degradation.

Another indirect correlation between the capitalist economic organization and the environmental degradation has to do with the fact that the capitalist countries are inherently imperialistic. Modern theories of imperialism generally shared the view that the slow economic growth, excessive inequality, and high levels of unemployment in developing countries result from the nonequal power relations between rich and poor countries [66]. Therefore, the economic problems of the least developed countries, particularly their indebtedness, make nature conservation to appear almost impossible. These countries, in order to acquire foreign currency so as to service their external debts, are trying to expand their exports and reduce imports. So these countries are intensifying the extraction of natural resources (eg, Amazon rainforests deforestation). Furthermore, poor countries, which are in a great need for funds, provide storage space for nuclear and toxic waste [67].

8.3.4 Three Case Studies

Japan: Minamata disease

Japan emerged from the World War II completely devastated. The rehabilitation and construction of the country was a vast unifying goal for the Japanese nation between 1950 and 1960. The rapid economic growth based on rapid industrialization resulted in widespread degradation of natural resources. Serious environmental problems became visible in the 1960s, when many diseases occurred because of the environmental degradation. Ui [68] provides a detailed chronicle of Minamata disease, which is related to intoxications from organic mercury wastes, generated by Chisso's company fertilizer factory. This disease spread to cities Kumamoto and Niigata and had, according to Japanese sources, 2239 casualties including 987 deaths.

Malaysia: the environmental mess of the Southeast Asia's "tiger" during 1970—80

Malaysia, as soon as political autonomy was granted in 1957, placed high priority on economic development. According to Sakarajasekaran [69], during 1971—73 the GDP growth rate registered 11% growth, compared to the planned objective of 6.8% growth, while the 1976—78 period showed a real GDP growth rate of 8.4%, compared to the average 8.2% real growth per annum anticipated for the 1976—80 period.[1] Nowadays, despite the Asian Financial Crisis in 1997, Malaysia's economy is one of the most competitive in Asia, ranking 6th in Asia and 20th in the world, higher than countries like Australia, France, and South Korea.

All these impressive achievements, as Hamzah Majid [70] since the year 1979 had mentioned, have been accompanied by detrimental effects on the natural environment. According to Goh Kim Chuan during the 21st year of independence, the government's priorities (ie, economic growth, social issues, etc.) left no space for environmental concerns. Moreover, it is a common phenomenon during national economies'

---

[1] According to: Treasury Economic Report (1978). Kuala Lumpur: Government Printers.

reconstruction period the natural resources to be considered vast and limitless and at the same time probable environmental consequences of development projects to be confronted as unnecessary distractions. As a result, considerable environmental deterioration in some localities and severe water pollution problems had occurred. Particularly indicative, for the tremendous environmental problems created due to Malaysian economic growth model through the years 1960–80, is the publication of Goh Kim Chuan [42].

Forest ecosystem damage

The effects of changing land use and logging activities have been diminishing forest acreage in Peninsular Malaysia. A survey of forest resources in the peninsula, in 1966, showed that about 9.1 million hectares out of the total land area of 13.2 million hectares were forested (primary forest cover), thus accounting for some 69% of the land.[2] In 1978 the percentage had decreased to 54.6% [71]. As far as the effects on rain forest animals are concerned, there had been no full study during the same period. Nevertheless, Chivers [72] estimated that 10,000 gibbons die each year due to loss of habitat.

Water pollution

Due to the mountainous terrain, as well as the high annual rainfall in Malaysia, there is an extensive network of rivers. During the 1970s Malaysia was the world's leading producer of natural rubber and at the same time the industrialization of the country had begun. According to Singh et al. [73] these industrial activities had resulted in the daily discharge of some 97 million liters of effluents into the rivers and streams. This volume of discharge was equivalent to a pollution load of 200 tonnes of biochemical oxygen demand (BOD) per day, which was equivalent to the organic pollution load from a population of about 4.5 million people. Another important source of water pollution was the effluents of oil palm factories. It had been estimated that for every tonne of effluent discharged from an oil palm factory, the BOD output is equivalent to that discharged from a population of 500 people. The total discharge from all oil palm mills in 1974 was estimated as a population equivalent of 8 million, and in 1978 it was estimated that the figure had doubled to almost 16 million population equivalent [74].

Air pollution

According to Goh [75], over half a million tonnes of air pollutants were released into the air over Peninsular Malaysia as a result of fuel combustion and burning of wood wastes in 1975.

---

[2] For further information, see: P.C. Lee, (1973). Multi-use management of West Malaysia's forest resources, In: *Biological resources and national development*, E. Soepadmo and K.C. Singh (eds). Kuala Lumpur: Malayan Nature Society.

Among these, a share of 45% was derived from transportation, 28% from furnaces and power generators, and about 28% from the burning of wood and wood waste.

South Korea: the miracle of rapid economic growth

The air pollution level in Seoul is among the highest in the world. A study in the 1980s concluded that 67% of rainfall was highly acidic for its residents. Sulfur dioxide emissions in Seoul were found to be five times higher than those of Taiwan and eight times than that from Tokyo. In 1989, the government found out that the water into 10 factory purification plants contained heavy metals such as cadmium and manganese at a level twice the officially permitted limit. The use of herbicides has increased 26 times between 1970 and 1985. According to studies conducted in the mid-1970s, the use of fertilizers in Korean agriculture was six times higher than that in the United States and 13 times higher than the average global. Finally, Korea has one of the highest levels of diseases related to professional employment. In particular, 2.66 out of 11 people suffer from diseases related to their professional employment compared to 0.70 in Taiwan, 0.93 in Singapore, and 0.61 in Japan [76].

8.4 Quantifying the Interactions Between Development and Environment

As it has already been stated, nowadays, air pollution, degradation of soil and water quality, desertification, biodiversity loss, global warming, etc. have become major environmental problems. Irrespective of the approach, if one aims at examining the causes and interdependencies between environmental degradation and the development process, as they were presented above, it is imperative to select appropriate indicators, which will illustrate both the progression of degradation and the improvement of environmental resources.

Given the aims of this section, among the great variety of indicators, the indicators used for monitoring the Millennium Development Goals (MDGs) concerning the environmental sustainability will be delineated in this chapter. These indicators cover a wide area of environment—development implications and comprise a basis for understanding their quantitative dimension. In chapter "Urban Environment," a more detailed analysis of the content and progress of the MDGs is presented. It should be also noted that the procedure of selecting an indicator is not an unambiguous process. It depends on data availability, as well as on the particular conditions of the problem and/or the thematic area, which this indicator is intended to describe. Besides, according to Mavraki et al. [77], there is no acceptable set of environmental indicators, being able to be applied at every different analysis level. This fact has led to the development of many different indicators, according to the case under study.

The indicators used to reflect the progress of the MDGs in the field of environmental sustainability—and in our view are the appropriate ones to quantify the interactions between the development models and the environment—are:

- *Proportion of land area covered by forest:* Land use change, principally deforestation, is responsible for the release of large amounts of carbon into the atmosphere. Deforestation in 2005 was accounted for an estimated 17% of global greenhouse gases (GHG) emissions, more than the entire transport sector [78]. Most of this is generated in developing countries. In recent years deforestation in Brazil and Indonesia has produced over half of all GHG emissions associated with land use change. Forest cover continues to decrease on a global scale. Between 1990 and 2005 the global surface of forests was reduced by 1.3 million km^2. Latin America and the Caribbean lost 7% of their forest during this 15-year period and Sub-Saharan Africa lost 9% [79].

- *CO_2 emissions, total, per capita, and per GDP (PPP):* GHG emissions have continued to increase since 1990. Deforestation and fossil fuel consumption primarily produce CO_2, while agriculture and waste are the main source of methane and nitrous oxide emissions. For the very poorest countries most GHG emissions come from agriculture and land use changes. When emissions from land use change are included, the top 10 emitters account for the two-thirds of CO_2 emissions—including China, India, and Brazil. In general the amount of GHG emitted per capita is far higher in the developed than in the developing countries [80].

- *Proportion of fish stocks within safe biological limits:* Developing countries are highly dependent on marine and freshwater fisheries. Fish provide 2.6 billion people with over 20% of their protein intake. Two-thirds of world fisheries production comes from fish capture. Together China, Peru, Chile, Indonesia, and India accounted for 45% of inland and marine fish catches in 2008 [79]. The share of overexploited fish populations has increased the last 40 years from 10% to 25% [79] and the most commercially successful species are fully exploited.

- *Proportion of total water resources used:* UN Water estimated in 2007 that by 2025 two-thirds of the world's population could be under conditions of water stress, defined as 1.700 m^3/person/year, the threshold for meeting the water requirements for agriculture, industry, domestic purpose, and energy. The International Water Management Institute also recently assessed global environmental water needs. It went beyond traditional calculations which compare water withdrawals to mean annual runoff, measuring the water needs at a river basin level and finding the amount of water needed to maintain ecosystem functionality [81].

- *Proportion of terrestrial and marine areas protected:* Only 12% of the planet is under some form of protection: about 18 million square kilometers of protected land and over 3 million square kilometers of protected territorial waters. Protected areas are also often poorly managed and suffer from pollution, irresponsible tourism, etc. [82].

- *Proportion of species threatened with extinction:* Measuring the diversity of animals, plants, and other organisms is inherently very difficult. Some progress has been made by the World Wildlife Fund (WWF) which summarizes changes in populations of vertebrate species in its Living Planet Index (LPI). This index tracks over 3600 populations of 1313 vertebrate species [83]. The LPI indicates a downward trend since 1970 with no signs of recovery [79].

It is worth mentioning that as a part of an evaluation of progress toward environmental MDGs, the World Bank has assembled and compared measures of the quality of national environmental policy and institutions. One of these, the Environmental Performance Index (EPI) takes broadly accepted targets for a set of 25 environmental indicators regarding: air pollution, water resources, biodiversity, productive natural resources, and climate change and ranks countries on the basis of their performance relative to these. Using the scores from 149 countries the 2008 EPI revealed that lower-income countries generally lag behind higher-income countries [84].

8.5 Development and Poverty Alleviation

8.5.1 Background and Questions

Welfare and poverty

Economic growth and development are supposed to beget welfare and well-being. One way to define the latter is through the level of command or access to resources or commodities [85]. Poverty, on the other hand, is considered "as a condition involving critical shortages of those elements" [85: 16]. In this sense, poverty statistics are indicative of the status that has emerged by the implementation of the selected developmental strategies and practices. What is more, not only income-based indicators, but also other nonincome dimensions of well-being/poverty are harnessed and are recognized as vital [86]. As McGillivray [87: 1] points out, it appears that there is a convergence toward the fact "that well-being and poverty are multidimensional and, in particular, that no single uni-dimensional measure adequately captures the full gamut of well-being achievement." This seems to be particularly true since the poor, apart from having little money and face shortage of quantifiable resources, enter into a more general situation of vulnerability such as deprivation in nutrition, limited access to health and education services, sense of impotence, etc. [85].

There is a series of multidimensional conceptualizations and indicators regarding well-being and poverty. The best known well-being indicator having been developed by the United Nations Development Programme (UNDP) is that of Human Development Index (HDI). The latter is "a composite measure that includes indicators along three dimensions: life expectancy, educational attainment, and command over the resources needed for a decent living" [88: 23]. But UNDP also publishes statistical data about poverty, attempting to approach it in a multifarious manner. The Multidimensional Poverty Index (MPI) constitutes such an indicator.

Poverty and hunger

Hunger is shaped by several factors, but poverty is deemed to be one of its most important determinants [89]. A large share of the income of poor households is spent on food—even for those engaged in farming; in extreme cases, though, very poor families are not even capable of buying and consuming "enough food to meet dietary requirements," a fact that "can have long-lasting impacts on labor productivity" and eventually lead to the hampering of development prospects [90: 74]. Moreover, hunger, like poverty, with which is closely related, is multidimensional [89]. More specifically, hunger—in the shape of both undernourishment and malnourishment—is approached by the concept of food insecurity.

So, in an attempt to overcome the drawbacks of hunger, food security is measured across its four dimensions, namely: availability, access, stability, and utilization [91]. As it is shown in practice, all of these dimensions matter and need to be considered in order to measure, estimate, and enhance food security. This is in particular true in cases where food insecurity in portrayed as solely an (economic/physical) access problem: yet, the relationship between food insecurity and access is not as straightforward as it is often assumed to be, given that other "exogenous" economic conditions and political or natural hazard events occasionally severely deteriorate the state of food security [90].

In any case, no matter which are the determinants of food insecurity, the latter "usually takes a huge toll on labour productivity, and thus perpetuates a vicious circle where food insecurity causes low labour productivity, low incomes and thus further food insecurity" [90: 68]. This vicious circle refers to what is called the "hunger trap" or the "agriculture—hunger—poverty nexus." It is essential that one understands that poverty is not only a determinant of hunger, but it also results from hunger. And the deeper insight of how these two injustices interconnect is a prerequisite in order to eradicate both [92]. As von Braun et al. [92] explicitly state:

> *Hunger, and the malnourishment that accompanies it, prevents poor people from escaping poverty because it diminishes their ability to learn, work, and care for themselves and their family members. If left unaddressed, hunger sets in motion an array of outcomes that perpetuates malnutrition, reduces the ability of adults to work and to give birth to healthy children, and erodes children's ability to learn and lead productive, healthy, and happy lives. This truncation of human development undermines a country's potential for economic development—for generations to come.*

Questions

In this respect, both poverty and hunger—being tightly intertwined—are shaping the welfare of a region or country, and thus, the level of development. So, what is the state of the world at the present time, in terms of welfare and development? In what ways have the given developmental strategies affected this worldwide state of affairs, or to what extent have some of the MDGs and World Food Summit (WFS) target been met? The answers can be partially derived by appealing to the pertinent UNDP statistics and their overall historical trends and geographic differentiations. Yet, a more comprehensive and illuminating image is to emerge by putting these figures and findings in the wider context and assessing them in a critical perspective.

8.5.2 Statistics and Analysis

Poverty

According to the most recent UNDP Reports, although there is a generic trend of progress in human development—in terms of HDI— there is also a great disparity across HDI group countries. More specifically, in 2012, the global average value of HDI was 0.694. The lowest in the rank was Sub-Saharan Africa with an HDI value of 0.475, followed by South Asia (0.558), while the group of countries of very high human development exhibited a value of 0.905 [88]. In 2013, the pertinent HDI values were 0.702 (globally), 0.502, 0.588, and 0.890, respectively [93]. Now, by examining the HDI trends for the periods of 1990–2000, 2000–08, and 2008–13 per region, it is shown that "[w]hile all regions are registering improvement, signs of a slowdown are emerging" [93: 33]. The same slowdown in growth is monitored in all four (low, medium, high, very high) human development groups [93]. These temporal changes in the slowdown are portrayed in Figs. 8.2 and 8.3.

Aside from this decreased growth, which is displayed by the utilization of the HDI, there are some significant findings pertaining to poverty indices. It is worth noticing that MPI has substituted HPI (Human Poverty Index), avoiding the latter's aggregate character and providing the more specific overlapping deprivations (health, education, living conditions) faced by households and individuals per geographic region, ethnic groups, etc. [94]. Now, by appealing to the MPI, it is revealed that the number of people who live in multidimensional poverty are much more than those who live on less than $1.25 a day (ie, than those below the extreme income poverty); more precisely, the former group is estimated to count about 1.56 billion people, whereas the latter comprises 1.14 billion people [88,95]. So, these MPI statistics do show in an empirical means that income alone cannot warrant a reliable estimator of poverty—something that the theoretical research has indicated as well. Another interesting finding is that the HDI group (low, medium, high) has to do something with this disparity: "The lower the HDI value, the larger the gap

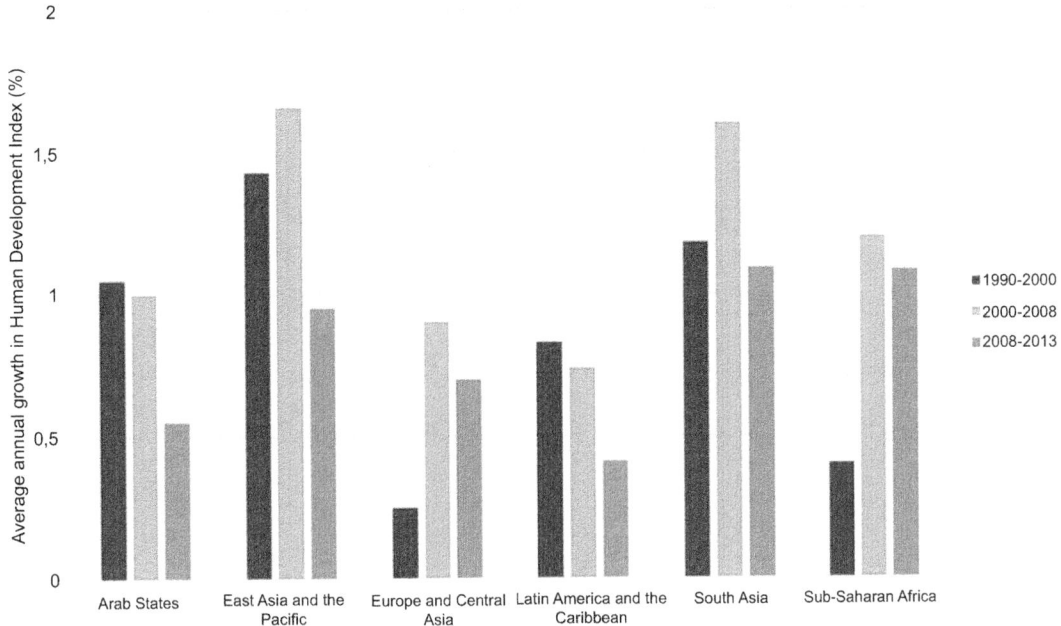

Figure 8.2

Trends of the Human Development Index from 1990 to 2013 per geographic region [93].

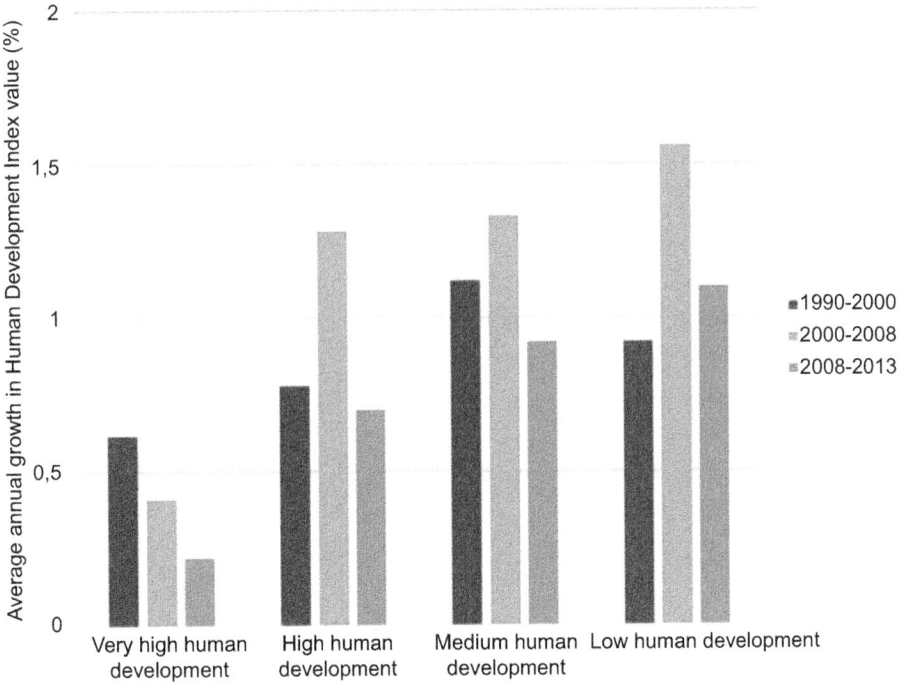

Figure 8.3

Trends of the Human Development Index from 1990 to 2013 for low, medium, high, and very high HDI groups of countries [93].

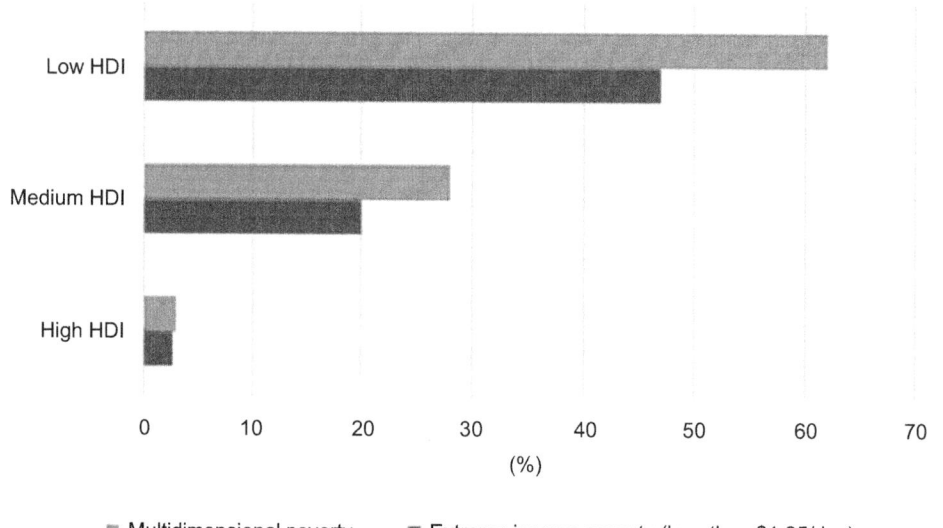

Multidimensional poverty   Extreme income poverty (less than $1.25/day)

Figure 8.4

Trends of the Human Development Index (HDI) from 1990 to 2013 for low, medium, high, and very high HDI groups of countries [88].

between income poverty and multi-dimensional poverty" (data refer to 2002–11) [88: 29] (Fig. 8.4). This arises to be consistent with the fact that the countries with higher HDI—and not solely higher Gross National Income (GNI) per capita—have the socioeconomic structures, are "equipped" with the appropriate institutions, and provide the social services that can mitigate/"alleviate" the multidimensional poverty. As a consequence—aside from the apparent fact that both types of poverty are dramatically lower in high-HDI countries than those of the low-HDI countries—the multidimensional poverty of high-HDI countries is slightly higher than their poverty income in absolute figures since the income does not play such a decisive role in countries with developed social services. Although this explanation seems rather plausible, one should consider that both types of poverty apply to a percentage of less than 5% of the total population of the high-HDI countries, so the consideration of this gap in relative terms may have some importance as well.

Hunger

The hunger statistics show that about 870 million people were undernourished around the world in 2010–12; 852 million of these people were residing in developing countries, comprising the 15% of the total population of these countries [90]. In 2011–13, the respective population numbers were 842 and 827 million people [96]. From 1990 to 1992, the proportion of chronically hunger people has declined from 23.5% to 14.3% in 2011–13 and to 13.5% in 2012–14 for the developing countries [91,96].

Despite this conspicuous reduction of the number of undernourished people during the past two and a half decades, there are several aspects of this decline that, if examined properly, do not offer so promising an image for the development of the posterity and all of the regions/countries worldwide. To elucidate, the progress in the undernourishment has been slowed down during the past decade in comparison to that had taken place in the 1990s [96]. Furthermore, this progress in the percentage of undernourished people is not that impressive in absolute numbers [91,96]. Besides, the geographical distribution of the current status of hunger is not that promising as well. According to FAO's 2014 Hunger Map [97], large regional disparities do not seem to be bridged: The chronically hungry people reside almost exclusively in developing countries; about one-quarter of the Sub-Saharan people are undernourished at the present time, while in Southern Asia the chronically hungry people are more than half a billion; contrariwise, only Latin America and South-Eastern Asia are those subregions that display some very significant amelioration in the status of undernourishment.

8.5.3 Millennium Development and World Food Summit Goals and Targets

Therefore, at this point we return to some of the initial questions: where are we with respect to United Nations standards and goals?

On the one hand, UN MDG 1 is about eradicating extreme poverty and hunger and tow sub-goals (targets) refer directly to the subject of this section:

- Target 1-a. "Halve, between 1990 and 2015, the proportion of people whose income is less than $1 a day" and
- Target 1-c. "Halve, between 1990 and 2015, the proportion of people who suffer from hunger."

On the other, the goal or target set by the 1996 World Food Summit (WFS) was "… to eradicate hunger in all countries, with an immediate view to reducing the number of undernourished people to half their present level no later than 2015."

Regarding the 1-a target, according to the UNDP [96], "[t]he extreme poverty rate has been halved, but major challenges remain." More specifically, the accomplishment of this goal is mainly attributed to the success of some of the most populous countries such as China, Brazil, and India to significantly reduce the number of their extremely poor people (China's, Brazil's, and India's poor people percentage declined from 60.2% to 13.1%, from 17.2% to 6.1%, and from 49.4% to 32.7%, respectively). In addition, in 2010, the vast majority (nearly two-thirds) of the extremely poor people reside in five countries (India (33%), China (13%), Nigeria (9%), Bangladesh (5%), and the Democratic Republic of the Congo (5%)) [96].

As for the 1-c target, "halving the proportion of undernourished people by 2015 is within reach" [91], "but major efforts are needed to achieve the hunger target globally by 2015" [96: 12]. Given the sharp geographical discrepancies in the prevalence of undernourishment, these efforts should be intensified in Sub-Saharan Africa and Southern and Western Asia [97]. Yet, whereas this goal appears achievable globally, the World Food Summit target of halving the number of undernourished people by 2015 cannot be accomplished by 2015 [91].

A general comment about these goals is that poverty and possibly chronic hunger reduction are attainable in relative terms. Yet, when it comes to their mitigation in absolute terms, they remain stubbornly high—especially in the case of undernourishment [90].

8.5.4 Notes on Rural Poverty

Generally, an area is characterized as rural, when it is located outside urban centers. Stricter, technical definitions of rural areas vary. In order to facilitate development policies, the European Union defines rural areas according to population density. According to this definition, when the population density of an area is below 300 inhabitants per square kilometers, this area is considered as rural [98]. In the United States, the Department of Agriculture Economic Research Services is based for its studies on counties, which are considered as rural when they include a combination of open countryside, rural towns with fewer than 2500 inhabitants and urban areas with less than 50,000 people. According to the World Bank (data based on national statistical offices and UN estimations for the year 2013[3]), at global level 3.34 billion people live in rural areas. This corresponds to a 47% share of the planet's population. The share has fallen by 5% since 2003 and by 9% since 1993, showing a continuous trend to urbanization. Rural areas of the world do not have homogenous characteristics.

A special case among rural areas is that of mountainous areas. Mountains cover about 25% of the earth's surface, and 12% of the human population are there. Mountains are considered particularly important for the planet [99]:

- They are the "water towers" of the world—about 50% of the global population depends on them for freshwater.
- They include fragile ecosystems rich in biodiversity due to remoteness and variety of climatic conditions.
- They are reserves of energy and mineral resources.
- They are islands of cultural diversity.

---

[3] Data available at the online database of the World Bank: http://data.worldbank.org/indicator/SP.RUR.TOTL/ countries.

Table 8.2: Percentage of Rural People Living in Extreme Poverty in Various Regions of the World [100].

Region	Percentage of Rural Population in Extreme Poverty Conditions (%)
East Asia	12
South Asia	50
Southeast Asia	8
Sub-Saharan Africa	30
North Africa and Middle East	<1
Latin America	1

Despite their importance and magnificence, mountain areas, in particular, and rural areas, in general, encounter severe poverty problems. Rural populations are considered the world's poorest and most disadvantaged. According to the International Fund for Agricultural Development (IFAD) [100], the largest segment of the world's poor are the 1-billion people living in rural areas, who are not able to cover their basic needs. In total, about 1.4 billion people around the world live in extreme poverty condition. So, more than the two-thirds of the world's population living in extreme poverty are inhabitants of rural areas. In Table 8.2, data related to rural population living in extreme poverty conditions are summarized. South Asia (mainly India) and Sub-Saharan Africa are the regions most exposed to extreme poverty. Generally, extreme poverty is defined as the situation, where a person lives with less than 1.25$ per day.

In 1990, more than half of the rural population (54%) in developing countries lived with less than 1.25$ a day and so, considered as extreme poor. By 2010 this share had dropped to 35%. This is, undoubtedly, an important step toward poverty eradication. High rates of economic growth in East Asia, and particularly in China, account for much of the decline in the rates of extreme poverty. However, the majority of people of rural areas, who were left-off living in extreme poverty, do not enjoy high living standards. Low income, food insecurity, insufficient infrastructure, analphabetism, and other negative factors still remain a threat for the rural areas of developing countries. The following characteristic statement of an African woman gives in a disarming manner the dimensions of poverty.

[Poverty] means the person is stuck. You cannot go anywhere or do anything to get out of the situation. You are not in a mood to rejoice, you can get rough with your children. You fear the future.

Abibatou Goudiaby, female, 21 years, Senegal[4]

---

[4] The words of Abibatou Goudiaby have been retrieved by http://www.un.org/en/globalissues/briefingpapers/ruralpov/quotes.shtml.

The problems faced by rural populations can be intensified under the influence of phenomena such as desertification and land degradation. It is estimated that about 1 billion people in the world are affected by land degradation, which mainly arises when fragile land is overexploited, in order to cover the demands of an expanding population. Excessive human activity leads also to desertification, which is additionally enhanced by climate change. Not only the growing need for food, but also the need for fuel lead to land degradation and desertification. People without access to energy services are obliged to consume biomass, in order to cook and warm their dwellings. This is an important factor of pressure to forest ecosystems. The major global challenge of climate change has also great effects on rural populations. In the case of people in semiarid and arid regions, as well as in mountain areas, agriculture takes place under marginal conditions. Hence, small changes in temperature and precipitation have great, direct impact on it. Taking into account that agriculture is the main means of survival for rural people, such changes could have dramatic effects.

The strong dependency of rural populations on the environmental conditions for their survival shows that there is a definite necessity for sustainable approaches to development in these areas. People facing extreme poverty conditions and suffering from chronic hunger are forced to struggle for their survival. This leads to adopting practices which ignore the environmental concerns, such as the already mentioned forest degradation for obtaining fuel, "slash-and-burn" agricultural techniques, overexploitation of water resources, etc. The environmental impact of poverty is partly due to the lack of economic resources and the inaccessibility to technological means and know-how. Therefore, apart from direct humanitarian aid, there is a dire necessity for integrated programs that will provide rural, poor communities with the appropriate means and knowledge, in order to escape from this vicious circle and become able to cover their essential needs without further environmental degradation. The environmental problems of the developing countries may be attributed—to some extent—to inequalities sustained by the dominant global economic policies. Hence, there is a moral obligation from the part of the "developed North" toward the "developing South" in the direction of diffusing resources, good practices, and policies.

As far as mountainous areas are concerned, according to the Food and Agriculture Organization (FAO), over 35% of their population faces food insecurity. Moreover, mountain populations are exposed to natural hazards (especially landslides and earthquakes); they face marginalization and do not have access to basic services. High-altitude areas have a series of specific characteristics (cold climate, limited space for agriculture, great distance from major urban and commercial centers, etc.), which require specialized approaches—mountain-specific strategies—for mountain development. Besides, since 1992 through a special chapter in Agenda 21 the need for sustainable mountain development was highlighted.

8.5.5 Lack of Access to Energy—Energy Poverty

Taking into account the great importance of energy services for modern societies, based on technology, in this subsection, some essential facts related to problems in energy supply and energy poverty are mentioned.

Living standards are closely related to energy sufficiency. There are approaches which connect social development itself with energy consumption. For instance, it is claimed that civilization is the figment of progress in machine manufacturing and intensive energy consumption [101]. According to the United Nations Development Programme, in order to ensure a satisfactory level of development, it is necessary not only to ensure sufficient quantities of energy, but also good quality of energy. In other words, households need to be at the higher steps of the "energy ladder" and be able to use electricity and/or fine liquid or gaseous fuels [102]. According to Reddy [103] the impact of energy on human development is related to the end-uses of energy and the tasks, which energy puts through. This explains the high importance that is attributed to access to electricity, since it has a multitude of end-uses: lighting, cooking, pumping, heating, cooling, etc. Reddy [103] also shows that in developing countries, small investments in the energy sector could produce dramatic improvements in the Human Development Index (HDI). With an increase of 100 W/person in electrical energy supply (a small fraction of the average electricity use in the industrialized countries), a radical improvement in quality of life could be achieved [104,105]. Energy consumption is not only related to quality of life improvement and poverty alleviation, but also with economic development. It is recognized that access to energy and, particularly, electrical energy is crucial for achieving the MDGs [106].

Despite the wide recognition of the importance of energy for development, globally, over 1.3 billion people live without access to electricity and 2.8 billion people without clean cooking facilities [107]. More than 95% of these people are either in sub-Saharan Africa or developing Asia and 84% are in rural areas, according to the International Energy Agency (IEA) [108]. At global level, the electrification rate is 80.5%, while in the world's rural areas this percentage is only 68%. More detailed data about access to electricity are given in Table 8.3. Between 1990 and 2010 the percentage of the world's population without access to electricity and clean cooking facilities has fallen by about 7%. However, in a period of 20 years this improvement is not particularly satisfying. Hence, the problem of lack of access to energy remains worrying and efforts for overcoming it should be intensified.

Even among developed countries, problems in energy supply are increasing. It is estimated that between 50 million and 125 million people in Europe do not have the possibility for adequate coverage of their energy needs [109]. Inadequate access to energy services is generally described either as energy or fuel poverty. Energy poverty is usually used for

Table 8.3: Access to Electricity in 2009 [108].

	Population Without Electricity (Millions)	Electrification Rate (%)	Urban Electrification Rate (%)	Rural Electrification Rate (%)
Africa	587	41.8	68.8	25.0
North Africa	2	99.0	99.6	98.4
Sub-Saharan Africa	585	30.5	59.9	14.2
Developing Asia	675	81.0	94.0	73.2
China and East Asia	182	90.8	96.4	86.4
South Asia	493	68.5	89.5	59.9
Middle East	21	89.0	98.5	71.8
Latin America	31	93.2	98.8	73.6
Developing countries	1314	74.7	90.6	63.2
World	1317	80.5	93.7	68.0

delineating problems related to dealing with energy services in the home, whereas fuel poverty is more often used for referring to the insufficient coverage of heating needs [110]. The concept of fuel poverty has not been defined yet with sufficient clarity. The United Kingdom is a country, which has adopted an official definition, legitimized in 1998. Under this definition, a household is considered to be fuel poor, when it has to spend more than 10% of its income on all domestic energy use, including appliance and heating. Apart from the United Kingdom, France, Slovakia, and Ireland have attempted to define the issue of energy poverty. The European Union through the European Economic and Social Committee (EESC) states the necessity of forming a common, general definition for energy poverty and establishing common indicators for it. This is the first step toward establishing integrated strategies for alleviating this important social problem. It is proposed that the definition given by the EESC in 2011 ("the difficulty or inability to ensure adequate heating in the dwelling and to have access to other essential energy services at a reasonable price") could be a basis for developing a common general definition of the problem.

Energy poverty—understood as inadequate coverage of energy needs—is, obviously, different than the total absence of energy services and its impacts are less severe. However, it still demands special attention. As shown in Table 8.4 [111], people at the risk of poverty in Europe are highly exposed to fuel poverty, too. In Table 8.4 three indices related to fuel poverty are used, namely arrears on utility bills, inability to keep the residence warm, and problems in the house attributed to inadequate heating. Energy supply problems seem to be acute in Central, Eastern, and Mediterranean EU countries, especially in Bulgaria, Hungary, Greece, and Cyprus, while in northern European countries the situation is far better. In Europe, as a whole, about 1 out of 10 people are considered to be energy poor and one-fourth of low-income people are energy poor. These facts,

Table 8.4: Percentage of People at the Risk of Poverty Affected by Fuel Poverty in European Countries [111].

Country	Arrears on Utility Bills (%)	Inability to Keep Home Adequately Warm (%)	Dwellings With Leakages and Damp Walls (%)	Country	Arrears on Utility Bills (%)	Inability to Keep Home Adequately Warm (%)	Dwellings With Leakages and Damp Walls (%)
Bulgaria	50.7	70.0	29.5	Estonia	20.0	9.6	30.3
Hungary	58.8	33.9	53.0	Belgium	14.0	18.8	26.2
Greece	54.4	47.6	21.0	Ireland	27.5	12.5	16.2
Latvia	39.5	35.1	43.3	France	17.8	15.2	22.1
Cyprus	25.9	50.6	34.6	Czech Republic	19.4	15.3	20.0
Slovenia	37.5	17.3	46.1	Spain	17.9	18.2	17.9
Italy	24.5	44.1	30.1	Slovakia	18.3	13.6	19.7
Romania	41.5	25.4	30.0	Netherlands	8.6	8.7	27.4
Lithuania	22.8	38.2	28.6	Germany	8.6	14.8	21.0
Portugal	14.5	43.0	28.4	Denmark	5.5	7.1	25.3
Croatia	40.9	21.8	19.9	Luxembourg	6.6	2.2	28.9
Poland	30.1	27.6	20.0	Austria	11.3	7.7	15.2
Malta	19.4	32.1	12.4	Finland	13.7	3.8	8.6
United Kingdom	20.3	19.4	21.4	Sweden	10.3	3.5	11.0

taking into account, the generally high living standards in Europe, show that there should be no complacency about the problem.

Insufficient energy supply poses serious threats for human health; inadequate heating increases winter mortality; and inability to use safe fuel (either for heating or for cooking) downgrades indoor air quality. Between 30% and 50% of excess winter mortality is attributed specifically to housing conditions and inadequate heating [112]. A special index (EWDI—Excess Winter Deaths Index) is used for monitoring winter mortality. This index indicates if the expected deaths in the winter are higher than in the rest of the year and is considered to be closely related to energy efficiency of houses and sufficient heating [111]. In Malta, Spain, and Portugal, in the period 2011−12, EWDI was over 30%. Even in Germany, EWDI exceeded 10% in the same time period. The World Health Organization considers indoor air pollution as one of the top 10 health risks, at a global level [113]. Furthermore, another aggravating factor for human health is time and effort to procure fuel, especially in cases of high dependency on firewood, like in sub-Saharan Africa. Mostly women, in rural sub-Saharan regions, have to carry about 20 kg of wood for an average distance of 5 km every day [114]. The abovementioned facts constitute additional causes for international mobilization toward alleviating problems in energy supply.

8.6 Development Policies and Useful Tools for development Planning
8.6.1 A Synopsis of Major International Policies

The fight against poverty is, probably, the most important target of international development policy, as it has already been noticed. Poverty alleviation is an essential step in the direction of defending the humanity's dignity and it is the prerequisite for creating development perspectives for the "damned of the earth," which now number about a billion, when speaking about extreme poverty. The United Nations through the MDGs gives the tone of international policies/actions against poverty, hunger, disease, and child mortality. In 2002 the Millennium Project was commissioned by the UN Secretary-General, aiming at the development of a concrete action plan for the world to achieve the eight MDGs, which are related to confronting poverty, hunger, and disease. The MDGs were set out in 2000, by the historical Millennium Summit of the UN. In this summit, leaders of a large number of nations adopted the UN Millennium Declaration that committed the nations to reduce poverty and set out certain targets, with a deadline of 2015. The MDGs and their basic context, in the form of 18 targets, are summarized below:

- Goal 1: Eradicate Extreme Hunger and Poverty
  - Target 1: Halve, between 1990 and 2015, the proportion of people whose income is less than $1 a day
  - Target 2: Halve, between 1990 and 2015, the proportion of people who suffer from hunger

- Goal 2: Achieve Universal Primary Education
  - Target 3: Ensure that, by 2015, children everywhere, boys and girls alike, will be able to complete a full course of primary schooling
- Goal 3: Promote Gender Equality and Empower Women
  - Target 4: Eliminate gender disparity in primary and secondary education, preferably by 2005, and in all levels of education no later than 2015
- Goal 4: Reduce Child Mortality
  - Target 5: Reduce by two-thirds, between 1990 and 2015, the under-five mortality rate
- Goal 5: Improve Maternal Health
  - Target 6: Reduce by three-quarters, between 1990 and 2015, the maternal mortality ratio
- Goal 6: Combat HIV/AIDS, Malaria, and Other Diseases
  - Target 7: Have halted by 2015 and begun to reverse the spread of HIV/AIDS
  - Target 8: Have halted by 2015 and begun to reverse the incidence of malaria and other major diseases
- Goal 7: Ensure Environmental Sustainability
  - Target 9: Integrate the principles of sustainable development into country policies and programs and reverse the loss of environmental resources
  - Target 10: Halve, by 2015, the proportion of people without sustainable access to safe drinking water and basic sanitation
  - Target 11: Have achieved by 2020 a significant improvement in the lives of at least 100 million slum dwellers
- Goal 8: Develop a Global Partnership for Development
  - Target 12: Develop further an open, rule-based, predictable, nondiscriminatory trading and financial system (includes a commitment to good governance, development, and poverty reduction both nationally and internationally)
  - Target 13: Address the special needs of the least developed countries (includes tariff- and quota-free access for least developed countries exports, enhanced program of debt relief for heavily indebted poor countries (HIPCs) and cancellation of official bilateral debt, and more generous official development assistance for countries committed to poverty reduction)
  - Target 14: Address the special needs of landlocked developing countries and small island developing states (through the Program of Action for the Sustainable Development of Small Island Developing States and 22nd General Assembly provisions)
  - Target 15: Deal comprehensively with the debt problems of developing countries through national and international measures in order to make debt sustainable in the long term
  - Target 16: In cooperation with developing countries, develop and implement strategies for decent and productive work for youth

- Target 17: In cooperation with pharmaceutical companies, provide access to affordable essential drugs in developing countries
- Target 18: In cooperation with the private sector, make available the benefits of new technologies, especially information and communications technologies

In 2008 a high-level event took place at the UN Headquarters. Governments, foundations, businesses, and civil society groups refreshed the efforts toward the achievement of the MDGs and announced new commitments in order to meet the targets. An amount of about $16 billion was agreed to be provided, in order to reach the MDGs, according to the announcements of the UN. Then, in 2010, another UN Summit focused on the MDGs adopted a global action plan for maintaining the course toward achieving the targets, especially the antipoverty ones. The action plan included also new commitments for women's and children's health and other initiatives. For accelerating progress related to women's and children's health, the states and the organizations that took part in the Summit allocated over $40 billion for 5 years. Arriving mid-2015, the following general deduction could be put forward: At global level, significant progress in achieving many of the MDGs has been made. However, great disparities across and even within countries are observed. It should not be overlooked that people pulled through extreme poverty and hunger still do not enjoy satisfying living conditions and so, the potential for making more progress is great. In Table 8.5, it has been attempted to gather the progress toward achieving the MDGs by combining the relevant reports [96,115]. It has been tried to keep the targets in their initial form (despite some changes made since their initial adoption), in order to be able to follow the whole course.

As far as Goal 8, is concerned, only Target 18 has been included in the Table as it has easily measurable results. Some key points regarding the some other targets of Goal 8 are summarized below [96]:

- Target 12: Great percentages of developing countries' exports enter the developing world duty free.
- Target 13: Not satisfactory financial aid is provided by the developed countries to the developing world.
- Target 15: Significant progress has been made but there are still various countries with very highly indebted.

In 1999, the income of the richest 5% of the world's population was 114 times higher than the income of the poorest 5% [116]. Between 1998 and 2008 the richest 1% of the world's population saw their real income increased by 60%. During the same period the incomes of the poorest 5% remained the same [117]. These facts show that we live in a world of raising inequalities. While the wealth of a small proportion of people steadily increases, the efforts for improving the living conditions of billions of people in the developing world have not been successfully completed. Almost 30 years after the publication of the

Table 8.5: Progress Over the Millennium Development Goals (MDGs) [96,115].

MDGs	Targets	Overall Progress	Regional Progress						
			Northern Africa	Sub-Saharan Africa	East Asia	South Asia	Southeast Asia	Central Asia/ Caucasus	Latin America
Goal 1	Target 1 Halve extreme poverty	Achieved Still great numbers of extreme poor	Achieved Low poverty	Not achieved Very high poverty	Achieved Moderate poverty	Achieved Very high poverty	Achieved Moderate poverty	Achieved Low poverty	Achieved Low poverty
	Target 2 Halve the proportion of people suffering from hunger	Not achieved Slower rates of hunger reduction than in the 1990s	Achieved Low hunger	Not achieved High hunger	Achieved Moderate hunger	Not achieved High hunger	Achieved Moderate hunger	Achieved Moderate hunger	Achieved Moderate hunger
Goal 2	Target 3 Achieve access of all children to primary education	Not achieved Important progress, still 10% of children in developing countries have no access	Achieved High enrollment	Not achieved Moderate enrollment	Not achieved High enrollment	Achieved High enrollment	Not achieved High enrollment	Not achieved High enrollment	Not achieved High enrollment
Goal 3	Target 4 Eliminate gender disparity in primary and secondary education	Achieved Progress should be made in some regions	Achieved Close to parity	Not achieved Close to parity	Achieved Parity	Achieved Parity	Achieved Parity	Achieved Parity	Achieved Parity
Goal 4	Target 5 Reduce by two-thirds the under-five mortality rate	Not achieved The rate has been halved. Accelerating pace in reducing child mortality	Achieved Low mortality	Not achieved High mortality	Achieved Low mortality	Not achieved Moderate Mortality	Achieved Low mortality	Achieved Low mortality	Achieved Low mortality

Continued

Table 8.5: Progress Over the Millennium Development Goals (MDGs) [96,115].—cont'd

MDGs	Targets	Overall Progress	Regional Progress						
			Northern Africa	Sub-Saharan Africa	East Asia	South Asia	Southeast Asia	Central Asia/ Caucasus	Latin America
Goal 5	Target 6 Reduce by three-quarters the maternal mortality ratio	Not achieved The ratio has fallen by 45%. In developing region 14 times higher maternal mortality than in developed	Not achieved Low mortality	Not achieved Very high mortality	Achieved Low mortality	Not achieved Moderate mortality	Not achieved Moderate mortality	Achieved Low mortality	Not achieved Low mortality
Goal 6	Target 7 Halt and reverse the spread of HIV/AIDS	Partially achieved The number of new infections has declined. Still great numbers of infections and alarmingly high levels of insufficient knowledge about HIV	Not achieved Low incidence	Achieved High incidence	Not achieved Low incidence	Achieved Low incidence	Not achieved Low incidence	Not achieved Low incidence	Achieved Low incidence
	Target 8 Halt and reverse the spread of malaria and tuberculosis	Partially achieved Millions of people still do not have access to public health programs	Not achieved Low mortality	Not achieved Moderate mortality	Achieved Low mortality	Achieved Moderate mortality	Achieved Moderate mortality	Achieved Low mortality	Achieved Low mortality

Goal 7	Target 9 Integrate the principles of sustainable development into country policies	Not achieved. Apart from the elimination of the use of ozone depletion substances, other crucial factors regarding environmental sustainability need great efforts to be achieved. CO_2 emissions present increase, renewable water sources become more and more scarce						
	Target 10 Halve the proportion of people without sustainable access to safe drinking water	Achieved Not all people gained access to water have access to safe water	Achieved High coverage	Not achieved Low coverage	Achieved High coverage	Achieved High coverage	Achieved Moderate coverage	Not achieved Moderate coverage	Achieved High coverage
	Target 11 Significant improvement in the lives of at least 100 million slum dwellers	Not achieved The number of slum dwellers is increasing due to accelerating urbanization	Achieved Moderate percentage of slum dwellers	Not achieved Very high percentage of slum dwellers	Achieved Moderate percentage of slum dwellers	Achieved High percentage of slum dwellers	Achieved High percentage of slum dwellers	–	Not achieved Moderate percentage of slum dwellers
Goal 8	Target 18 Increase the users of the Internet	Achieved Great increase in the Internet users. Need to expand the use in certain areas	Achieved High usage	Not achieved Moderate usage	Achieved High usage	Not achieved Moderate usage	Achieved High usage	Achieved High usage	Achieved High usage

historic document "Our common future" by the World Commission on Environment and Development, fundamental issues for the world's welfare remain pending. The MDGs should be fully achieved and without delay more efforts should be made for ensuring better perspectives, especially for people in the developing world. Otherwise, the famous aphorism of Karl Marx, "The rich get richer and the poor get poorer," will continue haunting humanity.

In 2010 the High-level Plenary Meeting of the General Assembly on the MDGs requested the Secretary-General to start processing the creation of a post-2015 development agenda. The Secretary-General established the UN System Task Team on the Post-2015 UN Development Agenda. This team brings together the efforts of more than 60 UN agencies and international organizations, in order to reach a set of sustainable development targets. The Task Team issued a first report to the Secretary-General in May 2012 regarding the post-2015 development agenda. In this report it is noted that the central challenge should be ensuring that globalization becomes a positive force for all the world's peoples. Some more detailed recommendations are summarized below [118]:

- The goals and targets of the post-2015 development agenda should be concrete and precise. This was the strength of the MDG framework. The MDGs should be reorganized according to four key dimensions: (1) inclusive social development, (2) inclusive economic development, (3) environmental sustainability, (4) peace and security.
- High level of policy coherence at the global, regional, national, and subnational levels will be required. There are no solutions, which fit to all cases. Hence, the post-2015 development agenda should include and encourage national policy design, as well as adaptation to local settings, under the overall vision and its principles.
- It is too early to define concrete goals and targets for the post-2015 UN development agenda. Various processes will need to be completed first. The outcome of and follow-up to the Rio+20 Conference on Sustainable Development will provide critical guidance.

The acceleration of progress toward the MDGs and the creation of a new global alliance, through the post-2015 development agenda, including more drastic measures for global development, are absolutely necessary for improving the current condition of our world. Toward this direction, courageous political decisions, sufficient financial resources, and integrated approaches to the global problems and challenges are required. Without radical changes in the global priorities, progress toward improving peoples' lives would always be deficient.

As already discussed, the vast majority of the world's poor live in rural areas. Great numbers of rural people in the developing world live under unbearable conditions and do not have access to basic services. The Food and Agriculture Organization (FAO) of the UN is the major international body which coordinates policies for rural populations.

According to the relevant Website: "Achieving food security for all is at the heart of FAO's efforts—to make sure people have regular access to enough high-quality food to lead active, healthy lives." The basic strategic framework of FAO, as well as the major dimensions of action, is included in Table 8.6 [119]. The strategic framework of FAO is implemented through regional programs/initiatives, which are coordinated by regional offices. In partnership with regional organizations, it is attempted to raise political commitment for specialized, country-level actions, in order to improve the capacities of governments and stakeholders. FAO creates and shares information about food, agriculture, and natural resources. Additionally, it plays a connector role, since it works with different partners and so, dialogue is facilitated. The programs run by the organization have the general aim of turning knowledge into action. Therefore, depending on the needs of its region and according to the strategic framework, certain actions are implemented. The programs implemented by FAO are funded both by the member countries (at a percentage of 41% in 2014) and by voluntary contributions (at a percentage of 59% in 2014). For 2014–15 the total budget of the organization is $2.4 billion. FAO and its actions have played a major role in making progress toward the achievement of the MDGs, especially in the fields of poverty and hunger alleviation.

Table 8.6: The Strategic Aims of FAO and Their Basic Specification [119].

Strategic Framework of the Food and Agriculture Organization (FAO)	
Eliminate hunger, food insecurity, and malnutrition	• **Policies, programmes, and legal frameworks**: Development of policies and legal framework with strong focus on hunger and malnutrition. • **Human and financial resources**: Efforts toward greater commitment and allocation of human and financial resources to support the implementation of policies. • **Governance, coordination mechanisms, and partnerships**: Creation of perspectives for stronger and more inclusive coordination across sectors and stakeholders. • **Evidence-based decision making**: Support to effective decision making, at the basis of food security information systems, enhanced tracking and mapping of actions, and improved impact assessment.
Make agriculture, forestry, and fisheries more productive and sustainable	• **Support practices that increase sustainable agricultural productivity**: Development of good practices, which increase productivity and save resources and share with decision makers. • **Provide information to support the transition to sustainable agriculture**: Gathering and share of information, development and sharing of analytical tools aiming at increasing productivity, and sustainable use of resources in agricultural systems.

Continued

Table 8.6: The Strategic Aims of FAO and Their Basic Specification [119].—cont'd

Strategic Framework of the Food and Agriculture Organization (FAO)	
Reduce rural poverty	• **Promote the transition to sustainable agriculture**: Help to countries, in order to evaluate the effectiveness of their strategies for sustainable agriculture. Supporting the development of policies and legal framework, which underpin the transition to sustainable agriculture. • **Advocate the adoption of international policies for productive and sustainable agriculture**: Urge countries to adhere to international agreements, which promote productive and sustainable agriculture. Support to the implementation of national laws. • **Improve opportunities for access to decent employment**: Improving the design of rural economic diversification policies, in order to promote decent work creation. Skills training for rural workers. Assistance in the application of international labor standards, ensuring occupational safety and health. • **Improve social protection systems**: Support synergies between social protection measures and food security. Support the development of national programs on social protection of the rural poor. Strengthen current social protection programmes for increasing their efficiency. • **Empower the rural poor gaining access to resources and services**: Strengthen rural organizations, including co-operatives. Improvement of rural infrastructure. Improvement of access of the poor to natural resources.
Enable inclusive and efficient agricultural and food systems	• **Improve the inclusiveness and efficiency of food systems**: Help governments to support the sustainable development of food systems and regulate plant and animal health, as well as food quality and safety. Documentation about food lost and waste and work on reducing it. • **Help strengthen public—private collaboration to improve smallholder agriculture**: Engagement of the food industry and nonprofit organizations in supporting smallholder farmers. Facilitation of investment increase in strengthening the food sector. Support governments to cooperate more effectively with the private sector. • **Improve the inclusiveness and efficiency of markets**: Gather and share information on market access and development. Strengthen financial mechanisms to support the growth of agriculture and food industries.
Increase the resilience of livelihoods to threats and crises	• **Help countries govern risks and crises**: Help countries develop strategies and plans for reducing and managing risks. Advocating for the mobilization of resources toward risk reduction for agriculture, food, and nutrition.

Table 8.6: The Strategic Aims of FAO and Their Basic Specification [119].—cont'd

Strategic Framework of the Food and Agriculture Organization (FAO)
• **Help countries watch to safeguard**: Development and share of mechanisms that monitor and warn about hazards and risks. Help countries prevent and mitigate risks. Development of strategies that reduce the impact of disasters. Help countries to make their agricultural systems withstand and recover from crises. • **Support countries to prepare and response**: Provide assistance so that humanitarian action protects the livelihoods of vulnerable farmers, fishers, herders, etc. during emergencies. Help toward ensuring that disaster response plans are coordinated at all levels.

As it was described in Section 8.5, the development perspectives of great shares of the world's population are downgraded, due to lack of access to energy and, especially, electricity. The UN Secretary-General, Ban Ki-Moon, has stated:

Energy is the golden thread that connects economic growth, social equity, and environmental sustainability […]. Widespread energy poverty condemns billions to darkness, to ill health, to missed opportunities. Energy poverty is a threat to the achievement of the Millennium Development Goals. It is inequitable and unsustainable. Children cannot study in the dark. Girls and women cannot learn or be productive when they spend hours a day collecting firewood. Businesses and economies cannot grow without power. We must find a way to end energy poverty.

The UN Secretary-General, in cooperation with the President of the World Bank, introduced an initiative called "Sustainable Energy for All" in 2011. This global initiative has three main pillars:

• providing universal access to modern energy services,
• doubling the global rate of improvement in energy efficiency, and
• doubling the share of renewable energy in the global energy mix.

It is aimed to meet the aforementioned targets by 2030.

In support of the initiative, the UN Foundation created a global Energy Access Practitioner Network, which has 2000 members in 170 countries. The Network focuses on removing market barriers to the delivery of energy services through the adoption of new technologies and innovative financial and business models. The UN Foundation is also collaborating with a series of public and private sector partners, in order to facilitate the initiative. As far as the steps toward achieving each goal are concerned, in Table 8.7 , the recommended actions are summarized. As it can be seen in Table 8.7, the first stage of the "Sustainable Energy for All" initiative is the creation of a robust data platform, which

Table 8.7: Areas of Action for Each Goal of the "Sustainable Energy for All" Initiative for Improvement of Global Energy Databases.

	Recommended Targeting of Effort Over Next 5 Years
Energy access	Improve energy questionnaires for global networks of household surveys
	Pilot country-level surveys to provide more precise measures of access to electricity and clean cooking
	Develop suitable access measures for heating
Energy efficiency	Integrate data systems on energy use and associated output measures
	Strengthen country capacity to collect data on sectoral intensities
	Improve data on physical activity drivers
	Improve data on energy efficiency targets, policies, and investments
Renewable energy	Improve data and definitions for bio-energy and sustainability
	Capture renewable energy used in distributed generation
	Capture renewable energy used off-grid and in microgrids
	Promote a more harmonized approach to target setting

will have the capability of monitoring global progress toward the specific objectives. Household surveys are the main source of data for energy access and national energy balances the basic tool for monitoring renewable energy use and progress in energy efficiency. Until now, it has been made possible to monitor between 126 and 181 countries, depending on the indicator. This corresponds to coverage of 96–98% of the world's population, a rather satisfying achievement.

Through the extensive monitoring framework, the establishment of certain quantitative targets for the initiative was made possible. These targets are listed in Table 8.8. Actual global investments in the areas covered by the initiative's objectives were estimated at about $400 billion in 2010. The additional necessary investments for achieving the

Table 8.8: Specific Targets of the "Sustainable Energy for All" Initiative.

	Objective 1		Objective 2	Objective 3
	Universal access to modern energy services		Doubling global rate of improvement of energy efficiency	Doubling share of renewable energy in global energy mix
Proxy indicator	Percentage of population with electricity access	Percentage of population with primary reliance on nonsolid fuels	Rate of improvement in energy efficiency	Renewable energy share in TFEC
Reference 1990	76	47	−1.3	16.6
Starting point 2010	83	59		18.0
Objective for 2030	100	100	−2.6	36.0

objectives are considered to be at least $600–800 billion per year. The access to energy-related expenses count for about 15% of the incremental costs. The greatest share of the expenses is attributed to energy efficiency and renewable energy. Box 8.2 contains an interesting successful example of a small-scale intervention for providing rural communities with green energy.

Within the developed world, lack of access to electricity is, practically, not a problem. Nevertheless, energy poverty, insufficient coverage of energy needs, becomes an intensifying issue. Action against energy poverty can be characterized as fragmentary, until now. For instance, some member states of the European Union proceeded to the adoption of national policies for alleviating energy poverty. Italy, Spain, and Greece have introduced low electricity tariffs for vulnerable customers. In Sweden, the social protection system undertakes the coverage of unpaid energy bills by customers with economic problems [120]. It can be claimed that better-coordinated actions should be adopted against energy poverty, aiming at protecting vulnerable citizens and avoid negative impacts on human health and the environment, caused by insufficient access to energy services.

Box 8.2 Successful Case Study of Action Toward Supplying Rural Areas in Peru With Energy

Successful example for sustainable energy supply: Micro-hydro revolving fund in Peru
"Practical Action"* has been using Inter-America Development Bank (IDB) funds in Peru since 1994, in order to implement a "Revolving Fund" of soft loans. The fund combined with technical assistance aims at constructing micro-hydro power plants in isolated rural areas. The fund consists of a financial model based on loans subsidized with technical assistance for individual clients (microrural entrepreneurs). It covers the installation of new systems, as well as the rehabilitation and/or repair of existing systems.

The amount of loans ranges from US $10,000 to $50,000, with an interest rate of 10%. The payback period is 1–5 years, and the grace period varies, depending on the client's situation. The types of guarantee vary according to the status of the client, collective or individual. In the case of collective clients, a positive cash flow should be demonstrated. In the case of individual clients, they must give collaterals for an amount equivalent to or greater than 30% of the loan received. The electromechanical equipment may form part of the guarantee.

To date, this model has allowed the conclusion of 22 loan contracts, amounting about $800 million, in total. The loans have enabled an additional installed capacity of over 1.5 MW to be put into operation in remote areas, which has provided 15,000 rural inhabitants with energy.

---
* Practical Action is an international organization was founded by radical economist and philosopher E.F. Schumacher over 45 years ago. It tries to use low cost, appropriate, small-scale development solutions for supporting people in helping themselves.

The MDGs include targets related to environmental sustainability. The content of the MDGs, themselves, show the strong interactions between environment and development. Apart from sufficient food, income, and decent housing conditions, good living standards require a clean, safe, and pleasing environment. Therefore, global efforts for protecting the environment and corresponding to major challenges, such as climate change, are absolutely necessary for ensuring a viable future for humanity. Unfortunately, there are major indices showing that we are, at global level, far away from a pathway respectful to the environment. For instance the world's average CO_2 emissions per capita were 14% higher in 2011 than in 1990 [121]. Despite the overall reduction in deforestation rates, in Latin America about 4 million hectares of forests are lost every year [122]. The depletion rates of the world's water reserves have not stopped increasing since 1960 [123]. Therefore, global concern about environmental issues should be intensified and international policies have to be more fruitful.

The global environmental agenda is set by the United Nations Environment Programme (UNEP), which can be characterized as the leading global environmental authority. UNEP focuses on the following thematic priorities, which contain particular goals. In Table 8.9 the thematic priorities of UNEP and their goals are listed. For each thematic priority global, regional and national programs are coordinated by UNEP, in order to achieve the specific targets. The activities of UNEP are concentrated on the developing world.

Table 8.9: Thematic Priorities and Goals of UNEP.

Priorities of UNEP	
Climate change	• **Adaptation to climate change**: Countries are helped to reduce their vulnerability and use ecosystem services to build natural resilience.
	• **Mitigating climate change**: Support in making sound policy, technology, and investments that lead to GHG emission reductions, with a focus on renewable energy sources and energy efficiency.
	• **Reducing emissions from deforestation**: Incentives for developing countries to reduce emission from forested lands.
	• Enhancing knowledge and communication.
Disasters and conflicts	• **Disaster risk reduction**: Prevention and reduction of the impacts of natural hazards on vulnerable communities and countries through sustainable natural resource management.
	• **Assessment**: UNEP conducts field-based scientific assessments to identify the environmental risks to human health, livelihoods, and security following conflicts, disasters, and industrial accidents.
	• **Recovery**: In the aftermath of a crisis, implementation of environmental recovery programmes through field-based project offices to support long-term stability and sustainable development.
	• **Cooperation for peacebuilding**: It is aimed to use environmental cooperation to transform the risks of conflict over resources into opportunities for peace.

Table 8.9: Thematic Priorities and Goals of UNEP.—cont'd

	Priorities of UNEP
Ecosystem management	• **Making the case**: Promotion of the ecosystem management approach and explanation of its advantages for development. • **Restoration and management**: UNEP develops and tests tools and methodologies for national governments and regions to restore and manage ecosystems. • **Development and investment**: Providing help to national governments, in order to integrate ecosystem services into development planning and investment decisions.
Environmental governance	• **Sound science for decision making**: Reviewing global environmental trends and emerging issues and bringing these findings to policy forums. • **International cooperation**: UNEP helps countries cooperate to achieve agreed environmental priorities and supports efforts to implement new international environmental laws and standards. • **National development planning**: Promoting integration of environmental sustainability into regional and national development policies. • **International policy setting and technical assistance**: UNEP works with countries and other stakeholders to strengthen their laws and institutions, helping them achieve environmental goals, targets, and objectives.
Chemicals and waste	• **Scientific assessments**: Global assessments of the environmental impacts of harmful substances and awareness raising. • **Legal instruments**: UNEP assists governments to develop appropriate policy and control systems for harmful substances. • **National implementation**: UNEP provides the tools, methodologies, and technical assistance to help countries design, finance, and implement national programs, which improve assessment and management of harmful substances and hazardous waste. • **Monitoring and evaluation**: Promotion of best practice for monitoring and evaluating the progress of national programmes.
Resource efficiency	• **Assessing critical trends**: Assessments and reports on trends in how resources are extracted, processed, consumed, and disposed of. • **Building capacity for policy action**: Work with partners from government, city authorities, and the research community to develop policy tools and instruments that lead to more resource-efficient societies. • **Seizing investment opportunities**: UNEP forges expert networks and industry partnerships. These collaborations help small and large businesses adopt resource-efficient technologies. • **Stimulating demand for resource-efficient goods and services**: UNEP develops consumer and procurer information tools, market incentives, and public–private initiatives to promote sustainable lifestyles and value chains.

Apart from the general directions given by UNEP and the relevant projects, there are numerous regional and national policies for the environment. In the United States, the Environmental Protection Agency (EPA) is a major organization which aims at protecting the environment and health of citizens. EPA develops and enforces regulations on environmental issues, gives grants for implementing environmental programs, and studies environmental issues through a big network of laboratories and monitoring stations. EPA is activated to the whole range of environmental issues, namely: Air, Chemicals and Toxics, Climate Change, Emergencies, Greener Living, Health and Safety, Land and Cleanup, Pesticides, Waste, Water. EPA receives generous funding and this is a major cause for the high impact of its activities and its global reputation. In 2014 the EPA's costs of operation were $8.5 billion [124]. It should be noted that the budget of UNEP for the same year was $630 million. The mission of the two organizations is different. However, taking into account that UNEP coordinates environmental programs worldwide, the great difference in the budgets is an indicator of the particularly heterogeneous exercise of environmental policy among the world.

Finally, it is worth mentioning a central policy of the European Union, aiming at confronting climate change. This is the well-known "climate and energy package 2020," which contains the "20-20-20" target setting three objectives for 2020:

- a 20% reduction in EU GHG emissions from 1990 levels,
- raising the share of EU energy consumption produced from renewable resources to 20%, and
- a 20% improvement in the EU's energy efficiency.

Although the "20-20-20" target is based on the will for environmental protection and climate change alleviation, it is a typical example of environment−development interactions. More specifically, all EU member states, according to the 2009/28/EC Directive, were obliged to set binding national targets for raising the share of renewables in their energy consumption. Obviously, a great amount of investments is needed, in order to achieve these targets and hence, the climate and energy package has strong implications on development across the European Union.

It seems that Europe has been particularly successful in parts of its environmental policy—compared to other regions of the world—and the "20-20-20" has played a positive role toward this success. Between 1995 and 2011, the European Union has been the only region globally that achieved reduction of CO_2 emissions, at a percentage of 5%. The overall share of renewables in 2012 within the the European Union was 14.1%, compared to 6% in 1997 [125].

The strong environment and development interactions may affect the EU policies after 2020. Ambitious targets have been set by the the European Union for 2030, such as increasing the share of renewable to 27% and achieving 30% energy savings. However, recession/crisis phenomena are still present within the the European Union and they, de facto, set restrictions in environmentally friendly strategies, which demand high investment costs.

8.6.2 Tools for Supporting Environment/Development Issues and Policies

Environmental Impact Assessment

The multifariousness of the environment—perceived as a whole—has already been noted. Furthermore, the inevitability of the human action(s) and intervention within this multidimensional environment has also been clearly discussed. That is, the developmental process, involving human activities, projects, and works, is viewed as an inescapable necessity.

However, this necessity—being attributed to the development-related activities—is not to be accompanied by the potential adverse environmental implications that these activities entail. One kind of dealing with this situation is through some means of preestimating these implications. The pertinent history goes back to the enactment of the National Environmental Policy Act of 1969,[5] where the newly created tool (then) of *Environmental Impact Assessment (EIA)* was envisioned to be of use for federal agencies with the aim of integrating the socioeconomic aspects of development with environmental concerns [126]. Since then, "EIA systems have been set up worldwide and become a powerful environmental safeguard in the project planning process" [127]. In the context of the European Union, the EIA has been introduced with the European Union Directive (85/337/EEC) on Environmental Impact Assessments[6] and has been amended through time (eg, Directive 2011/92/EU).

In a rather generic definition, EIA is "the process of identifying the environmental consequences of human activities, before those activities begin" [128: 1]. More specifically, EIA's main focus is to anticipate the important consequences from proposed projects on biophysical (soils, flora, fauna, etc.), built (settlements, infrastructures, etc.), socioeconomic (education, health care, recreation services, etc.), and cultural systems (art, beliefs, etc.). A much more detailed definition is furnished by Lawrence ([129: 7]—different levels of bullet/indentation added by the authors) [129], whereby EIA is treated as:

"a systematic process of:

- Determining and managing (identifying, describing, measuring, predicting, interpreting, integrating, communicating, involving and controlling) the

---

[5] National Environmental Policy Act, 42 United States Code Sections 4321–4370.
[6] Council Directive 85/337/EEC on the Assessment of the Effects of Certain Public and Private Projects on the Environment (1985-06-27) from Eur-Lex. Available at: http://eur-lex.europa.eu/LexUriServ/LexUriServ.do?uri=CELEX:31985L0337:EN:HTML.

- • Potential (or real) impacts (direct and indirect, individual and cumulating, likelihood of occurrence) of
  - Proposed or existing human actions (projects, plans, programs, legislation, activities) and their alternatives on the
    - Environment (physical, chemical, biological, ecological, human health, cultural, social, economic, built and interrelations)"

In practice, there are several key stages involved in the EIA process: from the initial stages of Project Preparation, Screening and Scoping—to the Environmental Studies, Review of Adequacy of the Environmental Information, Consideration of the Environmental Information by the Competent Authority before Making Development Consent Decision, the Announcement of Decision and the Post-Decision Monitoring (to mention only some of these stages) [130].

Yet, despite the formalities suggested and the considerable attention that it appears to be given to the environmental consequences that emerge from various proposed human actions, projects, etc., the EIA is not a panacea. There are several caveats compromising the reliability and scientific adequacy of the EIA process. Wright et al. [131] stress out the fact that the limited time for the completion of EIAs to meet legislative deadlines, the cost-reduction strategies, and a lack of impartiality degrade the results of the process. In such an approach—which is usually far from being scientifically rigorous— a habitual treatment of topics of the utmost importance is favored and cultivated, "perpetuating previous misconceptions and analytical flaws"; instances of negligence of nonlethal/nonsignificant environmental impacts, along with instances of dismissal of the cumulative effects that nonsignificant impacts may induce are indicative of the frivolous and, from time to time, opportunistic character of the EIAs [131: 72]. Although one could assert that such problems occur only in practice, the way our socioeconomic systems are structured—promoting the call for rapid and inexpensive solutions—makes it plausible to claim that these problems shift even toward the very core of the EIA process. The major pitfall here is the false sense of security and of environmental safeguarding that the abiding by the legislative imperatives and completion of EIAs entail. Since our security and well-being is conditioned to a great extent by the health of our environment [132], the naïve reliance on (socially constructed) conventions which do not intend to describe the actual environmental conditions in a genuinely scientific manner is deemed at least insufficient. The acknowledgment that "financial constraints are real logistical challenges and that workloads are continually increasing" along with the revealing of "ever-more complex relationships between human activities and their consequences [...] and the impact on animals, species or habitats" [131: 75] does not justify or legitimize an easy way-out, that is adjusting EIAs to our own desires.

Remote sensing for environmental applications

The abundant, multifarious, and extremely variable—temporally and spatially—physical, chemical, biological, and human components of the Earth's environment involve highly complex interactions and feedbacks, rendering our environment a highly dynamic system [133]. Managing the entirety of the biophysical and human environment requires, at first, the inventorying of a variety of elements that constitute it. Yet, the vastness and the inconstancy of these elements call for methods and techniques that enable us to acquire information about the cover of the Earth surface without being in direct contact with it.

Toward this direction, *Remote Sensing* can be of significant aid. As Jensen [134] states: "Remote Sensing is the art and science of obtaining information about an object without being in direct physical contact with the object" and it "can be used to measure and monitor important biophysical characteristics and human activities on Earth." Typically, the electromagnetic radiation carries the input information in the Remote Sensing system, while the output information of such a system can be either an image directly representing the scene being viewed, or it can occur by further analysis and interpretation [135]. In practice, this information acquired by Remote Sensing methods provide us with several capabilities which range from: (1) observation and identification of natural features or patterns, (2) analysis and measurement of various biological and physical variables, (3) mapping the geographic location, spatial extent, and distributions of physical features and of biophysical variables, (4) monitoring of how these mapped features and variables vary over time (and space), up to the (5) support of decision making by resorting to all the previous capabilities [136].

The previously mentioned capabilities encompass various aspects of the biophysical environment such as vegetation and soil cover, geologic composition and structure, water surfaces' and hydrology's characteristics, etc., but they also extend to the human or built environment. In the following, we refer to some of the Remote Sensing applications which typically harness satellite imagery and address issues pertaining to these various aspects of the environment.[7]

Managing the natural/biophysical environment

There are several Remote Sensing applications pertaining to the management of the natural (biotic and abiotic) environment:

Biotic Environment: Remote sensing methods are often utilized to furnish spatial, temporal, and thematic information about living organisms. This information can range from the detection of large mammals' occurrence to the identification of larger features

---

[7] It should be noted that we refer only to aspects of the environment located *on* the Earth's Surface. Thus, we do not include the observation of what is above its surface (eg, clouds, aerosols, etc.).

habitat types, land covers, and landscape patterns [136]. Knowledge about vegetation species, spatial patterns, and modifications in phenological cycles can be derived by several vegetation indices/transformation processes and landscape ecology metrics based on remote sensing inputs and digital image analyses [134,137]. Spatial patterning quantification in landscapes and especially in forests has been a core matter of much of ecological, forestry, and management-related research, while "the technological and conceptual advances in remote sensing have shaped the way landscape ecologists [...] conduct research," ie, "using landscape metrics or landscape pattern indices" [138: 174]. Moreover, applications regarding the more generic issue of managing the biological diversity "include land cover and land use change, grassland conditions, oil-spill cleanup, wildfire fighting, monitoring of fire scars in tropical forests, postfire recovery, and changes in fragmentation patterns" [136: 9]. In an even more generic sense, as Franklin [139: 1] states, Remote Sensing has been developed as a foundation for a transdisciplinary approach to biodiversity and wildlife management and this new approach relies "on the capability to derive multispectral views of environment at multiple spatial and temporal scales, which are readily integrated with other forms of data, including Global Positioning Systems (GPS) and Geographical Information Systems (GIS)" in order to produce suitable models and visualizations and ultimately "explore solutions to previously intractable environmental science and management problems."

Aquatic Environment: Another important domain of Remote Sense applications for environmental management refers to water. Needless to offer an argumentation, information about its quality, quantity, extent, and geographic distribution is vital. Water (or hydrologic) parameters can be collected via in situ point measurements, but, typically, they cannot afford statistically significant distribution geovisualizations because these point measurements are usually sparse and insufficient to infer regional geographic patterns [140]. According to Jensen [134], the most important hydrologic parameters or variables pertain mainly to the: surface extent of water bodies, water (biochemical) components, water-surface temperature, water depth, snow and ice-surface cover, cloud cover, precipitation and water vapor—and it is hard to obtain regional information about these crucial variables by means of in situ point observations. However, by harnessing the merits of Remote Sensing, it is possible to map river channel morphology and in-stream habitat [141], to identify "spatial–temporal patterns of snow cover across large areas in inaccessible terrain, providing useful information on a critical component of the hydrological cycle" [142] or to detect changes in wetlands using time-series vegetation indices derived from multitemporal satellite imagery [143]—to mention only a few out of a multitude of applications for inland water observation and monitoring. Besides, the investigation of the oceans' dynamic processes—ie, the study of their changes in terms of their biological, chemical, and physical properties and in terms of the oceans' interactions the atmosphere, cryosphere, and land—can be now realized on a global level by dint of the utilization of satellite Remote Sensing methods [144].

Rocks and Soils: The soil, geological, and geomorphological aspects of the environment of an area significantly affect the properties of the latter. Remote sensing methods can be employed to identify the "chemical composition of rocks and minerals that are on the Earth's surface and not completely covered by dense vegetation" [134: 507]. More recent papers cover the literature review on geological remote sensing approaches, satellite imagery selection, and applications [145] or suggest methods for the enhanced detection of lithological and hydrothermal alteration patterns and regional mineral exploration [146]. Moreover, the spatial structure or the 3D features of "the Earth's surface formed by natural processes" [134: 529] that is, geomorphology and landforms, respectively, can be extracted, classified, and mapped by means of Remote Sensing and GIS methods/ techniques [147,148]. Remotely sensed images can also aid in measuring the soil properties and in soil mapping in an essential means—as well as in landform mapping [149]. Notwithstanding the "theoretical" importance and value of registering such kind of information, lithological/geological, landform/geomorphological, and soil mapping do have many practical bearings that are able to affect entities other than the nonliving ones, ie, living organisms and the humans. For instance, Bocco et al. [150,151] have explored the contribution of geomorphologic mapping resulting from the integrated Remote Sensing/GIS approaches in enabling/enhancing natural resource management and in ameliorating land use planning in developing countries.

Actually, the majority of the applications of Remote Sensing for all the previous domains can and usually has further implications on the human environment, and, specifically, on people. However, there are some applications that focus directly in observing and monitoring the built environment.

Managing the human/built environment

Urban Growth and Sprawl: One of the most essential issues relating to the built environment is the delineation of its spatial extent and the registering of its development—through time. The study of urban growth is subsumed in the subdiscipline of urban geography and its main focus is on cities and towns and on how they expand in physical and demographic terms [152]. Gatrell and Jensen [153] stress out the significance and effectiveness of the Remote Sensing methods in exhibiting the interactions between people and their urban environments. More precisely, assets of greatly relying on Remote Sensing are the rapidity of image acquiring over extended areas, the capability of getting large multitemporal data sets, the advancements in digital processing and analysis, and the potential of integration with GIS, etc. [154]. In a more specific sense, Donnay et al. [155: 7] ascribe to the Remote Sensing the task of "detection, identification, and analysis of urban features" in order to produce data sets pertaining to:

1. "the location and extent of urban areas;
2. the nature and spatial distribution of different land use categories within urban areas;
3. the primary transportation networks and related infrastructure;
4. various census-related statistics and socio-economic indicators;
5. the 3-D structure of urban areas for telecommunications (inter-visibility) and Environmental Impact Assessment (EIA) studies; and
6. the ability to monitor changes in these features over time."

In practice, Remote Sensing data enable us to detect and measure features related to the land use/land cover and morphology of the urban area [152,156,157]. Elements pertaining to the land use/land cover are "'urban' pixels that form the basis of many remote sensing analyses consist typically of developed and impervious surfaces (pixels) that include built structures, concrete, asphalt, runways, and buildings" [152: 4], whereas features related to the morphology refer to parameters such as shape, density, and texture [156,157]. Yet, since urban areas may also encompass a multitude of nonimpervious, nonbuilt pixels such as parklands and urban forests while at the same time may rule out developments concerning other land-uses [158], and because of the distinct and complex spatial structures and functions in urban areas, Bhatta [152] labels the latter as urban ecosystems or landscapes. In this respect, an urban ecosystem or landscape can be seen to "integrate physical, social, economic, ecological, environmental, infrastructure and institutional subsystems [while] urban growth and sprawl is an outcome of change in performance/ functioning of these subsystems" [152: 6].

Therefore, maybe the most significant advantage of Remote Sensing data in urban applications is the monitoring of urban development "[…] to determine the type, amount, and location of land conversion," especially in cases "of rapid land-use changes where the updating of information is tedious and time-consuming via traditional surveying and mapping approaches." [152: 51]. Another very important technical issue in urban Remote Sensing is the spatial resolution of the image data [159]. In this perspective, major advances in both the temporal and the spatial resolution of satellite imagery (large time-series of images and image resolutions of less than 1 m respectively) over the last years, now furnish a high potential in the management of the built environment.

The pertinent literature teams with research studies regarding the utilization of Remote Sensing to monitor urban expansion and land-use changes [160–162]. The study of the urban expansion is a matter of high significance by itself. Nonetheless, there are specific cases where this expansion is so rapid and "irregular"—leading to the encroachment of large proportions of farmland and forest, traffic problems, and other environmental problems [152]. The geospatial patterns accompanying urban development "out of control" are deemed as urban sprawl [163]. The harnessing of Remote Sensing data and the implementation of respective methods and techniques do can and should aid in managing this special issue of sprawling which entangles a multitude of detrimental environmental effects for the urban landscape.

Enhancing (Human) Livelihoods by Managing the Nonhuman Environment: There are some other domains where Remote Sensing can contribute indirectly to the human well-being and development. In the previous sections, the issues of poverty and hunger have been addressed as being closely intertwined with human development. Food security and famine, in particular, are dimensions that have long ago been considered: Since 1986, the US Agency for International Development (USAID) has developed and promoted the Famine Early Warning System (FEWS) "to provide information on food security of communities in semi-arid regions so that such a widespread catastrophe does not occur again" [164: 3]. Granted that (governmental) strategies and policies of early and proper intervention can facilitate the detachment of the linkage between the climate extremes and famine [165], it is vital that we employ methods of acquiring and processing relevant data and information in handy manners.

"The development of remote sensing systems to monitor environmental conditions has offered for the first time a way to monitor current climate variations over entire continents for very little expense" [164: 4]. This development has occurred in a variety of ways, enhancing its potential to support agricultural management, thus allowing "a more accurate analysis and interpretation of the Remote Sensing data in terms of: type of crop, soil type, state of growth, and presence of disease" [166]. It should be noticed, though, that the elements and phenomena pertaining to the state and the evolution of agriculture, food security, and famine are multifarious and are interrelated in intricate means. Valuable information about environmental conditions, in general, can be gained through the combination of remotely sensed data and other data types for estimating the impacts on the agriculture-related issues and farmer's livelihoods [167]. More specifically, "[s]igns of imminent famine can be identified by combining remotely sensed data, precipitation data, and surface water data to characterize and model hazards threatening vulnerable livelihoods" [167: 18].

As mentioned above, field-based hydrological data are point-based and dispersed, not being able to provide sufficient regionalized information. Similarly, hydrometeorological data are sparsely distributed in space and are not capable of furnishing information in (near) real time and, thus, of effectively predicting famine, but, fortunately, "[r]emotely sensed data, combined with numerical modeling and GIS, are important tools for FEWS NET" [167: 18]. For instance, the research paper of Ratnasari and Kusumawardani [168] focuses on how food insecurity can be estimated in spatial terms through (1) remotely sensed image manipulation—approximating crop failure potential level and the erosion potential—and (2) GIS utilization—determining the erosion potential due to rainfall and slope that can be obtained from spatial interpolation of point-based measurements regarding rainfall and elevation.

GIS integration for environmental applications

Thus far, it has occurred that Geographical Information Systems are closely associated with Remote Sensing in addressing various environmental issues. There are several definitions for GISs; yet, according to a relatively broad one, a GIS is an integrated system of effective capture, storage, retrieval, processing, analysis, and visualization of information pertaining to affairs that are inherently geographical, depending on the specified requirements of various users [169–171]. In this respect, it appears that GIS serves in incorporating Remote Sensing data, methods, and techniques—when it comes to environmental applications. Since Remote sensing is a tool extracting valuable information about geographic objects, entities, or areas, it provides a type of information that "functions in harmony with other spatial data-collection techniques or tools of the mapping sciences, including cartography and GIS" [152: 10]. In other words, remotely sensed data afford "thematic and metric information" (about diversified domains of interest—soils, geomorphology, hydrology, land cover, etc.)—"making it ready for input into GIS" [135: 9].

Such theses regarding integration have been endorsed in an even stronger and explicit means by Rokos [172: 151]: "the most appropriate methods and techniques for the systematic, holistic and dynamic observation and monitoring of the global changes are those that can integrate the capabilities of the Remote Sensing and of the Land and Environment Information Systems." In order to "read" the environment as a holistic, global system, one needs to adapt models—abstractions or simplifications of the reality—that furnish insights about both its biophysical and socioeconomic dimensions [171,173,174]. Now, by fusing GISs and Remote Sensing in the context of environmental modeling, Skidmore [171: 5] cites some examples of applications, including:

> *monitoring of deforestation, agro-ecological zonation, ozone layer depletion, food early warning systems, monitoring of large atmospheric-oceanic anomalies such as El Nino, climate and weather prediction, ocean mapping and monitoring, wetland degradation, vegetation mapping, soil mapping, natural disaster and hazard assessment and mapping, and land cover maps.*

Recent literature further endorses such kind of integration by dint of its being rather necessary and suitable both in theory and in practice [168,175,176]. In any case, one should bear in mind the multiple benefits of Remote Sensing for environmental applications—as demonstrated in the previous sections. Yet, it is the GISs' analytical capacity that enables the decision making, planning, and management in environmental-related issues [171].

8.7 Alternative Concepts and Pathways

8.7.1 Ethical–Philosophical Issues

The present section is not dedicated to advocate a certain ethical/moral thesis, or even to outline and evaluate such kind of theses. However, when it comes to the election and/or the assessment of development approaches and strategies, it appears more than necessary to draw upon crucial issues of environmental ethics and philosophy.

Future generations

The modern concept of sustainable development stems from the UN Brundtland Report. In the latter, it has been explicitly stated that "[h]umanity has the ability to make development sustainable—to ensure that it meets the needs of the present without compromising the ability of future generations to meet their own needs" [23: 8]. Through this definition, the issue of future generations arises. In other words, the crucial question emerges: Do we have *any* obligations to the posterity with reference to our development strategies and their subsequent long-term environmental impacts? Moreover, of what kind are these obligations (if we have any) toward them and, thereof, should we regulate our actions and steer our options taking into consideration the needs and desires of the future generations?

There is a widespread assumption whereby the advancement of science and technology is propelling progress in a manner that future generations will benefit. Yet, this assumption according to which one need not worry about the posterity is faulty and problematic since current trends of success (eg, in terms of productivity and of food availability) are to be abated in the future (eg, due to the forthcoming shortage of freshwater and land degradation) [177]. Therefore, granted that there are some absolute resource-based limits to the progress and to its capacity to sustain the needs of future generations, this technological optimist perspective is shown to fail. Even worse (for the technological optimist), just "[b]ecause of advances in science and technology, the current generation may bear a greater burden of moral responsibility toward its successors than that of any previous generation" [178: 444]. In this respect, progress by itself not only does not resolve the core matter of the sustainability—delivering us from the burden of weighing our present choices—but it does further charge us with the grave moral responsibility of caring for our successors. As Partridge [179] puts it, the state of both being able to predict the future outcomes and having the capacity to affect them set us to the position of being morally responsible.

But things are getting more complicated when one is wondering *what* our obligations to future generations are. Certainly, concerns about intragenerational equity and our obligations to the today's world's poor in specific "are [important] issues in the environmental ethics, because our reaction to nutritional needs in an expanding human population affects the total impact of human beings on the environment" [180: 11]. Herein, we take for granted such obligations to our contemporaries. Yet, contrary to the ordinary belief that our responsibilities to them is analogous to the responsibilities to future generations, philosophers "recognize that future persons have a moral, epistemological, and even ontological status that is radically different from the status of our contemporaries" [181: 444]. The future persons' contingent existence, their potential (active/passive) rights, and the nonreciprocal relationship between us and them are indicative of this different status [181]. Nevertheless, this stark discrepancy in their status should not pose insurmountable barriers in coming up with strategies that have a practical bearing. For instance, Partridge, and Callicott and Frodeman [181,182] propose some policy guidelines for sustainable future such as: "Leaving 'Enough and as Good,'" "Do No Harm," and "Doing Well by Doing Good."

Responsibilities to nonhumans

So far, the interest of assessing the environmental ethics has been on the well-being of our contemporaries or of future generations. Yet, aside from considering the impacts/benefits on human beings from environmental degradation/conservation, which is concentrating on *anthropocentric concerns*, there are also *nonanthropocentric concerns* involved in the domain of environmental ethics. The latter concerns are "about nonhuman lives and live forms for their sake rather than for our own" [180: 13]. Traditionally, ethics "was focusing on the obligations of humans to other humans, because until now it was deemed that only human beings and the granting of their interests have an inherent (intrinsic) value" [180: 38] —while other living and nonliving entities have only an utilitarian value. But an expanding shift appears to have been/be taking place regarding the spatiotemporal boundaries of ethics. As Nash [181] theorizes, ethics has been historically evolved from merely attending to the individual interests toward encompassing the whole human race at the present time; according to his hypothesis, this ethical perspective is to be broadened even to the point of integrating all the living and nonliving things, ultimately enveloping the universe as a whole.

Irrespective of how extreme such a perspective may (seem to) be, certain criteria are required for a being or entity to be included in the "ethical/moral community," ie, to be morally considerable. As it can be reasonably presumed, the criteria for such moral considerability "are multiple, complex, and sometimes intensely disputable" [183: 42]. One criterion for moral considerability is the existence of benefit or interest (eg, welfare) for an entity [182]. Another criterion is that of complexity—in terms of the level/degree of the organization of matter. Now, humans clearly satisfy both criteria. But, to tell that living beings other than humans are excluded from having interests because these nonhuman interests differ from the human ones equals to the adoption of a sheer anthropocentric, or the *basic human chauvinism*, principle [183,184]. So, in a rather broad sense, animals and plants might be said to satisfy these two criteria, albeit, in more strict terms, this is not the case for the latter, owing to the fact that they do not have desires, intentionality, and self-consciousness [183,185]. However, the argument whereby intentionality is not a prerequisite for the existence of interest or benefit—eg, "a tree [...] can see its 'interest' for continuing and unencumbered existence to be satisfied" [183: 43]—can be extrapolated to nonliving entities of nature like rocks, mountains, rivers, and stalactites [186]. Such entities can be alleged to promote "their own good" by dint of organizing and structuring the matter that are consisted of, but the case of ascribing to the organization of organic/biological matter a higher moral status than to the structuring of inorganic materials can be condemned as a kind of biological chauvinism or bio-chauvinism [187]. A milder version for assigning an intrinsic value to nonhuman entities is related to the line of reasoning that they are potentially morally (or aesthetically) considerable, simply because they exist [188,189].

Further remarks

Even from this elementary examination of these two (human and nonhuman) dimensions of the environment, the difficulty and the perplexity of deciding about our moral obligation toward the posterity and the nonhuman entities is apparent. Certainly, there is a series of environmental ethics' issues, but Papadimitriou [190] subsumes them under four basic conceptual dualisms that arise in the pertinent domain, namely: anthropocentricism versus nonanthropocentricism, instrumental versus intrinsic values, individualistic versus holistic perspectives, and shallow/deep ecology. As for the first case of dualism, it is pointed out that one cannot just discard nonanthropocentric concerns neither in principle, nor in practice. However, it is worth noticing that, as Norton [191] suggests, it may not be necessary to embrace altogether the intrinsic value of the environment in a venture to preserve it as a whole. In this sense, it might be possible for the human race to live in harmonic coexistence with the rest of the nature without adopting a biocentric perspective, but by merely being compelled to conform to the ideal of harmonic coexistence through proper social structuring and mechanisms [183,194].

This weak version of anthropocentrism does not comply with deep ecological principles; nonetheless, it is in line with more holistic approaches. The WCED's declarations for intra- and intergenerational equity strategies have emerged within the stream of weak anthropocentrism. Yet, the problem of environmental ethics remains both in principle and in practice. How can we ensure the respect toward the environment in a holistic manner merely by means of socially oriented coercion, that is, without appealing to the intrinsic value of nonhuman entities (including the intrinsic value that is possibly derived by their interrelations) and to a more radical reformation of the human conscience and behavior based on the awareness of the fundamental unity between the self and the nature? Could it be that it is solely the standard Judeo-Christian tradition and Western ethical theory that have promoted this "human chauvinism" [186,192], or that our very (human) nature urges us to conceive the rest of our environment as inferior to ourselves, shaping behaviors that may incur destruction and annihilation? As it seems, such clear-cut distinctions cannot grasp the multifariousness of the matter, albeit there are historical cases that can verify the former or the latter ends of this "moral gamut."

These crucial moral-philosophical questions comprise the departing point for the rest of this section. In the remainder, we proceed with the critical evaluation of the dominant political and socioeconomic structuring and patterns that are intermingled with the current status and trends for the policies regarding the relationship between environment and development. Moreover, we adduce and describe some alternative concepts and pathways in an attempt to put forward alternative development strategies which may be beneficiary for the rest of humanity, the posterity, and the environment as a whole.

8.7.2 An Alternative Development Paradigm

In this final section the framework of an alternative development paradigm, as well as a proposed planning methodology according to it are presented.

There are many ways to categorize development thinking through time. Potter et al. [12: 82] suggested here that four major approaches to the examination of development theory can be recognized. The four approaches are:

- the classical–traditional approach,
- the historical–empirical approach,
- the radical–political economy–dependency approach, and
- alternative and bottom-up approaches.

Development theories and strategies have been many and varied, with new approaches generally being added alongside existing ones. Classical and neoclassical economic approaches generally stress the need for unrestrained, polarized growth and of letting the market decide for itself. Neoliberalism as a generic development paradigm stems from the New Right and emphasizes what is seen as the continuing need for market liberalization and for the economy to be market- and performance-driven. Historical models give a normative impression of the degree to which in the past, since mercantilism and colonialism, development has been highly uneven and spatially polarized. Both dependency (radical) approaches and alternative/another development can be seen as direct critiques of modernization theory. Thus, the economic growth paradigm of the 1950s was challenged by socialist and environmentally oriented paradigms in the 1960s and 1970s, respectively [12].

While much alternative development thinking makes a diffuse impression, this has gradually been giving way to a sharper and more assertive positioning on account of several trends [1: 89]:

- The enormous growth of NGOs in numbers and influence generates a growing demand for strategy and, therefore, theory.
- The importance of environmental concerns and sustainability has weakened the economic growth paradigm and given boost to alternative and ecological economics.
- The glaring failure of several development decades further unsettles the mainstream paradigm of growth.
- The growing challenges to the Bretton Woods institutions lead to the question of whether these criticisms are merely procedural and institutional (for more participation and democratization) or whether they involve fundamentally different principles.

Based on the previous approaches, a country is considered "developed" even if it has an increasingly large percentage of citizens living below poverty line, unemployed, underemployed, and marginalized; even if the quality of social services, education health, and security is constantly degrading; even if the private interests gradually replace even the most obvious government responsibilities; whether insecurity, alienation, racism, xenophobia, nationalism in its most hideous form, prostitution, drugs, crime, and interweaving economic and political interests grow bigger; even if ignorance, illiteracy, inequality of any type, and the progressive decline of the multidimensional educational and cultural aspect of education in a one-dimensional technical—vocational training and retraining are getting more and more alarming.

Many scientists, economists, sociologists, biologists, engineers, philosophers have tried and are currently attempting to challenge this approach from different perspectives, directly or indirectly, through their noninstitutional works, without having managed to establish a coherent alternative theory on development. Thus, it appears as a sine qua non priority the need of forming a mutually acceptable interdisciplinary and integrated theory for development, which is keeping up with the multidimensional and comprehensive nature of physical and socioeconomic reality [11]. The challenge is to define a development mode that not only combines social needs, economic needs, and environmental needs, but also gives a meaning to the lives of the individuals in a society [193]. In this context, Rokos' proposal perceives development as "a 'better' balance of social and human relations and land use systems, of production, employment, distribution and consumption, according to the values and choices of the forces in power as these coexist combatant and interact at the natural environment, each time according to the particular social dynamic and average social consciousness" [11].

Rokos [11] considers that development can only be an organic "whole" of the multidimensional and complex relations; interactions and interdependencies of economic, social, political, cultural, and technical/technological energetics; and efforts to achieve this evaluative/axiological "better" balance that is different each time for people and societies. Thus, the objective concept of the "whole" of development cannot be fragmented in the particularism of the thematic/sectoral forms (economic, agricultural, industrial, etc.), ignoring the impact of their activities on the natural (ie, biophysical) environment and the political, social, and cultural field. It cannot be contained in a certain room, in a state, or a region, ignoring what this means for the broader international and planetary surrounding. The organic "whole" of development cannot be restricted in the time limits of a small or large, one-dimensional or multidimensional, national or supernational planning or project; neither sustainable development can be characterized as integrated because of the partiality of its conception since its main foundation is economic competitiveness and sustainability of companies.

That is why Rokos [11] envisions a Worth-Living Integrated Development (WID), namely that kind of development which exists simultaneously and over time at global, supernational, national, regional, and local level, in economy, society, policy, culture, and technology. This kind of development only exists in dialectical harmony and always with respect toward humanity, its age-old noble values, and the "whole" of the natural and cultural environment, in which man fits peacefully and creatively as integral but not as the dominant part. Integrated Development can only exist:

- when citizens of the world understand and believe that development cannot be only "economic" or even state centric, partially centric, Eurocentric, or bank centric, but concerns all aspects of life and is their duty and responsibility to take their lives into their own hands and fight for it,
- when citizens understand that economy, society, and politics have no value without morality, and
- when they take the actual share of responsibility for the present state of things, as actors, accomplices, or just spectators of events of a truncated social, economic, and political democracy.

WID should be always in dialectical harmony and with respect toward human beings and their natural and cultural environment, in which they behave as their integral parts and not as owners, "investors," and exploiters [11,194]. And, of course, WID presupposes respect and equality among human beings. And this is the radical difference between sustainable and Worth-living Integrated Development, since the supporters of sustainable development are, in the best case, used for the fiercest exploitation of human beings by other human beings, in the name of progress and neoliberal globalization in the era of the absolute dominance of rapacious markets [195].

The alternative way for a WID involves the principles of interdisciplinary and integrated approach, study, research, analysis, and treatment of the objectively complex and multidimensional environmental and development problems in their political, cultural, economic, technical/technological dimension. This requires as instruments a common code of communication, of sciences, of perceptions, beliefs and ideas, and also a free, emancipated consultation, documentation, acceptance, agreement, and commitment on the way to a WID as a way-out of the total, global, and multidimensional crisis. This alternative path is based on the conscious, sensitive, responsible, active citizens (producers, creators, workers, scientists, professionals) and their freely coordinated collectives, to their cooperatives and collective initiatives, partnerships and actions: of social solidarity, of integrated land consolidations, integrated production units, standardization, processing, distribution, and exchange of local products of certified quality and institutions of free education, culture, research, and technology in constant collaboration with universities and institutes of technology.

Consequently, Rokos [196] has developed an interdisciplinary and holistic methodology concerning development in accordance with the philosophy, principles, and values of WID. The processes and phases of actions needed in the proposed methodology follow the steps listed below:

*   Stakeholders of local development initiatives should document, agree, and accept in an interdisciplinary and holistic way the principles, values, goals, actions, and practices of WID as the optimum choice. The core of the methodological approach is the definition of the purpose, philosophy, principles, and values of development theory, in a dialectical relation with the quantities and qualities of the elements constituting natural and socioeconomic reality, as well as with the relevant problems, the real potentials, and objective constraints. The purpose of development, according to Rokos [196] should be the simultaneous—in space and time—appropriate economic, social, political, cultural, and technical/technological development, which will be performed in a dialectical harmony with human, as the latter is historically, peacefully, and creatively integrated to natural and cultural environment, as an organic and integral part and not as a master, owner, or exploiter.
*   The necessary Integrated Surveys research and studies of the natural and socioeconomic reality of a region and the systematic monitoring of its changes through time should be conducted. The research should also take into account the dynamics of the multidimensional relationships, interdependences, and interactions between nature and society. The right choice of parameters that determine the economic, social, political, cultural, technical/technological, and natural reality of the study area should be made. Furthermore the adequate methods for recording and monitoring should be applied. The more accurate, real, and reliable the data are, the more valid and safely documented the development policies can be. These basic studies should be organically combined and

supplemented with specific on-site sampling corollaries, researches, and scrutiny, evaluating at the same time local people's indigenous expertise and wisdom.

• The elements/data that constitute the natural and socioeconomic reality of the area should be properly analyzed, evaluating and studying the quantities, qualities, and dialectical relations of the real problems, the objective potentials (perspectives), and constraints for Integrated Development. Proposals for thoroughly carried out researches and studies concerning specific areas or natural resources, confrontation of special issues, as well as the analysis of possible financing sources should be documented.

• Alternative scenarios for Integrated Development of the specific region should be formulated and documented. These scenarios should utilize the certain conditions of reality, confront problems according to local constraints and perspectives, specifying basic development directions—axes (ie, guidelines). They should also have a specific spatial reference and take into account the perceptions, desires, and suggestions of the stakeholders, residents, and emigrants, investigating not only the possible financing sources, but also the availability of the latter to take action and responsibilities, to contribute in a positive way in all levels for the realization of each scenario. The interdisciplinary evaluation of advantages and disadvantages for each scenario as well as between the different scenarios is also required.

• Predicting and evaluating the balance of positive and negative elements of each scenario. Choosing the optimum scenario, according to the purpose of development research, the time frame for their implementation, the social acceptance of individual actions, the constraints, internal and external, which multidimensionally affect the implementation of each scenario,

• The balance between positive and negative elements of each scenario for the Integrated Development of the region should be evaluated and then the optimum scenario should be chosen. This step should take into account the purpose of the research, the time frame for the implementation, social acceptance of individual actions, and internal or external objective constraints (intraregional, interregional, national, European, and international). All of the above factors can and will definitely affect in a multidimensional way, the sizes and qualities as well as the implementation of the outcomes of each scenario.

• Implementation of the optimum Integrated Development scenario for the region, constant monitoring, and feedback. This step includes the integration of all the positive elements from other alternative scenarios in the optimum scenario, constant monitoring and scrutiny of the implementation as well as realization of possible problems, investigation of sizes and qualities, evaluation of their importance, and finally taking over the right feedback initiatives through collective participatory processes.

8.7.3 Alternative Pathways in Practice

It is not easy to find successful cases of development based on alternative approaches in a world dominated by the markets. "Alternative islets" cannot easily be expanded to a wider geographical space without major social changes and/or changes in the productive processes. However, studying them is useful and can enhance the dialogue for overcoming the problems of the current, global development policies.

Energy cooperatives: sustainable energy solution in Costa Rica

Energy cooperatives in Costa Rica (a country with very high shares of renewable energy by the way) mainly focus on rural electrification. Rural electrification in the country is advanced, with over 98% of the nation's population having access to electric power. This is something exceptional for Central America. This achievement related in particular to the cooperatives which have been active since the 1960s. Four cooperatives (Coopelesca, Coope Alfaro Ruiz, Coope Guanacaste, and Coopesantos) operate in the rural regions. The primary objective of these cooperatives is to achieve levels of rural electrification in accordance with the requirements of the law on the participation of rural electrification cooperatives. The four cooperatives are completely self-sustaining and create a surplus on their operations. They are also continually expanding their scope and range of consumer services, for example, into telecommunications, by reinvesting the surplus. The cooperative model in energy production is a very good alternative example for overcoming energy poverty in the developed world. Three of the country's electricity cooperatives—Coopelesca, Coope Guanacaste, and Coopesantos—jointly own a wind farm. A federation of energy cooperatives, Conelectricas R.L., was established in 1989 and is involved on behalf of its members in power generation investment and operations, strategic services and policy advocacy, and various technical services. The energy cooperatives of Costa Rica supply energy to 150,000 customers in rural areas, a really great number. The profit-driven energy companies in a liberalized energy market are rarely interested/involved in rural electrification, since opportunities for profit are found mainly in urban and industrialized areas. In contrast to this situation, cooperative schemes can form the solution for rural electrification.

Marinaleda: radical solutions at the heart of the developed world

Marinaleda is a small town in Andalusia region, in Spain. Since 1979, the town is led by a charismatic major, Juan Manuel Sánchez Gordillo. After a decade of occupations and hunger strikes the citizens of Marinaleda won a 1200-hectare farm from the Duke of Infantado. This property was just one of many instances in Spain of vast estates with arable land fenced off from the area's surrounding, usually starving, population. Villagers of Marinaleda walked 10 miles, every day, to occupy the Duke's land. The police every day evicted them and they returned, peacefully, the next day. In 1991, the Andalusian government compensated the Duke with an undisclosed sum, and gave it for the people of Marinaleda. They planted the Duke's land, which previously grew nonlabor-intensive crops like sunflowers, with labor-intensive crops like olive trees. The land's exploitation is based on a total cooperative model. The idea was based on the following simple thought: the more labor required, the more jobs would be created. Once the olive trees were grown, an already labor-intensive process, they then had to be processed into oil. This requires a processing plant, and the employment of more workers. The necessary infrastructure was created by the surplus of the agricultural production, which was reinvested. The reinvestment of surplus without chasing profits is ever since the basis of Marinaleda's productive model. The town has nowadays an unemployment rate of 5%, whereas Spain has an unemployment rate of 27%. The cooperative model of Marinaleda is further supported by simple social welfare practices. For example, when a team goes to the farm, other teams stay back in the village for taking care of the children. The farm, known as El Humoso, sells its products internationally. The major claims that even if Marinaleda exists in a capitalist world, by proving that "we can work for reasons other than money" an act of subverting capitalism itself is realized.

References

[1] J.N. Pieterse, Development Theory, SAGE Publications, London, 2009.

[2] OECD, Glossary of Statistical Terms. Gross Domestic Product (GDP), 2002. http://stats.oecd.org/glossary/detail.asp?ID=1163 (accessed 02.05.15).

[3] F. Braudel, Grammar of Civilizations (A. Alexakis, Transl.), National Bank of Greece Cultural Foundation, Athens, 2001 (in Greek).

[4] D. Harvey, A Brief History of Neoliberalism, Oxford University Press, Oxford, 2005.

[5] Encyclopaedia Britannica, Environmentalism. History of the Environmental Movement, 2015. http://www.britannica.com/EBchecked/topic/189205/environmentalism/224631/History-of-the-environmental-movement (accessed 02.05.15).

[6] UN, General Assembly of the United Nations. Sustainable Development, 2015. http://www.un.org/en/ga/president/65/issues/sustdev.shtml (accessed 03.05.15).

[7] H. Hove, Critiquing sustainable development: a meaningful way of mediating the development impasse? Undercurrent 1 (1) (2004) 48–54.

[8] B. Hettne, Thinking About Development, Zed Books, London, 2009.

[9] G. Esteva, Development, in: W. Sachs (Ed.), The Development Dictionary. A Guide to Knowledge as Power, Zed Books, London & New Jersey, 1992.

[10] G. Rist, The History of Development: From Western Origins to Global Faith, third ed., Zed Books, London & New Jersey, 2008.

[11] D. Rokos, From 'Sustainable' to Worthliving Integrated Development, Livanis Publications, Athens, 2003 (in Greek).

[12] R. Potter, T. Binns, J.A. Elliott, D. Smith, in: Geographies of Development. An Introduction to Development Studies, third ed., Pearson Education Limited, New York, 2008.

[13] D. Gasper, The Ethics of Development, Edinburgh University Press, Edinburgh, 2004.

[14] A. Thomas, Development as practice in a liberal capitalist world, J. Int. Dev. 12 (6) (2000) 773–787.

[15] D. Rokos (Ed.), Environment and Development. Dialectical Relations and Interdisciplinary Approaches, Alternative Editions, Athens, 2005 (in Greek).

[16] C.A. Talmage, Development, in: A.C. Michalos (Ed.), Encyclopedia of Quality of Life and Well-Being Research, Springer Reference, Netherlands, 2014.

[17] E.F. Schumacher, Small Is Beautiful, 1980 (F. Choidas, O. Tremi, Transl.). Glaros, Athens. (in Greek).

[18] K. Willis, Theories and Practices of Development, Taylor & Francis e-Library, 2005.

[19] M. Horkheimer, Th W. Adorno, Dialectic of Enlightenment, 1996 (L. Anagnostou, Transl.). Nissos, Athens.

[20] W. Sachs, The Development Dictionary: A Guide to Knowledge as Power, Zed Books, London, 1992.

[21] J. Rapley, Understanding Development Theory and Practice in the Third World, third ed., Lynne Rienner Publishers, Boulder, 2007.

[22] D. Rokos, The integrated development of mountainous areas in times of 'crisis'. Seventeen years of the N.T.U.A. M.I.R.C. contribution, in: 6th Interdisciplinary Interuniversity Conference "The Integrated Development of Mountainous Areas". National Technical University of Athens, 16−19 September 2010, Metsovo, Greece, 2010 (in Greek).

[23] WCED − World Commission on Environment and Development, Our Common Future, Oxford University Press, Oxford, 1987.

[24] G. Atkinson, R. Dubourg, K. Hamilton, M. Munasinghe, D. Pearce, C. Young, Measuring Sustainable Development: Macroeconomics and the Environment, Edward Elgar, Cheltenham, 1997.

[25] I.U.C.N, Caring for the Earth. A Strategy for Sustainable Living, 1991 (Gland, Switzerland).

[26] K. Magliveras, The development of international environmental law from the Stockholm convention to the Kyoto protocol, in: M. Kaila, E. Theodoropoulou, A. Dimitriou, G. Xanthakou, N. Anastasatos (Eds.), Environmental Education, Research Findings and Educational Planning, Atrapos, Athens, 2005 (in Greek).

[27] M. Mawhinney, Sustainable Development. Understanding the Green Debates, Blackwell Science, Hoboken, 2002.

[28] J. Blewitt, Understanding Sustainable Development, Earthscan, London, 2008.

[29] A.D. Basiago, Methods of defining 'sustainability', Sustain. Dev. 3 (1) (1995) 109−119.

[30] P. Bifani, Medio Ambiente y Desarrollo Sostenible, AIEPALA, Madrid, 1999.

[31] L.A. Rios Osorio, M.O. Lobato, D.C. Xavier Alvarez, Debates on sustainable development: towards a holistic view of reality, environment, Dev. Sustain. 7 (2005) 501−518.

[32] D. Reid, Sustainable Development: An Introductory Guide, Earthscan, London, 1995.

[33] B. Hopwood, M. Mellor, G. O'Brien, Sustainable development: mapping different approaches, Sustainable Dev. 13 (2005) 38−52.

[34] M. Salomone, Cities, sustainable, in: A.C. Michalos (Ed.), Encyclopedia of Quality of Life and Well-being Research, Springer Reference, Netherlands, 2014.

[35] J. O'Neill, A. Holland, A. Light, Environmental Values, Routledge, London, 2008.

[36] Scottish Executive Social Research, Sustainable Development: A Review of International Literature, 2006. http://www.gov.scot/Publications/2006/05/23091323/1 (accessed 08.05.15).

[37] M. Parker, V. Fournier, P. Reedy, The Dictionary of Alternatives: Utopianism and Organization, Zed Books, London, 2007.

[38] S.R. Devkota, Is strong sustainability operational? An example from Nepal, Sustainable Dev. 13 (2005) 297−310.

[39] D.W. Pearce, E. Barbier, Blueprint for a Sustainable Economy, Earthscan, London, 2000.

[40] D.W. Pearce, G.D. Atkinson, Capital theory and the measurement of sustainable development: an indicator of "weak" sustainability, Ecol. Econ. 8 (1993) 103−108.

[41] S. Serafy, In defense of weak sustainability: a response to Beckerman, Environ. Values 5 (1996) 75−81.

[42] M.C. Gutés, The concept of weak sustainability, Ecol. Econ. 17 (1996) 147−156.

[43] S.W. Verstegen, J.C. Hanekamp, The sustainability debate: idealism versus conformism. The controversy over economic growth, Globalizations 2 (3) (2005) 349−362.

[44] D.W. Orr, Ecological Literacy: Education and the Transition to a Postmodern World, State University of New York Press, Albany, 1992.

[45] T. O'Riordan, The politics of sustainability, in: R.K. Turner (Ed.), Sustainable Environmental Economics and Management: Principles and Practice, Belhaven Press, London, 1993.

[46] R. Kates, T. Parris, A. Leiserowitz, Sustainable development? Goals, indicators, values and practice, Environment 47 (3) (2005) 8–21.

[47] T. Hartman, The Last Hours of Ancient Sunlight, Harmony Books, New York, 1998.

[48] G.K. Chuan, Environmental impact of economic development in Peninsular Malaysia: a review, Appl. Geogr. 2 (1) (1982) 3–16.

[49] N. Hanley, J. Shogren, B. White, Environmental Economics in Theory and in Practice, Oxford University Press, New York, 1997.

[50] P.R. Portney, R.N. Stavins (Eds.), Public Policies for Environmental. Protection, second ed., 2000 (Resources for the Future, Washington, DC).

[51] J. O'Connor, Capitalism, nature, socialism: a theoretical introduction, Capitalism Nat. Socialism 1 (1) (1988) 11–38.

[52] P. Burkett, Marx and Nature: A Red and Green Perspective, St. Martin's, 1999 (New York).

[53] A. Vlachou, Nature, Capital and Society, Kritiki Publishing, Athens, 2007 (in Greek).

[54] D.W. Pearce, Environmental Economics, Longman, New York, 1976.

[55] R.K. Turner, D. Pearce, I. Bateman, Environmental Economics, John Hopkins University Press, Baltimore, 1993.

[56] M. Jacobs, The limit to neoclassicism: towards an institutional environmental economics, in: M. Redclift, T. Benton (Eds.), Social Theory and the Global Environment, Routledge, London, 1994.

[57] C. Leadbeater, The Weightless Society, Texere, New York, 2000.

[58] A. Mol, Globalization and Environmental Reform, MIT Press, Cambridge, 2001.

[59] T. Panayotou, Economic growth and the environment, in: Economic Survey of Europe, No. 2 (Secretariat of the Economic Commission for Europe, Edition), Economic Commission for Europe, UN, New York and Geneva, 2003, pp. 45–72.

[60] G. Economides, A. Philippopoulos, Growth enhancing policy is the means to sustain the environment, Rev. Econ. Dyn. 11 (1) (2008) 207–219.

[61] T. Everett, M. Ishwaran, G.P. Ansaloni, A. Rubin, Economic Growth and the Environment, Department for Environment Food and Rural Affairs (DEFDRA) Evidence and Analysis Series, 2010.

[62] J.B. Foster, Ecology and Capitalism, Metaixmio Publishing, Athens, 2003 (in Greek).

[63] J. Schumpeter, The instability of capitalism, in: R.V. Clemence (Ed.), Essays of J.A. Schumpeter, Addison-Wesley, Reading, MA, 1951.

[64] R. Bahro, Avoiding Social and Ecological Disaster, Getaway Books, Bath, 1994.

[65] W.S. Jevons, The coal question: an inquiry concerning the progress of the nation, and the probable exhaustion of our coal-mines, third ed. rev, Macmillan, London, 1906.

[66] M. Gillis, D.H. Perkins, M. Roemer, D.R. Snodgrass, Economics of Development, Typothito, Athens, 2011 (in Greek).

[67] P. Montague, Philadelphia Dumps on the Poor, Rachel's Environment & Healthy News, 1998. # 595.

[68] J. Ui, Minamata Disease, Industrial Pollution in Japan, The United Nations University, Tokyo, 1992.

[69] A. Sakarajasekaran, Industrial Production and Environmental Protection in Conflict? One World Only: Industrialisation and Environment, International Forum Friedrich Ebert Stiftung, 1973, pp. 190–200. Tokyo 2/11/1973-1/12/1973.

[70] A.M. Hamzah, Towards environmental management: the Malaysian experience, in: C. MacAndrews, C.L. Sien (Eds.), Developing Economies and the Environment: The Southeast Asian Experience, McGraw-Hill, Singapore, 1979.

[71] Anon, Country reports to ASEAN seminar on tropical rainforest management, Malays. For. 41 (2) (1978) 82–120.

[72] D.J. Chivers, The Siamang in Malaya – A Field Study of a Primate in Tropical Rainforest. (Contribution to Primatology vol. 4), Karger, Basel and New York, 1974.

[73] H.M. Singh, et al., Rubber factory discharges and their impact on environmental quality, in: Unpublished Paper Presented at the Symposium on Crises in the Malaysian Environment, Penang, 1978.

[74] Factories and Machinery Department, Palm Oil Processing: Effluent Treatment. 1 & 2, Government Printers, Kuala Lumpur, 1975.

[75] G.K. Seng, Air pollution control management programme, in: Unpublished Paper Presented at the Symposium on Crises in the Malaysian Environment, Penang, 1978.

[76] W. Bello, S. Rosenfeld, Dragons in Distress, Springer, San Francisco, 1992 (Institute of Food and Development Policy).

[77] D. Mavraki, A. Sitara, A. Loukatos, Environmental indicators: the case of Romania, in: HELECO 2005, Technical Chamber of Greece, Athens, 2005 (in Greek).

[78] World Bank, World Development Report 2010, World Bank, Washington, DC, 2009.

[79] World Bank, Global Monitoring Report 2008, World Bank, Washington, DC, 2008.

[80] WRI, Climate Analysis Indicators Tool (CAIT) Database 5.0, World Resource Institute, Washington, DC, 2007.

[81] V. Smakhtin, C. Revenga, P. Doell, Taking into Account Environmental Water Requirements in Global − Scale Water Resources Assessments, vol. 2, IWMI, 2004.

[82] UN, The Millennium Goals Report, United Nations, New York, 2009.

[83] WWF, Living Planet Report 2006, World Wildlife Fund, Geneva, Switzerland, 2006.

[84] D. Hulme, The making of the millennium development goals: human development meets result based management, in: An Imperfect World., Working Paper 16, Brooks World Poverty Institute, 2007.

[85] R. Nallari, B. Griffith, Understanding Growth and Poverty: Theory, Policy, and Empirics, World Bank Publications, 2011.

[86] M. McGillivray, A. Shorrocks, Inequality and multi-dimensional well-being, Rev. Income Wealth 51 (2005) 193−199.

[87] M. McGillivray (Ed.), Inequality, Poverty and Well-Being. Studies in Development Economics and Policy, Macmillan, Palgrave, 2006.

[88] UNDP, Human Development Report 2013. The Rise of the South: Human Progress in a Diverse World, United Nations Development Programme, New York, 2013.

[89] FAO, FAO Statistical Yearbook: World Food and Agriculture, FAO, Rome, 2012.

[90] FAO, FAO Statistical Yearbook: World Food and Agriculture, FAO, Rome, 2013a.

[91] FAO, IFAD, WFP, The State of Food Insecurity in the World − Strengthening the Enabling Environment for Food Security and Nutrition, FAO, Rome, 2014.

[92] J. von Braun, M.S. Swaminathan, M.W. Rosegrant, Agriculture, Food Security, Nutrition and the Millennium Development Goals, International Food Policy Research Institute (IFPRI), 2004.

[93] UNDP, Human Development Report 2014. Sustaining Human Progress: Reducing Vulnerabilities and Building Resilience, United Nations Development Programme, New York, 2014a.

[94] UNDP, What is the multidimensional poverty index. http://hdr.undp.org/en/content/what-multidimensional-poverty-index, 2015 (accessed 20.04.15).

[95] World Bank, An update to World Bank's estimates of consumption poverty in the developing world. Briefing note. http://siteresources.worldbank.org/INTPOvCALNET/Resources/Global_Poverty_Update_2012_02-29-12.pdf, 2012 (accessed 20.04.15).

[96] UNDP, The Millennium Development Goals Report 2014, United Nations Development Programme, New York, 2014b.

[97] FAO, FAO Hunger Map, FAO, Rome, 2014.

[98] EU, Rural Development in the European Union. Statistical and Economic Information, Report 2010, 2010, http://ec.europa.eu/agriculture/agrista/rurdev2010/RD_Report_2010.pdf (accessed 20.04.15).

[99] M.F. Price, Mountain Geology, Natural History and Ecosystems, Voyageur Press, Stillwater, USA, 2002.

[100] IFAD, Dimensions of Rural Poverty, 2015. http://www.ruralpovertyportal.org/region (accessed 20.04.15).

[101] D. Pimentel, M. Pimentel, Food, Energy and Society, Taylor & Francis Group, Boca Raton, USA, 2008.

[102] UNDP, Energy and the Challenge of Sustainability. World Energy Assessment, United Nations Development Programme, New York, 2000.

[103] A.K. Reddy, Energy technologies and policies for rural development, in: T.B. Johansson, J. Goldemberg (Eds.), Energy for Sustainable Development, United Nations Development Programme, New York, 2002.

[104] J. Goldemberg, T.B. Johansson, A.K. Reddy, R.H. Williams, Basic needs and much more with one kilowatt per capita, Ambio 14 (4/5) (1985) 190−200.

[105] S. Chakravarty, A. Chikkatur, H. de Coninck, S. Pacala, R. Socolow, Sharing global CO_2 emission reductions among one billion high emitters, Proc. Natl. Acad. Sci. U.S.A. 106 (29) (2009) 11884−11888.

[106] UNDP and World Bank, Energy Services for the Millennium Development Goals, United Nations Development Programme and World Bank, New York, 2005.

[107] N. Katsoulakos, L. Papada, D. Kaliampakos, The problem of energy poverty in mountainous areas, in: IISA 2014−5th International Conference on Information, Intelligence, Systems and Applications. Chania, Greece, 7−9/7/2014, 2014, http://dx.doi.org/10.1109/IISA.2014.6878794.

[108] IEA, World Energy Outlook 2011, International Energy Agency, Paris, 2011.

[109] EPEE - European Fuel Poverty and Energy Efficiency, Project Fact Sheet −Intelligent Europe, 2009. http://www.fuel-poverty.org/documents/epee_factsheet.pdf (accessed 10.02.15).

[110] B. Boardman, Fixing Fuel Poverty: Challenges and Solutions, Earthscan, London, 2010.

[111] BPIE, Alleviating Fuel Poverty in the EU. Investing in Home Renovation, a Sustainable and Inclusive Solution, Buildings Performance Institute Europe, Brussels, 2014.

[112] W.R. Keatinge, G.C. Donaldson, E. Cordioli, M. Martinelli, A.E. Kunst, J.P. Mackenbach, S. Nayha, I. Vuori, Heat related mortality in warm and cold regions of Europe: observational study, Br. Med. J. 321 (7262) (2000) 670−673.

[113] C. Liddell, C. Morris, Fuel poverty and human health: a review of recent evidence, Energy Policy 38 (6) (2010) 2987−2997.

[114] A.D. Sagar, Alleviating energy poverty for the world's poor, Energy Policy 33 (11) (2005) 1367−1372.

[115] UN, Millennium Development Goals: 2014 Progress Chart, United Nations, New York, 2014.

[116] UNDP, Human Development Report 2002, United Nations Development Programme, Oxford University Press, New York, 2002.

[117] B. Milanovic, Global Income Inequality by the Numbers: In History and Now − An Overview, The World Bank Development Research Group, Poverty and Inequality Team, 2012.

[118] UN System Task Team, Realizing the Future We Want for All, 2012 (Report to the Secretary-General. New York).

[119] FAO, Our Priorities. The FAO Strategic Objectives, Brochure, 2013b. Version 1.1.

[120] S. Bouzarovski, S. Petrova, R. Sarlamanov, Energy poverty policies in the EU: a critical perspective, Energy Policy 49 (2012) 76−82.

[121] EDGAR − Emission Database for Global Atmospheric Research, CO_2 Time Series 1990−2011 Per Capita for World Countries, 2011. http://edgar.jrc.ec.europa.eu/overview.php?v=CO2ts_pc1990-2011 (accessed 05.05.15).

[122] FAO, Global Forest Resources Assessment (FRA) 2010. Key Findings, Food and Agriculture Organization, Rome, 2010.

[123] Y. Wada, L.P.H. van Beek, C.M. van Kempen, J.W.T.M. Reckman, S. Vasak, M.F.P. Bierkens, Global depletion of groundwater resources, Geophys. Res. Lett. 37 (20) (2010).

[124] EPA, Agency Financial Report. Fiscal Year 2014, Environmental Protection Agency, Washington, DC, 2014.

[125] EC, EU Energy in Figures. Statistical Pocketbook 2014, Publications Office of the European Union, Luxembourg, 2014.

[126] E.A. Blaug, Use of the environmental assessment by federal agencies in NEPA implementation, Environ. Prof. 15 (1) (1993) 57−65.

[127] R. Therivel, P. Morris (Eds.), Methods of Environmental Impact Assessment, UCL Press, 1995.

[128] R.K. Morgan, Environmental Impact Assessment: A Methodological Approach, Kluwer Academic Publishers, Dordrecht, 1998.

[129] D.P. Lawrence, Environmental Impact Assessment: Practical Solutions to Recurrent Problems, John Wiley and Sons, 2003.

[130] EC, Guidance on EIA: EIS Review, 2001. European Communities, http://ec.europa.eu/environment/archives/eia/eia-guidelines/g-review-full-text.pdf.

[131] A.J. Wright, S.J. Dolman, M. Jasny, E.C.M. Parsons, D. Schiedek, S.B. Young, Myth and momentum: a critique of environmental impact assessments, J. Environ. Prot. 4 (2013) 72−77.

[132] Millennium Ecosystem Assessment, Ecosystems and Human Well-Being: Biodiversity Synthesis, World Resources Institute, Washington, DC, 2005. Also available: http://www.unep.org/maweb/documents/document.354.aspx.pdf (accessed 08.05.15).

[133] B. Alonso, F. Valladares, International efforts on global change research, in: E. Chuvieco (Ed.), Earth Observation of Global Change the Role of Satellite Remote Sensing in Monitoring the Global Environment, Springer, 2008.

[134] J.R. Jensen, Remote Sensing of the Environment: An Earth Resource Perspective, Pearson Prentice Hall, 2007.

[135] Q. Weng, Remote Sensing and GIS Integration: Theories, Methods, and Applications, McGraw-Hill, New York, 2010, p. 416.

[136] N. Horning, J. Robinson, E. Sterling, W. Turner, S. Spector, Remote Sensing for Ecology and Conservation: A Handbook of Techniques, Oxford University Press, 2010.

[137] K. McGarigal, S.A. Cushman, E. Ene, FRAGSTATS v4: Spatial Pattern Analysis Program for Categorical and Continuous Maps, Computer Software Program Produced by the Authors at the University of Massachusetts, Amherst, 2012. Available at: http://www.umass.edu/landeco/research/fragstats/fragstats.html (accessed 25.04.15).

[138] S.E. Gergel, New directions in landscape pattern analysis and linkages with remote sensing, in: M.A. Wulder, S.E. Franklin (Eds.), Understanding Forest Disturbance and Spatial Pattern: Remote Sensing and GIS Approaches, CRC Press, 2007.

[139] S.E. Franklin, Remote Sensing for Biodiversity and Wildlife Management, McGraw-Hill, 2010.

[140] S. Kröger, R.L. Law, Sensing the sea, Trends Biotechnol. 23 (5) (2005) 250−256.

[141] C.J. Legleiter, D.A. Roberts, W.A. Marcus, M.A. Fonstad, Passive optical remote sensing of river channel morphology and in-stream habitat: physical basis and feasibility, Remote Sens. Environ. 93 (4) (2004) 493−510.

[142] W.W. Immerzeel, P. Droogers, S.M. De Jong, M.F.P. Bierkens, Large-scale monitoring of snow cover and runoff simulation in Himalayan river basins using remote sensing, Remote Sens. Environ. 113 (1) (2009) 40−49.

[143] L. Chen, Z. Jin, R. Michishita, J. Cai, T. Yue, B. Chen, B. Xu, Dynamic monitoring of wetland cover changes using time-series remote sensing imagery, Ecol. Inf. 24 (2014) 17−26.

[144] D. Tang, G. Levy, Introduction, in: D. Tang (Ed.), Remote Sensing of the Changing Oceans, Springer, 2011.

[145] F.D. Van der Meer, H.M. Van der Werff, F.J. van Ruitenbeek, C.A. Hecker, W.H. Bakker, M.F. Noomen, M. van der Meijde, E. John, M. Carranza, J. Boudewijn de Smeth, T. Woldai, Multi-and hyperspectral geologic remote sensing: a review, Int. J. Appl. Earth Obs. Geoinformation 14 (1) (2012) 112−128.

[146] R.L. Langford, Temporal merging of remote sensing data to enhance spectral regolith, lithological and alteration patterns for regional mineral exploration, Ore Geol. Rev. 68 (2015) 14−29.

[147] S.J. Walsh, D.R. Butler, G.P. Malanson, An overview of scale, pattern, process relationships in geomorphology: a remote sensing and GIS perspective, Geomorphology 21 (3) (1998) 183−205.

[148] N.J. Schneevoigt, L. Schrott, Linking geomorphic systems theory and remote sensing: a conceptual approach to Alpine landform detection (Reintal, Bavarian Alps, Germany), Geogr. Helv. 61 (3) (2006) 181−190.

[149] V.L. Mulder, S. De Bruin, M.E. Schaepman, T.R. Mayr, The use of remote sensing in soil and terrain mapping—a review, Geoderma 162 (1) (2011) 1−19.

[150] G. Bocco, M. Mendoza, A. Velázquez, Remote sensing and GIS-based regional geomorphological mapping—a tool for land use planning in developing countries, Geomorphology 39 (3) (2001) 211–219.

[151] G. Bocco, A. Velázquez, C. Siebe, Using geomorphologic mapping to strengthen natural resource management in developing countries. The case of rural indigenous communities in Michoacan, Mexico, Catena 60 (3) (2005) 239–253.

[152] B. Bhatta, Analysis of Urban Growth and Sprawl from Remote Sensing Data, Springer-Verlag, Berlin, Heidelberg, 2010.

[153] J.D. Gatrell, R.R. Jensen, Sociospatial applications of remote sensing in urban environments, Geogr. Compass 2 (3) (2008) 728–743.

[154] B. Bhatta, Remote Sensing and GIS, Oxford University Press, New York and New Delhi, 2008.

[155] J.P. Donnay, M.J. Barnsley, P.A. Longley, Remote Sensing and Urban Analysis: GISDATA 9, CRC Press, 2003.

[156] C.J. Webster, Urban morphological fingerprints, Environ. Plann. B 22 (3) (1995) 279–297.

[157] T.V. Mesev, P.A. Longley, M. Batty, Y. Xie, Morphology from imagery: detecting and measuring the density of urban land use, Environ. Plann. A 27 (5) (1995) 759–780.

[158] S. Martinuzzi, W.A. Gould, O.M.R. Gonzalez, Land development, land use, and urban sprawl in Puerto Rico integrating remote sensing and population census data, Landsc. Urb. Plann. 79 (2007) 288–297.

[159] R. Welch, Spatial resolution requirements for urban studies, Int. J. Remote Sens. 3 (2) (1982) 139–146.

[160] J. Xiao, Y. Shen, J. Ge, R. Tateishi, C. Tang, Y. Liang, Z. Huang, Evaluating urban expansion and land use change in Shijiazhuang, China, by using GIS and remote sensing, Landsc. Urb. Plann. 75 (1) (2006) 69–80.

[161] B. Bhatta, S. Saraswati, D. Bandyopadhyay, Quantifying the degree-of-freedom, degree-of-sprawl, and degree-of-goodness of urban growth from remote sensing data, Appl. Geogr. 30 (1) (2010) 96–111.

[162] Z. Zhu, C.E. Woodcock, J. Rogan, J. Kellndorfer, Assessment of spectral, polarimetric, temporal, and spatial dimensions for urban and peri-urban land cover classification using Landsat and SAR data, Remote Sens. Environ. 117 (2012) 72–82.

[163] B. Zhang, Study on Urban Growth Management in China, Xinhua Press, Beijing, 2004.

[164] M.E. Brown, Famine Early Warning Systems and Remote Sensing Data, Springer Science & Business Media, 2008.

[165] B. Wisner, P. Blaikie, T. Cannon, I. Davis, At Risk, second ed., Taylor and Francis Books Ltd, Wiltshire, 2004.

[166] P. Geerdeers, Remote Sensing Advances: ACP Countries Can Do More − Knowledge for Development, 2013. Available at: http://knowledge.cta.int/Dossiers/S-T-Issues/Remote-sensing-and-GIS/Feature-articles/Remote-sensing-advances-ACP-Countries-can-do-more (accessed 06.05.15).

[167] Committee on the Earth System Science for Decisions about Human Welfare: Contributions of Remote Sensing, Geographical Sciences Committee, National Research Council, Contributions of Land Remote Sensing for Decisions About Food Security and Human Health: Workshop Report, The National Academies Press, 2007.

[168] D.S. Ratnasari, P. Kusumawardani, Spatial modelling for food vulnerability using remote sensing data and GIS (Study case in Klungkung Regency, Bali), Proc. Environ. Sci. 24 (2015) 15–24.

[169] M.F. Goodchild, Geographic information systems in undergraduate geography: a contemporary dilemma, Oper. Geogr. 8 (1985) 34–38.

[170] D.J. McGuire, M.F. Goodchild, D.W. Rhind, Geographical Information Systems, Longman Scientific and Technical, New York, 1991.

[171] A. Skidmore, Environmental Modelling With GIS and Remote Sensing, CRC Press, 2003.

[172] D. Rokos, The contribution of remote sensing and integrated land and environment information systems in the study and monitoring of global changes, in: D. Rokos (Ed.), From 'Sustainable' to Worthliving Integrated Development, Livanis Publications, Athens, 1993 (in Greek).

[173] E.P. Odum, Ecology, Holt Rinehart and Winston, London, 1975.

[174] J.N.R. Jeffers, An Introduction to Systems Analysis: With Ecological Applications, Edward Arnold, London, 1978.

[175] L.M. Rebelo, C.M. Finlayson, N. Nagabhatla, Remote sensing and GIS for wetland inventory, mapping and change analysis, J. Environ. Manage. 90 (7) (2009) 2144−2153.

[176] I.R. Hegazy, M.R. Kaloop, Monitoring urban growth and land use change detection with GIS and remote sensing techniques in Daqahlia governorate Egypt, Int. J. Sustainable Built Environ. 4 (1) (2015) 117−124.

[177] P.S. Wenz, Environmental Ethics Today, Oxford University Press, New York, 2001.

[178] J.B. Callicott, R. Frodeman, Encyclopedia of Environmental Ethics and Philosophy, Macmillan Reference, USA, 2009.

[179] E. Partridge, Posterity and the "strains of commitment", in: T.-C. Kim, J.A. Dator (Eds.), Creating a New History for Future Generations, Institute for the Integrated Study of Future Generations, Kyoto, 1994, pp. 263−278. Also available at: http://gadfly.igc.org/papers/strains.html.

[180] A. Georgopoulos, Environmental Ethics, 2002. Gutenberg (in Greek).

[181] R.F. Nash, The Rights of Nature: A History of Environmental Ethics, University of Wisconsin Press, 1989.

[182] R. Elliot, Environmental ethics, in: P. Singer (Ed.), A Companion to Ethics, Blackwell, 1991.

[183] R. Routley, Is there a need for a new, an environmental, ethic? Proc. Fifteenth World Congr. Philos. 1 (1973) 205−210.

[184] R. Routley, V. Routley, Human chauvinism and environmental ethics, in: D. Mannison, M.A. McRobbie, R. Routley (Eds.), Environmental Philosophy, Australian National University, Research School of Social Sciences, Canberra, 1980.

[185] J. Feinberg, The rights of animals and future generations, in: W.T. Blacksonet (Ed.), Philosophy and Environmental Crisis, The University of Georgia Press, Athens, GA, 1974.

[186] R. Elliot, Faking nature, in: R. Elliot (Ed.), Environmental Ethics, Oxford University Press, Oxford, 1995.

[187] H. Rolston III, Environmental Ethics: Duties to and Values in the Natural World, Temple University Press, Philadelphia, 1988.

[188] A. Carlson, Nature and positive aesthetics, Environ. Ethics 6 (1) (1984) 5−34.

[189] E.C. Hargrove, Foundations of Environmental Ethics, Prentice Hall, New Jersey, 1989.

[190] E. Papadimitriou, Nature and ethics, in: I. Modinos, I. Efthimiopoulos (Eds.), Ecology and Environmental Sciences, Interdisciplinary Institute of Environmental Surveys, Athens, 1998 (in Greek).

[191] B.G. Norton, Environmental ethics and weak anthropocentrism, Environ. Ethics 6 (2) (1984) 131−148.

[192] L. White Jr., The historical roots of our ecologic crisis, Science 155 (3767) (1967) 1203−1207.

[193] C.J. Koroneos, D. Rokos, Sustainable and integrated development—a critical analysis, Sustainability 4 (2012) 141−153.

[194] D. Rokos, The integrated development of mountainous areas. Theory and practice, in: D. Rokos (Ed.), Proceedings of the 3rd Interdisciplinary Interuniversity Conference: The Integrated Development of Mountainous Areas. Theory and Practice. National Technical University of Athens, Metsovion Interdisciplinary Research Center, Metsovo, 7−10.6.2001, Metsovo Conference Center, Alternative Editions, Athens, 2004 (in Greek).

[195] E. Michailidou, D. Rokos, Greek mountainous areas: the need for a worthliving integrated development, in: Regional Studies Association Annual International Conference, Newcastle, UK, 17−20 April 2011, 2011.

[196] D. Rokos, The integrated development of Epirus. Problems, potentials, limitations, in: A.A. Livanis (Ed.), Proceedings of the 4th Interdisciplinary Interuniversity Conference of the N.T.U.a. And the N.T.U.A. M.I.R.C. The Integrated Development of Epirus, Metsovo Conference Center, Metsovo, 23−26 September 2004, 2007, pp. 138−153 (Athens, in Greek).

Index

'*Note:* Page numbers followed by "f" indicate figures and "t" indicate tables.'